Exam-Oriented
Practical
ANATOMY

Exam-Oriented
Practical
ANATOMY

Exam-Oriented Practical ANATOMY
A Student's Manual

Tapan Kumar Jana
MBBS DGO MS FAIMS
Associate Professor
Department of Anatomy
Calcutta National Medical College
Kolkata, West Bengal, India

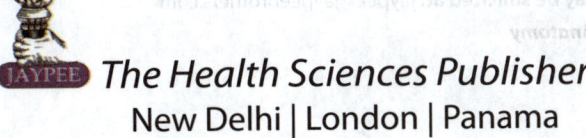

The Health Sciences Publisher
New Delhi | London | Panama

 **Jaypee Brothers Medical Publishers (P) Ltd**

Headquarters
Jaypee Brothers Medical Publishers (P) Ltd
4838/24, Ansari Road, Daryaganj
New Delhi 110 002, India
Phone: +91-11-43574357
Fax: +91-11-43574314
Email: jaypee@jaypeebrothers.com

Overseas Offices

J.P. Medical Ltd
83 Victoria Street, London
SW1H 0HW (UK)
Phone: +44 20 3170 8910
Fax: +44 (0)20 3008 6180
Email: info@jpmedpub.com

Jaypee-Highlights Medical Publishers Inc.
City of Knowledge, Bld. 235, 2nd Floor, Clayton
Panama City, Panama
Phone: +1 507-301-0496
Fax: +1 507-301-0499
Email: cservice@jphmedical.com

Jaypee Brothers Medical Publishers (P) Ltd
17/1-B Babar Road, Block-B, Shaymali
Mohammadpur, Dhaka-1207
Bangladesh
Mobile: +08801912003485
Email: jaypeedhaka@gmail.com

Jaypee Brothers Medical Publishers (P) Ltd
Bhotahity, Kathmandu, Nepal
Phone: +977-9741283608
Email: kathmandu@jaypeebrothers.com

Website: www.jaypeebrothers.com
Website: www.jaypeedigital.com

© 2018, Jaypee Brothers Medical Publishers

The views and opinions expressed in this book are solely those of the original contributor(s)/author(s) and do not necessarily represent those of editor(s) of the book.

All rights reserved. No part of this publication may be reproduced, stored or transmitted in any form or by any means, electronic, mechanical, photocopying, recording or otherwise, without the prior permission in writing of the publishers.

All brand names and product names used in this book are trade names, service marks, trademarks or registered trademarks of their respective owners. The publisher is not associated with any product or vendor mentioned in this book.

Medical knowledge and practice change constantly. This book is designed to provide accurate, authoritative information about the subject matter in question. However, readers are advised to check the most current information available on procedures included and check information from the manufacturer of each product to be administered, to verify the recommended dose, formula, method and duration of administration, adverse effects and contraindications. It is the responsibility of the practitioner to take all appropriate safety precautions. Neither the publisher nor the author(s)/editor(s) assume any liability for any injury and/or damage to persons or property arising from or related to use of material in this book.

This book is sold on the understanding that the publisher is not engaged in providing professional medical services. If such advice or services are required, the services of a competent medical professional should be sought.

Every effort has been made where necessary to contact holders of copyright to obtain permission to reproduce copyright material. If any have been inadvertently overlooked, the publisher will be pleased to make the necessary arrangements at the first opportunity.

Inquiries for bulk sales may be solicited at: jaypee@jaypeebrothers.com

Exam-Oriented Practical Anatomy

First Edition: **2018**
ISBN: 978-93-86150-95-0

Printed at Sanat Printers

Dedicated to
My Parents
Late Asutosh Jana
and
Late Namita Jana

who have raised me on uncompromising principles and infused in me a habit of hard work and meticulous attention to details. They have taught me never to forget my humble origins and to live a life of simplicity and discipline.

Preface

It gives me great pleasure to present the book *Exam-Oriented Practical Anatomy* to all my students past, present and future. In my long tenure as a teacher of anatomy, I have found that 1st year medical students find the subject of anatomy both thrilling and terrifying. The enjoyment of the subject lies in the exploration of human body structure by dissections, histology, radiology and other lectures and practical classes. However, the terror of the subject lies in its huge extent, the innumerable facts, figures and relations that have to be memorized and the difficult level of examinations. The oral and practical examinations in particular have been a tough ordeal for many aspiring students. There has been a long-standing need for presenting the practical aspects of anatomy in a simple and organized manner.

In the present book, I have attempted to organize the content of practical anatomy in such a way as to take the tedium out of the subject. The information is presented in a simple and systematic manner which is easy to understand and recapitulate. The important facts presented in the book are useful to medical students of most Indian universities and institutions. Some aspects of the subject need depth of understanding. I have provided simple and straightforward explanations for these items.

The book is organized into four sections: **Window dissections**—these have been presented according to the standard dissection items taught in most medical colleges in India. A brief idea of the incisions, superficial and deep structures have been provided along with supplementary questions and answers that may be expected in the examinations. **Surface anatomy**—accurate and easily reproducible landmarks have been used to give a precise account of important points and lines representing important body structures along with further questions and answers to be expected in the examination. **Histology**—a brief outline of the standard staining procedures and microscopy is followed by brief notes on most of the relevant slides taught in the undergraduate course. Important points are followed by relevant questions and their appropriate answers. **Radiological anatomy**—standard views of X-rays covering the trunk and extremities have been provided along with detailed discussions of the features, to be studied in each film.

I hope the book will serve the needs of the students for whom it is meant and be welcomed by the medical fraternity. I wish my students all the best not only for their study of anatomy, but also for the long and difficult years which lie ahead of them to achieve success as doctors in society.

Tapan Kumar Jana

Acknowledgments

It is never an easy task to compile a list of people who have helped me in my growth and maturity as doctor and a teacher. It must include at least, all the people who stood by me from my infancy to adulthood, from high school to medical college, my teachers, colleagues, students, administrators, friends, relatives and my family members.

Writing a comprehensive treatise on 'Practical Anatomy' is a daunting task which cannot be taken lightly. To write such a book I recognize the suggestions and supports that I received from my teachers, colleagues, students, publisher and well-wishers.

However, it would be quite remiss of me if I fail to acknowledge the particular help and encouragement I received from Professor Sibani Mazumdar, Head, Department of Anatomy, Calcutta National Medical College, Kolkata, West Bengal, India. I extend my sincere and heartfelt thanks to Professor Bijon Chandra Dutta, Head, Department of Anatomy, Silchar Medical College, Assam, India, for his constructive suggestions in writing this book. I would like to acknowledge the assistance of Dr Arunabha Tapadar, Associate Professor, Malda Medical College, Malda, West Bengal. I should not forget to mention the name of my wife Dr Susmita Jana and my beloved son Anubhav Jana, for their constant inspiration and morale boosting. I am especially indebted to my postgraduate students, who with their incisive questions have helped to further sharpen my knowledge on the subject.

Last but not least, I owe much of my endeavor to my students, past and present, for showing me their points of view, their lacunae in the understanding of human anatomy and a valuable insight on how it feels like to be on the other side of the table in an examination.

I hope my humble work can take the terror out of the subject and make it pleasant and palatable.

Contents

SECTION 1: WINDOW DISSECTIONS

1. **Window Dissections: Introduction** 3
   - Study of Anatomy *3*
   - Dissection *3*
   - Relative and Descriptive Terms in Anatomy *4*
   - Different Terminologies *5*
   - Structures Encountered in Dissection *7*
   - Procedure and Steps of Dissection *20*

2. **Window Dissections: Upper Limb (Superior Extremity)** 30

   Lesson 1: Introduction to Superior Extremity *30*

   Lesson 2: Clavipectoral Fascia (Pectoral Region) *34*
   - Steps of Dissection *34*
   - Attachment and Distribution of the Fascia *36*
   - Structures Piercing the Fascia *37*
   - Muscles Related to Clavipectoral Fascia *38*

   Lesson 3: Triangular and Quadrangular Space *42*
   - Steps of Dissection *42*
   - Muscles Related to Triangular and Quadrangular Space *45*

   Lesson 4: Axilla (Armpit) *48*
   - Steps of Dissection *48*
   - Identification of Different Nerves of Brachial Plexus *53*
   - Cords of Brachial Plexus *54*
   - Muscles of Axilla *54*

   Lesson 5: Front of Arm (Brachium) *60*
   - Steps of Dissection *60*
   - Principal Neurovascular Bundle *64*
   - Muscles Related to Front of the Arm *65*
   - Important Landmarks at the Middle of the Arm *65*

Lesson 6: Cubital Fossa 68
- Steps of Dissection 69
- Muscles Related to Cubital Fossa 73

Lesson 7: Front of the Forearm 75
- Steps of Dissection 75
- Facts to be Noted 81
- Muscles Related to Front of Forearm 81

Lesson 8: Palm of the Hand 86
- Steps of Dissection 86
- Muscles on the Palm 94

Lesson 9: Back of the Arm 102
- Steps of Dissection 103

Lesson 10: Back of the Forearm 107
- Steps of Dissection 107
- Muscles Related to Back of the Forearm 112

Lesson 11: Dorsum of the Hand 115
- Steps of Dissection 116

3. Window Dissections: Lower Limb (Inferior Extremity) 121

Lesson 1: Introduction to Inferior Extremity 121

Lesson 2: Femoral Triangle 125
- Steps of Dissection 125
- Muscles Related to Femoral Triangle 131

Lesson 3: Adductor Canal 143
- Steps of Dissection 143

Lesson 4: Anterolateral Compartment of the Leg 148
- Steps of Dissection 149
- Muscles Related to Anterolateral Compartment of Leg 153

Lesson 5: Dorsum of the Foot 158
- Steps of Dissection 158
- Muscles Related to the Dorsum of the Foot (Intrinsic Muscles of the Dorsum) 161

Lesson 6: Gluteal Region 165
- Steps of Dissection 165
- Muscles Related to Gluteal Region 170

Lesson 7: Back of the Thigh 178
- Steps of Dissection 178
- Muscles of the Back of the Thigh 180

Lesson 8: Popliteal Fossa *184*
- Steps of Dissection *184*

Lesson 9: Back of the Leg (Posterior Crural Region) *192*
- Steps of Dissection *192*
- Muscles Related to the Back of the Leg *197*

Lesson 10: Sole of the Foot *204*
- Steps of Dissection *204*
- Muscles Related to the Sole of the Foot *210*

4. Window Dissections: Abdomen — 217

Lesson 1: Introduction to Abdomen *217*
- Boundaries of the Abdomen *217*
- Contents *218*
- Relationship of the Abdomen to Other Regions *218*
- Topographical Divisions of the Abdominal Wall *219*

Lesson 2: Inguinal Canal *221*
- Steps of Dissection *221*

Lesson 3: Rectus Sheath *231*
- Steps of Dissection *231*

Lesson 4: Exposure of Kidney from Back *240*
- Steps of Dissection *240*

5. Window Dissections: Thorax — 246

Lesson 1: Introduction to Thorax *246*
- Thoracic Cage *246*
- Inlet of Thorax *246*
- Outlet of Thorax *248*

Lesson 2: Dissection of Intercostal Space (Upper Intercostal Spaces) *248*
- Steps of Dissection *248*

6. Window Dissections: Head and Neck — 254

Lesson 1: Introduction to Head and Neck *254*

Lesson 2: Face *255*
- Extent of Face *255*
- Steps of Dissection *255*
- Muscles Related to Face *261*
- Muscles of Mastication *262*

Lesson 3: Anterior Triangles of Neck *274*
- Steps of Dissection *276*
- Muscles Related to Anterior Triangle of Neck *284*

Lesson 4: Posterior Triangles of Neck 298
- Steps of Dissection 299
- Muscles Related to Posterior Triangle of Neck 304

SECTION 2: SURFACE ANATOMY

7. Surface Anatomy: Upper Limb (Superior Extremity) 317

Lesson 1: Points 317
- Head of Radius 317
- Head of Ulna 319
- Styloid Process of Radius 320
- Styloid Process of Ulna 320
- Pisiform Bone 321
- Tip of Coracoid Process 323
- Acromial Angle 325
- Hook of Hamate 326
- Bifurcation of Brachial Artery 326
- Beginning of Brachial Artery, Radial Nerve, Median Nerve and Ulnar Nerve 327

Lesson 2: Lines 330
- Radial Nerve in the Back of the Arm 330
- Ulnar Nerve in Forearm 331
- Axillary Artery 331
- Brachial Artery 333
- Radial Artery in the Forearm 334
- Ulnar Artery in Forearm 335
- Superficial Palmar Arch 337
- Flexor Retinaculum 338

8. Surface Anatomy: Lower Limb (Inferior Extremity) 341

Lesson 1: Points 341
- Adductor Tubercle 341
- Tuberosity of Navicular 342
- Medial Malleolus 343
- Lateral Malleolus 344

Lesson 2: Lines 345
- Popliteal Artery 345
- Anterior Tibial Artery 347
- Posterior Tibial Artery 348
- Arteria Dorsalis Pedis 350

- Tibial Nerve in Popliteal Fossa *351*
- Common Peroneal Nerve *353*
- Deep Peroneal (Anterior Tibial) Nerve *355*

9. Surface Anatomy: Abdomen 357

Lesson 1: Points *357*
- Cardiac Orifice *357*
- Pyloric Orifice *359*
- Fundus of Gallbladder *361*
- Appendicular Orifice *363*
- McBurney's Point *364*
- 4th Lumbar Spine *365*
- Origin of Celiac Artery *367*
- Origin of Superior Mesenteric Artery *368*
- Duodenojejunal Flexure *369*

Lesson 2: Lines *371*
- Fundus of Stomach *371*
- Lesser Curvature of Stomach *373*
- Lower Border of Liver *374*
- Root of the Mesentery *376*
- Kidney from Back *377*

10. Surface Anatomy: Thorax 380

Lesson 1: Points *380*
- Tip of 9th Costal Cartilage *380*
- Sternal Angle *382*
- Apex of Heart *383*
- Tracheal Bifurcation *384*

Lesson 2: Lines *387*
- Anterior Border of Left Lung *387*
- Right Border of Heart *389*
- Left Border of Heart *391*
- Arch of Aorta *393*
- Superior Vena Cava *394*

11. Surface Anatomy: Head and Neck 396

Lesson 1: Points *396*
- Supraorbital Notch *396*
- Bifurcation of Common Carotid Artery *398*
- Arch of Cricoid Cartilage *400*
- Spine of 7th Cervical Vertebra *401*
- Nasion *402*

- Infraorbital Foramen 403
- Thyroid Prominence (Laryngeal Prominence) 404
- Tips of Greater Cornu of Hyoid 406

Lesson 2: Lines 407
- Isthmus of Thyroid Gland 407
- Lateral Lobe of Thyroid Gland 408
- Frontal Air Sinus 411
- Parotid Duct 412
- Right Common Carotid Artery 414
- Internal Carotid Artery 415
- Internal Jugular Vein 418
- External Jugular Vein 419
- Facial Artery in the Face 421
- Spinal Accessory Nerve 422
- Palatine Tonsil 425

SECTION 3: HISTOLOGY

12. Histology: Introduction — 431

Lesson 1: Microscope 431

Lesson 2: Preparation of Tissue for Histological Study 433
- Steps 433

Lesson 3: Procedure of Hematoxylin and Eosin Staining 435
- Hematoxylin and Eosin (H&E) Staining Protocol 435

Lesson 4: Epithelial Tissue (Epithelium) 436

13. Histology: Musculoskeletal System — 442

Lesson 1: Compact Bone 442

Lesson 2: Skeletal Muscle 449

Lesson 3: Cardiac Muscle 456

14. Histology: Blood Vascular System — 460

Lesson 1: Arteries 460

Lesson 2: Veins 465

15. Histology: Gastrointestinal System — 469

Lesson 1: Tongue 469

Lesson 2: Esophagus 475

Lesson 3: Stomach 477

Lesson 4: Duodenum *482*

Lesson 5: Jejunum and Ileum *485*
- Histological Comparison between Duodenum, Jejunum and Ileum *485*

Lesson 6: Appendix *486*

Lesson 7: Rectum (Large Intestine) *488*

16. Histology: Liver and Pancreas (Both Exocrine and Endocrine Types of Gland) — 492

Lesson 1: Liver *492*

Lesson 2: Pancreas *501*

17. Histology: Salivary Glands (Exocrine Glands) — 508

Lesson 1: Parotid Gland *508*

Lesson 2: Submandibular Gland *512*

Lesson 3: Sublingual Gland *515*

18. Histology: Endocrine Glands — 517

Lesson 1: Thyroid Gland *517*

Lesson 2: Adrenal Gland *521*

19. Histology: Lymphatic System — 526

Lesson 1: Lymph Node *526*

Lesson 2: Thymus *532*

Lesson 3: Spleen *538*

Lesson 4: Palatine Tonsil *542*

20. Histology: Respiratory System — 545

Lesson 1: Trachea *545*

Lesson 2: Lungs *549*

21. Histology: Urinary System — 558

Lesson 1: Kidneys *558*

Lesson 2: Ureter *571*

Lesson 3: Urinary Bladder *574*

22. Histology: Male Reproductive System — 577

Lesson 1: Testis *577*

Lesson 2: Vas Deferens *584*

Lesson 3: Prostate *585*

23. Histology: Female Reproductive System — 588

Lesson 1: Uterus 588
Lesson 2: Uterine Tubes 591
Lesson 3: Ovary 593
Lesson 4: Mammary Gland 599

24. Histology: Nervous System — 603

Lesson 1: Spinal Cord 603
Lesson 2: Cerebellum 612

25. Histology: Integumentary System — 620

Lesson 1: Skin 620

26. Histology: Identification of Histological Slides At a Glance — 634

SECTION 4: RADIOLOGICAL ANATOMY

27. Radiological Anatomy: Introduction — 641

General Considerations 641

28. Radiological Anatomy: Upper Limb (Superior Extremity) — 647

Lesson 1: Shoulder Region 647
Lesson 2: Elbow Region 652
Lesson 3: Region of Wrist and Hand 657

29. Radiological Anatomy: Lower Limb (Inferior Extremity) — 662

Lesson 1: Hip Region 662
Lesson 2: Knee Region 668
Lesson 3: Ankle and Foot 674

30. Radiological Anatomy: Abdomen — 680

Lesson 1: Plain X-rays 680
- Abdomen 680
- Lumbosacral Spine 683

Lesson 2: Contrast X-rays 688
- Barium Meal X-ray (Contrast) 688
- Barium Follow Through 691
- Barium Enema 692

- Intravenous Pyelography *695*
- Hysterosalpingography *698*
- Choleycystogram *700*

31. Radiological Anatomy: Thorax — 703

Lesson 1: Plain X-rays *703*

Lesson 2: Contrast X-rays *709*
- Barium Swallow of Esophagus *709*

32. Radiological Anatomy: Head and Neck — 712

Lesson 1: Plain X-rays of Head and Neck Region *712*

Index — *731*

SECTION 1

Window Dissections

Section Outline

- ❖ Window Dissections: Introduction
- ❖ Window Dissections: Upper Limb (Superior Extremity)
- ❖ Window Dissections: Lower Limb (Inferior Extremity)
- ❖ Window Dissections: Abdomen
- ❖ Window Dissections: Thorax
- ❖ Window Dissections: Head and Neck

Chapter 1

Window Dissections: Introduction

STUDY OF ANATOMY

Anatomy is the science which deals with the macroscopic (gross) and microscopic (cytological and histological) study of any animal. Our subject matter is to study human body which we call it '**human anatomy**'. It is one of the important basic subjects in medical science. So, we should learn the subject from the very begining. The term 'Anatomy' has been derived from the Greek Word **ana—'up' + tomia—'cutting' means 'cutting up'**. There are various methods in our hands by means of which anatomical studies can be carried out. Our knowledge in this discipline is mostly based on studies on the cadaver (dead body). In cadaveric anatomy we can study the subject in two ways—systematically and regionally. We study the different systems of human body such as locomotor system, visceral system (splanchnology), cardiovascular system, nervous system, integumentary system and the organs of special senses in systematic anatomy. In regional anatomy, the whole body is subdivided into five regions, such as—head and neck, superior extremity, thorax, abdomen and inferior extremity. Various structures (bones, muscles, vessels, nerves, lymphatics, joints, etc.) of a particular region of the body are studied together with their relation to one another in regional anatomy. In this book, the dissections of different parts of the body will be discussed regionally.

DISSECTION

(Latin, dissect—dissecare 'cut up')

Dissection is the 'cutting up' of a body methodically in order to expose and study of its internal parts. Dissection forms the base of anatomical knowledge. We need cadavers or dead bodies for dissection. These cadavers help us to gain knowledge. The value of the gift that the cadavers provide for us cannot be measured. So the dissector should take proper care and should be well conscious regarding the proper use of the cadavers. The cadavers are kept in preservative fluid. If a part or whole body is allowed to dry, the texture and appearance of the organs of the body will be altered. Therefore, one should expose only those parts of the body to be dissected. The dissected part should be kept moist by warpping it with cotton soaked in formalin or at least water.

Though the study of anatomy is carried out by dissection on dead preserved body, the student should remember that the purpose of such studies is to allow him to visualize the living body in action. The method of learning will be more meaningful provided the student approaches it with an enquiring mind.

Common Instruments for Dissection

A student should have few personal dissection instruments while entering into a dissection hall. The instruments are as follows (Fig. 1.1):

i. **Scalpel:** It is a small surgical knife usually having a convex edge. It is consisting of a handle made of metal and a sharp blade of 3.5 to 4 cm long. The blade is removable and can be replaced by a new blade. Scalpels are not recommended for general dissection because the sharp edge of the knife may cut small structures. The knife is usually reserved for cutting the skin and the dense layer of deep fascia.

ii. **Forceps:** It is a two-bladed instrument with a handle for compressing or grasping tissues. Two pairs of forceps are needed in dissection. One pair should have blunt and rounded tips and the gripping surfaces should be corrugated. Other pair of forceps should have teeth at its tips for gripping tissue. These forceps are known as toothed forceps.

iii. **Scissors:** Two pairs of scissors are required. A large and heavy pair of about 15 cm long for transection (cutting in two in the transverse plane) and blunt dissection. The second pair is a small pair of scissors with two sharp points for sharp dissection of delicate structures. In this context, one should know what blunt or sharp dissection is. **Blunt dissection** is the dissection accomplished by separating tissues along natural cleavage lines without cutting. It is done with a probe, fingers or scissors. **Sharp dissection** is accomplished by incising tissues with a sharp edge. It is done by using a scalpel and small fine scissors.

iv. **Probe:** A long, slender instrument for exploring structures. The tip of this primary dissecting tool is slightly bent and blunt to tear connective tissue.

v. **Artery forcep (hemostat):** It is used for holding structures. It has powerful grasping capacity. Two types of hemostat are usually recommended. One is medium-sized straight or curved variety and other one is small, straight or curved (mosquito forcep).

vi. **Tissue forcep (of Allis):** It is a powerful grasping tool having multiple teeth at the tips of the blades. This instrument is sometimes needed to grasp tough structures.

vii. **Hand lens:** Though it is not absolutely essential, it is especially useful as an aid to bridging the gap between gross (macroscopic) and microscopic anatomy.

RELATIVE AND DESCRIPTIVE TERMS IN ANATOMY

When encountering a structure during dissection, one should be aware of its position, its relationship to other structures and so on. Structures are described with the body in anatomical position. The **anatomical position** (Fig. 1.2) of a person means the standing erect position with eyes looking forward, feet together, arms by the sides with the palms of the hands facing forward. Though during dissection the cadaver lies on the table in supine or prone position, one should correlate the structures in its anatomical position.

Window Dissections: Introduction

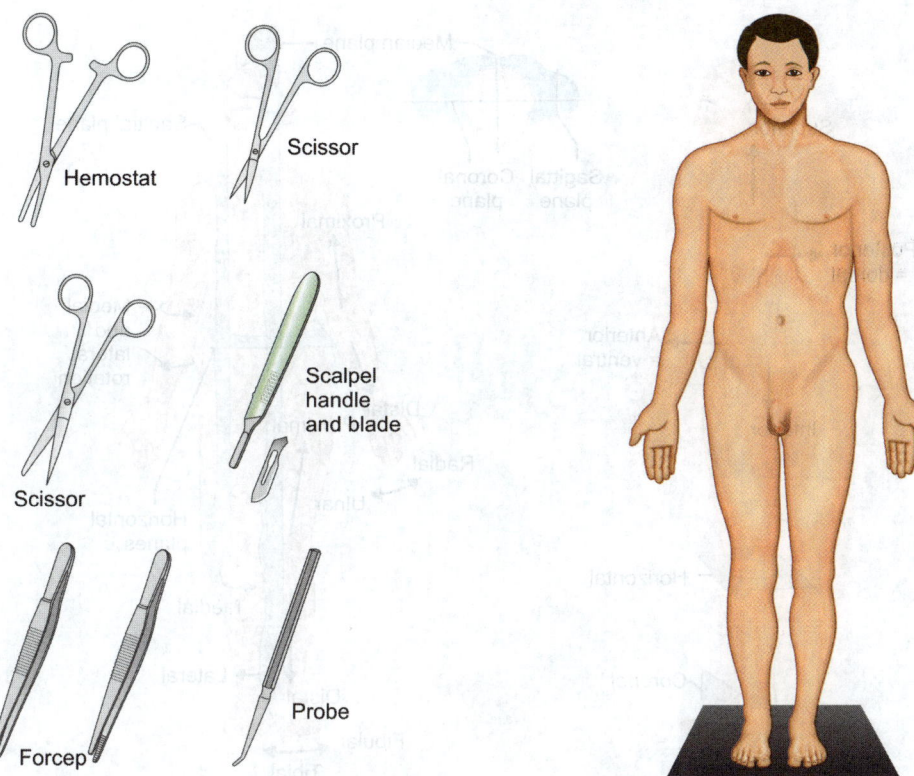

Fig. 1.1: Instruments for dissection

Fig. 1.2: Anatomical position

Anatomical planes (Fig. 1.3): For better understanding of the relations of the deeper structures, section is made through the different parts of the body. This line of section is called the **plane of section**.

The different planes are:
  i. **Median plane or mid-sagittal plane:** Median means in the middle. The median plane is an imaginary vertical anteroposterior plane that passes through the sagittal suture dividing the body into two apparently equal halves, right and left. A sagittal plane passes through any part of the body parallel to the median plane.
 ii. **Coronal plane:** It is an imaginary vertical side to side plane that cuts the sagittal plane at right angle.
iii. **Transverse or horizontal plane:** This plane cuts the long axis of the body or part of the body at right angle.

DIFFERENT TERMINOLOGIES

- **Anterior or ventral:** Situated at or directed towards the front of the body.
- **Posterior or dorsal:** Looking towards the back.
- **Medial:** Nearer the median plane.

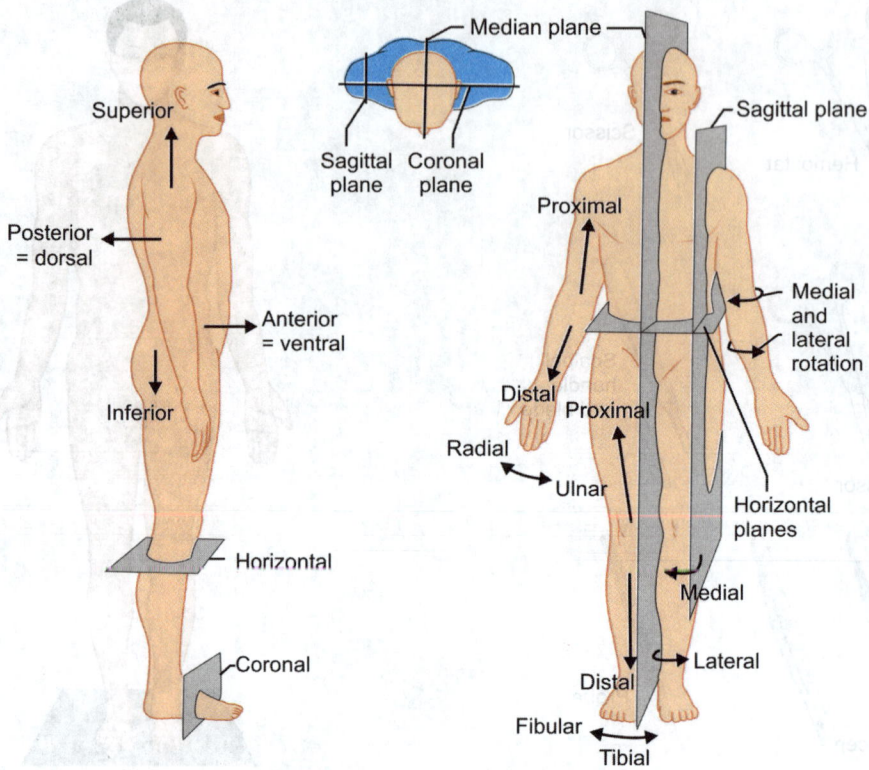

Fig. 1.3: Anatomical planes of the body

- **Lateral:** Away from the middle line of the body.
- **Median:** On the middle line of the body.
- **Superior:** Looking upwards, i.e. nearer the head of upright body.
- **Inferior:** Looking downwards, i.e. nearer the feet.
- **Cranial or cephalic or rostral:** Close to the head or headwards.
- **Caudal:** Towards the tail or away from the head.
- **Proximal:** Nearest to the point of reference.
- **Distal:** Farther from any point of reference.
- **Ipsilateral or homolateral:** Refers to the same side of the body.
- **Contralateral:** Refers to the opposite side of the body.
- **Superficial:** Close or nearer to the skin.
- **Deep:** Further from the skin.
- **Supine:** Lying with the face upwards, or on the dorsal surface.
- **Prone:** Lying face downwards.
- **Interior:** Inside.
- **Exterior:** Outside.

- **Invagination (Latin vagina = sheath):** It is defined as inward bulging of the wall of a cavity.
- **Evagination:** Outward bulging of the wall of a cavity.
- **Raphe:** The line of union of the halves of various symmetrical parts.
- **Bursa (Latin bursa = a purse):** A synovial bursa is a closed sac differentiated out of areolar tissue. It is a lubricating device and by diminishing friction it allows free movement. Accordingly, **bursa may be of three types: (1) Subcutaneous, (2) Subtendinous and (3) Articular**.
- **Synovial sheath:** A tubular bursa that envelops a tendon. It is required only where a tendon is subjected to friction or pressure.
- **Anastomosis:** Communication between vessels by collateral channels.
- **Tributary (vein):** A small vein flowing into a larger vein.
- **Fossa:** A hollow or depressed area (e.g. cubital fossa, popliteal fossa).
- **Retinaculum:** A structure (band of connective tissue) that retains an organ or tendons or other tissues in place.

Some Movements

- **Flexion:** Bending anteriorly or the act of bending.
- **Extension:** The movement by which the two ends of any jointed part are drawn away from each other.
- **Adduction (Latin ad = to):** The movement towards the median plane or towards the axial line of a limb (in case of digits).
- **Abduction (Latin ab = from):** The movment away from the median plane or from the axial line of a limb (in case of digits).
- **Rotation:** The process of turning around an axis. In the limbs, lateral and medial rotations refer to the direction of movement of the anterior surface.
- **Circumduction (Latin circum = around):** Circular movement of a limb, i.e. movement of flexion, abduction, extension and adduction in sequence, thereby describing a cone.
- **Pronation:** The act of turning the palm backward (posteriorly).
- **Supination:** The act of turning the palm forward (anteriorly).
- **Inversion:** The act of turning the sole (plantar aspect) of the foot inward.
- **Eversion:** The act of turning the sole of the foot outward.
- **Protract (Latin pro = forward):** To move forward.
- **Retract:** To move backward.

STRUCTURES ENCOUNTERED IN DISSECTION

Skin
(Latin cutis = skin, Greek derma = skin)
This is the first layer to come across in dissection in any parts of the body.

Functional Aspect of Skin

The outer covering of the body is the skin. It is a waterproof envelope preventing the evaporation and escape of tissue fluids. It acts as an excretory organ and regulator of body temperature. It is the most extensive sense organ of our body. Its surface area in an average adult is 1.7–2 m^2.

The skin presents a number of lines all over the body. These are **cleavage lines or Langer's line**. These lines are formed due to the collagen bundles of the dermis. In the trunk and neck the cleavage lines are horizontal but in the limbs they are mostly longitudinal. Any surgical inscision along these lines produces thin and linear scar because of better union.

Structural Aspect of Skin (Figs 1.4A to C)

Structurally, the skin consists of a superficial nonvascular epithelial component, the **epidermis** and a deeper vascular connective tissue component, the **dermis**. The skin thickness (epidermis + dermis) varies between 0.3 mm and 3 mm. Somewhere the skin is thick and somewhere it is thin. This thick or thin skin refers to the thickness of the cornified layer of epidermis (stratum

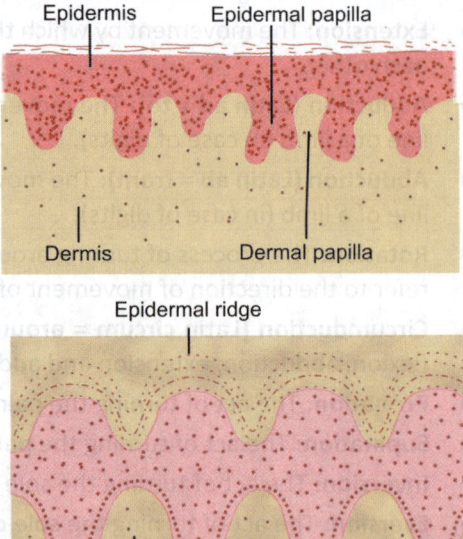

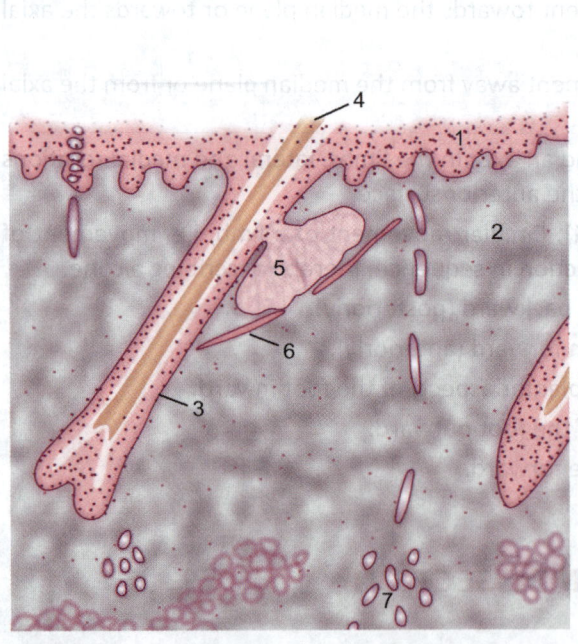

Fig. 1.4A: Structure of skin: (1) Epidermis, (2) Dermis, (3) Hair follicle, (4) Hair, (5) Sebaceous gland, (6) Arrector pili, (7) Sweat gland

Fig. 1.4B: Epidermal and dermal papillae

Window Dissections: Introduction

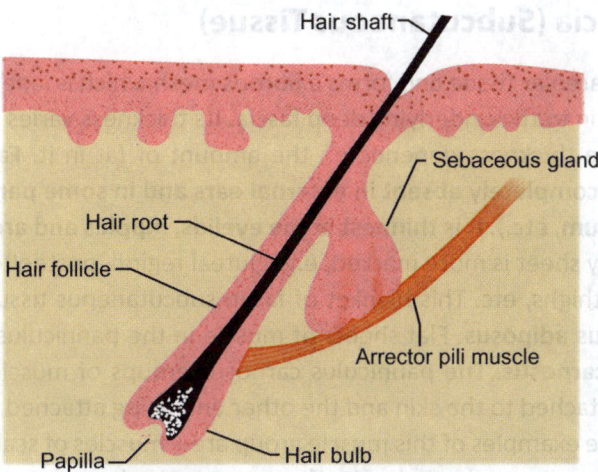

Fig. 1.4C: Structure of hair follicle

corneum). Where the stratum corneum is thick, paradoxically the dermis is thin and vice versa. Microscopically, **the epidermis is lined by keratinized (cornified) stratified squamous epithelium.** The cellular components of epidermis arranged from superficial to deep are **(i) stratum corneum (ii) stratum lucidum (iii) stratum granulosum (iv) stratum spinosum and (v) stratum basale**. The basement membrane of epidermis lies at dermoepidermal junction. The epidermis gets its nutrition by diffusion from nearest capillary loops of the dermis, as epidermis is an avascular layer. Deep to epidermis, the dermis (corium) is consisting of bundles of white fibers and elastic fibers, blood vessels, nerve fibers, lymphatics, etc.—all embedded in ground substance. It is in general 1–2 mm thick except on palms and soles where it is thicker. Structurally, dermis consists of a superficial **(i) papillary layer and (ii) deep reticular layer**.

The dermis contains various skin appendages like hair follicles, sebaceous and sweat glands. However, **sebaceous glands are absent on palms and soles**. Sweat glands are present all over the body. Smooth muscles of the dermis in connection with hair follicles are called **arrectores pilorum** (pili) muscles that pass obliquely from the epidermis to the slanting surface of the hair follicle. Contraction of these muscles causes erection of hairs. However, these muscles are absent in facial and axillary hairs, eyebrows and eye lashes, etc. **Arrector pili muscle is innervated by efferent autonomic fibers (sympathetic fibers).**

Developmental Aspect of Skin

The epidermal part of skin is developed from surface **ectoderm**. The skin appendages like nail, hair follicle, sebaceous and sweat glands are specialized derivatives of the epidermis. So they are also ectodermal in origin. One of the important cells of epidermis is melanocytes (melanin) on which color of the skin depends. It is also ectodermal in origin. The dermis, on the other hand, is **mesodermal** in development.

Superficial Fascia (Subcutaneous Tissue)

It is a layer of loose areolar tissue and forms a fibrous mesh which is laden with fat. It connects the dermis of the skin to the underlying deep fascia. Its thickness varies from region to region of the body and the thickness depends on the amount of fat in it. **Fat (adipose tissue) in superficial fascia is completely absent in external ears and in some parts of external genital organs (penis, scrotum, etc.). It is thinnest in the eyelids, nipples and areola of the breasts**. In some areas this fatty sheet is more marked, e.g. gluteal region, postdeltoid region, mammary gland, front of the thighs, etc. This blanket of fat in subcutaneous tissue deep to the skin is called the **panniculus adiposus**. Flat sheets of muscle in the panniculus adiposus are known as the **panniculus carnosus**. The panniculus carnosus groups of muscles are those muscles whose one end is attached to the skin and the other end being attached usually to deep fascia or bone. Some of the examples of this muscle group are—muscles of scalp and face, platysma, palmaris brevis, corrugator cuti ani, dartos of scrotum and subareolar muscle of the nipple. In the superficial fascia small blood vessels, lymph vessels with a few lymph nodes and nerves of the skin are found.

Deep Fascia

Deep fascia is an inelastic membrane of fibrous tissue that invests the structures deep to the superficial fascia. Deep fascia is well-marked in the neck and limbs where the fibers are arranged chiefly in circular fashion. Even in the neck it is arranged in three layers—investing, pretracheal and prevertebral. It is wrapped around the muscles, vessels and nerves like a bandage. Deep fascia sends partition or septa between the muscles and groups of muscles. These septa called intermuscular septa firmly anchored to the periosteum. The deep fascia never passes freely over the bone, rather firmly attached to periosteum. The deep fascia widely varies in thickness. It is thickened where muscles are attached to it. Somewhere it is thickened to form **retinacula** which keeps the tendons of muscles in position and prevents from bow stringing. Such retinacula are found around the wrist (**wrist band**) and ankle as flexor and extensor retinacula. However, **true deep fascia is absent in thorax, abdomen and face and in the ischiorectal fossa.** The structures (muscles, vessels, nerves) which are deep to deep fascia are called deep structures and superficial structures (superficial veins, subcutaneous lymph vessels or nerves) are superficial or outside the deep fascia. The deep fascia is pierced or perforated by these vessels and nerves. Deep fascia is very sensitive too. The muscles when contract compress the veins within and between them, thereby playing an important role in venous return from extremeties (antigravity flow). **The deep fascia is sometimes named according to their position in the body such as fascia lata (thigh), fascia cruris (leg), plantar aponeurosis (sole of foot), brachial fascia (arm), antebrachial fascia (forearm), palmar aponeurosis (palm of hand), fascia coli (neck), etc.**

Muscles

(Musculus, Latin = Mus, meaning Mouse)
Some muscles resemble a mouse in gross appearance and the tendons of these muscles represent their tails. **Muscles are of three varieties:** (i) Skeletal/striated/voluntary/striped, (ii) Visceral/smooth/non-striated/involuntary/unstriped and (iii) Cardiac muscles. We are concerned with the skeletal muscles because we come across mainly these muscles during dissection. **A skeletal muscle** consists of group of muscle fasciculi and covered by an areolar membrane called **epimysium** (fascia covering a muscle) (Fig. 1.5). Each muscle fasciculi is covered by another areolar membrane called **perimysium** (Fig. 1.5) and consists of a group of muscle fibers. Each muscle fiber again consists of a group of myofibrils and covered by another connective tissue (delicate layer of loose areolar tissue) called **endomysium** (Fig. 1.5). Basically, the term 'muscle fiber' and the 'muscle cell' are same thing and considered to be the structural unit of a muscle. Each muscle cell or fiber consists of a cell membrane (sarcolemma), protoplasma (sarcoplasma) and multiple nuclei.

The two chief component parts of a skeletal muscle are **(i) Fleshy and (ii) fibrous** (Fig. 1.6). The fibrous component may be **tendon** or **aponeurosis**. The junction between fleshy and tendinous part is known as myotendinous junction where both the parts are contiguous but not continuous. However, their connective tissue covering is continuous with one another. A tendon or aponeurosis is inelastic, noncontractile and relatively avascular, but it is immensely powerful. During contraction of a muscle, there is no change of the length of the tendon but the fleshy part of a muscle shortens between a third and a half of the resting length. The power of a muscle depends on the number and diameter of its fibers.

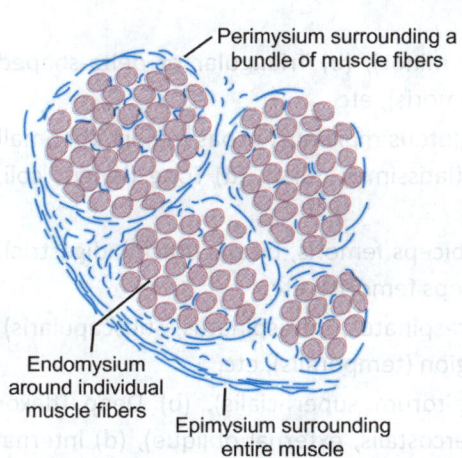

Fig. 1.5: Epimysium, perimysium and endomysium

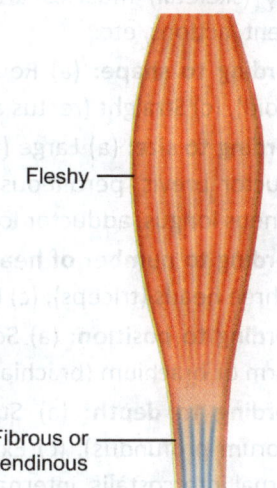

Fig. 1.6: Component parts of a skeletal muscle

Histologically, tendon and aponeurosis are similar. **A tendon** is a cord-like structure where the collagen fibers predominate and run parallel to one another. On other hand, **an aponeurosis** is sheet like, broad and membranous. Tendons are supplied only by the sensory nerves. But the fleshy part of a muscle is supplied by a mixed nerve having both motor and sensory filaments in 3:2 ratios. The nerve and blood vessels of a muscle enter its deep surface at an area which is less mobile. This oval area of entry of nerve and vessels is known as **neurovascular hilus**.

Origin and insertion of a muscle is a popular term found in anatomy. Conventionally, the fixed end (usually fleshy end) of a muscle is its origin and the movable end is its insertion (tendinous end). Usually, the upper attachment is the origin and the lower attachment is insertion. In the limbs, the proximal attachment is origin and the distal attachment is usually called insertion. When the two attachments of a muscle (origin and insertion) become closer to each other; it is conventionally described as the **'action' of a muscle**. Due to action of a muscle, it shortens due to contraction. To act on a joint (movement of joint) one muscle should cross at least one joint. Multiple joints may be crossed by a single muscle. Distal to the joint, the muscle may be inserted at the proximal (biceps brachii), middle (coracobrachialis) or distal (brachioradialis) end of a bone. During contraction or shortening, though the origin remains fixed and the other end (insertion) moves, the reverse can occur where the insertion remain fixed and the origin moves. This depends on circumstances and varies with most muscles. Recently, the term **'attachment'** is gaining popularity as an alternative for both origin and insertion. Proximal attachment for origin and distal attachment for insertion.

Nomenclature of Skeletal Muscles

Voluntary (skeletal) muscles are named according to their shape, size, position, depth, attachment, actions, etc.

 i. **According to shape:** (a) Round (teres major et minor), (b) Triangular or delta-shaped (deltoid), (c) Straight (rectus abdominis, rectus femoris), etc.
 ii. **According to size:** (a) Large (adductor magnus, gluteus maximus, psoas major), (b) Small (adductor brevis, peroneous brevis), (c) Broad (latissimus dorsi), (d) Long (longus coli, peroneus longus, adductor longus), etc.
 iii. **According to number of heads:** (a) Two heads (biceps femoris, biceps brachii, digastric), (b) Three heads (triceps), (c) Four heads (quadriceps femoris), etc.
 iv. **According to position:** (a) Scapular region (supraspinatus, infraspinatus, subscapularis), (b) Arm or brachium (brachialis), (c) Temporal region (temporalis), etc.
 v. **According to depth:** (a) Superficial (flexor digitorum superficialis), (b) Deep (flexor digitorum profundus), (c) External (external intercostalis, external oblique), (d) Internal (internal intercostalis, internal oblique), etc.
 vi. **According to attachments:** (a) Styloid process to hyoid bone (stylohyoid), (b) Coracoids process to brachium or arm (coracobrachialis), (c) Sternum and clavicle to mastoid process (sternocleidomastoid), etc.

Window Dissections: Introduction

vii. **According to action:** (a) Extensor, (b) Flexor, (c) Adductor, (d) Abductor, (e) Rotator.

viii. **According to function:** (a) Dilators (dilator nares), (b) Compressors (compressor nares), (c) Depressor (depressor labii inferioris), (d) Elevators (levator labii superioris), (e) Constrictor (constrictor pupillae), (f) Invertors, (g) Evertors, etc.

ix. **According to compositiom:** Semimembranosus, semitendinosus.

x. **According to contrasting feature:** Pectoralis major/pectoralis minor; gluteus maximus/gluteus minimus; Teres major/teres minor; Zygomaticus major/zygomaticus minior, etc.

xi. **According to direction of fibers:** (a) Oblique (obliquous capitis superior et inferior, external oblique, (b) Transverse (transversus abdominis, transversus thoracis), (c) Straight (rectus femoris).

According to the action of skeletal muscles, they are grouped as follows:

i. **Prime movers:** The muscle or a group of muscles which are directly responsible for causing a particular movement are called prime movers. Biceps brachii is an example of prime mover causing flexion of elbow joint.

ii. **Antagonists:** The muscles which oppose the action of prime movers and initiate opposite movement are called antagonists, e.g. triceps is the antagonist of biceps brachii.

Sometimes, the prime movers and antagonists' contract together to stabilize or fix a particular joint to allow movements at the distal joints. These group of muscles are called **fixator.** During abduction of shoulder by deltoid the scapula is fixed by fixators.

iii. **Synergists:** When the two muscles act together they perform a particular movement, but they are different in action while acting individually. These are synergists. Synergists prevent unwanted movement. Flexor carpi ulnaris causes adduction and flexor carpi radialis causes abduction of wrist when they act individually. But when they act simultaneously they cause flexion of wrist joint.

Gravity plays a very important role in **paradoxical action** of a muscle. When a prime mover of one particular movement helps the opposite movement by active lengthening of its fibers against gravity it is called **paradoxical action**. For example, **biceps brachii** which is a prime mover for flexion of elbow joint controls extension of the elbow. During sitting from standing position the **quadriceps femoris** which is an extensor of knee, acts as the flexor. The **deltoid** (abductor) helps in adduction of shoulder joint by lengthening against gravity.

Vessels

A vessel is a channel for carrying fluid, such as blood (**blood vessels**) or lymph (**lymph vessels**). Blood vessels comprise of: (i) arteries, (ii) capillaries and (iii) veins. Lymph vessels consist of: (i) lymphatic capillaries, (ii) collecting vessels or lymphatic vessels and (iii) lymph trunks.

Blood vascular system is a closed system of tubular passages which convey blood from the heart and circulate to the different parts of the body and again return to the heart. So the system starts from large arteries and ends in large veins. **Heart (left ventricle)** → large artery (aorta) → medium and small-sized artery → arteriole → terminal arteriole → meta-arteriole → capillaries or sinusoids → venules → small and medium sized veins → large vein (vena cava) → **heart (right atrium)**.

Structurally, the wall of an artery is consisting of three layers (Figs 1.7A and B). From outside inwards these coats or layers are (i) Tunica adventitia, (ii) Tunica media and (iii) Tunica intima. **Tunica adventitia** is the outermost and strongest of all coats. It is made up of connective tissue in which collagen fibers predominate. **Tunica media** is the thickest of all coats. It is made up of smooth muscles and elastic tissue. Depending upon the predominance of type of tissue in tunica media, the **artery are of two types:** (a) **Elastic artery** where tunica media consists mainly of elastic tissue and less muscle tissue (e.g. aorta, pulmonary trunk,

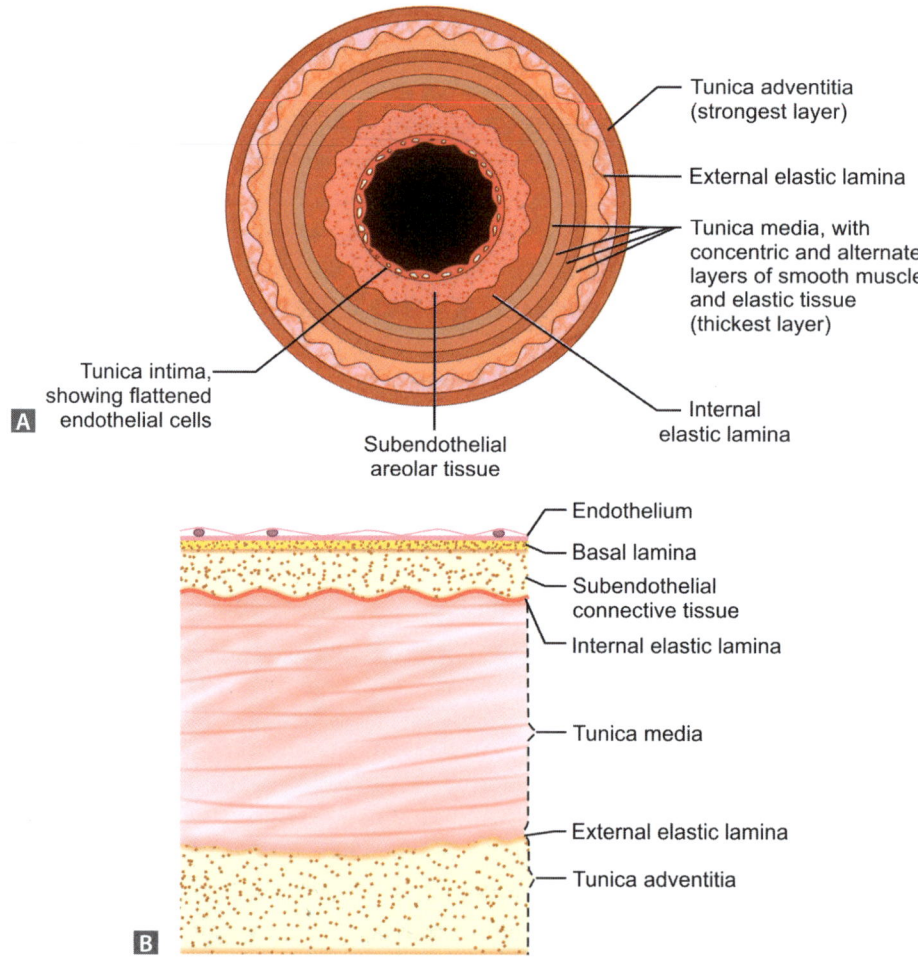

Figs 1.7A and B: (A) Cross-section of an artery; (B) Coats (layers) of an arterial wall

Window Dissections: Introduction

carotid artery, subclavian artery, etc.). (b) **Muscular artery** where muscle fiber predominates (e.g. all other distributing arteries). **Tunica intima** is the innermost coat. It is lined by **endothelium** which is a layer of flattened endothelial cells.

The structure of veins is similar to that of arteries. But the wall of a vein is thinner than a same sized artery. Unlike arteries, tunica adventitia of large vein is thicker than tunica media. Veins accompany arteries except in superficial fascia and they have the same tree-like pattern like arteries. A **tributary** is a stream (vein) flowing into a larger vein. As an artery may have multiple branches, a vein may also have many tributaries. Depending on the position of the veins, they are either superficial or deep. The superficial veins run in the superficial fascia and they are not accompanied by corresponding arteries.

On the other hand, the deep veins run deep to the deep fascia and accompanied by arteries. In the forearm and leg, the deep veins are arranged in pairs and run along the sides of the arteries. These types of veins are known as **venae comitantes** which help in venous return by the transmitted pulsation of the arteries towards the heart. Counter current heat exchange may occur between these arteries and veins.

Some veins called **perforating** or connecting veins which perforate the deep fascia and connect the superficial veins with the deep veins. Most of the veins of the body contain valves that allow unidirectional flow of blood towards heart and prevent regurgitation in opposite direction. Such valves are formed by the reduplication of tumica intima (endothelium) and made up of two semilunar cusps. Valves are more numerous in limbs and are commonly placed just before the mouth of a tributary.

Another commonly used term in relation to blood vessels is **anastomoses**. It means communication between vessels by collateral channels. Communications between adjacent arteries are called **interarterial** anastomosis and direct communication between arteriole and venule are called **arteriovenous anastomosis**. **End arteries** (Fig. 1.8) are those which do not anastomose with neighboring arteries, i.e. they do not form any precapillary anastomosis (e.g. central artery of retina, appendicular artery, etc.). The walls of large and medium-sized vessels (artery and veins) particularly the adventitia and media coats are supplied by small arteries called **vasa vasorum**. The small vessels and tunica intima of larger vessels are supplied by diffusion from blood in their lumen.

Blood vessels also have rich nerve supply. These are autonomic nerves (sympathetic) which consist of both efferent and afferent fibers.

Lymph vessels carry lymph (Latin lympha = pure, clear water) or tissue fluid from different parts of the body and convey into the blood vascular system (vein).

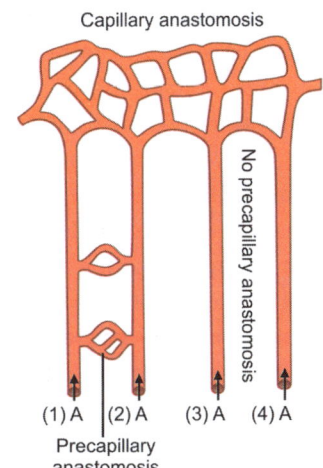

Fig. 1.8: End artery: A—arteries; (1 + 2) ordinary arteries with precapillary anastomosis; (3 + 4) end arteries without precapillary anastomosis

Though lymph vessels are more plentiful than veins, they tend to accompany veins and drain corresponding territories. So, the lymphatic system is auxiliary to venous system. The lymph vessel also consists of superficial and deep sets like that of veins in the limbs. The superficial set is accompanied by superficial veins but the deep set of lymph vessels are accompanied by arteries. Lymph vessels also have valves, more numerous than veins. Because of these valves the vessels are beaded in appearance. Though the vessels proper have plenty of valves, the lymph capillaries which begin blindly in the tissue spaces are valveless.

However, **lymph capillaries** are absent in epidermis, cornea, articular hyaline cartilage, brain, spinal cord, bone marrow, liver lobule, splenic pulp, etc. Blood capillaries are also absent in certain areas such as epidermis of skin, cornea, articular hyaline cartilage, hair and nails. **Lymph nodes** are bean-shaped structures which interrupt the flow of lymph and act as filters for lymph and factories for lymphocytes. The shape and size may also vary. They are usually situated along the blood vessel and arranged in groups such as cervical group, mediastinal group, axillary group, pre and para-aortic group, iliac group, inguinal group, popliteal group of lymph nodes and so on. **There are about 800 lymph nodes in human body.**

Nerves

A nerve is a delicate structure comprising a collection of nerve fibers that convey impulses between a part of the central nervous system and some other body region. Nerve cells or neurons are the structural and functional units of the nervous system (Fig. 1.9).

A cell body (soma) and processes or neurites (axons and dendrites) are the main parts of a neuron. Aggregations of cell bodies outside the brain and spinal cord are known as **ganglia**. Isolated masses of grey matter (cell bodies of neurons) present in the nervous system are referred to as **nuclei**. The axon of a neuron forms the structural basis of the nerve fiber. A single nerve fiber is surrounded by very finest fibril (connective tissue) called endoneurium which is visible only with electron microscope. Adjoining nerve fibers are held together by their endoneurium to form bundles or nerve **fasciculi.** Each fasciculus is surrounded by another layer of connective tissue called **perineurium**. Several fasiciculi are held together and surrounded by a layer of connective tissue called **epineurium**. In these connective tissue coverings the larger nerves have also their own nerves called **nervi nervorum** (vasa vasorum in larger arteries). In the epineurium of peripheral nerves there are free anastomoses of small arteries which are derived from the regional larger arteries. These vessels supplying the nerves are called **vasa nervorum** or more precisely **arterio nervorum**. The corresponding veins in

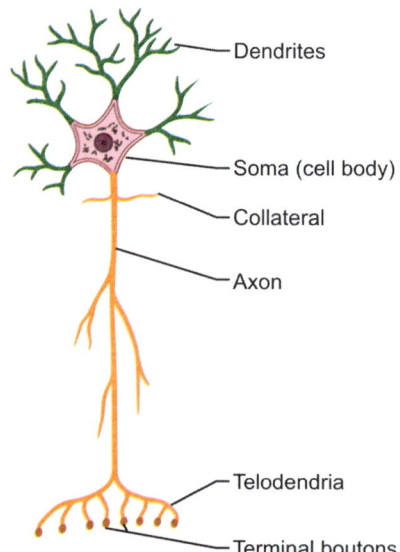

Fig. 1.9: Structure of a neuron

the epineurium of nerves have a similar pattern. These are called **venae nervorum**. But there is a **blood-nerve barrier** in the peripheral nerves. The circulating blood is separated from the nerve fibers by this barrier. This barrier is formed by tight junctions between the endothelial cells of the capillaries (nonfenestrated) in nerves and perineurium.

Nerves may be white or gray in color. The whiteness of the nerves is due to its myelin sheath. Accordingly, nerve fibers may be **myelinated** or **nonmyelinated**. Nonmyelinated nerves are gray in color. Myelin sheath is a white material of varying thickness consisting of alternate concentric layers of lipids and proteins. It is formed by the Schwann cells in peripheral nerves and by the oligodendrocytes in the nerves of central nervous system. The white matter of the brain, different nerve tracts in central nervous system (CNS), most of the somatic nerves (somatic fibers of less than 1 μm in diameter are nonmyelinated) and preganglionic fibers of autonomic nervous system are myelinated. Myelination increases the velocity of conduction of nerves. Around the myelin sheath there is a thin but tough membrane which is formed by the outer cell membrane and thin layer of cytoplasm of Schwann cell with its peripheral nuclei called **neurilemma sheath**. **However, this sheath is absent in the nerves of central nervous system and these nerves canot regenerate after injury due to the absence of this layer.**

There are some general principles of nerve supply. If a structure or part of the body is supplied by a nerve in the embryonic period (intrauterine period) it never alters thereafter. If the structure migrates in the developing fetus, it drags the nerve with it. The conspicuous example is the phrenic nerve (C_3, C_4, C_5) supplying the diaphragm which migrates from the cervical region to form abdominothoracic partition, dragging its nerve from the cervical segments (C_3, C_4, C_5) of spinal cord. One thing is to be kept in mind that in a few cases, muscle of a limb receives double nerve supply. Generally, the flexor muscles receive the second supply from the nerves of extensor compartment. This is due to fact that embryologically the muscle was developed in extensor compartment of the fetal limb and drags its nerve with it, but for functional reasons the muscle lies in the flexor compartment along with the nerve of flexor compartment. It is examplified by the nerve supply of brachialis which is supplied by both musculocutaneous nerve (flexor compartment) and radial nerve (extensor compartment).

Each spinal nerve (total 31 pairs) divides into anterior (ventral) rami and posterior (dorsal) rami (Fig. 1.10). **The posterior rami never form plexus.** It divides into a medial and lateral branch and supplies muscles and skin on the back. **But the larger anterior rami form somatic nerve plexus—cervical, brachial, lumbar and sacral.** In all vertebrates including human the muscle and skin (somatic supply) of the limbs are supplied by somatic plexus (e.g. brachial plexus to upper limb; lumbar and sacral plexus to lower limb).

The constituents of plexuses divide into anterior and posterior divisions to supply the flexor and extensor compartments of the limb respectively. The branches of the spinal nerves (peripheral nerves) supplying the limb muscles which are the derivatives of somatic nerve plexuses are not purely motor since they also contain sensory or afferent nerve fibers. So, these are mixed nerves with motor and sensory component in 3:2 ratios. The peripheral spinal nerves supply both skeletal muscle and skin (cutaneous). The amount of muscle innervated by a single nerve or one segment of spinal cord is called **myotome**. A **dermatome** is the area of skin supplied by a single spinal nerve. The succeding and preceding dermatomes are supplied by their corresponding succeding and preceding spinal nerves. Due to **overlaping** of adjacent dermatomes (Figs 1.11A and B), some area of skin is supplied by two adjacent consecutive spinal nerves. So, interruption of a single spinal nerve due to injury does not produce anesthesia. Flowchart 1.1 shows the classification of nervous system.

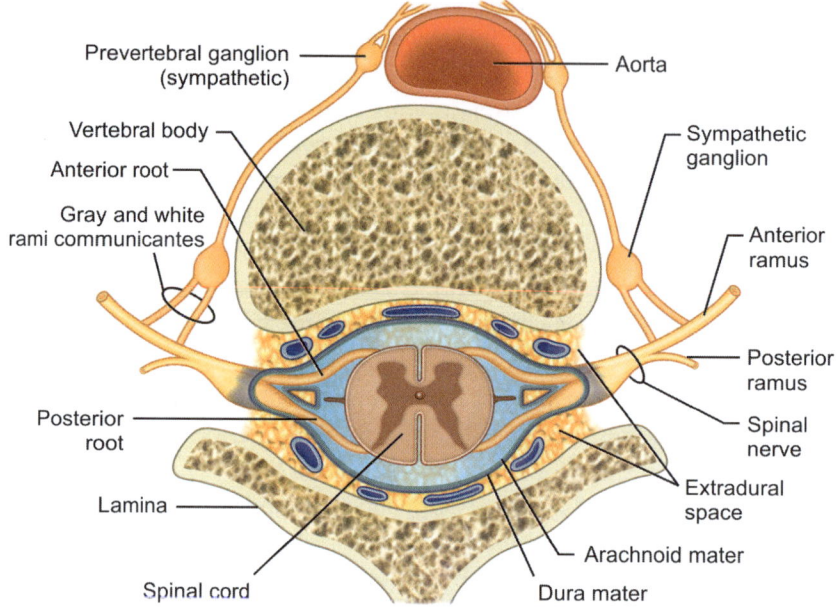

Fig. 1.10: Divisions of a spinal nerve

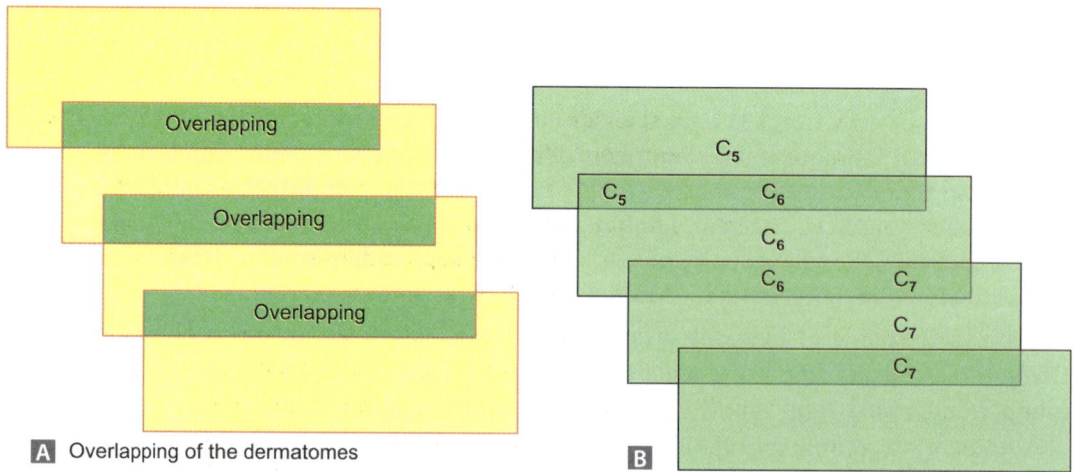

Figs 1.11A and B: Overlapping of dermatomes

Window Dissections: Introduction

Flowchart 1.1: Classification of nervous system

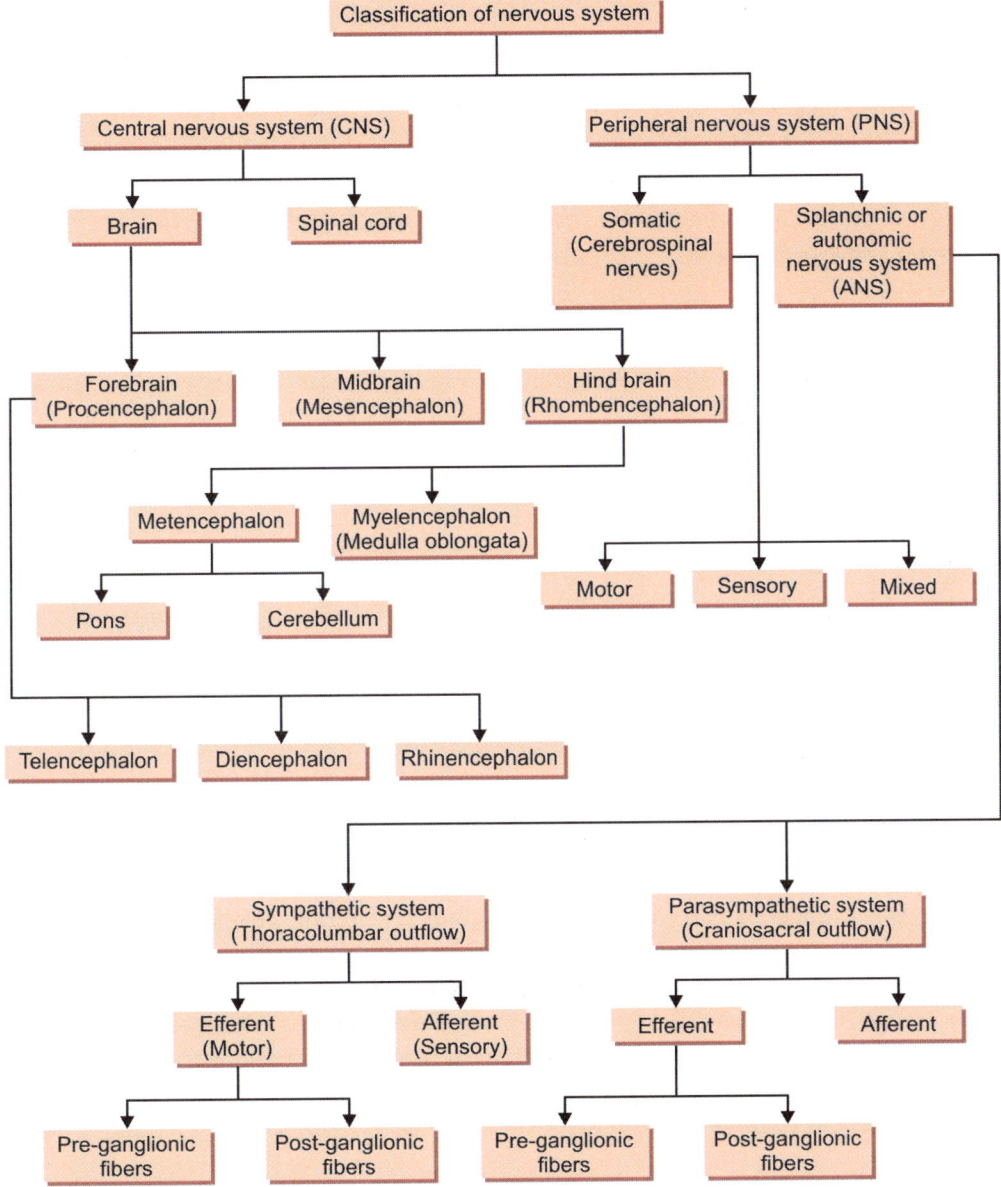

So, from the above discussion it is clear that the posterior rami of spinal nerves supply the extensor muscles of the vertebral column and skull and overlying skin. The anterior rami with or without forming plexus supply all other muscles of the trunk and limbs along with the skin at the sides and front of the neck and body **(anterior rami of thoracic spinal nerves ($T_2 - T_{12}$) do not form plexus, instead they form intercostal nerves).**

Ligaments

Ligaments are the thickenings of general mass of fascia and appear in response to tensile strength. It is strong, inelastic white fibrous tissue, collagen fibers being predominant and connects the bones at joints, serving to support and strengthen them. But some ligaments are a double layer of peritoneum and few are cord-like remnants of fetal tubular structures which become nonfunctional after birth.

Bursa

It is a fluid-filled sac or sac-like cavity lined by a smooth synovial membrane and appears in places where friction would otherwise occur. When a muscle, tendon or skin slide over bone or fascia, friction occurs between them and the sac-like bursa appears between them to reduce friction.

Joints

The site of junction or union between two or more bones is called a joint. On the basis of the connecting medium between the bones the joints may be (i) **Fibrous joint** (immovable), (ii) **Cartilaginous joint** (slightly movable) and (iii) **Synovial joint** (freely movable). Bones may articulate by their ends (long bones), by margins (flat bones) or by surfaces (flat or irregular bones). Flowchart 1.2 shows the classification of joints.

PROCEDURE AND STEPS OF DISSECTION (FIGS 1.12A TO C)

Position of the cadaver is important to dissect a particular part of body. To dissect an anterior part of body, the cadaver is placed in supine position and prone position is required for dissection on posterior aspect of the body. Some alteration in the position is done for proper exposure of the structures. Neck is extended and rotated for dissection of triangles of the neck; arm is abducted for axillary dissection, etc. The following steps are followed in dissection in most of the parts of the body with few exceptions in some specific regions.

Incision: It is the first step of dissection. It is very important to decide the line of incision so that the structures in the deep can be exposed properly. It is decided by locating the soft tissue and bony landmarks.

Skin is incised with the scalpel along the specified line of incision. Skin thickness varies from region to region. So, the force is applied to a certain extent without incising the deeper layers. In certain places, the skin and superficial fasica are incised together and reflected as a single sheet. But usually, skin and superficial fascia are incised and reflected separately. Skin removal is done carefully. After the skin incision is made, skin is held at the intersection of two incision lines by toothed forcep and by the help of the scalpel blade the skin is raised. Once it

Window Dissections: Introduction

Flowchart 1.2: Classification of joints

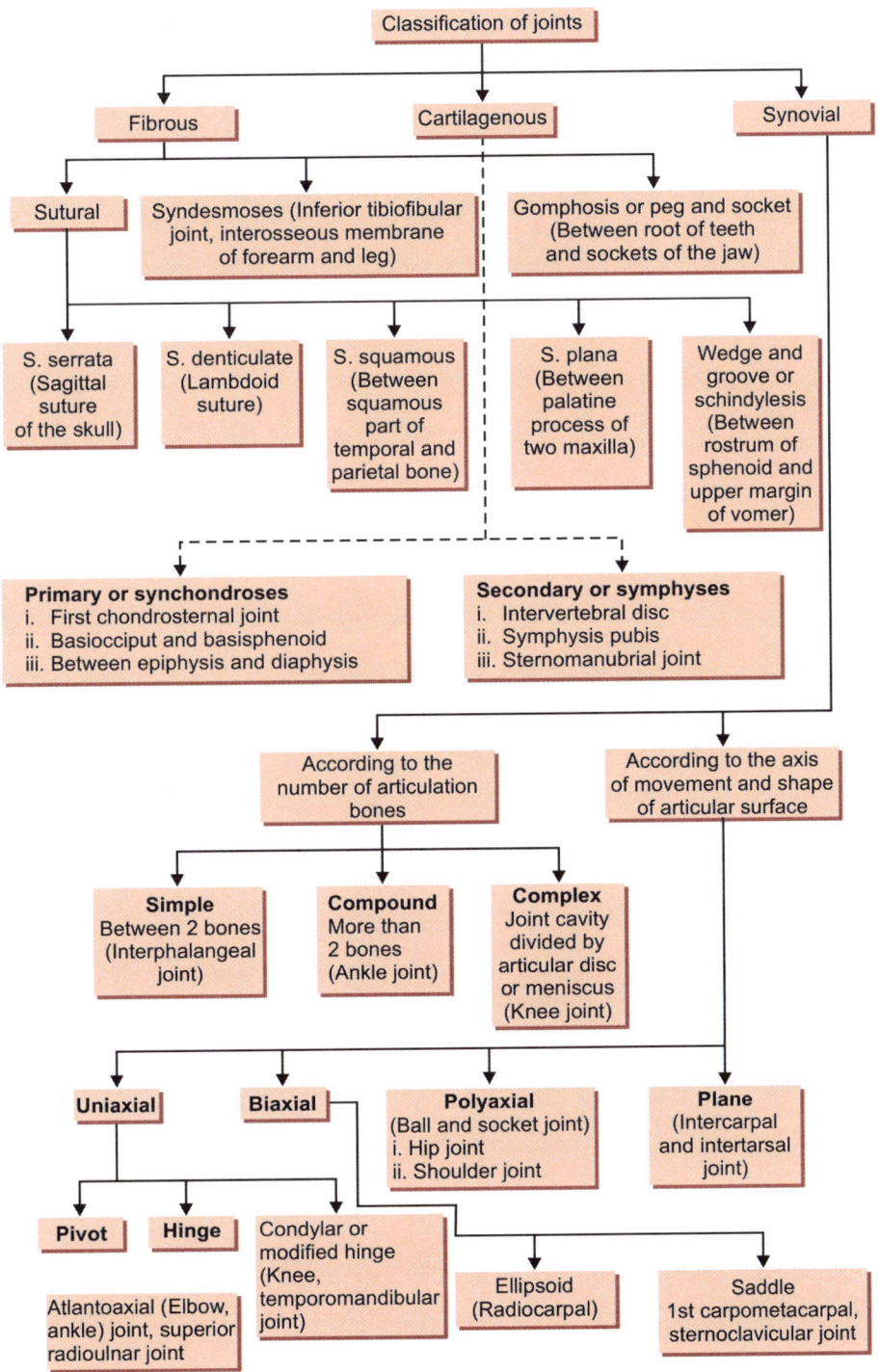

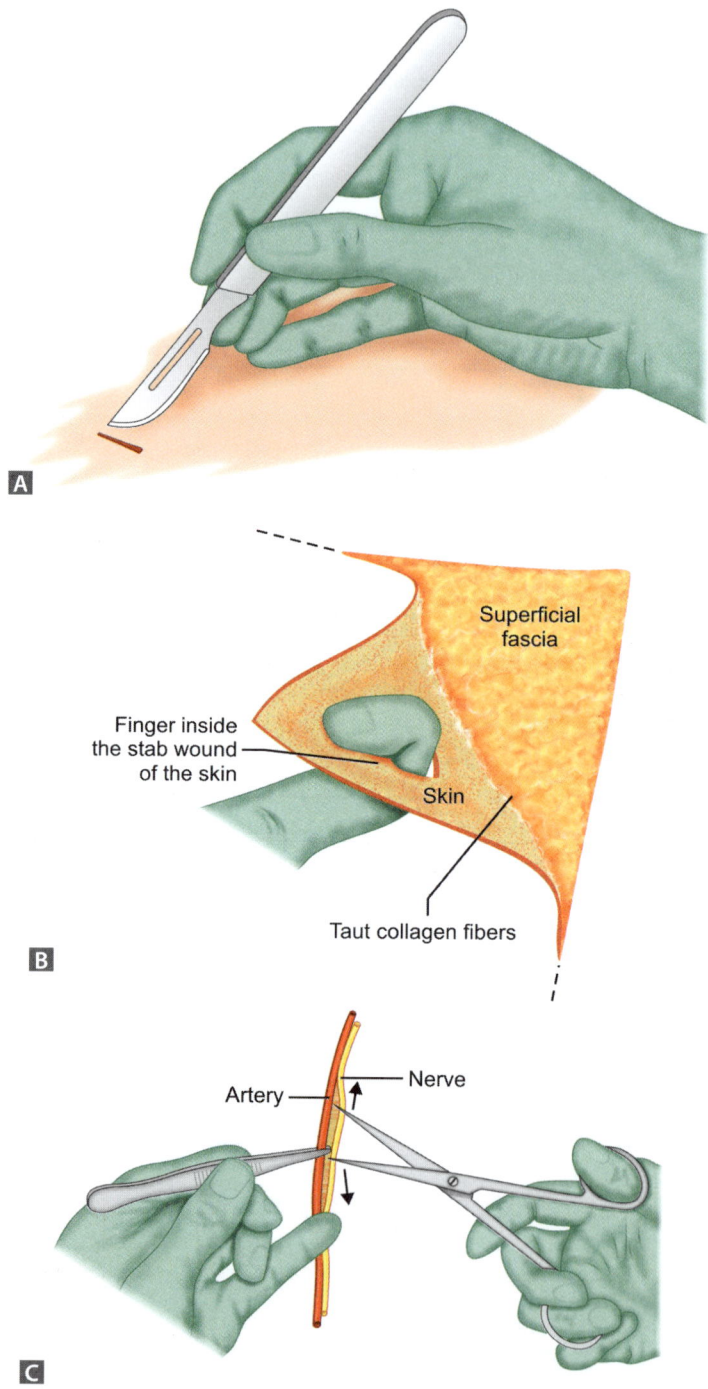

Figs 1.12A to C: (A) Skin incision by using the scalpel; (B) Exposure of superficial fascia; (C) Dissection to expose an artery and nerve

is raised, the flap of skin is retracted and collagen fibers are cut with the scalpel blade and the skin flap is reflected.

Superficial fascia comes into view after skin reflection. This fascia (subcutaneous tissue) contains fat, cutaneous nerves and superficial blood vessels. The nerves are usually accompanied by a small artery and one or more small veins. Sometimes larger veinss are also found in superficial fascia. These veins pierce the deep fascia and drain into deep veins. The fascia is then incised and reflected like that of skin. The fascia may also be separated from the deep fascia by blunt dissection. In this way, the blood vessels and nerves entering the superficial fascia through deep fascia are well-visualized. These structures are traced and studied.

The deep fascia is then cut at one point and scissor is introduced inside the cut area and by separating the blades of scissor this dense fibrous sheet is freed from the underlying structures. Then the whole length of the fascia is cut and reflected sideways. During this process of dissection, the deep fascia can also be made free from the deep structures by blunt dissection. The fascia is seen to be pierced by cutaneous nerves running from nerves trunks to supply the skin. The superficial vessels also pierce this matrix of fibrous connective tissue. Deep to deep fascia the main structures are muscles, large arteries with their branches deep veins with tributaries and nerves with branches.

Muscles form the main bulk of the deep structure. Every muscle is covered by a delicate membrane called epimysium. Muscles are separated from each other by blunt dissection using fingers. As far as practicable each muscle should be exposed properly and lifted from its bed. Then the muscle is traced proximally and distally to find out its attachments onto the bone, because this will determine the function of the muscle. Neurovascular bundles enter into the muscle. Look for the bundle on its surfaces and follow the vessels and nerves that are forming the bundle upto the main nerve trunk and vessels from which they arise. Clean the whole area by using forceps and fine scissors. Sometimes blunt dissection is required.

Try to find out the different branches of the main nerve trunk and branches of the main artery and also the tributaries of the large deep veins. In most of the cases, arteries are accompanied by veins. Variations in the arrangements of these structures are also to be kept in mind. **Blood vessels** (arteries and veins) are hollow tubular structures and they contain blood in living body. In the cadaver, they are identified by rolling them between the fingers and lumen must be felt. The arteries are thick-walled, but the veins are thin walled and wider than the arteries. The veins may contain clotted blood and the wall looks dark whereas the arterial lumen is empty and sometimes decomposed air bubbles appear within its lumen. Veins are more numerous than arteries. Sometimes the vein and its tributaries are stripped (removed) from the dissection field to see the artery and other related structures more clearly by blunt dissection using a probe. **The nerve** is not having a lumen. It is whitish in color. So, it is well differentiated from vessels which have lumen and may contain liquid blood or blood clots (veins). But a nerve may be mistaken for tendon of a muscle (median neve for palmaris longus tendon). They can be differentiated by the dull appearance of the nerve whereas a tendon is glistening and avascular.

PROBABLE QUESTIONS AND ANSWERS

1. What do you mean by midsagittal plane?

Ans. It is an imaginary vertical anteroposterior plane that passes through the sagittal suture dividing the body into two equal halves.

2. What is coronal plane?

Ans. It is an imaginary vertical side to side plane that cuts the sagittal plane at right angle.

3. What is the anatomical position of a person?

Ans. The anatomical position of a person means the standing erect position with eyes looking forward, feet together, arms by the sides with palms of the hands facing forward.

4. What are the major layers of skin?

Ans. Two layers:
   i. Outer nonvascular epidermis.
   ii. Inner vascular connective tissue component dermis.

5. What is the thickness of the skin (Epidermis + Dermis)

Ans. The thickness varies between 0.3 mm and 3 mm.

6. What is cleavage lines (Langer's line)?

Ans. Cleavage lines are those lines that are found in the skin all over the body and formed due to the collagen bundles of the dermis. In the trunk and neck, these lines are horizontal but in the limbs they are mostly longitudinal.

7. What is arrectores pilorum?

Ans. Arrectores pilorum is the muscle of the skin that passes obliquely from the epidermis to the slanting surface of the hair follicle. Contraction of these muscles causes erection of hairs. These are supplied by sympathetic nerve.

8. Name the sites where arrectores are absent?

Ans. Facial and axillary hairs, eyebrows and eyelashes, etc.

9. What are the appendages of skin?

Ans. Nails, hair follicles, sebaceous glands and sweat glands.

10. What is the developmental source of skin?

Ans. Epidermis and skin appendages (specialized derivatives of epidermis) are ectodermal and dermis of the skin is mesodermal in origin.

Window Dissections: Introduction

11. What is the lining epithelium of skin?

Ans. Skin is lined by keratinized (cornified) stratified squamous epithelium.

12. What is superficial fascia?

Ans. Superficial fascia is a layer of loose areolar tissue and forms a fibrous mesh which is laden with fat. It connects the dermis of the skin to the underlying deep fascia.

13. What is panniculus adiposus?

Ans. The blanket of fat in subcutaneous tissue deep to the skin is called panniculus adiposus.

14. What is panniculus carnosus?

Ans. Flat sheets of muscle in the panniculus adiposus are known as the panniculus carnosus. The panniculus carnosus group of muscles are attached to the skin at one end and to the deep fascia or skin at the other end.

15. Give examples of panniculus carnosus group of muscles.

Ans. Example: platysma, palmaris brevis, corrugator cuti ani, dartos muscle of scrotum, muscles of scalp and face, etc.

16. What is deep fascia?

Ans. Deep fascia is an inelastic membrane of fibrous tissue that invests the structures deep to the superficial fascia.

17. What is retinaculum?

Ans. Somewhere, the deep fascia is thickened and forms retinacula which keeps the tendon of muscles in position and prevents bow-stringing.

18. What do you mean by wrist band?

Ans. Retinacula around the wrist joint are called wrist band. These bands prevent bow-stringing of muscle.

19. Name the areas where deep fascia is absent.

Ans. Abdomen, thorax, face and ischiorectal fossa.

20. What do you mean by deep structures?

Ans. The structures which are deep to deep fascia are called deep structures (e.g. muscles, deep veins, arteries and nerves).

21. What are the different kinds of muscles?

Ans. Three kinds:
  i. Skeletal muscle
  ii. Smooth muscle
  iii. Cardiac muscle.

22. What are the chief component parts of a skeletal muscle?

Ans.
  i. Fleshy part
  ii. Tendon or aponeurosis.

23. What are the differences between a tendon and an aponeurosis?

Ans. Though histologically both are similar, a tendon is a cord-like structure where the collagen fibers predominate; on the other hand, an aponeurosis is sheet-like, broad and membranous.

24. What do you mean by origin and insertion of a muscle?

Ans. Conventionally, the fixed and fleshy end of a muscle is its origin. Usually, the upper or proximal attachment is origin. On the other hand, the movable and tendinous end is its insertion. Usually, the lower or distal attachment is insertion.

25. What do you mean by the 'action of a muscle'?

Ans. When the two attachments (origin and insertion) of a muscle become closer to each other, it is conventionally described as the 'action of a muscle'. Due to action of a muscle, it shortens due to contraction.

26. What is prime mover? Give example.

Ans. Prime movers are those muscles or group of muscles which are directly responsible for causing a particular movement of a joint. Biceps brachii is an example of prime mover causing flexion of elbow joint.

27. What are antagonists? Give example.

Ans. The muscles which oppose the action of prime movers and initiate opposite movement are called antagonists, e.g. triceps is the antagonist of biceps brachii.

28. What is meant by synergists?

Ans. When the two muscles act together, they perform a particular movement, but they are different in action while acting individually. These are synergists.

Window Dissections: Introduction

29. What is the paradoxical action of muscle? Examplify.

Ans. When a prime mover of one particular movement helps the opposite movement by active lengthening of its fibers against gravity is called the paradoxical action of a muscle. The deltoid (abductor) helps in adduction of shoulder joint by lengthening against gravity.

30. What are the coats (structural layers) of a blood vessel (artery/veins)?

Ans. From outside inwards, the coats are tunica externa, tunica media and tunica interna.

31. What is meant by anastomosis?

Ans. Anastomosis means communication between vessels by collateral channels.

32. What is a tributary of a vein?

Ans. A tributary is a smaller vein flowing into a larger vein.

33. What do you mean by venae comitantes?

Ans. In the forearm and leg, the deep veins are arranged in pairs and run along the sides of the arteries. These types of veins are called venae comitantes.

34. What is the function of venae comitantes?

Ans.
i. Venae comitantes help in venous return towards the heart by the transmitted pulsation of the arteries.
ii. Counter-current heat exchange may occur between these arteries and veins.

35. What is meant by end artery? Give example.

Ans. End arteries are those which do not anastomose with neighboring arteries, i.e. they do not form any precapillary anastomosis, e.g. central artery of retina, appendicular artery, etc.

36. How do you differentiate an artery from a vein in cadavers?

Ans.
i. The arteries are thick-walled. The veins are thin-walled but wider than the corresponding artery.
ii. The veins may contain clotted blood and the wall looks dark, whereas the arterial lumen is empty and sometimes decomposed air bubbles appear within its lumen.

37. Why does a peripheral nerve look white?

Ans. The whiteness of the nerve is due to its myelin sheath. Myelin sheath is a white material of varying thickness consisting of alternate concentric layers of lipids and proteins.

38. Which structural component of a nerve is required for its regeneration after injury?

Ans. Neurolemna sheath of a nerve is required for its regeneration following injury. But this sheath is absent in the nerves of central nervous system.

39. How do you differentiate a nerve from a blood vessel?

Ans.
i. A blood vessel is having a lumen which can be felt by rolling it between fingers, but the nerve is not having such lumen.
ii. A vessel (vein) may contain blood clots, but a nerve does not.

40. What is vasa vasorum?

Ans. The smaller arteries which supply the walls of large and medium-sized vessels are called vasa vasorum.

41. What do you mean by dermatome?

Ans. A dermatome is the area of skin supplied by a single spinal nerve.

42. What is meant by myotome?

Ans. The amount of muscle innervated by a single nerve or one segment of spinal cord is called myotome.

43. What is bursa?

Ans. A bursa is a closed synovial lubricating sac and allows free movement by diminishing friction.

44. What are ligaments?

Ans. Ligaments may be the thickenings of general mass of fascia or fold of a double layer of peritoneum or cord-like remnants of fetal tubular structures.

45. What is a joint?

Ans. A joint or articulation means union between two or more bones.

46. What are the types of joints?

Ans. Three types:
   i. Fibrous (inmovable)

Window Dissections: Introduction

ii. Cartilaginous (slightly movable)
iii. Synovial (freely movable)

47. Name the various types of movements of a joint.

Ans. Flexion, extension, abduction, adduction, circumduction, rotation (medial and lateral), gliding.

48. What are the different axis of movements?

Ans. Anteroposterior axis (abduction and adduction), transverse axis (flexion and extension) and vertical axis (rotation).

CHAPTER 2

Window Dissections: Upper Limb (Superior Extremity)

- Introduction to Superior Extremity
- Clavipectoral Fascia (Pectoral Region)
- Triangular and Quadrangular Space
- Axilla (Armpit)
- Front of Arm (Brachium)
- Cubital Fossa
- Front of the Forearm
- Palm of the Hand
- Back of the Arm
- Back of the Forearm
- Dorsum of the Hand

LESSON 1: INTRODUCTION TO SUPERIOR EXTREMITY

The upper limb hangs on each side of the trunk of the body with the palms of the hands facing forwards. In fetal life, the limbs project outwards at right angles to the body. Axial lines extend from the trunk onto the limbs. **Axial line** is said to be the line of junction of two dermatomes supplied from discontinuous spinal levels. Though adjacent dermatomes overlap, no overlaping occurs across the axial lines. **The anterior axial line** extends from the sternal angle → second costal cartilage → front of the limb → wrist. **The posterior axial line** runs from c_7 spine (vertebral prominens) → back of the arm up to the level of insertion of deltoid. The thumb and radius form the **preaxial** border, and ulna and the little finger form the **postaxial** border in the upper limb (Fig. 2.1). On assuming the anatomical position, the upper limbs undergo axial rotation of 90° laterally bringing the palm to face anteriorly.

The upper limb has the power of grasping or holding the objects in hands and has wide range of movements at shoulder joint. The force is transmitted from the upper limb to the axial skeleton via the bones of the upper limb. Axial skeleton includes: (1) skull bones, (2) vertebrae, (3) ribs and (4) sternum. **The line of force (weight) transmission is as follows** (Fig. 2.2):

  i. **Carpal bones** → Radius (via radio → carpal or wrist joint) → maximum force transmitted to ulna (via interosseous membrane) → humerus (via elbow or humeroulnar joint) → head of the humerus → shoulder joint → glenoid cavity of scapula → coracoid process of scapula → coracoclavicular ligament → clavicle (junction of medial 2/3rd and lateral 1/3rd) → **Either,** sternoclavicular joint → **sternum (axial bone) or,** costoclavicular ligament → **ribs (axial bone).**
  ii. Another route of transmission may also be possible, i.e. from humerus → head of humerus → acromion process with coracoacromial ligament → acromioclavicular joint → clavicle → **sternum (axial bone).**

Window Dissections: Upper Limb (Superior Extremity)

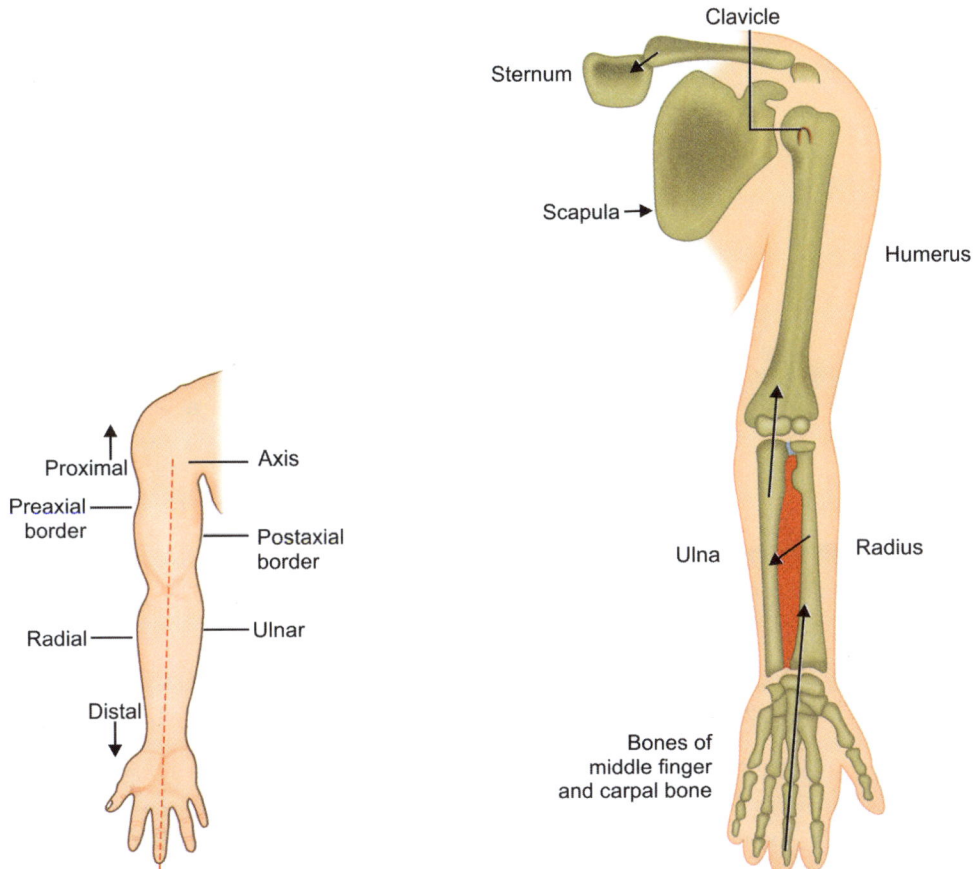

Fig. 2.1: Preaxial and postaxial borders of upper limb

Fig. 2.2: Line of force transmission

As the clavicle plays an important role in weight transmission, it is called modified long bone though the other features of a typical long bone are lacking. For descriptive purpose, the upper limb is mainly divided into four regions such as:

1. **Shoulder region** which includes:
   a. Pectoral region.
   b. Scapular region.
   c. Axilla or armpit.

   The bones of the **shoulder girdle (pectoral girdle)** are clavicle or collar bone (Fig. 2.3) and scapula or shoulder blade (Fig. 2.4). These bones connect the upper limb bones with axial bones. **A limb girdle is defined as a bone or bones that connect a limb to the axial skeleton.**

2. **Arm or brachium** which extends from shoulder to elbow is having only one bone called humerus (Fig. 2.5).

3. **Forearm or antebrachium** extends from elbow to wrist joint and the bones of this region are radius (Fig. 2.6) and ulna (Fig. 2.7).

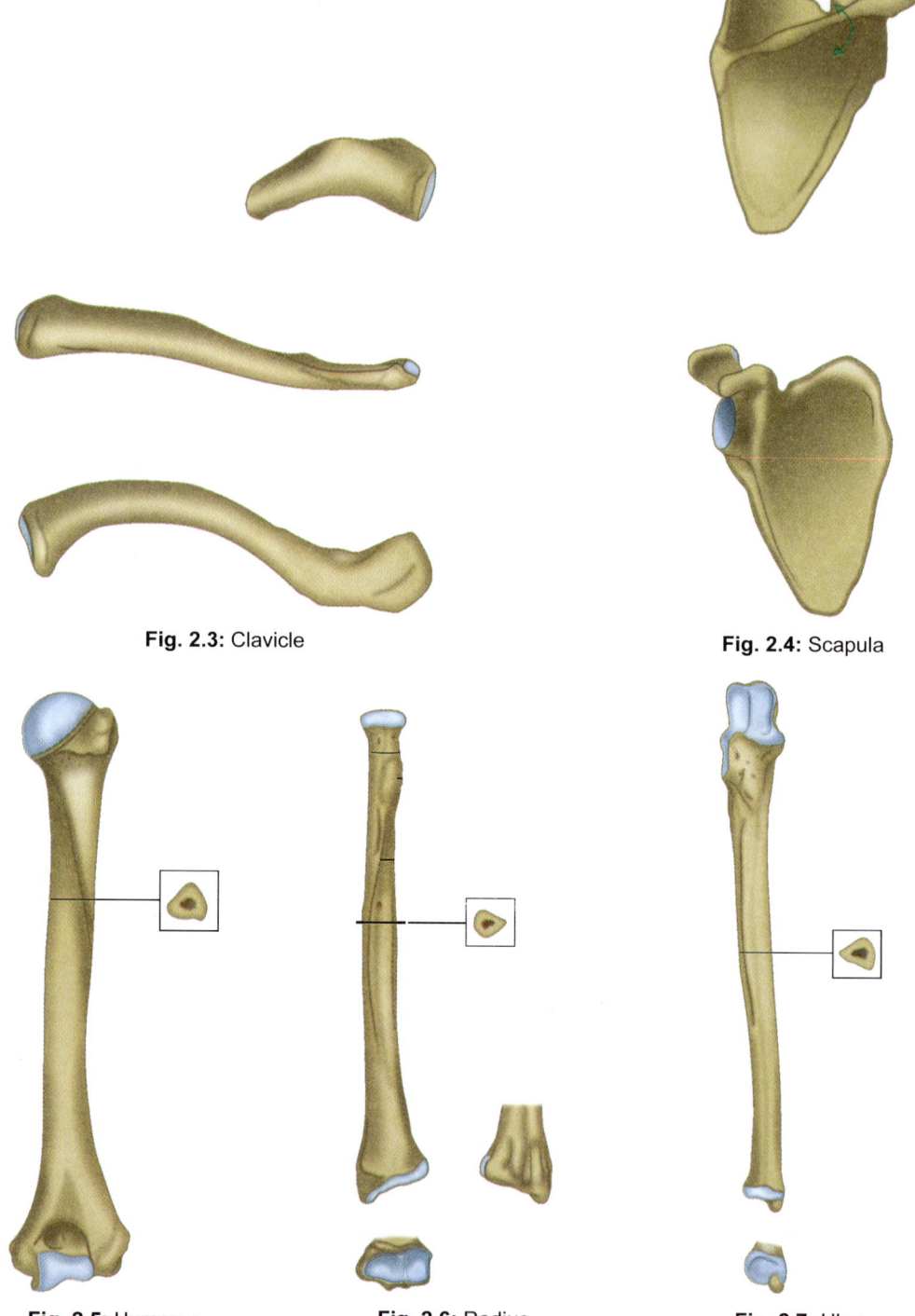

Fig. 2.3: Clavicle

Fig. 2.4: Scapula

Fig. 2.5: Humerus

Fig. 2.6: Radius

Fig. 2.7: Ulna

Window Dissections: Upper Limb (Superior Extremity)

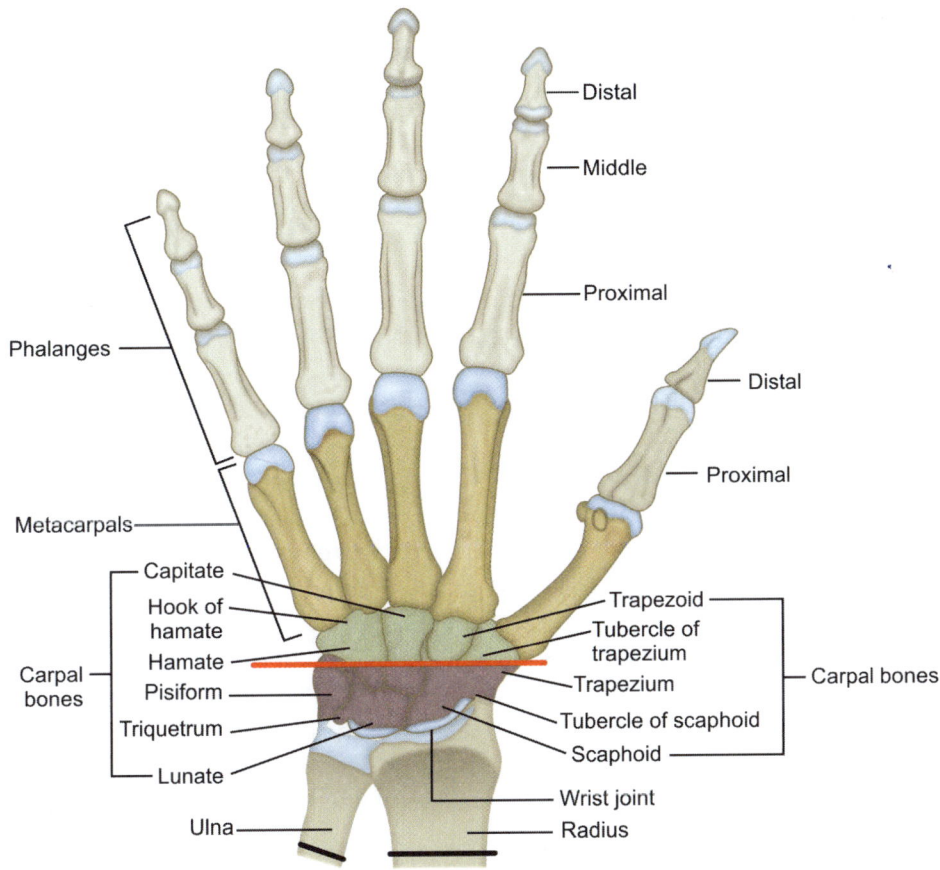

Fig. 2.8: Skeleton of hand

4. **Hand:** Skeleton of hand consists of (Fig. 2.8):
    a. Carpal bones (8):
        i. Proximal row (4)—scaphoid, lunate, pisiform, and triquetral.
        ii. Distal row (4)—trapezium, trapezoid, capitate and hamate.
    b. Metacarpal bones (5) named as 1st, 2nd, 3rd, 4th, and 5th (from lateral to medial).
    c. Phalanges (14)—3 for each finger (proximal, middle and distal) except thumb—2 (proximal and distal).

The deep strctures of these above regions are ensheathed by a jacket of common tubular connective tissue membrane called **deep fascia**. This deep fascia only changes its name from region to region such as axillary fascia (axilla), brachial fascia (arm), antebrachial fascia (forearm), palmar aponeurosis (palm of hand), etc.

LESSON 2: CLAVIPECTORAL FASCIA (PECTORAL REGION)

Clavipectoral fascia is a strong membranous sheet of fascia which extends from the inferior surface of the clavicle above to the pectoralis minor muscle below; hence the name clavipectoral fascia. This fascia lies deep to the pectoralis major muscle and encloses the pectoralis minor muscle.

STEPS OF DISSECTION

1. **Postion of the cadaver:** The cadaver is allowed to lie on the dissection table in supine posture with arm abducted at right angle to the body. The upper extremity is placed on a wooden block.
2. **Skin incision** (Fig. 2.9) (Exposure of upper three intercostal spaces):
   a. A vertical incision starts from the mid point of the jugular notch down the sternum in the median plane up to the level of the fourth costal cartilage (A–B)
   b. Second incision starts form the jugular notch (point A) and runs along the clavicle up to the tip of acromion process of scapula laterally (point C).
   c. Third incision passes obliquely from point B (sternum) and runs along the anterior axillary fold up to the junction of upper one-fourth and lower three-fourths of the arm (point D). Along these above incisions the skin is cut up to the depth of superficial fascia and the skin flap is reflected upwards and laterally. Superficial fascia is exposed.
3. **Superficial fascia:** In the superficial fascia, search for the **medial, intermediate and lateral branches of supraclavicular nerve (C_3, C_4).** Anterior cutaneous vessels and nerves are also found in this fascia. Anterior cutaneous branches of intercostal nerves become cutaneous after piercing anterior intercostal membrane of external intercostalis muscle. In this particular dissection, the anterior cutaneous branches of the first, second and third intercostal spaces are found.

   Then the superficial fascia is cut along the skin incision and reflected like that of skin. Deep fascia (pectoral fascia) is exposed.
4. **The deep fascia** covering the surface of pectoralis major muscle is called **pectoral fascia.** This fascia is continuous with the axillary fascia (fascia of the axillary floor). This pectoral fascia (deep fascia) is also continuous with the periosteum of clavicle and sternum and then passes laterally over the deltopectoral groove to become continuous with fascia covering deltoid muscle.

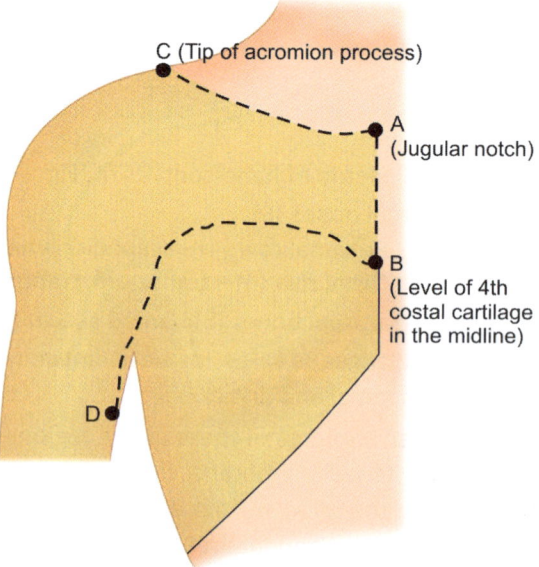

Fig. 2.9: Incision for clavipectoral fascia

The deep fascia is then cut in the same line of skin incision and reflected. Deep structures are exposed.

5. **Deep structures**: After reflecting the deep fascia, identify the both heads (clavicular head and sternocostal head) of pectoralis major muscle and deltoid muscle. Using fingers for blunt dissection, define the lateral border of pectoralis major and medial border of deltoid muscle in the upper part of the arm, thereby identify the deltopectoral groove. In this groove, look for the cephalic vein. This vein drains into the axillary vein. Occasional lymph nodes are also found in this groove. Deltoid branch of acromiothoracic artery also runs in this groove.

6. **In the next step of dissection**, cut across the clavicular head of pectoralis major muscle with the help of scissors and reflect it downwards and laterally towards its insertion into the humerus. During reflection, the vessels and nerves that are entering its deep surface are to be preserved. It is to be mentioned that the lateral pectoral nerve and pectoral branch of thoracoacromial artery enter the deep surface of the clavicular head of pectoralis major. Deep to the pectoralis major muscle the subclavius muscle, clavipectoral fascia and the pectoralis minor muscle are found. If these structures are not properly exposed, detach the sternocostal head of the pectoralis major from its attachment to the sternum using scissors and reflect laterally. The medial pectoral nerve after piercing pectoralis minor muscle enters the deep surface of pectoralis major. So palpate the deep surface of the muscle to palpate the pectoral nerves and preserve them (Fig. 2.10).

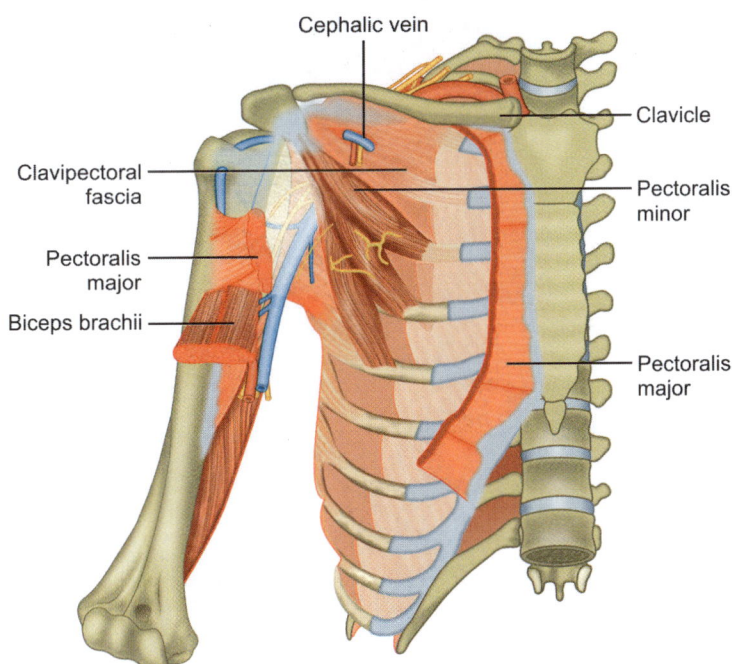

Fig. 2.10: Clavipectoral fascia

7. Clavipectoral fascia is exposed. Clean the fascia and study it. This fascia passes both superficial and deep to the subclavius muscle above and pectoralis minor muscle below.

ATTACHMENT AND DISTRIBUTION OF THE FASCIA (FIGS 2.11A AND B)

Above: It splits into two layers which enclose the subclavius muscle. Both the superficial and deep layers are attached to the margins of the subclavian groove on the inferior surface of the clavicle. Then the deep layer continues above to become continuous with the deep cervical fascia covering the inferior belly of omohyoid in the neck.

Below: It splits to enclose the pectoralis minor muscle. At the lower border of the muscle the fascia again unites and then becomes continuous with the axillary fascia forming the **suspensory ligament of the axilla**. This maintains the depression of the floor of the axilla.

Laterally: It is attached to the root of the coracoid process of scapula and blends with the coracoclavicular ligament.

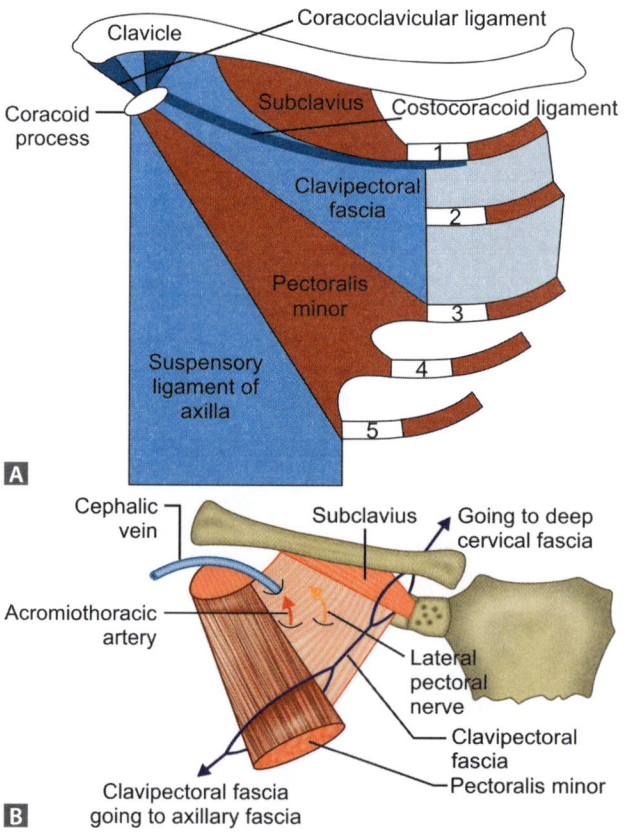

Figs 2.11A and B: Extent of clavipectoral fascia

Medially: It is attached to the first rib and blends with the fascia covering the first and second intercostal space.

This fascia is related superficially to the pectoralis major muscle and deep to it axillary vessels and cords of brachial plexus are noted.

The clavipectoral fascia is not uniform in its thickness throughout its entire extent. It is very thick and strong in the upper part extending from the root of the coracoid process of scapula laterally to the first costal cartilage medially. This part is known as **costocoracoid membrane**. The lower portion of the fascia is thin and delicate.

STRUCTURES PIERCING THE FASCIA (FIG. 2.12)

1. **Acromiothoracic artery (thoracoacromial artery):** It is a branch of second part of axillary artery. After piercing the fascia, it divides into four branches such as:
    a. **Acromial branch** runs under the deltoid.
    b. **Clavicular branch** runs medially and supplies the subclavius muscle and sternoclavicular joint.
    c. **Deltoid branch** runs in the deltopectoral groove.
    d. **Pectoral branch** runs between the pectoralis major and the pectoralis minor muscle and supplies both of them.
2. **Acromiothoracic vein:** It drains into the axillary vein.
3. **Lateral pectoral nerve** (C_5, C_6, C_7): It is a branch of lateral cord of brachial plexus and supplies both the pectorals.

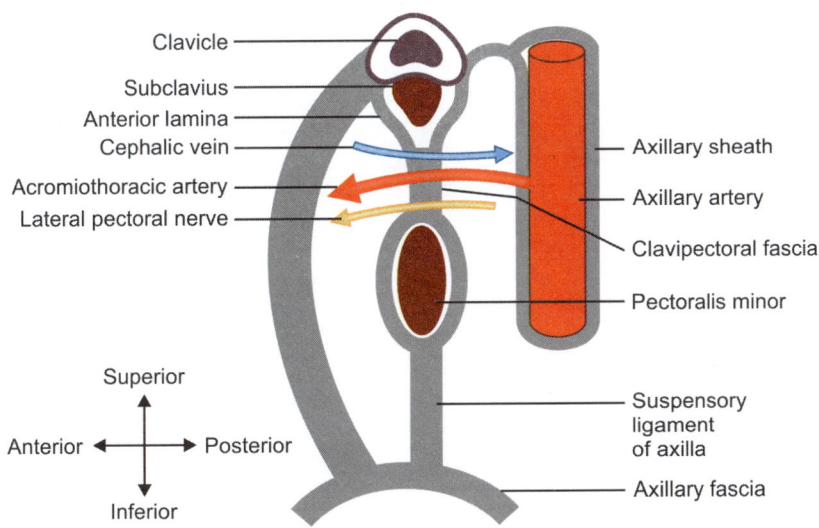

Fig. 2.12: Structures piercing clavipectoral fascia

4. **Cephalic vein:** It is the preaxial, superficial vein of the upper limb. It starts from the dorsal venous arch and ends in the axillary vein after piercing the clavipectoral fascia.
5. **Some lymph vessels**.

These above mentioned structures are thoroughly cleaned and dissected out by using fine forceps, fine scissors and blunt dissection. In fact, these structures pierce the costocoracoid part of the fascia.

The branches of the thoracoacromial arteries are not properly visible here. To identify the branches, use scissors to detach the pectoralis minor muscle from its origin from 3rd to 5th ribs and reflect it upwards towards its insertion to the coracoid process of scapula. Then clean the area and define the branches of the artery.

MUSCLES RELATED TO CLAVIPECTORAL FASCIA (REFER FIGS 2.10 AND 2.11A)

Muscles	Origin	Insertion	Nerve supply	Action
Petoralis major	**Clavicular head:** Anterior surface of medial half of clavicle **Sternocostal head:** • Lateral part of anterior surface of sternum up to 6th costal cartilage • 2nd to 6th costal cartilage	Lateral lip of inter tubercular sulcus of humerus	Medial and lateral pectoral nerve (C_5-T_1)	• Adduction and medial rotation of shoulder joint • **Clavicular head:** Flexion of shoulder joint • **Sternocostal head:** It helps in extension • Accessory muscle of inspiration
Pectoralis minor	3rd to 5th ribs excluding costal cartilages	Medial border and upper surface of coracoid process of scapula	Medial pectoral (pierces the muscle) and lateral pectoral nerve	• Protraction of scapula • Depression of shoulder
Subclavius	Junction of the 1st rib and its costal cartilage	Groove on the under surface of clavicle	Nerve to subclavius (Branch from upper trunk of brachial plexus) carrying the fibers from C_5, C_6	• Depresses the clavicle • Stabilizes sternoclavicular joint
Deltoid (Fig. 2.13)	• Whole length of the lower lip of the crest of the spine and lateral border of acromial process of scapula • Lateral one-third of clavicle	Deltoid tuberosity of humerus	Axillary nerve (C_5, C_6)	• **Acromial fibers:** Abduction of shoulder joint • **Clavicular fibers:** Flexion, adduction and medial rotation • **Posterior fibers:** Extension, adduction, lateral rotation

Window Dissections: Upper Limb (Superior Extremity)

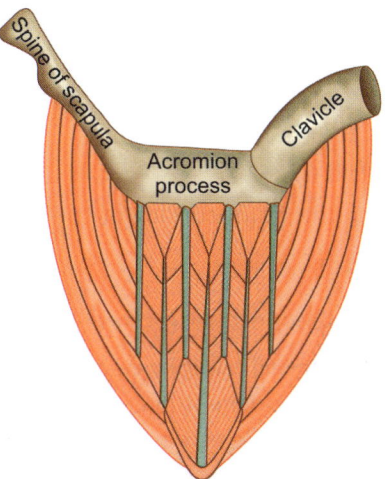

Fig. 2.13: Origins of deltoid

Summary

1. Incision of skin, superficial fascia as in Figure 2.9.
2. Structures found in superficial fascia: Twigs of medial and intermediate branch of supraclavicular nerve (C_3, C_4).
3. Deep fascia (pectoral fascia) are cut and reflected.
4. Deltopectoral groove between pectoralis major and deltoid are defined with its contents (cephalic vein, deltoid and acromiral branch of acromiothoracic artery, deltopectoral lymph node).
5. Clavipectoral fascia, pectoralis minor and subclavius muscle are exposed deep to pectoralis major.
6. Clavipectoral fascia is a strong membranous sheet of fascia extending from the inferior surface of the clavicle to pectoralis minor. This fascia is pierced by:
    i. Cephalic vein
    ii. Lateral pectoral nerve
    iii. Thoracoacromial vessels
    iv. Lymphatics
7. Both pectoralis major and minor are supplied by both lateral and medial pectoral nerves. Subclavius muscle is supplied by nerve to subclavius (C_5, C_6) from upper trunk of brachial plexus.

PROBABLE QUESTIONS AND ANSWERS

1. **What is clavipectoral fascia?**

Ans. Clavipectoral fascia is a strong membranous sheet of fascia extending from the inferior surface of the clavicle above to the pectoralis minor muscle below.

12. What are the muscles of the pectoral region?

Ans. Pectoralis major, pectoralis minor and subclavius muscle.

13. What is pectoral fascia?

Ans. The deep fascia covering the outer surface of the pectoralis major muscle is called pectoral fascia.

14. What is the boundary of deltopectoral groove?

Ans. Deltopectoral groove is bounded laterally by the clavicular origin of deltoid muscle and medially by the clavicular origin of pectoralis major muscle. The base is formed by the union of above two muscles. The floor is formed by the clavipectoral fascia and the roof is formed by the deep fasicia and the skin. This groove is situated in the upper part of the front of the arm.

15. What are the contents of the deltopectoral groove?

Ans. The contents are:
   i. Terminal portion of cephalic vein
   ii. Deltoid and acromial branches of acromiothoracic artery and
   iii. Occasionally, lymph nodes.

16. What are the origins and insertions of pectoralis major and pectoralis minor?

Ans. **Pectoralis major**
   a. Origin:
      i. **Clavicular head** takes origin from the anterior surface of medial half of clavicle
      ii. **Sternocostal head** takes origin from the lateral part of anterior surface of manubrium and body of sternum, cartilages of 2nd to 6th ribs.
   b. Insertion: Lateral lip of the bicipital groove of the humerus.

   Pectoralis minor:
   a. Origin: Outer surface of 3rd, 4th and 5th ribs close to their costal cartilages.
   b. Insertion: Medial border and upper surface of coracoid process of scapula.

17. What are the nerve supply of these muscles?

Ans. Both the pectoralis major and the pectoralis minor muscles are supplied by the lateral and medial pectoral nerves.

18. What is the nerve supply of subclavius muscle?

Ans. It is supplied by the nerve to subclavius which is a branch from the upper trunk of the brachial plexus (C_5 and C_6).

9. Where does the clavipectoral fascia exist?

Ans. It is found deep to the pectoralis major muscle and extends from the lower border of subclavius muscle to the upper border of pectoralis minor.

10. Trace the clavipectoral fascia.

Ans. **Superiorly**, the fascia encloses the subclavius muscle and attached to the margins of the subclavian groove on the inferior surface of the clavicle.

Inferiorly, it splits to enclose the pectoralis minor muscle and at the lower border of that muscle it reunites and continues as the axillary fascia forming the suspensory ligament of the axilla.

Medially, it is attached to the first rib and blends with the fascia covering the 1st and 2nd intercostal space.

Laterally, it is attached to the root of the coracoid process of scapula and blends with the coracoclavicular ligament.

11. What is costocoracoid ligament?

Ans. The upper portion of the fascia extending from the root of the coracoid process of scapula laterally to the first costal cartilage medially is very thick and strong. This part of the fascia is known as costocoracoid ligament or costocoracoid membrane.

12. What are the structures that pierce the clavipectoral fascia?

Ans. The **structures piercing the fascia are:**
   i. Acromiothoracic artery, branch of 2nd part of axillary artery.
   ii. Acromiothoracic vein, draining into the axillary vein.
   iii. Cephalic vein, ending in the axillary vein.
   iv. Lateral pectoral nerve (C_5, C_6, C_7)
   v. Lymph vessels.

13. What happens if the collection of pus occurs beneath the clavipectoral fascia?

Ans. The pus may pass into the axilla under the axillary fascia and point in the axilla. If not drained is time, the pus may follow the axillary and the subclavian vessels and may point in the neck under the deep cervical fascia. The pus may trickle down the arm following the course of the axillary vessels and the nerves. From the neck it may pass down into the mediastinum in the thorax.

14. What is deltopectoral triangle?

Ans. The deltopectoral groove in the infraclavicular fossa is also known as deltopectoral triangle because of its triangular outline.

LESSON 3: TRIANGULAR AND QUADRANGULAR SPACE

Triangular space: Two in number—upper triangular space and lower triangular space.
a. **Upper triangular space:** It is a three-sided intermuscular space bounded above by the **teres minor**, below by the **teres major**, laterally by the **long head of triceps** and the apex is formed by the convergence of teres major and teres minor at the lateral border of scapula (Fig. 2.14).
b. **Lower triangular space:** It is an intermuscular space bounded above by **teres major**, medially by long head of **triceps** and laterally by the shaft of humerus (Fig. 2.14).

Quadrangular space: It is a four-sided intermuscular space on the posterior aspect of shoulder bounded above by the **subscapularis**, capsule of shoulder joint and **teres minor** from before backwords, below by **teres major**, medially by **long head of tricep**s and laterally by **surgical neck of humerus** (Fig. 2.14).

STEPS OF DISSECTION

1. **Position of the cadaver:** The cadaver is allowed to lie on the dissection table in the prone position with the upper limb abducted to 45° and place a wooden block under the chest. Identify the inferior angle, spine and acromion process of scapula.

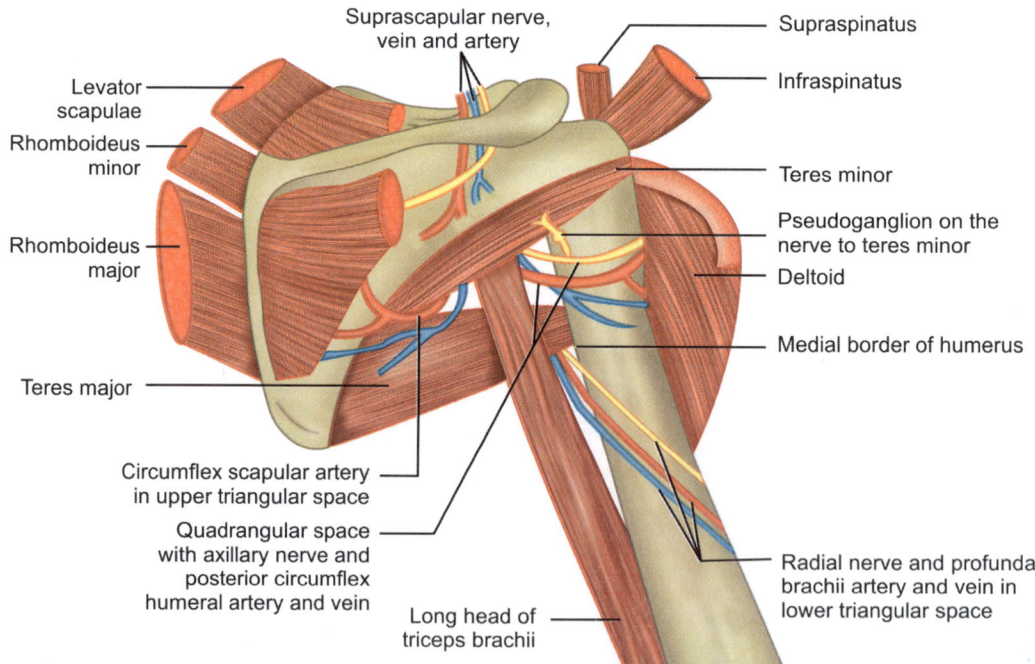

Fig. 2.14: Boundaries and contents of triangular and quadrangular space

Window Dissections: Upper Limb (Superior Extremity)

2. **Skin incision (Fig. 2.15):**
   a. A vertical incision starts from the midpoint of the scapular spine (A) down to the inferior angle of the scapula (B)
   b. Second incision starts from the midpoint of scapular spine (A) and runs laterally and obliquely along the spine up to the tip of the acromian process (C).
   c. Along these above incisions the skin is cut up to the depth of superficial fascia and the skin flap is reflected laterally and downwards.
3. **Superficial fascia:** In the superficial fascia, search for the branches of the posterior division of the supraclavicular nerve and dorsal cutaneous nerve. Then the superficial fascia is cut along the skin incision and reflected like that of skin. Deep fascia is exposed.
4. **Deep fascia:** The deep fascia is then cut in the same line of skin incision and reflected. Deep structures are exposed.
5. **Next** step is to identify the muscles that are exposed. Using blunt dissection the borders of the deltoid muscle are defined. The proximal attachments of deltoid muscle from the scapula and clavicle are detached as close to the bones as possible using a scalpel. The distal attachment of this muscle onto the deltoid tuberosity of the humerus should be left as it is. Reflect the muscle laterally, taking care not to tear the vessels and nerves running along its deep surface.
6. **Below** the deltoid muscle the **infraspinatus** muscle is visible. Just lateral to the infraspinatus the **teres minor** is identified. The **subscapularis** muscle is found in front of infraspinatus. Using a probe the borders of teres minor muscle are cleaned and defined.

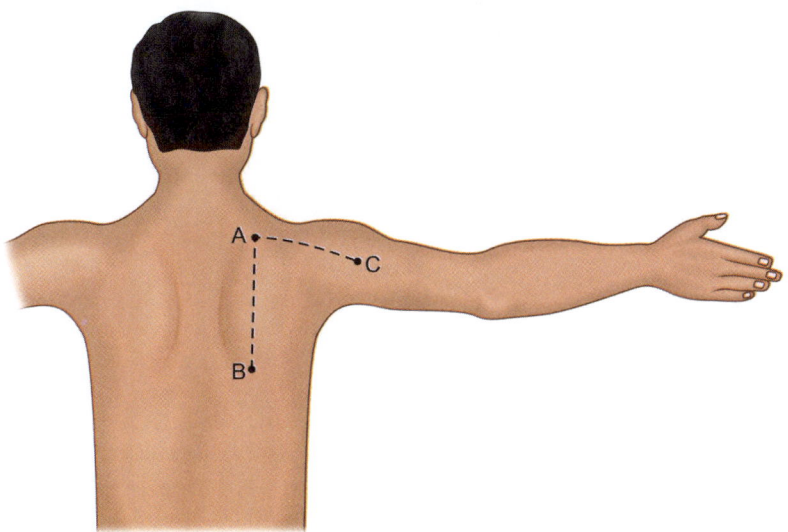

Fig. 2.15: Quadrangular-triangular space (skin incision)

7. **Just** below the teres minor a muscle belly passes laterally to the humerus across the posterior fold of axilla, dorsal to which lies **latissimus dorsi**—this is **teres major**. Clean and define the borders of teres major.
8. **Identify** the long head of the **triceps brachii** muscle. Observe that the long head of the triceps brachii passes anterior to the teres minor muscle and posterior to the teres major muscle.
9. **Now,** all the muscles related to quadrangular and triangular space are cleaned and defined **(Fig. 2.16)**.
10. **Observe** the axillary (circumflex) nerve and the posterior circumflex humeral artery and vein on the deep surface of the deltoid muscle near its attachment to the humerus. Clean the nerve and vessels using a probe and trace them around the surgical neck of the humerus.
11. **Push** your finger paralled to the nerve and vessels to open the **quadrangular space.** Now define the borders of this space (refer Fig. 2.14)
    a. Superior border: Inferior border of teres minor.

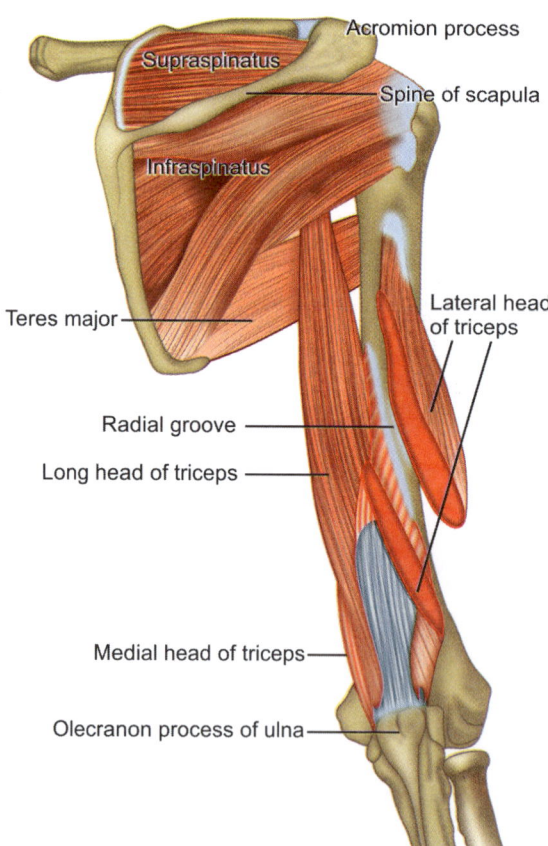

Fig. 2.16: Muscles related to triangular and quadrangular space

b. Lateral border: Surgical neck of the humerus.
c. Medial border: Long head of the triceps muscle.
d. Inferior border: Superior border of teres major muscle.

The structures that traverse this space are axillary nerve and posterior circumflex humeral artery (refer Fig. 2.14).

12. To dissect out the upper triangular space, clean and identify the muscles—teres minor, teres major and long head of biceps brachii. Boundaries of the triangular space are defined as follows (refer Fig. 2.14):
    - Superior border: Inferior border of teres minor.
    - Lateral border: Tendon of long head of triceps.
    - Inferior border: Superior border of teres major.

    Note that circumflex scapular artery and its accompanying vein may be found within this space.

13. **Another** triangular space known as **lower triangular space** is also identified below and lateral to the upper triangular space. This space contains **radial nerve** and **arteria profunda brachii** (refer Fig. 2.14).

> **Triangle of Auscultation**
> This a triangular area on the back close to the chest wall bounded below by the upper horizontal border of **latissimus dorsi**, medially by lateral border of **trapezius** and laterally by vertebral border of **scapula** and part of **rhomboideus major**. The floor of this triangle is formed by the 6th intercostal space without having any overlying muscle. The apex of lower lobe of each lung lies beneath this triangle and cardiac orifice of stomach lies deep to this triangle on left side.

MUSCLES RELATED TO TRIANGULAR AND QUADRANGULAR SPACE

Muscles	Origin	Insertion	Nerve supply	Action
Teres minor	Upper 2/3rd on the dorsal surface of lateral border of scapula	Lowest impression of greater tubercle of humerus	Posterior branch of axillary nerve (C_5, C_6) where it presents a **pseudoganglion**	• Lateral rotation of the shoulder joint • Stabilization of shoulder joint
Teres major	Inferior angle of scapula	Medial lip of bicipital groove	Lower subscapular nerve	Medial rotation and extension of shoulder joint
Triceps brachii	• **Long head:** Infraglenoid tubercle • **Lateral head:** Posterior surface of humerus above spiral groove • **Medial head:** Posterior surface of humerus below spiral groove	Posterior part of upper surface of olecranon process of ulna	Radial nerve	Extension of elbow

> ### Summary
> 1. Incision of skin, superficial fascia and deep fascia as in Figure 2.15.
> 2. Structures found in superficial fascia: Branches from the posterior division of supraclavicular nerve and dorsal cutaneous nerve.
> 3. Deep structures (deep to deep fascia):
>   Boundaries and contents of triangular and quadrangular space:
>
>   **Upper triangular space:**
>   - *Above:* Teres minor (posterior branch of axillary nerve).
>   - *Below:* Teres major (lower subscapular nerve)
>   - *Laterally:* Long head of triceps (radial nerve).
>   - *Contents:* Circumflex scapular artery and vein.
>
>   **Lower triangular space:**
>   - *Above:* Teres major (lower subscapular nerve).
>   - *Medially:* Long head of triceps (radial nerve).
>   - *Laterally:* Shaft of humerus.
>   - *Contents*: **Radial nerve** (branch of posterior cord of brachial plexus) and profunda brachii vessels.
>
>   **Quadrangular space:**
>   - *Above:* Subscapularis (upper and lower subscapular nerve).
>
>   Teres minor (posterior branch of axillary nerve)
>   - *Below:* Teres major (lower subscapular nerve).
>   - *Medially:* Long head of triceps (radial nerve).
>   - *Laterally:* Surgical neck of humerus.
>
>   *Contents:* Axillary nerve (C_5, C_6), a branch of posterior cord of brachial plexus and posterior circumflex humeral vessels.

PROBABLE QUESTIONS AND ANSWERS

1. Of the two spaces, which one is lateral?

Ans. Quadrangular space is lateral and upper triangular space is medial.

2. Which structure intervenes between these spaces?

Ans. Long head of triceps.

3. Which muscle intervenes between quadrangular space and lower triangular space?

Ans. Teres major.

4. What is the relative position between the two triangular spaces?

Ans. Lower triangular space is diagonally opposite the upper triangular space.

Window Dissections: Upper Limb (Superior Extremity)

5. What is the other name of the quadrangular space?

Ans. Quadrilateral space, because it is having four sides and four angles.

6. Which circumflex humeral artery is longer ?

Ans. Posterior circumflex humeral artery is longer than anterior circumflex artery.

7. Which nerve is also known as circumflex nerve?

Ans. Axillary nerve, because it winds round the surgical neck of the humerus.

8. Which structure accompanies the posterior circumflex humeral artery?

Ans. The anterior branch of axillary nerve (not the posterior branch) accompanied by the posterior circumflex humeral artery winds round the posterior surface of the surgical neck of the humerus under cover of deltoid.

9. What is the first branch of axillary nerve?

Ans. An articular twig to the shoulder joint is the first branch from the trunk of the axillary nerve while passing through the quadrangular space.

10. What are the branches of the axillary nerve?

Ans. Anterior and posterior branch, both supply **deltoid** muscle. Posterior branch, in addition supplies **teres minor** muscle.

11. What is the fate of posterior branch of axillary nerve?

Ans. The posterior division is continued as upper lateral cutaneous nerve of arm.

12. Which nerve gives off the lower lateral cutaneous nerve of arm?

Ans. Radial nerve in the spiral groove.

13. What is the peculiarity of posterior division of axillary nerve?

Ans. It is having a **pseudoganglion** (pseudo = false, ganglion = a group of nerve cell bodies).

14. What are the other nerves where do you find pseudoganglion?

Ans. Posterior interosseous nerve (back of the wrist), lateral terminal branch of deep peroneal nerve (dorsum of foot) and median nerve (hand).

15. Does the axillary nerve obey the Hilton's law?

Ans. Yes, because it supplies the shoulder joint and innervates muscles (deltoid, teres minor) and skin (upper lateral cutaneous nerve of the arm) overlying the joint.

16. Which artery gives rise to circumflex humeral artery?

Ans. Anterior and posterior circumflex humeral arteries are the branches of third part of axillary artery.

17. What happens in fracture of surgical neck of humerus?

Ans. Loss of power of abduction (due to paralysis of deltoid) and loss of cutaneous sensation over he lower part of deltoid.

18. What is root value of axillary nerve?

Ans. C_5, C_6 (from the posterior cord of brachial plexus).

19. From which artery the circumflex scapular artery arises?

Ans. Subscapular artery (branch of 3rd part of axillary artery).

20. What are the boundaries and contents of the triangular and quadrangular spaces?

Ans. Discussed within this chapter.

LESSON 4: AXILLA (ARMPIT)

The axilla is a pyramidal-shaped space (Fig. 2.17) betwen the pectoral muscles, upper part of the arm, the scapula and the lateral thoracic wall. It is a region of passage for vessels and nerves that course from the root of the neck into the upper limb. It presents an apex called **cervico-axillary canal**, base and four walls (anterior, posterior, medial and lateral). The main contents of the axilla are the axillary sheath, brachial plexus, axillary vessels and their branches, lymphatics (lymph nodes and lymphatic vessels), muscles, fat and connective tissue.

STEPS OF DISSECTION

1. **Position of the cadaver:** The cadaver is allowed to lie in supine position with the arm abducted to about 90°:The upper extremity is placed on a wooden block.
2. **Skin incision (Fig. 2.18):**
    a. Incision starts from (point A) the anterior axillary fold at the level of the fourth rib (2nd rib corresponds to the sternal angle) and runs along the anterior fold of axilla up to the junction of upper one-fourth and lower three-fourths of the arm (point B).
    b. 2nd incision starts from point A and carry it backwards along the lateral thoracic wall up to the posterior axillary fold (point C).
    c. 3rd incision begins from point B and carries it along the medial surface of upper arm up to the posterior axillary fold (point D).

Window Dissections: Upper Limb (Superior Extremity)

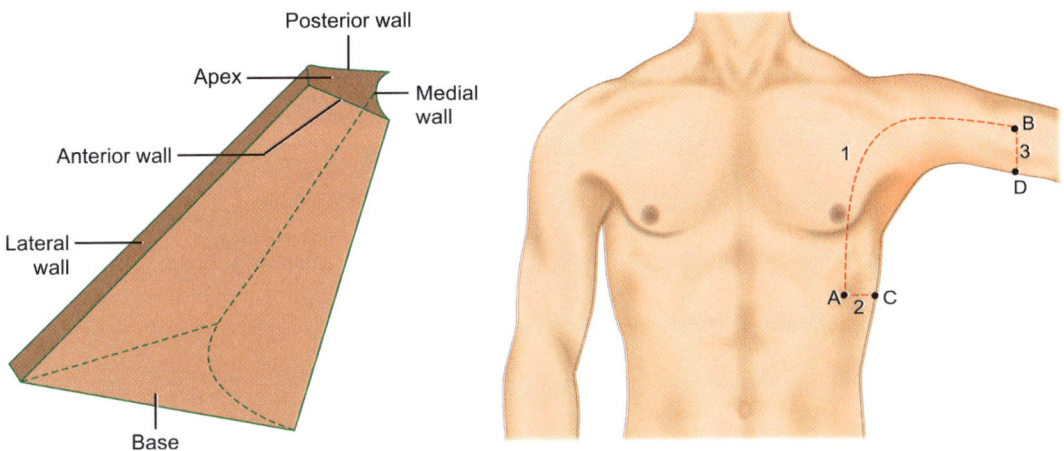

Fig. 2.17: Shape of axilla

Fig. 2.18: Skin incision of axilla

The skin flap is reflected backwards. The superficial fascia is exposed.

3. **Superficial fascia:** The superficial fascia is incised in the same line of skin incisions and reflected in the same way. In this layer lymph nodes (glands) and abundant fats are available. Remove lymph glands and fat carefully without destroying any important structure. Lateral cutaneous branches of second and third intercostal nerves are available in the superficial fascia. Dissect carefully to find out the lateral cutaneous branch of second intercostal nerve (**intercostobrachial nerve**) and clean it and preserve it.

4. **Deep fascia (axillary fascia):** Deep fascia is exposed after removing the fats and lymph glands. The deep fascia forms the base or floor of the axilla. It is continuous with the **pectoral fascia** (deep fascia on the surface of the pectoralis major muscle). It extends from the anterior to posterior axillary folds is supported from above by the suspensory ligament of axilla. The deep fascia is then cut in the same line of skin incision and reflected backwards.

5. The contents and 4 walls of the axilla are exposed. The next step will be to clean and identify the structures forming the different walls (boundaries) of the axilla (**Figs 2.19A and B**).
    Let us clean and identify the walls one by one.
    a. **Medial wall:** The medial wall of the axilla means the upper part of the lateral thoracic wall. This wall is mainly formed by the **serratus anterior** muscle. Identify the upper four digitations (total eight digitations are there) of this muscle and upper 4–5 ribs and their intercostal spaces. The serratus anterior muscle is followed posteriorly towards the medial margin of the scapula using fingers only. On the superficial surface of this muscle, find out a nerve which is running vertically and giving branches to the serratus anterior muscle. Use a probe to free the nerve and follow it superiorly as far as possible towards the apex of the axilla. This nerve is the **long thoracic nerve** (**nerve of Bell**).

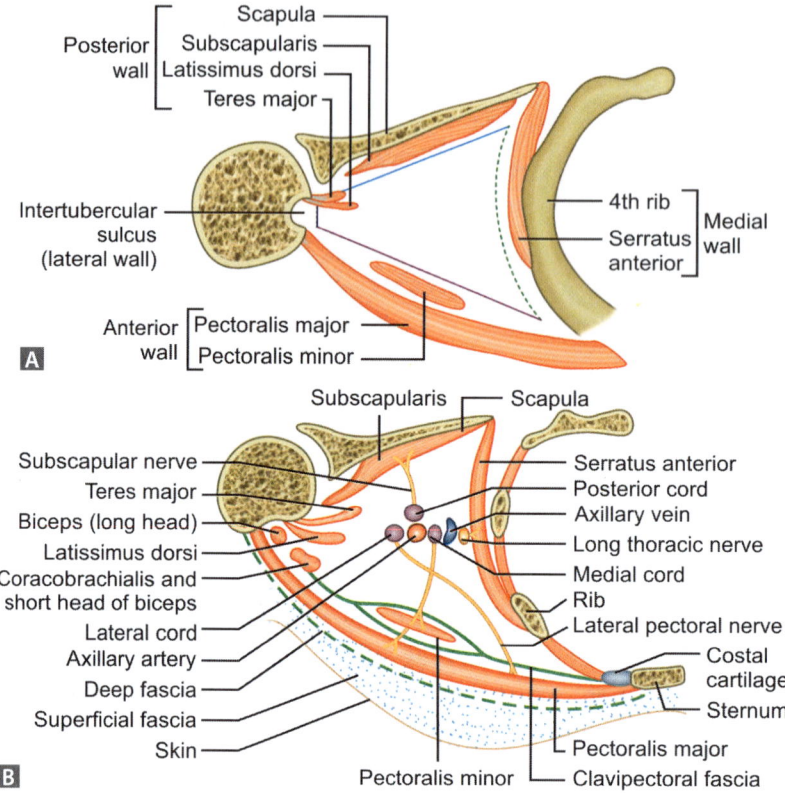

Figs 2.19A and B: Walls of axilla

b. **Lateral wall:** The lateral wall of the axilla is formed by the medial suface of the upper one-fourth of the humerus. Muscles related are **coracobrachialis** and upper part of **biceps brachii** muscle. Nerve is **musculocutaneous** nerve.

   Identify and clean the musculocutaneous nerve which enters the **coracobrachialis** muscle to supply. *It is the most lateral terminal branch of the brachial plexus.*

c. **Anterior wall:** The anterior wall of the axilla is formed by the **pectoralis major** (superficial plane) and **pectoralis minor, subclavius, clavipectoral** fascia (deep plane). The anterior axillary fold is formed by the spirally arranged lower border of pectoralis major muscle. Clean and identity these muscles. The pectoralis major is to be reflected to expose the pectoralis minor, clavipectoral fascia and subclavius.

d. **Posterior wall (Fig. 2.20):** The posterior wall is formed by the **subscapularis, latissimus dorsi** and **teres major** muscles from above downwards. Identify these muscles and clean them. Identify and clean the **subscapular nerves** that arise from the posterior cord of brachial plexus. Verify that nerves run in the loose connective tissue on the anterior surface of the subscapularis muscle.

Window Dissections: Upper Limb (Superior Extremity)

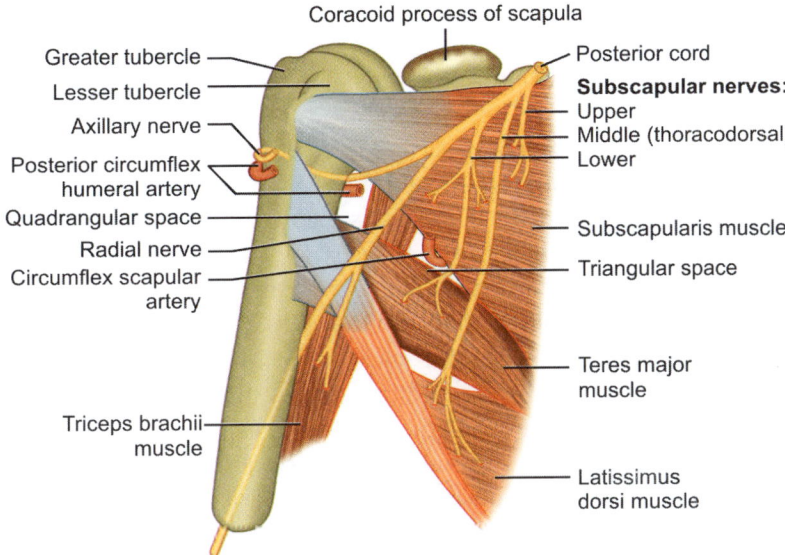

Fig. 2.20: Posterior wall of axilla

The **upper subscapular nerve** lies at the upper part of the subscapularis muscle. Clean and identify it. It innervates the subscapularis muscle.

The **middle subscapular nerve** (thoracodorsal nerve) lies on latissimus dorsi and supplies it.

The **lower subscapular nerve** lies at the lower border of subscapularis and supplies it and **teres major** muscle.

6. **Apex** of the axilla is directed above and triangular in shape. It communicates with the neck through the **cervicoaxillary** canal which transmits cords of brachial plexus and axillary vessels enclosed in a sheath called **axillary sheath** which is derived from prevertebral fascia. Efferent subclavian lymph trunks extend from the apical group of axillary nodes through the cervicoaxillary canal.
7. **Contents:** The main contents of the axilla are axillary sheath which surrounds the axillary vessels and brachial plexus, axillary lymphatics and lymph nodes, considerable amount of fat and connective tissue, sometimes axillary tail of the breast in female.

    The next approach will be to expose the contents of the axilla by careful dissection of the important structures.

    a. **Axillary sheath** is a connective tissue structure that surrounds the axillary vessels and brachial plexus. It extends from the lateral border of the first rib to the inferior border of the teres major muscle. It is located about 2 cm inferior to the coracoid process. Anterior suface of the axillary sheath is now opened to reveal and follow the course of axillary vessels using scissors and forceps.

b. **Axillary vein** is identified within the axillary sheath by its relation to the axillary artery. **The vein lies medial to the artery. Medial cutaneous nerve of forearm** lies between the artery and the vein, whereas the **medial cutaneous nerve of the arm** lies medial to the vein. The axillary vein is formed at the lateral border of the teres major muscle by the joining of two brachial veins and ends at the lateral border of the first rib to be continuous with the subclavian vein. To get a clear view, the smaller tributaries of the vein may be removed. Since the veins follow the corresponding branches of the artery, their loss is of no significance. If necessary, the axillary vein may be removed to enhance dissection of the arteries and nerves in the axilla (Grants). Note the presence of lymph nodes that are associated with the veins.

c. **Axillary artery:** Axillary artery is the main artery of the upper limb. It begins at the outer border of the first rib as a continuation of the **subclavian artery** and ends at the lower border of the teres major muscle where it changes its name to **brachial artery**. The axillary artery is surrounded by **brachial** plexus. The brachial plexus must be retracted and preserved during dissection of the axillary artery. The artery lies lateral to the axillary vein. **Radial nerve** lies posterior to the artery but the **ulnar nerve** lies between and behind the artery and vein. Identify the three parts of the axillary artery (Fig. 2.21). The branching pattern of the artery may vary. The branches are named according to their distribution rather than by their origin.

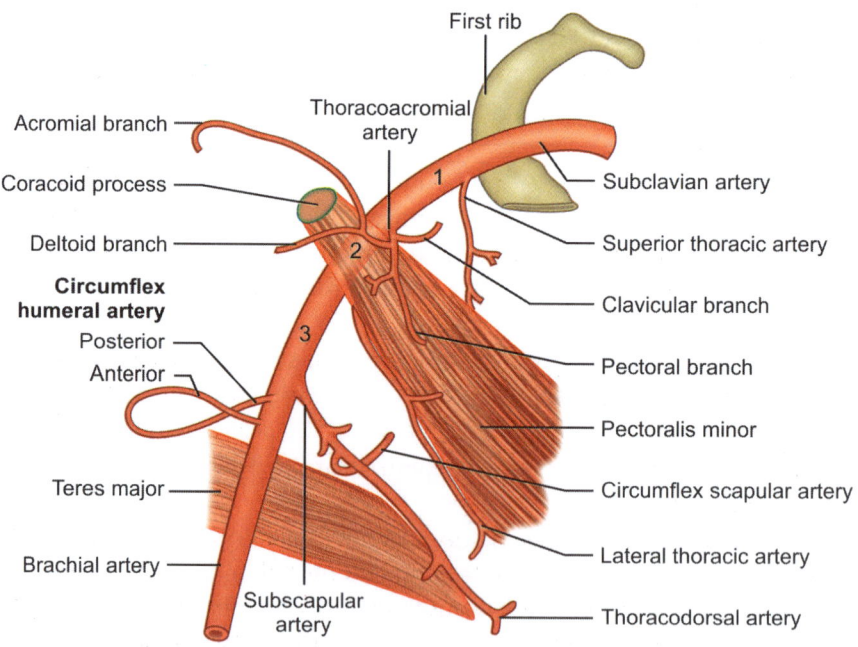

Fig. 2.21: Branches of axillary artery

Axillary artery	Extent	Branches
1st part	From outer border of first rib to upper border of petoralis minor	Superior thoracic artery
2nd part	Deep (posterior) to the pectoralis minor	• Lateral thoracic artery • Acromiothoracic artery
3rd part	From distal border of pectoralis minor to inferior border of teres major muscle	• Subscapular artery • Anterior circumflex humeral artery • Posterior circumflex humeral artery

d. **Nerves:** The nerves that are found in the axilla are the **infraclavicular part of brachial plexus** and its branches. The supraclavicular part is found in the dissection of the neck. The brachial plexus is formed by the ventral rami of C_5 to T_1 spinal nerves in the root of the neck and then passes into the axilla through the cervico axillary canal (apex of the axilla).

The plexus is having three cords. The cords lie posterior to the first part of axillary artery. The names of the three cords (lateral, medial and posterior) are given according to their relationship with the 2nd part of the axillary artery, while the main nerves arising from the cords surround the third part.

IDENTIFICATION OF DIFFERENT NERVES OF BRACHIAL PLEXUS

- **Musculocutaneous nerve (C_5, C_6, C_7):** It is the most lateral terminal branch of the brachial plexus and lies on the lateral aspect of the axillary artery. It pierces the coracobrachialis muscle and passes laterally between the biceps brachii in front and brachialis muscle behind.
- **Lateral cord:** Musculocutaneous nerve is traced proximally to find out the lateral cord, because musculocutaneous nerve is the terminal branch of lateral cord.
    - **Lateral root of median nerve** is another branch of lateral cord which lies on the lateral aspect and then on the anterior aspect of the axillary artery to join the medial root (branch of medial cord) to form **median nerve**.
    - **Median nerve (C_5–T_1):** It lies on the anterior aspect of the axillary artery. The two roots of median nerve, median nerve, musculocutaneous nerve and ulnar nerve (branch from medial cord) form the letter **'M'** in front of the 3rd part of the axillary artery.
- **Medial cord:** If the medial root of median nerve is traced proximally, the medial cord is identified. It lies on the medial side of the 2nd part of the axillary artery.
    - **Medial root of median nerve:** It lies medial to the axillary artery and then comes in front of it to join the lateral root and form median nerve.
    - **Ulnar nerve (C_7, C_8, T_1):** Medial cord continues distally as ulnar nerve. It lies behind and between the axillay artery and vein.
    - **Medial cutaneous nerve of arm:** It lies medial to the axillary vein.
    - **Medial cutaneous nerve of foream:** It lies between the axillary artery and the axillary vein.
- **Posterior cord:** Retract the axillary artery, the lateral cord and the medial cord. This procedure exposes the posterior cord of the brachial plexus. Branches of posterior cord that are found in the axilla are radial nerve, axillary nerve, thoracodorsal and subscapular nerves.

- **Radial nerve (C_5–T_1):** It is a thick nerve and lies posterior to the axillary artery. Use blunt dissection to clean the radial nerve and it is seen that it leaves the axilla running posterior to humerus.
- **Axillary nerve (C_5, C_6):** It is also a thick nerve. Use blunt dissection to clean the axillary nerve. It lies behind the axillary artery and passes immediately backwards through the quadrangular space accompanied by the posterior circumflex humeral artery (branch of 3rd part of the axillary artery).
- **Thoracodorsal and subscapular nerves:** Described earlier. Lateral pectoral nerve (branch of lateral cord) and medial pectoral nerve (branch of medial cord) are mainly found in the pectoral region, Hence, not described in axilla.

CORDS OF BRACHIAL PLEXUS

Lateral cord (Branches—3)	Medial cord (Branches—5)	Posterior cord (Branches—5)
1. Lateral pectral 2. **Musculocutaneous (C_5, C_6, C_7)** 3. Lateral root of median nerve	1. Medial pectoral 2. Medial cutaneous nerve of the arm (C_8, T_1) 3. Medial cutaneous nerve of the forearm (C_8, T_1) 4. **Ulnar nerve (C_7, C_8, T_1)** 5. Medial root of median nerve	1. Upper subscapular (C_5, C_6) 2. Thoracodorsal C_6, C_7, C_8 (middle subscapular) 3. Lower subscapular (C_5, C_6) 4. Axillary nerve (C_5, C_6) 5. Radial nerve (C_5, C_6, C_7, C_8, T_1)

MUSCLES OF AXILLA

Anterior Wall

Muscles	Origin	Insertion	Nerve supply	Action
Pectoralis major	• **Clavicular head:** Anterior surface of medial half of clavicle • **Sternocostal head:** Lateral part of anterior surface of sternum up to 6th costal cartilage • 2nd to 6th costal cartilage	Lateral lip of intertubercular sulcus of humerus	Medial and lateral pectoral nerve (C_5–T_1)	• Adduction and medial rotation of shoulder joint • Clavicular head: Flexion of shoulder joint • Sternocostal head: Helps in extension • Accessory muscle of inspiration
Pectoralis minor	3rd to 5th ribs excluding costal cartilages	Medial border and upper surface of coracoid process of scapula	Medial pectoral (pierces the muscle) and lateral pectoral nerve	• Protraction of scapula • Depression of shoulder

Window Dissections: Upper Limb (Superior Extremity)

> **Rectus Sternalis Muscle**
> Present occasionally, fibers are disposed vertically in parasternal region in front of pectoralis major. It is believed that it is a derivative of superficial part of rectus abdominis muscle and supplied segmentally by intercostal nerve.

Posterior Wall

Muscles	Origin	Insertion	Nerve supply	Action
• Subscapularis	• Subscapular fossa (costal surface)	Lesser tubercle of the humerus	Upper and lower subscapular nerve (C_5–C_7)	• Medial rotation of arm • Stabilizes the shoulder joint
• Latissimus dorsi	• Spinous process and supraspinous ligaments of all lumbar, all sacral and lower six (T_7–T_{12}) thoracic vertebrae. • Outer lip of iliac crest • Thoracolumber fascia	Floor of the intertubercular sulcus (bicipital groove) of humerus	Thoracodorsal nerve (C_6–C_8)	Acts on shoulder joint in: • Adduction • Extension • Medial rotation
• Teres major	Inferior angle of scapula	Medial lip of bicipital groove of humerus	Lower subscapular nerve (C_5–C_7)	• Medial rotation • Extension of shoulder joint

Medial Wall

Muscles	Origin	Insertion	Nerve supply	Action
Serratus anterior (Fig. 2.22)	Upper border and outer surface of upper eight ribs (R_1–R_8) (8 digitations)	Medial border (costal surface) of scapula	Long thoracic nerve or nerve of Bell (C_5–C_7)	Protraction and rotation of scapula

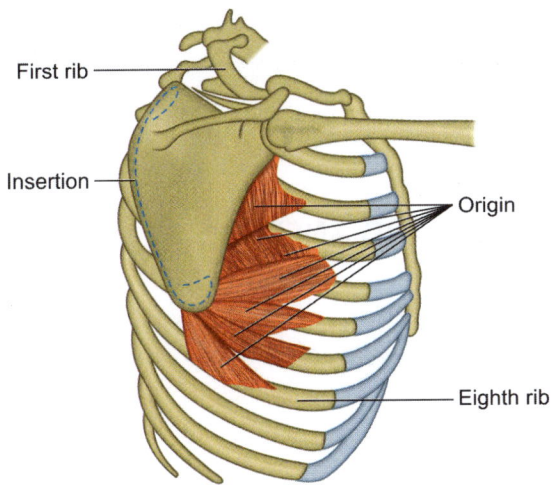

Fig. 2.22: Serratus anterior on the medial wall of axilla

Lateral Wall

Muscles	Origin	Insertion	Nerve supply	Action
Coracobrachialis	Coracoid process of scapula	Medial surface of the shaft of the humerus	Musculocutaneous nerve (C_5–C_7)	Adduction and flexion of shoulder joint
Biceps brachii	• **Long head:** Supraglenoid tubercle • **Short head:** Coracoid process of scapula	Radial tuberosity	Musculocutaneous nerve (C_5–C_7)	• Flexion of elbow • Supination

Axillary arch: It is a muscular slip arising from the latissimus dorsi in the posterior axillary fold and joins the tendons of pectoralis major, coracobrachialis or biceps. This arch is present only in about 7% subjects. It crosses the axilla in front of the vessels and nerves.

Summary
1. Incision of skin, superficial fascia, deep fascia (axillary fascia) as in Figure 2.18.
2. Structures found in superficial fascia: Intercostobrachial nerve, lateral cutaneous branch of 3rd intercostal nerve, fat and lymphatics.
3. Contents of the axilla: Axillary vessels and cords of brachial plexus within axillary sheath, lymph nodes and fat.
4. 4 walls of axilla: Medial wall: Upper 4 digitations of serratas anterior muscle (long thoracic nerve—C_3, C_4, C_5), upper 3–4 intercostal spaces.
   – Lateral wall: Biceps and coracobrachialis (both supplied by musculocutaneous nerve).
   – *Anterior wall:* Pectoralis major et minor (lateral and medial pectoral nerve), subclavius.
   – *Posterior wall:* Subscapularis (upper and lower subscapular nerve) latissimus dorsi (Thoracodorsal nerve) and teres major (lower subscapular nerve).
5. Three cords of brachial plexus (lateral, medial and posterior are named according to their relationship to the 2nd part of axillary artery.
6. Axillary artery begins as continuation of subclavian artery at the outer border of 1st rib and ends as brachial artery at the distal border of teres major muscle and divided into three parts by the intervening pectoralis minor.
7. Cervicoaxillary canal (apex of axilla): Bounded by clavicle (anteriorly), scapula (posteriorly) and first rib (medially).

PROBABLE QUESTIONS AND ANSWERS

 **1. What is cervicoaxillary canal?**

Ans. The apex of the axilla is known as cervicoaxillary canal.

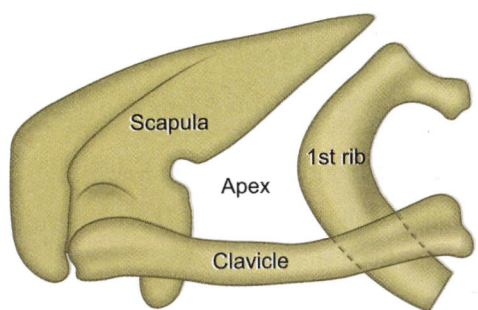

Fig. 2.23: Cervicoaxillary canal

2. What are its boundaries?

Ans. Triangular in shape, bounded in front by clavicle, behind by superior border of scapula and medially by outer border of the first rib (Fig. 2.23).

3. What are the structures that are passing through the apex?

Ans. Axillary vessels, cords of brachial plexus, efferent lymph vessels and long thoracic nerve (C_5, C_6, C_7).

4. Which structure forms the base of the axilla?

Ans. Axillary fascia or deep fascia of axilla extending from the anterior to posterior axillary folds forms the base of the axilla.

5. What is the support of the base?

Ans. Base is supported by the suspensory ligament of axilla which maintains the depression of the floor of the axilla.

6. What is the suspensory ligament of axilla?

Ans. At the upper border of pectoralis minor muscle, the clavipectoral fascia splits to enclose and reunites at the lower border of that muscle to be continued as the suspensory ligament of axilla.

7. How the axillary folds are formed?

Ans. i. The anterior axillary fold is formed by the spirally arranged lower border of pectoralis major muscle.
ii. The posterior fold is formed by the latissimus dorsi and teres major muscle.

8. How many walls of the axilla are there?

Ans. 4 walls—anterior, posterior, medial and lateral.

9. What are the muscles found in the posterior wall?

Ans. Subscapularis, latissimus dorsi and teres major muscles from above downwards.

10. Where do you find the long thoracic nerve of bell?

Ans. Along the medial wall of the axilla, deep to the fascia covering serratus anterior muscle and behind the midaxillary line.

11. What is intercostobrachial nerve and where do you find it?

Ans. Usually, the lateral cutaneous branch of intercostal nerve divides into an anterior branch and a posterior branch. But the lateral cutaneous branch of second intercostal nerve does not divide. This undivided lateral cutaneous branch of second intercostal nerve is called intercostobrachial nerve. It is found to pierce the medial wall of the axilla.

12. Why it is so named?

Ans. It is so named because this branch of **intercostal** nerve passes laterally after piercing the medial wall of axilla to supply the posteromedial part of the arm (**brachium**). It joins with the medial cutaneous nerve of the arm. Irritation of the nerve may cause radiation of pain from medial wall of axilla (lateral chest wall) towards upper part of medial aspect of the arm.

13. What is the nerve supply of subscapularis?

Ans. It is supplied by both upper and lower subscapular nerves. Both are the branches of posterior cord of brachial plexus.

14. Where from the suprascapular nerve arises?

Ans. It arises from the upper trunk of brachial plexus at **Erb's point**.

15. What is Erb's point?

Ans. It is the point of union of 6 nerves—roots of C_5, C_6, anterior and posterior divisions of upper trunk, suprascapular nerve and nerve to subclavius (Fig. 2.24).

16. What structure accompanies the subscapular artery?

Ans. Thoracodorsal nerve (nerve to latissimus dorsi).

17. What are the muscles supplied by lower subscapular nerve?

Ans. Lower subscapular nerve supplies both the subscapularis and teres major, whereas upper subscapular nerve supplies only the subscapularis muscle.

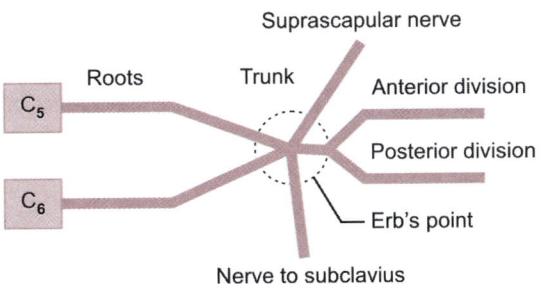

Fig. 2.24: Erb's point

18. What happens if the long thoracic nerve gets injured?

Ans. It produces winging of scapula due to paralysis of serratus anterior muscle and unopposed action of the rhomboideus and middle fibers of trapezius.

19. What is axillary sheath?

Ans. It is a fascial envelope derived from prevertebral layer of deep cervical fascia enclosing the proximal part of axillary artery and vein together with the brachial plexus.

20. How is the axillary vein formed?

Ans. It is formed by the union of basilic vein and venae comitantes of brachial artery at the lower border of teres major.

21. Is the axillary vein a content of axillary sheath?

Ans. It is said that the axillary vein is not a content of the axillary sheath in order to allow expansion of the vein during increased venous return **(RJ Last)**.

22. Which structure divides the axillary artery into three parts?

Ans. The pectoralis minor crosses in front of the artery and divides it into three parts: Proximal to the muscle (1st part), behind the muscle (2nd part), distal to the muscle (3rd part).

23. What are the continuations of different cords of brachial plexus?

Ans. Posterior cord continues as **radial nerve** (C_5–T_1) medial cord continues as **ulnar** nerve (C_7, C_8, T_1) and lateral cord continues as **musculocutaneous nerve** (C_5, C_6, C_7).

24. Why the cords are so named?

Ans. They are so named because of their relative position to the second part of the axillary artery.

25. How is the median nerve formed?

Ans. It is formed by the union of lateral root (C_5, C_6, C_7) of lateral cord and medial root (C_8, T_1) of medial cord.

26. Where from the C_7 fibers of ulnar nerve is derived?

Ans. Some fibers from the C_7 conveyed by the median nerve handed over to the ulnar nerve. This fibers from C_7 supply the flexor carpi ulnaris muscle in the forearm.

27. What is the mode of insertion of pectoralis major in bicipital groove?

Ans. Before insertion into the lateral lip of bicipital groove, the fibres bend up on itself to form a U-shaped lamina. So, it is bilaminar—anterior and posterior; clavicular fibers form another lamina, so trilamilar.

LESSON 5: FRONT OF ARM (BRACHIUM)

The arm is the upper part of the upper limb between the elbow and shoulder and it is enveloped by a sleeve of tough connective tissue (deep fascia of the arm or **brachial fascia**) that is continuous at its proximal end with the pectoral fascia, the axillary fascia and the deep fascia covering the deltoid and latissimus dorsi. Distally, the brachial fascia is continuous with the deep fascia of the forearm **(antebrachial fascia)**.

STEPS OF DISSECTION

1. **Position of the cadaver:** Supine position with the arm rotated laterally and elbow extended.
2. **Skin incision (Fig. 2.25):**
   a. A transverse incision in front of the arm at the junction of upper one-fourth and lower three-fourths (A–B).
   b. 2nd incision is given transversely between the two epicondyles of the humerus (C–D).
   c. 3rd incision given vertically from the mid point of the upper (E) and lower (F) incisions (E–F).

   Reflect the flap of skin laterally and medially. Superficial fascia is exposed.
3. **Superficial fascia:** In it some important superficial structures are found.
   a. **Cephalic vein** on the lateral aspect passing upwards.

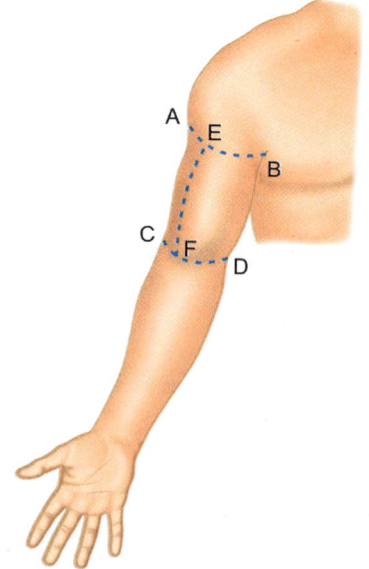

Fig. 2.25: Skin incision—front of the arm

b. **Basilic vein** on the medial aspect, piercing the deep fascia at the middle of the arm becomes deep vein and forms axillary vein after joining with the venae comitantes of brachial artery.
c. **Medial cutaneous nerve of the arm.**
   - Superficial fascia is reflected in the line of skin incision.
   - Deep fascia is exposed.
4. **Deep fascia:** The deep fascia of the arm (brachial fascia) is cut vertically on its anterior surface using scissors and forceps. Fingers are introduced inside the deep fascia to separate it from the underlying muscles. Proceed medially and laterally from the incision and note the presence of **medial and lateral intermuscular septum** extending from the deep surface of the deep fascia up to the medial and lateral sides of the humerus; thus creates flexor (anterior) compartment and an extensor (posterior) compartment for the muscles of the arm.
   Reflect the deep fascia like that of the skin and superficial fascia. Deep structures are exposed.
5. **After** reflecting the deep fascia, **biceps brachii** muscle is visible. Find the neurovascular bundle of the arm immediately deep to the deep fascia, medial to the biceps brachii muscle. Lift this muscle forwards using the fingers. One must be careful not to damage the **musculocutaneous nerve** which is passing deep to the biceps brachii from medial to lateral side of the arm. Now separate the three muscles in the anterior compartment of the arm using fingers and forceps. If necessary, a scalpel or scissor is used to incise the deep fascia at the border of the two adjoining muscles and then slide fingers up and down the adjoining borders to free one muscle from the other. One should keep in mind that many neurovascular bundles run between adjoining muscle bellies (Fig. 2.26).

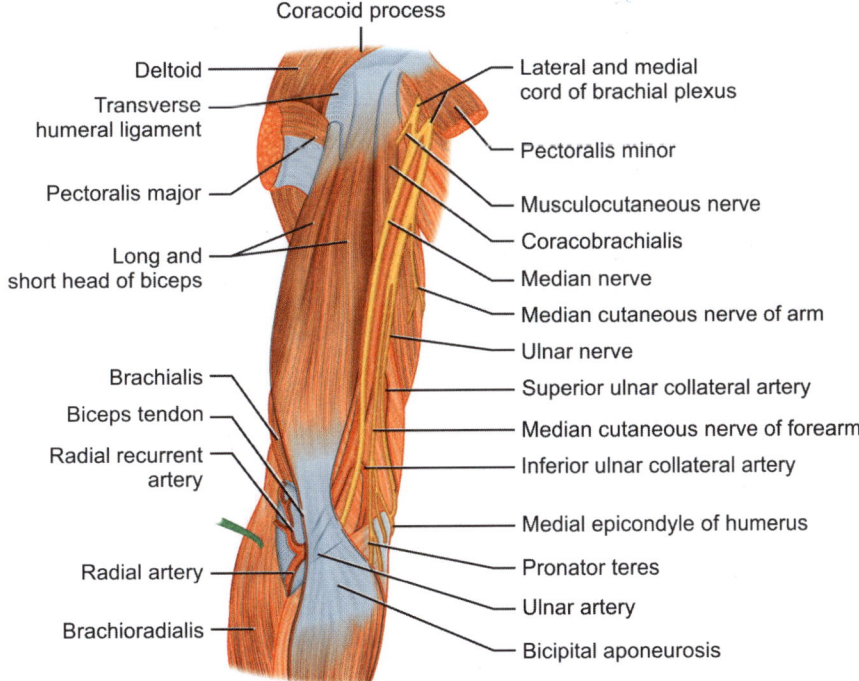

Fig. 2.26: Neurovascular bundle in front of arm

6. Transect the biceps brachii muscle (by scissor) about 5 cm above the elbow without damaging the musculocutaneous nerve. Reflect the two portions of biceps brachii muscle proximally and distally (Figs 2.27A and B).
7. **Brachialis muscle** is observed. Remove the fascia from brachialis and define **brachioradialis** and **extensor carpi radialis longus** muscle (both are forearm muscles). Try to find out a branch from **radial nerve** which supplies these three muscles (lateral 3rd of brachialis, brachioradialis and extensor carpi radialis longus).
8. **Musculocutaneous** nerve (Fig. 2.28) is cleaned proximally and distally. Proximally, it emerges from the **coracobrachialis** muscle and supplies it. Then it follows a course through the plane of loose connective tissue between the **biceps** and **brachialis** muscle. After the nerve gives off its muscular branches to **biceps brachii** and medial 2/3rd of **brachialis** it continues distally as the **lateral cutaneous nerve** of the **forearm** in front of the lateral epicondyle of the humerus after piercing the deep fascia.
9. **Coracobrachialis** muscle which lies on the medial side of biceps brachii is cleaned and found to be pierced by the musculocutaneous nerve. A slender branch of medial cord of brachial plexus known as **medial cutaneous nerve of arm** passes along the medial side of the brachial vessels and then it pierces the deep fascia.
10. **Medial cutaneous nerve** of the forearm, branch of medial cord of brachial plexus descends downwards between the brachial artery and brachial vein and at the middle of the arm it pierces the deep fascia.

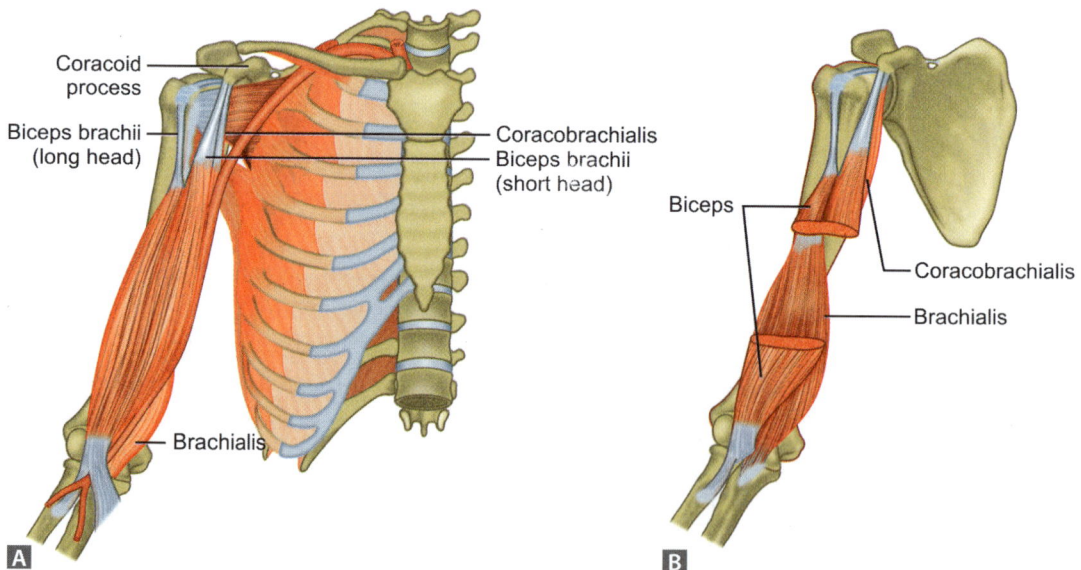

Figs 2.27A and B: (A) Front of arm (before transection of muscles); (B) Front of arm (after transection of muscles)

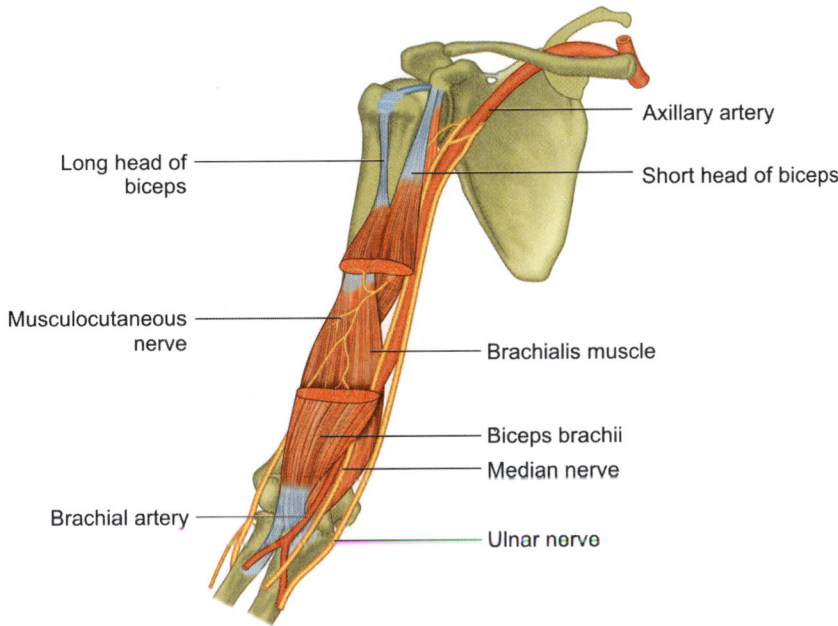

Fig. 2.28: Musculocutaneous nerve lying in front of brachialis muscle

11. **Median nerve** (C_5-T_1), formed by lateral root (from lateral cord) and medial root (from medial cord) descends downwards passing in front of the brachial artery from its lateral to medial side. Note that **median nerve is the only large structure to cross the anterior surface of the brachial artery.**
12. **Ulnar nerve** (C_7, C_8 and T_1) descends downwards between the brachial artery and vein. At the middle of the arm it pierces the **medial intermuscular septum** and passes distally in the posterior compartment of the arm to enter the forearm by passing over the posterior surface of the medial epicondyle of the humerus.
13. **Brachial artery and vein (Fig. 2.29)** are dissected out and cleaned. The artery is traced from its beginning at the lower border of the **teres major** muscle upto its bifurcation into **radial and ulnar artery**. The brachial artery has **three** named branches in the arm.
    a. Arteria profunda brachi (accompanied by **radial** nerve).
    b. Superior ulnar collateral (accompanied by **ulnar** nerve).
    c. Inferior ulnar collateral (arises 3 cm above the medial epicondyle and passes in front of it).
    Others:
    – Muscular.
    – Nutrient (arises near the middle of the humerus).

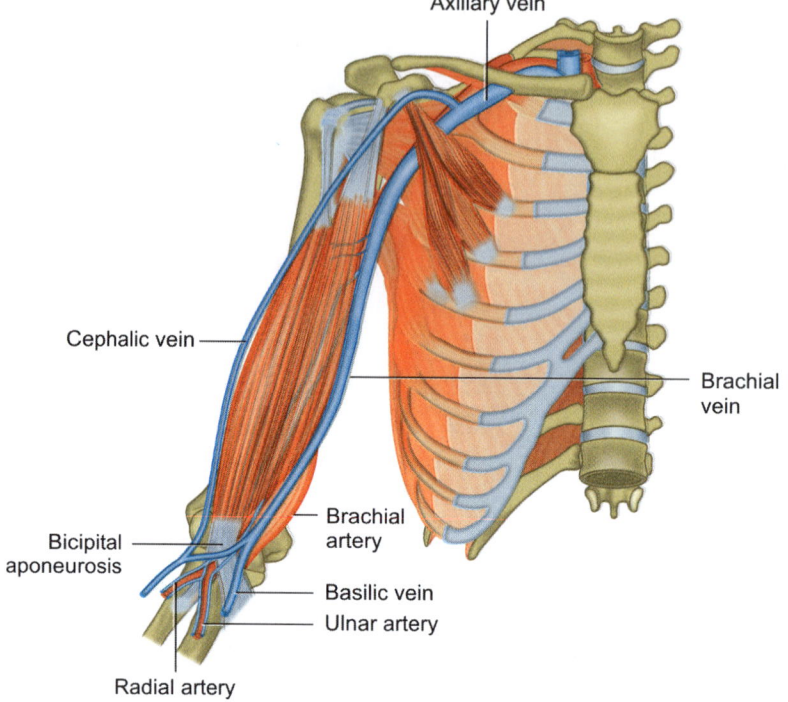

Fig. 2.29: Brachial artery and vein

PRINCIPAL NEUROVASCULAR BUNDLE

This bundle consists of artery (brachial artery), veins (basilic and brachial veins) and nerves (median, ulnar, radial and medial cutaneous nerve of the forearm). The radial nerve is the first to leave the bundle and accompanies arteria profunda brachii in the spiral groove on the posterior surface of the humerus. The ulnar nerve is the next to leave the bundle. Then, the basilic vein and medial cutaneous nerve of forearm pass into the superficial fascia. Thus, the bundle contains only the median nerve and brachial vessels in the lower third of the arm to lie just medial to the tendon of biceps at the elbow.

Note that the brachial artery lies medial to the biceps brachii tendon and close to the shaft of the humerus at the lower part of the arm. So, while taking a blood pressure reading, the brachial artery is compressed at this location.

MUSCLES RELATED TO FRONT OF THE ARM

Muscles	Origin	Insertion	Nerve supply	Action
Biceps brachii	• **Longhead:** Supraglenoid tubercle • **Short head:** Coracoid process of scapula	Radial tuberosity	Musculocutaneous nerve	• Flexion of elbow • Supination
Brachialis	Anterior surface of distal one-half of the humerus	• Tuberosity of ulna • Anterior surface of coronoid process of ulna	• Medial larger part by musculocutaneous nerve • Lateral smaller part by radial nerve	Flexion of elbow
Coraco-brachialis	Coracoid process of scapula (medial to the origin of short head of biceps)	Medial surface of the shaft of the humerus	Musculocutaneous nerve	• Adduction • Flexion of the shoulder joint

Coracobrachialis represents adduction muscle of the arm and homologous with the adductors of thigh. In some mammals, it has three heads of origin. Upper two heads fused to originate from tip of coracoid process of scapula enclosing musculocutaneous nerve between them. The third head is usually suppressed in man and represented by a fibrous band called **Struther's ligament** extending from the supratrochlear spur of the lower part of anteromedial surface of humerus to medial epicondyle deep to which may pass the **median nerve** or **brachial artery.**

IMPORTANT LANDMARKS AT THE MIDDLE OF THE ARM

1. Insertion of coracobrachialis and deltoid muscle.
2. Two heads of biceps brachii joint.
3. Nutrient foramen of the humerus.
4. Median nerve crosses the front of brachial artery from lateral to medial side.
5. Basilic vein and medial cutaneous nerve of forearm pierce the deep fascia.
6. Ulnar nerve and superior ulnar collateral artery pierce the medial inter-muscular septum and reach the posterior compartment of the arm.

Summary

1. Incision of skin, superficial fascia, deep fascia (brachial fascia) as in Figure 2.25.
2. Structures found in superficial fascia: **Cephalic vein** (laterally), **basilic vein** (medially) and **medial cutaneous nerve of arm** (from medial cord).
3. Deep structures (deep to deep fascia):
   a. **Muscles:** Both the heads of biceps brachii and coracobrachialis (both supplied by musculocutaneous nerve), brachialis (musculocutaneus nerve—medial part; radial nerve—lateral part).
   b. **Nerves:** Musculocutaneous nerve (lateral cord) median nerve (lateral and medial cord) ulnar nerve (medial cord), medial cutaneous nerve of arm and forearm.
   c. **Vessels:** Brachial vessels (brachial artery with venae comitantes).
4. Brachial artery begins at the lower border of teres major as a continuation of 3rd part of axillary artery and ends at the level of neck of radius by dividing into radial and ulnar artery in the cubital fossa.

PROBABLE QUESTIONS AND ANSWERS

1. How the arm is divided into anterior and posterior compartment?

Ans. The deep fascia of the arm (**brachial fascia**) projects into its interior as medial and lateral intermuscular septum which are attached to the medial supracondylar ridge and lateral supracondylar ridge respectively.

2. What are the structures that are pierced by medial intermuscular septum?

Ans.
   i. Ulnar nerve in the middle of the arm.
   ii. Superior ulnar collateral artery.
   iii. Posterior branch of inferior ulnar collateral artery above the medial epicondyle.

3. What are the structures that are pierced by lateral intermuscular septum?

Ans.
   i. Radial nerve.
   ii. Radial collateral branch of arteria profunda brachii.

4. Name the muscles that are arising from the lateral intermuscular septum in addition to their bony origin?

Ans. Brachioradialis (BR) and extensor carpi radialis longus (ECRL).

5. Name the muscles of the anterior compartment of the arm.

Ans.
   i. Coracobrachialis—medial aspect of the upper arm.
   ii. Biceps brachii and brachialis—front of the arm.
   iii. BR and ECRL—in the lower part.

6. What is origin of coracobrachialis?

Ans. Tip of coracoid process of scapula medial to the origin of short head of biceps brachii muscle.

7. Where do the two heads of biceps brachii join to form a single tendon?

Ans. About 7 cm above the elbow joint.

8. How does the biceps look?

Ans. Biceps tendon is flat; muscle is elongated fusiform; arises by two tendinous heads (short and long).

9. How does the biceps insert?

Ans. The tendon of biceps undergoes twisting so that the anterior surface of the tendon becomes lateral and inserted into the posterior part of radial tuberosity and separated from the anterior part of radial tuberosity by a bursa.

10. To what extend, the long head of biceps accompanies the synovial sheath?

Ans. The synovial sheath extends upto the surgical neck of humerus.
[**Note that the origin of long head is intracapsular but extrasynovial**].

11. How does the coracobrachialis get its nerve supply?

Ans. Musculocutaneous nerve (C_5, C_6, C_7) branch of lateral cord of brachial plexus supplies the muscle before piercing it and after piercing it.

12. Where do you find musculocutaneous nerve?

Ans. After piercing the coracobrachialis, the nerve passes laterlly between biceps brachii and brachialis, then it pierces the deep fascia to be continued as lateral cutaneous nerve of forearm.

13. What are muscles supplied by musculocutaneous nerve?

Ans. Coracobrachialis, both heads of biceps brachii and medial part of brachialis (lateral part supplied by radial nerve).

14. What is the relation between median nerve and brachial artery in the arm?

Ans. In the upper part of the arm, the median nerve is lateral to the artery; at the middle of the arm, the nerve crosses in front of the artery from lateral to medial side.

15. How does the ulnar nerve appear in the forearm from the arm?

Ans. The ulnar nerve piercing the medial intermuscular septum in the middle of the arm appears on the back and lies is a groove on the posterior surface of the medial epicondyle and then appears in the forearm between the two heads of flexor carpi ulnaris muscle.

16. How many times the radial nerve appears in front of the arm?

Ans. Two times:
   i. **In the upper part of the arm:** It lies behind the brachial artery and enters into the spiral groove accompanied by the arteria profunda brachialis.
   ii. **In the lower part of the arm:** It reappears for the second time after piercing the lateral intermuscular septum and runs in the interval between the brachioradialis and extensor carpi radialis longus on the lateral side and brachialis on the medial side.

17. Where does the radial nerve end?

Ans. Radial nerve ends by dividing into a superficial (cutaneous) and a deep (posterior interosseous) branch on reaching the front of the lateral epicondyle.

18. Is there any branch of median nerve in the arm?

Ans. No, except a branch to pronator teres.

19. What are the structures that lodge in intertubercular sulcus?

Ans.
  i. Long head of biceps.
  ii. Ascending branch of anterior humeral circumflex artery.

20. What is the paradoxical action of biceps?

Ans. Gradual relaxation of the biceps brachii muscle and brachialis muscle (both are flexors of the elbow) helps in extension of the elbow. This is called the **action of paradox.**

21. How do you explain the double nerve supply of brachialis?

Ans. In fetal life, this muscle consists of two portions, one occupies the flexor compartment with its supply by the nerve of flexor compartment, i.e. musculocutaneous nerve and the other occupies the extensor compartment with its supply by the nerve of that compartment, i.e. radial nerve.

Branches of Brachial Artery

1. Arteria profunda brachii **(largest and first branch)**
2. Superior ulnar collateral.
3. Inferior ulnar collateral.
4. Nutrient artery to humerus.
5. Muscular: Muscles of the anterior compartment of the arm.
6. Terminal: **Radial and ulnar.**

LESSON 6: CUBITAL FOSSA

Cubitus (Latin)—**Elbow**; Fossa—**Depression** (hollow).

Cubital fossa is the triangular depression on the anterior aspect of the elbow. Triangle is formed by (Fig. 2.30):

- Base: An imaginary line drawn between the two epicondyles of the humerus.
- Medial arm: Pronator teres muscle.
- Lateral arm: Brachioradialis muscle.

Window Dissections: Upper Limb (Superior Extremity)

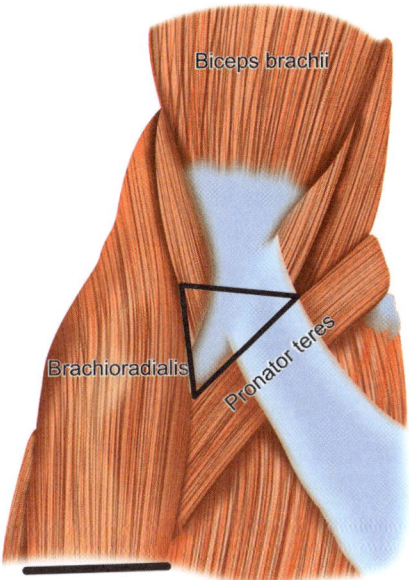

Fig. 2.30: Cubital fossa (marked by a triangle)

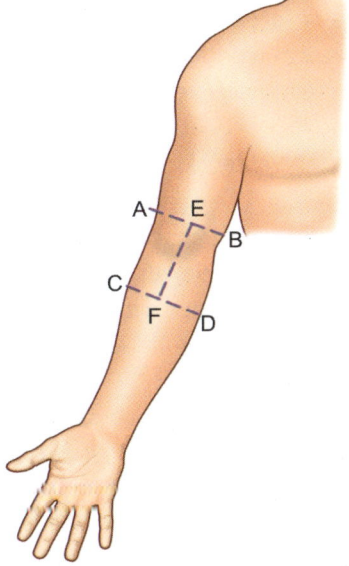

Fig. 2.31: Skin incision

- Apex: Meeting of lateral border (arm) and medial border (arm).

STEPS OF DISSECTION

1. **Position of the cadaver:** The cadaver is allowed to lie in supine position with the forearm extended and supinated. The upper limb is placed preferably on a wooden block.
2. **Skin incision (Fig. 2.31):**
   a. Identify the two epicondyles of the humerus. A transverse incision is given about one finger above the two epicondyles (A–B).
   b. Another transverse incision at the junction of the upper one-third and lower two-thirds of the front of the forearm (C–D)
   c. Third incision extends from the midpoint of proximal incision (E) to midpoint of distal incision (F).

   The skin flaps are reflected side ways. The superficial fascia is exposed.
3. **Superficial fascia (Fig. 2.32):** The superficial fascia contains the superficial veins of upper limb and cutaneous nerves of the forearm. The veins are **cephalic vein**, **basilic vein** and

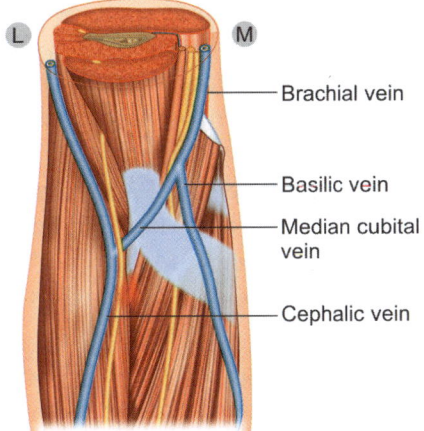

Fig. 2.32: Superficial structures of cubital fossa

median cubital vein connecting the former two veins. The nerves are **medial cutaneous nerve of forearm** (branch of medial cord of brachial plexus) and **lateral cutaneous nerve of forearm** (continuation of musculocutaneous nerve of lateral cord). Clean and find out these superficial structures. To gain access to deeper structures, it may be necessary to cut the median cubital vein and retract it medially and laterally. The superficial fascia is then cut in the line of skin incision and reflect it. Deep fascia is exposed.

4. **Deep fascia:** The deep fascia is known as antebrachial fascia which is continuous proximally with the brachial fascia (deep fascia of the arm). **Bicipital aponeurosis** is attached to the antebrachial fascia. Cut this aponeurosis near the biceps brachii tendon and reflet it medially. Do not disturb the brachial artery deep to this aponeurosis.

> ### Bicipital Aponeurosis (Figs 2.33A and B)
> It is the broad extension from the medial side of the biceps tendon that attaches to the antebrachial fascia. It forms a secondary insertion of biceps (primary insertion is radial tuberosity) through which biceps is indirectly attached to the posterior border of the ulna. Along with antebrachial fascia it forms the roof of the cubital fossa. It protects the **brachial artery** and **median nerve** from injury during venipuncture (of **median cubital vein**).

Reflect the deep fascia in the line of skin incision.

5. The triangle is exposed. The boundary and contents of the triangle are cleaned. Remove any fat that may be obstructing the proper view of the structures.
6. **Note** that the **brachioradialis** muscle which forms the lateral boundary of the triangular fossa overlaps the **pronator teres** muscle (medial border) at the apex. The floor is formed by the **brachialis** (superomedially) and **supinator** muscle (inferolaterally). To expose the supinator properly, **brachioradialis** is retracted laterally. Two nerves are lying on supinator muscle. The **radial nerve** descends downwards lying on the muscle and **posterior interosseous nerve** (deep branch of radial nerve) goes on to the posterior compartment of the forearm passing between the two strata (superficial and deep) of the supinator muscle.
7. Then three main structures of the cubital fossa are identified and followed them from the arm. The **tendon** of biceps passes backwards between the radius and ulna to reach the posterior part of the tuberosity of radius for insertion in a twisted manner. The **brachial artery** divides into radial and ulnar artery opposite the neck of the radius in this region. The **radial artery** leaves the cubital fossa at the apex and the **ulnar artery** passes deep to the deep (ulnar) head of pronator teres. The **median nerve** supplies **pronator teres** and leaves the fossa by passing between the two heads (superficial or humeral and deep or ulnar) of pronator teres. **So the deep head of pronater teres separates the median nerve from the ulnar artery (Fig. 2.34).** Note the relative positions of these three structures— biceps **T**endon is lateral, brachial **A**rtery is intermediate and median **N**erve is medial **(TAN from lateral to medial) (Figs 2.35A and B).**

Window Dissections: Upper Limb (Superior Extremity)

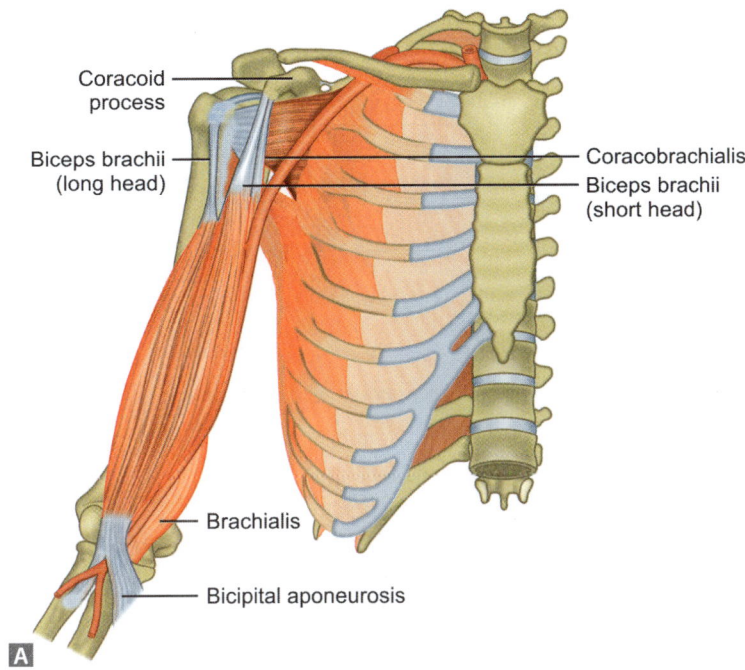

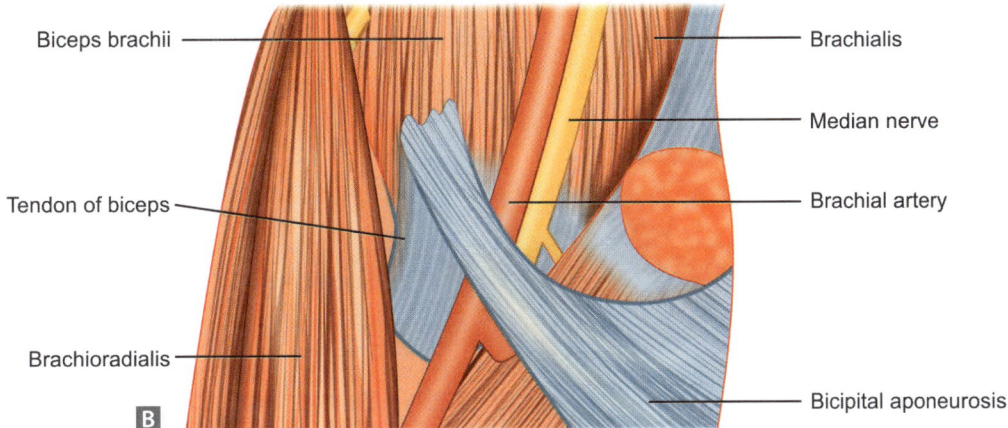

Figs 2.33A and B: Bicipital aponeurosis and deep structures of cubital fossa

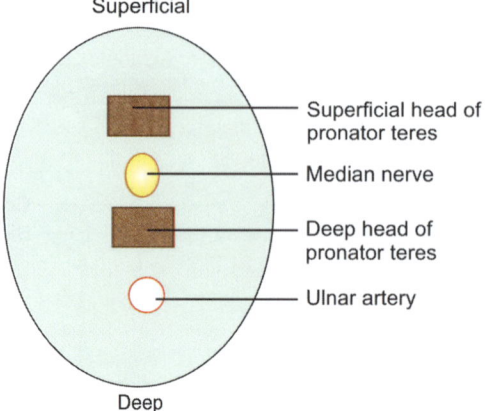

Fig. 2.34: Median nerve and ulnar artery

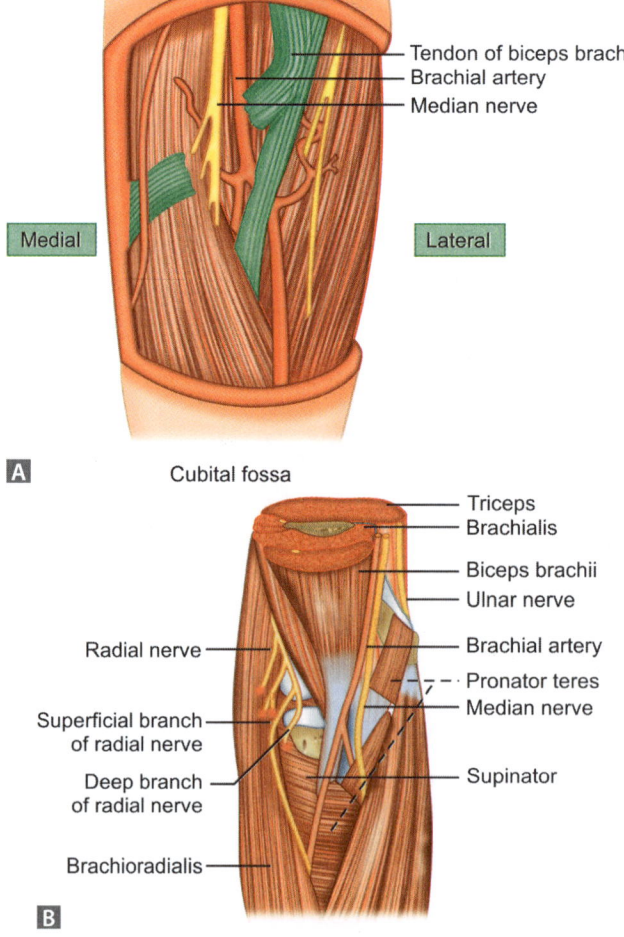

Figs 2.35A and B: Contents of cubital fossa

Window Dissections: Upper Limb (Superior Extremity)

> **Median cubital vein** is used for intravenous injection, blood transfusion, cardiac catheterization because:
> i. It is fixed by the perforating veins and does not slip away.
> ii. It is separated from the brachial artery and median nerve by bicipital aponeurosis—so no chance of injury.

MUSCLES RELATED TO CUBITAL FOSSA

Muscles	Origin	Insertion	Nerve supply	Action
Pronator teres (medial boundary)	• **Humeral head**: Medial epicondyle of humerus • **Ulnar head**: Medial border of coronoid process of ulna	Middle of the lateral surface of radius	Median nerve	• Pronation • Flexion of elbow
Brachioradialis (lateral boundary)	Lower part of lateral supracondylar line of humerus	Styloid process of radius	Radial nerve	Flexion of elbow (best in midprone position)
Supinator (floor)	• Lateral epicondyle of humerus (humeral head) • Supinator crest of ulna (ulnar head)	Distal to the radial tuberosity of radius	Posterior interosseous nerve (before passing between two heads of supinator)	Supination in extended elbow
Brachialis (floor)	Anteromedial and anterolateral surface of the lower half of the shaft of the humerus	• Anterior surface of coronoid process of ulna • Tuberosity of ulna	• Medial larger part by musculocutaneous nerve • Lateral small part by radial nerve	Flexion of elbow

> **Summary**
> 1. Incision of skin, superficial fascia and deep fascia (antebrachial fascia) as in Figure 2.30.
> 2. Structures found in superficial fascia:
>   a. **Laterally: Cephalic vein** and lateral cutaneous nerve of forearm.
>   b. **Medially: Basilic vein** and medial cutaneous nerve of forearm.
>   c. **Median cubital** vein connecting the above two veins superficial to bicipital aponeurosis.
> 3. Deep fascia and bicipital aponeurosis are cut and reflected.
> 4. **Boundaries of the triangular cubital fossa**:
>   a. Laterally: **Brachioradialis**.
>   b. Medially: **Pronator teres**.
>   c. Base: Line joining the two epicondyles.
>   d. Apex: Convergence of above two muscles.
>   e. Floor: **Brachialis and supinator**.
> 5. **Contents**: TAN (lateral to medial) (Figs 2.35A and B)—**tendon of biceps brachii, brachial artery, median nerve** and a short course of radial nerve at the suprolateral angle of the fossa between brachioradialis and brachialis muscle.

PROBABLE QUESTIONS AND ANSWERS

1. What is bicipital aponeurosis?

Ans. It is a fibrous expansion from the medial border of the biceps tendon extending downwards and medially to attach with the upper part of the posterior border of ulna.

2. What are the contents of the cubital fossa?

Ans. The contents (from lateral to medial) are **T**endon of biceps, brachial **A**rtery and median **N**erve (TAN).

3. Is the radial nerve is a content of the cubital fossa?

Ans. The radial nerve is not actually a content of the fossa because it is overlapped by the brachioradialis muscle which forms the lateral boundary of the fossa. So, the nerve can be seen by retracting the muscle laterally.

4. Where does the radial nerve divide?

Ans. The radial nerve divides in front of the lateral epicondyle of the humerus into a superficial branch and a deep branch.

5. How does the radial nerve enter into the back of the forearm?

Ans. The deep branch of radial nerve passes between the two heads of supinator muscle and appears in the extensor (back) compartment of the forearm.

6. How do the median nerve and ulnar nerve enter into the forearm?

Ans. The median nerve passes between the two heads of pronator teres and the ulnar nerve passes between the two heads of flexor carpi ulnaris to enter into the forearm.

7. How dose the ulnar artery leave the cubital fossa?

Ans. The ulnar artery leaves the cubital fossa passing downwards and medially deep to the ulnar head (deep head) of pronator teres. The deep head of pronator teres separates the ulnar artery (passing deep to it) from the median nerve passing between the two heads of pronator teres.

8. What is the level of bifurcation of the brachial artery?

Ans. About 1 cm below the elbow joint and opposite the neck of the radius it divides into the radial and ulnar artery.

Window Dissections: Upper Limb (Superior Extremity)

9. What is the only branch of median nerve in the arm?

Ans. Nerve to pronator teres.

10. What structure do you get after cutting the deep head (ulnar head) of pronator teres?

Ans. The ulnar artery which is seen lying on the flexor digitorum profundus.

11. How do you feel brachial artery in front of the elbow in living?

Ans. In flexion of the elbow the biceps tendon is being taut and felt. Just medial to this tendon the pulsation of the brachial artery is felt. This helps as standard method of recording of blood pressure.

12. Which muscle is known as supinator longus?

Ans. Biceps brachii is also called supinator longus as it causes supination.

LESSON 7: FRONT OF THE FOREARM

Forearm is the region between the elbow and the wrist and enveloped by a sleeve of connective tissue called, **antebrachial fascia** or deep fascia of the forearm. The forearm is divided by the intermuscular septa, the interosseous membrane and the forearm bones together into an anterior (flexor) compartment and a posterior (extensor) compartment. **The flexor muscles are more massive and powerful than the extensor muscles because of their antigravity action.** The tendons of these muscles are retained in position due to the deep fascial thickening at the wrist (**flexor** and **extensor retinacula**).

STEPS OF DISSECTION

1. **Position of the cadaver:** Supine with the upper limb abducted. Supinate the hand forcefully and hold it in position tying with the wooden block/dissection table. The wrist and elbow are extended.
2. **Skin incision (Figs 2.36A and B):**
    a. A transverse incision joining the two epicondyles of the humerus (A–B)
    b. 2nd incision (C–D) joins the styloid process of radius and ulna.
    c. A third vertical incision (E–F) joining the mid-points of these two transverse lines.

    But if the cubital fossa already has been dissected, the first transverse incision (A–B) should be given at lower level, usually at the junction of upper 1/3rd and lower 2/3rd of the forearm.

    Reflect the skin flap side ways. Superficial fascia is exposed.

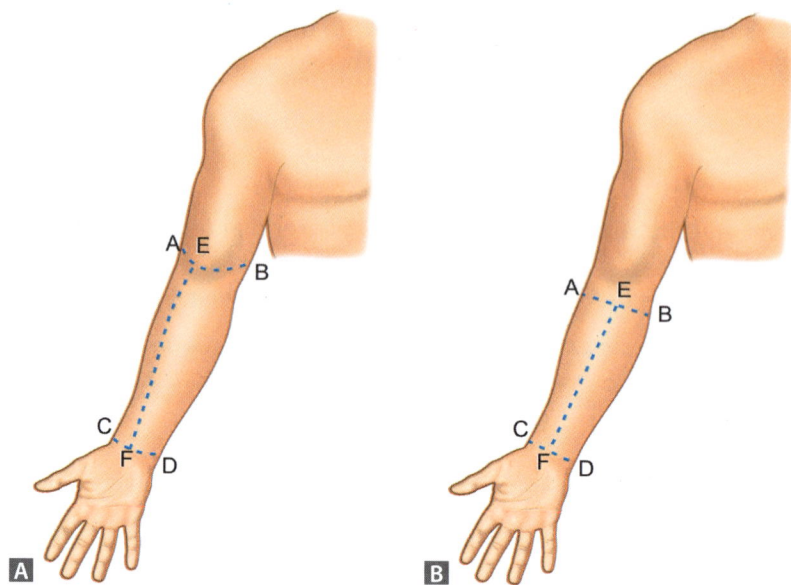

Figs 2.36A and B: Skin incision—front of the forearm

3. **Superficial fascia:** In the superficial fascia, anterior branch of medial and lateral cutaneous nerve of forearm are found only at the lower part. Care should be taken to preserve the **cephalic** and **basilic vein** during removal of the remnants of the superficial fascia. Deep to this layer deep fascia is found.
4. **Deep fascia (antebrachial fascia):** The deep fascia is cut along the line of skin incision. It is separated from the muscles that lie deep to it using a probe or fingers. The fascia is then detached from its attachments to the radius and ulna. Then it is reflected medially and laterally like that of skin.
5. **Deep structures (Fig. 2.37): Deep** to the deep fascia, the structures available are the muscles, vessels and nerves. The muscles that are found beneath the deep fascia are arranged in two groups' **superficial muscles** and **deep muscles**.

   Separate one muscle from another. To do this use a scalpel to incise the deep fascia at the border of two adjoining muscles and slide your fingers up and down the adjoining borders to free one muscle from the other. One thing must be kept is mind that many neurovascular bundles run between these adjoining muscles.
6. **Clean** the superficial group of flexor muscles (five in number) using blunt dissection. These muscles are **pronator teres** (PT), **flexor carpi radialis** (FCR), **palmaris longus** (PL), **flexor carpi ulnaris** (from lateral to medial side) and **flexor digitorum superficialis** (FDS) deep to FCR and PL (Figs 2.38A and B).

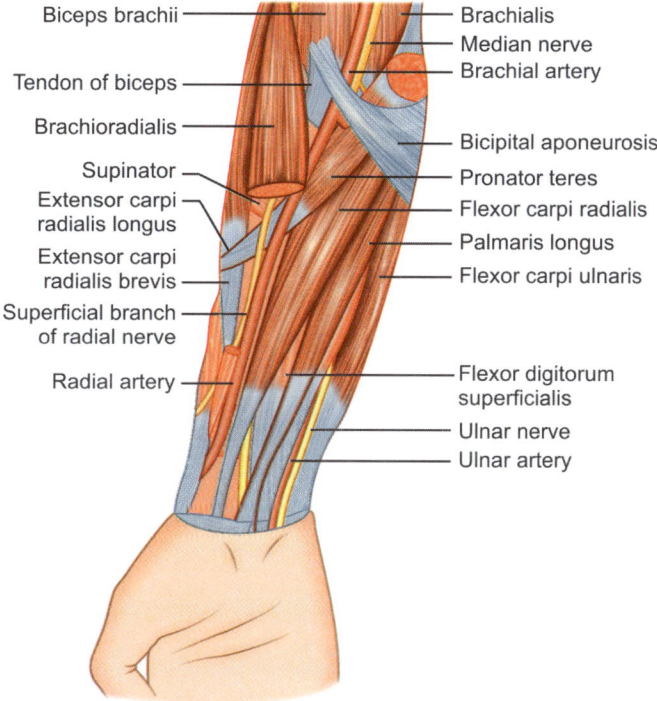

Fig. 2.37: Deep structures of anterior compartment of forearm

Now, separate the tendons of the superficial group of flexor muscles from one another. Most laterally and at the highest level lies the pronater teres and most medially lies the flexor carpi ulnaris. Medial to flexor carpi radials is the palmaris longus tendon which joins the palmar aponeurosis passing superficial to the flexor retinaculum. **It is believed that the palmar aponeurosis is the degenerated distal part of palmaris longus.** Flexor digitorum superficialis is identified by its position and large fleshy belly and ends in four tendons.

7. **Between** the superficial muscle group and deep muscle group, identify the **neurovascular bundles** which consist of median nerve, ulnar nerve, radial nerve and their accompanying vessels **(Figs 2.39A to C).**
8. **Ulnar nerve** enters into the front of the forearm between the two heads of flexor carpi ulnaris. Follow it distally. In its course, it supplies **flexor carpi ulnaris** (superficial muscle) and medial half of **flexor digitorum profundus** (deep muscle). Just proximal to the flexor retinaculum, the **ulnar artery** and **ulnar nerve** become superficial between the tendons of FCU and FDS. Here, **the nerve lies medial to the artery.** Proximally, the ulnar artery passes deep to the ulnar head of pronator teres. Before passing into the hand both the ulnar artery and nerve (palmar branches) pierce the deep fascia proximal to the flexor

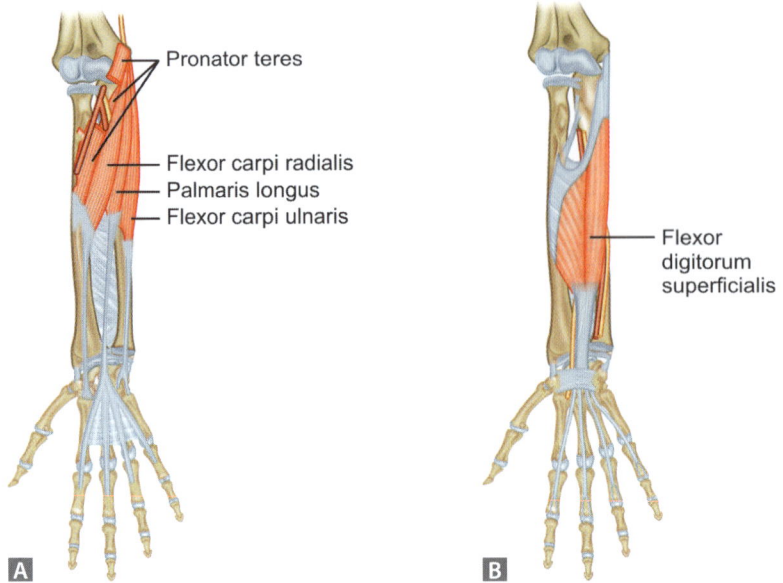

Figs 2.38A and B: (A) Superficial muscles of front of forearm; (B) Front of forearm—flexor digitorum superficialis

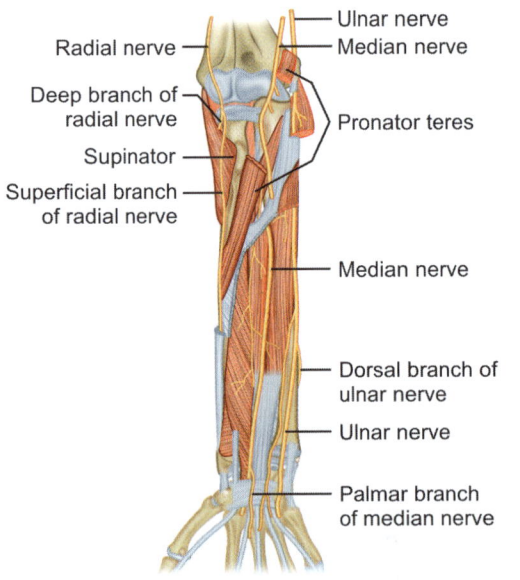

Fig. 2.39A: Nerves of front of forearm

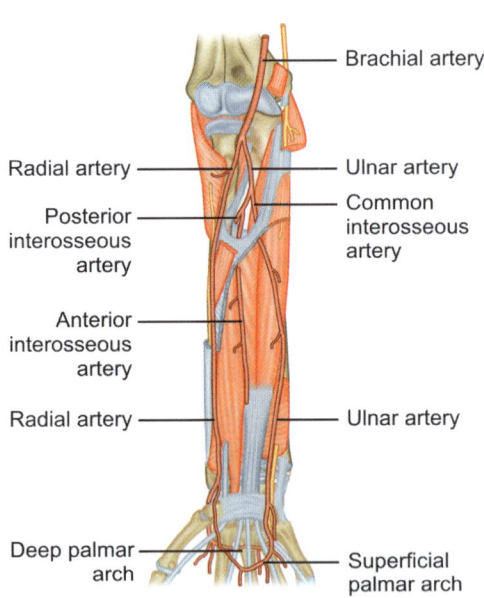

Fig. 2.39B: Arteries of front of forearm

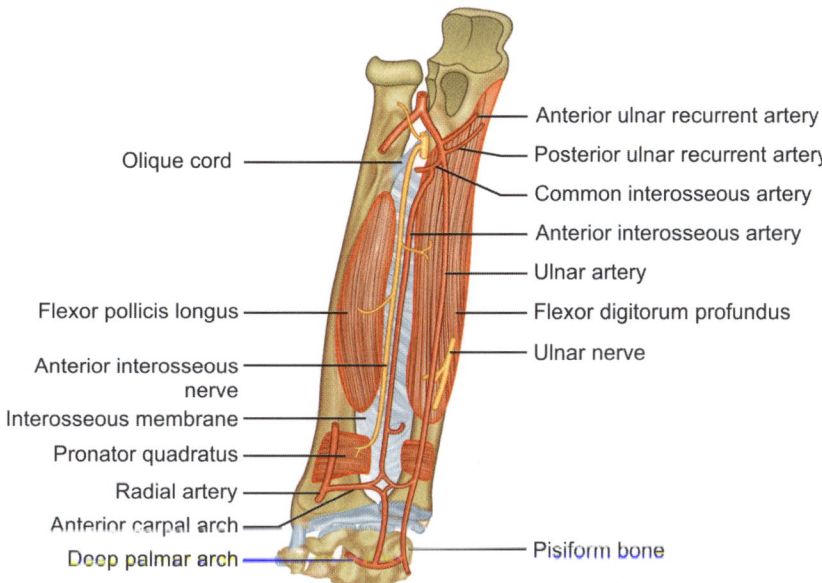

Fig. 2.39C: Neurovascular bundles of front of forearm

retinaculum and enter the hand superficial to the retinaculum, immediately lateral to the insertion of FCU to the pisiform bone. The dorsal branch of the ulnar nerve arises near the middle of the forearm.

9. **Radial nerve** (superficial branch) and **radial artery** are identified in the intermuscular plane between the superficial flexor group of muscles and extensor group of muscles on the lateral aspect of the forearm. Follow these structures proximally and distally. Proximally, the radial nerve divides into superficial branch and deep branch **(posterior interosseous nerve)** at the level of elbow. The deep branch passing through the supinator muscle appears in the posterior compartment of the forearm. The superficial branch courses distally on the deep surface of the brachioradialis muscle. Trace it to the distal 1/3rd of forearm and emerges on the dorsal side of the brachioradialis tendon to become a cutaneous nerve and it is observed that this nerve curls round the lateral border of the radius about 7.5 cm above the wrist and goes to the back of forearm.

10. **Median nerve** innervates most of the muscles of the anterior compartment of the forearm. In cubital fossa, it is medial to the brachial artery. Trace it downwards. It courses deep to the superficial group of flexor muscles after passing between two heads of pronator teres. For proper exposure of the nerve, cut the palmaris longus tendon about 3 cm proximal to the wrist and reflected it proximally. Similarly, flexor carpi radialis tendon is cut about 5 cm proximal to the wrist and reflected proximally. Then lift the medial edge of flexor digitorum superficialis and observe the median nerve to pass deep to the muscle. Use a probe to free the nerve from the loose connective tissue that lies between the superficial and deep group of forearm muscles. Note its nerve supply to these muscles. A **palmar**

branch arises from the nerve proximal to the wrist and enters the palm superficial to the flexor retinaculum. It gives off a branch in the forearm called **anterior interosseous nerve** which lies on the anterior surface of the interosseous membrance. The main trunk of the nerve becomes superficial between the flexor carpi radialis and palmaris longus before entering the palm deep to the flexor retinaculum through the **carpal tunnel.**

11. **The** humero-ulnar head of flexor digitorum superficialis is cut transversely and reflect the distal part of the muscle laterally to expose the **deep muscles of the forearm (Fig. 2.40),** median nerve and ulnar artery. The deep group of muscles are—**flexor digitorum profundus** (FDP), **flexor pollicis longus (FPL)** and **pronator quadratus** (PQ). The four tendons of FDP lie deep to the four tendons of flexor digitorum superficialis. Its lateral half is innervated by the median nerve.

12. **Retract** the **FDP** medially and **FPL** laterally to expose the **anterior interosseous nerve** (branch of median nerve) and **anterior interosseous artery** (branch of common interosseous artery, branch of ulnar artery) to lie in front of the interosseous membrane.

The **anterior interosseous artery** pierces the interosseous membrane at the upper border of pronator quadratus muscles and goes on to the posterior compartment of the forearm. On the other hand, the **posterior interosseous artery** passes on to the posterior compartment over the upper end of the interosseous membrane. **It is to be noted that the anterior interosseous artery supplies the deep group of flexor muscles, but the posterior interosseous artery supplies the extensor group of forearm muscles.**

13. **Pronator quadratus** (PQ) is a rectangular muscle that runs transversely from the anterior surface of the ulna to the radius in the distal part of the forearm. Retract the tendons of superficial and deep groups of flexor muscles to expose this quadrilateral muscle.

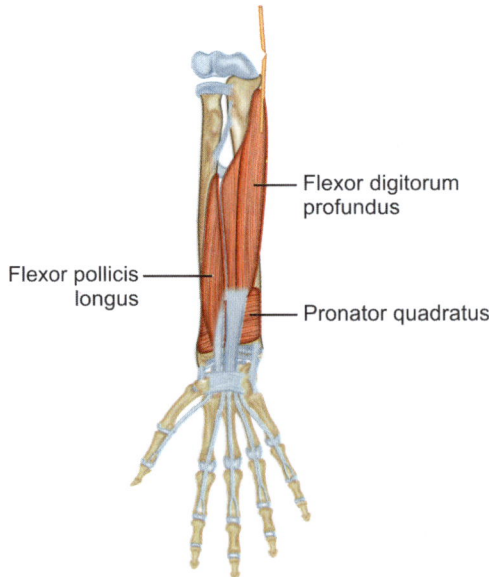

Fig. 2.40: Deep muscles of front of forearm

Window Dissections: Upper Limb (Superior Extremity)

> All the flexor groups of muscles in the anterior compartment of the forearm are innervated by the median nerve except flexor carpi ulnaris and medial half of flexor digitorum profundus (ulnar nerve).

FACTS TO BE NOTED

Muscles	From	To	Acting on
Long superficial flexor muscles (FCR, PL, FCU)	Humerus	Hand/digits	Elbow/wrist/digital joints
Deep flexor muscles (FPL, FDP)	Forearm bones	Digits	Wrist and digital joints
Intermediate flexor muscle (FDS)	Humerus and forearm bones	Digits	Elbow/wrist/digital joints
PT, BR, anconeus, supinator	Humerus	Forearm bones	Elbow/radioulnar joints
PQ, supinator	One forearm bone	Other forearm bone	Radioulnar joints

> **Note that** to enter into the forearm the **median nerve** passes between the two heads of **pronator teres**, **ulnar nerve** passes between the two heads of **flexor carpi ulnaris** and **radial nerve** passes between the two strata of **supinator muscle.**

MUSCLES RELATED TO FRONT OF FOREARM

Muscles	Origin	Insertion	Nerve supply	Action
Pronator teres	Medial epicondyle of humerus	Middle of the lateral surface of radius	Median nerve	• Pronation • Flexion of elbow
Flexor carpi radialis	Medial epicondyle of humerus	Base of 2nd and 3rd metacarpal bones	Median nerve	Flexion and abduction of wrist
Palmaris longus	Medal epicondyle of humerus	Flexor retinaculum and palmar aponeurosis	Median nerve	Weak flexor of wrist
Flexor carpi ulnaris	• **Humeral head**: Medial epicondyle • **Ulnar head**: Olecranon process	Pisiform, hamate and 5th metacarpal bone	Ulnar nerve	Flexion and adduction of wrist
Flexor digitorum superficialis	• **Humeroulnar head**: Medial epicondyle of humerus and coronoid process of ulna • **Radial head**: Anterior oblique line of radius	Sides of the shaft of middle phalanx of 2nd to 5th digits	Median nerve	Flexion of wrist, metacarpophalangeal and proximal interphalangeal joint

Contd...

Contd...

Muscles	Origin	Insertion	Nerve supply	Action
Flexor digitorum profundus (bulkiest muscle of forearm)	Upper ¾th of the anteromedial surface of the shaft of ulna and interosseous membrane	Palmar surface of the base of terminal (distal) phalanx of 2nd to 5th digits	**Median nerve** (anterior interosseous branch)—lateral part and **ulnar nerve**—medial part	Flexion of wrist and distal interphalangeal joint
Flexor pollicis longus	Anterior surface of radius below the oblique line	Palmar surface of base of distal phalanx of thumb (1st digit)	Median nerve (anterior interosseous branch)	Flexion of thumb
Pronator quadratus	Bony ridge on the anteromedial surface of lower ¼th of ulna	Anterior surface of lower ¼th of radius	Median nerve (anterior interosseous branch)	Pronation

---

Internervous Line of Forearm

The medial and lateral borders of the anterior (flexor) and posterior (extensor) compartments of the forearm are not crossed by motor nerves though they are crossed by cutaneous nerves. This is the line through which the deep parts of the forearm may be explored.

Summary

1. Incision of skin, superficial fascia and deep fascia as in Figures 2.36A and B.
2. Structures found in superficial fascia: Cephalic vein and lateral cutaneous nerve of forearm—laterally.
   Basilic vein and medial cutaneous nerve of forearm—medially.
3. Deep structures (deep to deep fascia or antebrachial fascia).
   a. **Muscles: Superficial group** (5)—common origin from medial epicondyle of humerus. They are (lateral to medial) PT, FCR, PL, FCU and FDS.
   b. **Deep group** (3) – FPL, FDP and PQ.
   c. **Nerves:** Median nerve and its anterior interosseous branch, ulnar nerve and superficial branch (cutaneous) of radial nerve only in the upper part (7.5 cm above the wrist it goes to the back of forearm).
   d. **Vessels:** Radial and ulnar artery, common interosseous branch of ulnar artery dividing into anterior and posterior branch.
4. All the muscles of the anterior compartment of the forearm are supplied by the median nerve except FCU muscle and medial half of FDP muscle which are supplied by the ulnar nerve.

PROBABLE QUESTIONS AND ANSWERS

1. What is antebrachial fascia?

Ans. The deep fascia of forearm is called antebrachial fascia.

2. What do you mean by wrist band?

Ans. Close to the wrist, the antebrachial fascia is thickened to from flexor and extensor retinaculum which retain the digital tendons in position. These retinaculum are called wrist band.

3. Why the muscles of anterior compartment of forearm are more massive than the muscles of posterior compartment?

Ans. The muscles of anterior compartment or flexor muscles are more massive because they work against gravity and act as **antigravity muscle**.

4. Name the superficial and deep muscles of the front of the forearm.

Ans. **Superficial muscles** (from lateral to medial) are 5 in number.
   i. Pronator teres
   ii. Flexor carpi radialis
   iii. Palmaris longus (absent in 15% cases)
   iv. Flexor carpi ulnaris
   v. Flexor digitorum superficialis.

Deep muscles are 3 in number:
   i. Flexion pollicis longus.
   ii. Flexor digitorum profundus.
   iii. Pronator quadratus.

5. Which is the most powerful and bulkiest muscle of the forearm?

Ans. **Flexor digitorum profundus** (FDP) is the most powerful and bulkiest muscle of the forearm and acting as a chief gripper.

6. What is the nerve supply of these muscles?

Ans. All the muscles of the flexor compartment are supplied by the interosseous branch of median nerve except medial part of flexor digitorum profundus and flexor carpi ulnaris which are supplied by ulnar nerve.

7. How do you explain that the palmaris longus is a degenerating muscle?

Ans. It is a degenerating muscle as it has a short belly and a long tendon. The degenerated distal part of palmaris longus is represented by the palmar aponeurosis.

8. Where does the ulnar nerve give off its dorsal branch?

Ans. The ulnar nerve gives off the dorsal branch about 5 cm above the styloid process of ulna.

9. Where does the superficial branch of radial nerve (cutaneous branch) cross the lateral border of the forearm?

Ans. This nerve goes to the back of the forearm deep to the brachioradialis muscle at a level 7 cm above the styloid process of radius.

10. How does the anterior interosseous nerve end?

Ans. The anterior interosseous nerve, branch of median nerve in the forearm passing deep to the pronator quadratus muscle, supplies it and ends by supplying the inferior radioulnar joint and wrist joint.

11. What is the main blood supply of the forearm?

Ans. The arterial supply for the forearm is chiefly derived from the common interosseous branch of the ulnar artery.

12. How does the posterior interosseous artery go to the back of the forearm from the anterior compartment?

Ans. It goes to the back through the gap between the upper border of interosseous membrane and oblique cord.

13. What is oblique cord?

Ans. Phylogenitically, the oblique cord represents the upper degenerated part of the flexor pollicis longus muscle. It is a thin fibrous band extending from the ulnar tuberosity to the radial tuberosity.

14. What is the largest branch of ulnar artery in the forearm?

Ans. Common interosseous artery.

15. Which is described as the intermediate layer of flexor muscles?

Ans. Flexor digitorum superficialis is often described as intermediate layer because it lies between the other four superficial group of flexor muscles and deep flexor muscles.

16. What is space of Parona?

Ans. It is a rectangular space in the lower part of the forearm bounded in front by long flexor tendons, behind by pronator quadratus, above by oblique origin of flexor digitorum superficialis and below up to the flexor retinaculum.

17. What is the importance of this space?

Ans. This space may be infected by the extension of infections of synovial sheaths of flexor tendons.

18. What are the supinators and pronators of the forearm?

Ans. **Supinators are:**
  i. Supinator.
  ii. Biceps brachii.
  iii. Brachioradialis.

Pronators are:
  i. Pronator teres.
  ii. Pronator quadratus, assisted by FCR and PL.

19. What are the joints responsible for supination and pronation?

Ans. Superior, middle and inferior radioulnar joints.

20. What is the direction of fibers of interosseous membrane?

Ans. In the upper part, the fibers slope downward and medially from radius to ulna. In the lower part, they are arranged in reverse direction.

21. What is the importance of knowing the fiber arrangement of the interosseous membrane?

Ans. i. It appears that such fiber arrangement withstands transmission of forces from hand to humerus through the radius, ulna and intact interosseous membrane.
ii. Radius articulates with the carpal bones (radiocarpal joint) and ulna with the humerus (humeroulnar joint).

22. What are the positions of the forearm where the interosseous membrance is stretched and relaxed?

Ans. i. Stretched in mid prone position.
ii. Relaxed in extremes of supination and pronation.

23. Which is the largest superficial flexor muscle in the forearm?

Ans. Flexor digitorum superficialis is the **largest** of all the superficial group of flexor muscles in the forearm.

LESSON 8: PALM OF THE HAND

The hand presents a palmar surface (palm) and a dorsal surface (dorsum). In contrary to the dorsum, the skin of the palmar surface is thick, devoid of hairs and sebaceous glands. The skin of the palm is connected to the underlying deep fascia (palmar fascia) by numerous fibrous septa thereby restricting the mobility of the skin for proper grip. The palmar surface of the hand is having two elevations—one medially called **hypothenar eminence** caused by underlying hypothenar group of muscles and laterally **thenar eminence** caused by underlying thenar group of muscles. The different varieties of the lines on the palmar skin are also to be noted. These are—**flexure lines, papillary ridges** and **Langer's line**.

STEPS OF DISSECTION

1. **Position of the cadaver and hand:** Cadaver—supine, forearm—supinated, wrist extended, clenched hand open forcefully and hold it open and preferably the finger tips are fixed by nails.
2. **Skin incision (Fig. 2.41):**
    a. A transverse incision (A–B) joining the styloid process of radius and ulna.
    b. Make another transverse incision at the level of the webs of the fingers (C–D).
    c. A longitudinal incision across the palm from the midpoint of first incision (A–B) to the tip of the middle finger (E–F).
    d. Make a longitudinal incision along palmar surface of the thumb (E–G).
    e. Make longitudinal incisions on the palmar surface of the digits (2–5) from the incision C–D.

   Reflect the skin flaps. During this procedure, proceed with caution on the digits. It is due to the fact that the subcutaneous tissue on the palmar surface of the digits is very thin and there are digital vessels, nerves and fibrous digital sheaths immediately deep to the skin.
3. **Superficial fascia:** In the superficial fascia, the structures found are:
    a. Palmar cutaneous branch of ulnar nerve.
    b. Palmar cutaneous branch of median nerve.
    c. **Palmaris brevis muscle (remnant of panniculus carnosus).**

   Palmaris brevis muscle is detached from palmar aponeurosis and reflect it medially along with its nerve supply from

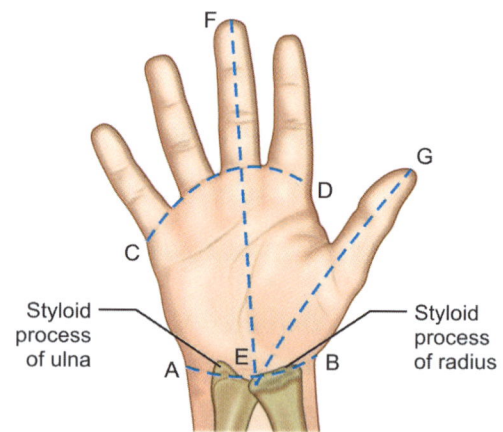

Fig. 2.41: Incision on palm

the ulnar nerve. Cut the superficial fascia and clean the fat from the palmar aponeurosis. Deep fascia is exposed.

4. **Deep fascia:** Deep fascia of the palm is known as **palmar fascia**. In the middle of the palm, the palmar fascia is thickened to form **palmar aponeurosis**. The palmar fascia is much thinner over the thenar and hypothenar eminences.

> **Palmar aponeurosis (Fig. 2.42)** is a thick, triangular portion of the deep fascia that lies in the central region of the palm with its apex at the flexor retinaculum and its base near the level of the heads of the metacarpals, spliting into four digital slips for the medial four fingers. Plantar aponeurosis of the foot presents **five digital slips**, whereas the palmar aponeurosis presents **four slips (digital slip to thumb is absent)**. This is because of allowing free movement of the thumb. Morphologically; the palmar aponeurosis is the degenerated primitive insertion of the palmaris longus tendon. **Although, the palmaris longus muscle may be absent, the palmar aponeurosis is always present.** The inflammatory contracture of the aponeurosis is known as **Dupuytren's contracture** which produces flexion of the fingers because the aponeurosis is attached to the proximal phalanges through the deep transverse metacarpal ligament.

Use a scalpel and by skinning motions detach the palmar aponeurosis from the underlying deep structure. Begin at the proximal end and proceed distally. Traction is applied to the **aponeurosis** using palmaris longus tendon. Cut the aponeurosis at the distal border of the flexor retinaculum and reflect it towards the roots of the fingers where it divides into

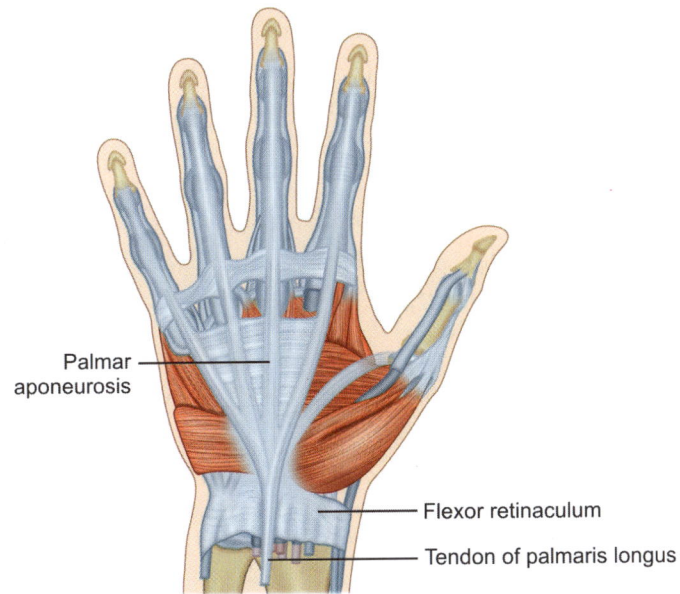

Fig. 2.42: Palmar aponeurosis

four slips. Do not cut too deeply as the superficial palmar arch is in contact with the deep surface of the aponeurosis.

5. **Deep structures:** Deep to the palmar aponeurosis **superficial palmar arch** is found. To dissect this palmar arch more clearly find out the ulnar artery which passes lateral to the pisiform bone with the ulnar nerve (**nerve is medial to artery**). Then in the palm it divides into a superficial branch and a deep branch. The **superficial palmar arch** is then formed by the superficial branch of the ulnar artery and by a smaller contribution from the superficial palmar branch of the radial artery. Clean the arch and its four palmar digital branches—**one proper palmar digital artery** to the medial side of little finger and other **three common palmar digital arteries.**

Dissect and find the **ulnar nerve** lateral to the pisiform bone and medial to the ulnar artery. The superficial branch of the ulnar nerve supplies cutaneous innervation to medial side of ring finger and either side of little finger **(1½ finger).**

Identify the **flexor retinaculum** (Fig. 2.43) in the proximal end of the palm between the thenar and hypothenar eminence. Cut this retinaculum to open the carpal tunnel. To cut through this retinaculum, insert a probe deep to the flexor retinaculum from proximal to distal end and use a scalpel to cut through the retinaculum over the probe (Fig. 2.44). Now the carpal tunnel is opened and the flaps of retinaculum are reflected sideways.

Look for the **synovial tendon sheaths** for flexor tendons deep to the retinaculum and extending into the palm (Figs 2.45A and B). The common flexor synovial sheath or **ulnar bursa** (Fig. 2.46) is found medially around four superficial (FDS) and four profundus (FDP) tendons. It extends about 2.5 cm proximal to the flexor retinaculum in the forearm and distally is continuous with the digital synovial sheath of the little finger. The **radial bursa** (Fig. 2.46) around the tendon of flexor pollicis longus (FPL) is found laterally and extends about 2.5 cm proximal to the flexor retinaculum and continuous with the digital synovial sheath of the

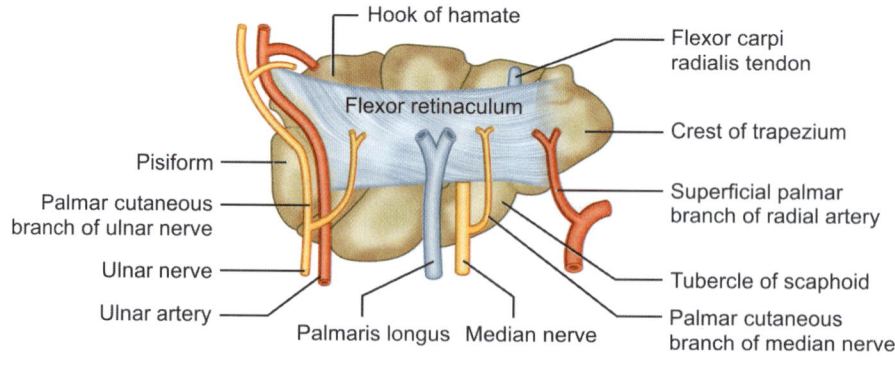

Fig. 2.43: Flexor retinaculum

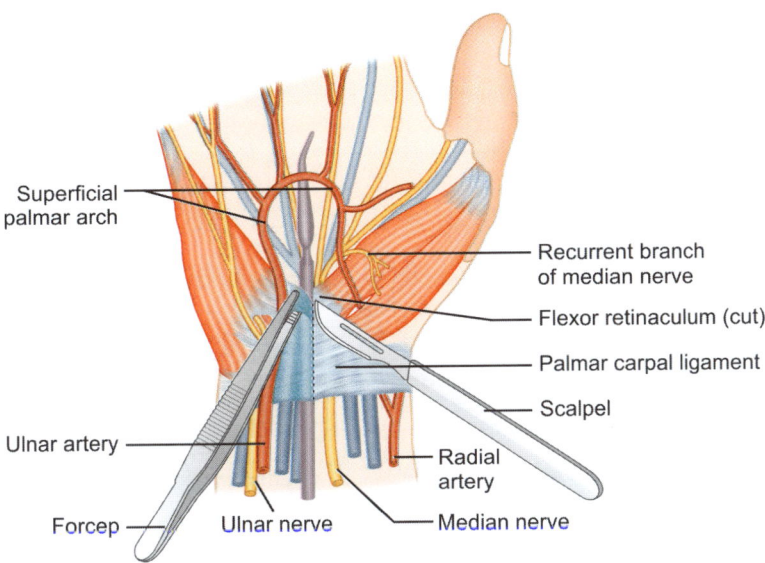

Fig. 2.44: Opening of flexor retinaculum

thumb. The radial bursa and ulnar bursa may communicate with each other within the carpal tunnel. So, **infection from the little finger may spread to the thumb or vice versa through this route**. Synovial sheaths of the index, middle and ring fingers are separate and extend proximally upto the heads of metacarpals. The **tendon of flexor carpi radialis** has its own synovial sheath. Synovial sheaths are difficult to demonstrate. However, to do this divide the loose connective tissue which surrounds the tendons of FDS, down to the tendon and note the smooth internal surface of the sheath and external surface of the tendon. Divide the fibrous flexor sheath longitudinally in all the five digits. **Understand** that the fibrous flexor sheaths are thickened tunnels of deep fascia within which the long flexor tendons slide in the digits. The sheath is thick over the phalanges and thinner at the level of joints to allow free flexion at the joints. Together with the phalanges each sheath forms an osteofascial canal for the long flexor tendons which prevents the tendons from springing away from the digits during flexion.

Transect the tendons of FDS close to the wrist and reflect the tendons distally. Observe the tendons of the FDP and four small muscles **(lumbricals)** arising from the four tendons of FDP (Figs 2.47A and B). Follow the lumbricals distally along the radial side of the corresponding metacarpophalangeal joint to be inserted into the dorsal surface of the bases of the middle and distal phalanx through the dorsal digital expansion. Transect the tendon of FDP close to the wrist and reflect distally. Note the nerves to lumbricals as they are reflected with the tendons. Now, note the distal attachments (insertion) of these long flexor tendons (FDS and FDP).

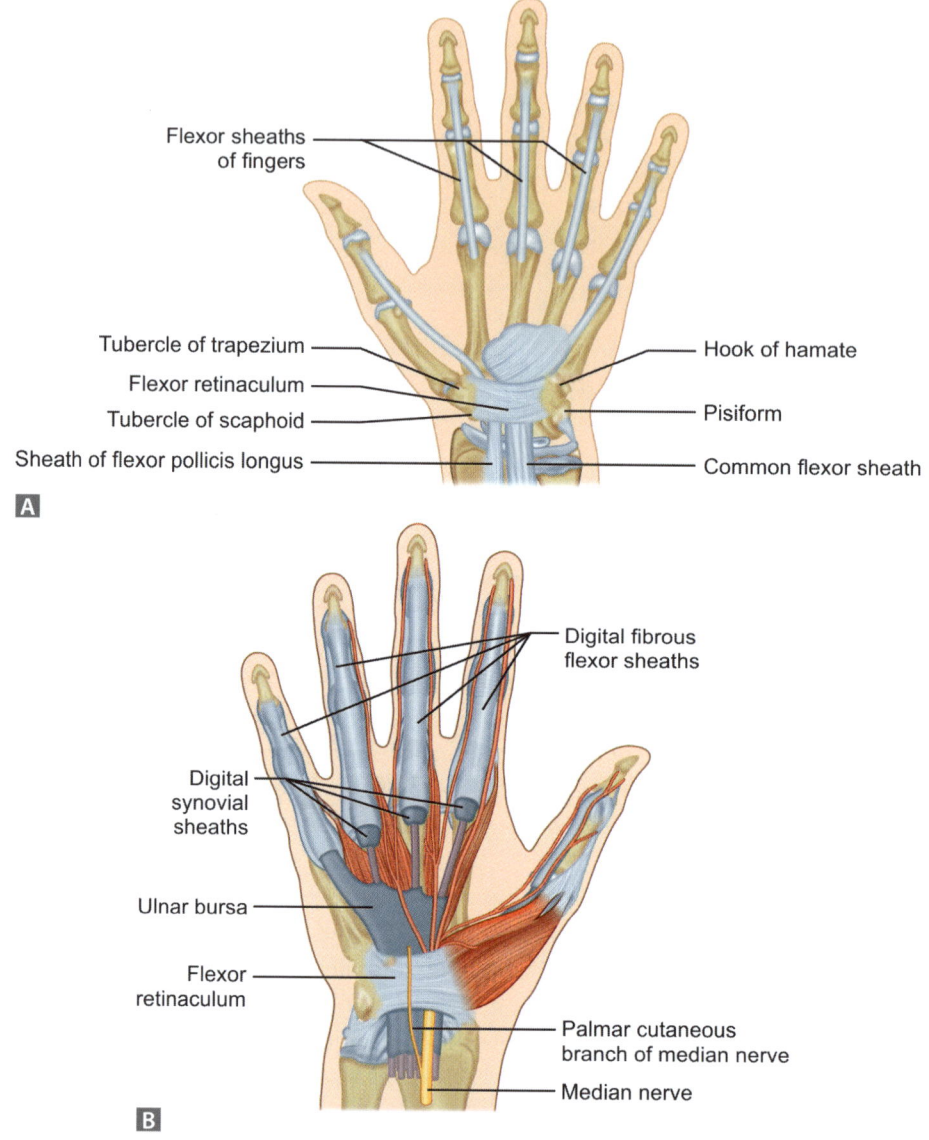

Figs 2.45A and B: (A) Fibrous flexor sheath; (B) Digital sheaths

> **Vinculae** (Fig. 2.48) are thin fibrous filaments which connect the phalanges and interphalangeal joints with the flexor tendons. They are located only in the digits. Within the digital synovial sheaths they are enclosed in synovial membrane and transmit blood vessels to the flexor tendons. Each digit presents two **vincula brevia** (triangular folds) and three **vincula longa** (filiform folds). **Vincula brevia** (1) Extends from FDS tendon to proximal phalanx and proximal IP joint; (2) FDP tendon to middle phalanx and distal IP joint. **Vincula longa** (1 and 2) extend from the superficial tendon to the proximal end of proximal phalanx and (3) profundus tendon to distal end of proximal phalanx.

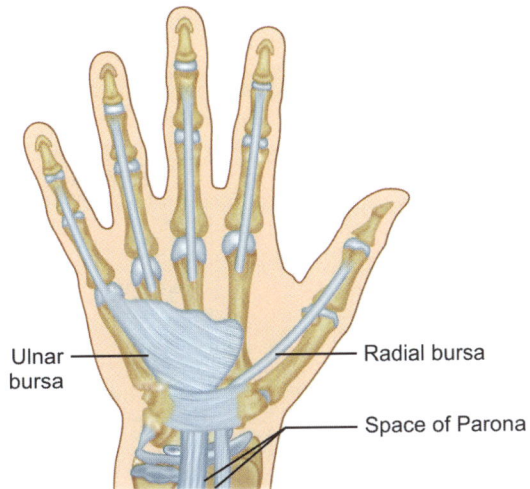

Fig. 2.46: Bursa in hand

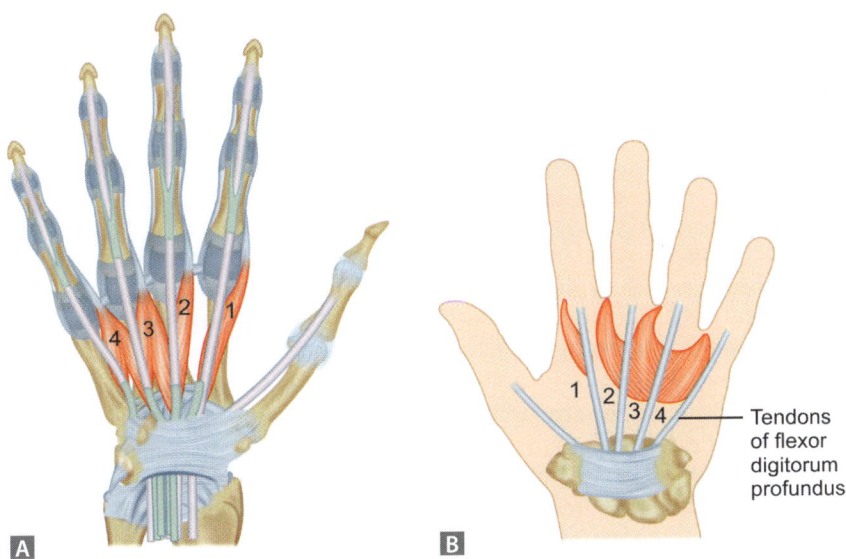

Figs 2.47A and B: Lumbricals

Pick up the tendon of FDS and note that it splits into two slips at the level of metacarpophalangeal joint to enclose the corresponding tendon of FDP and then inserted on the palmar aspect of the middle phalanx after reuniting dorsal to the tendon of profundus. The profundus tendon passes through the splitting fibers of superficialis tendon and inserted into the palmar surface of the base of the terminal phalanx. So, **FDS tendon is attached to the middle phalanx, whereas FDP tendon is attached to the distal phalanx.** Follow the **flexor pollicis tendon** (FP) distally for its insertion into the distal phalanx of the thumb.

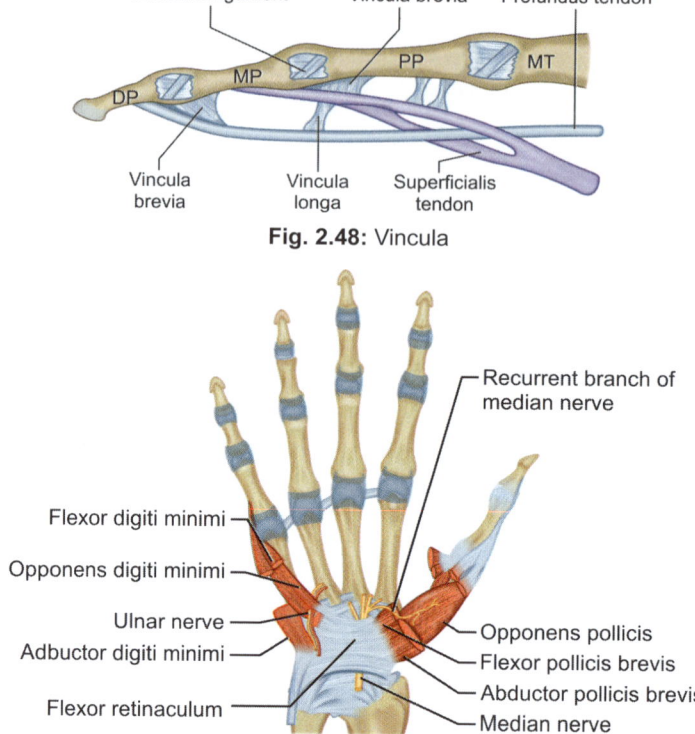

Fig. 2.48: Vincula

Fig. 2.49: Thenar and hypothenar muscles

Next step of dissection will be to dissect the lateral **(thenar)** eminence and medial **(hypothenar)** eminence of the palm. Remove the palmar fascia off the thenar and hypothenar muscles. Be careful to preserve the recurrent branch of **median nerve** within the thenar muscles and **ulnar nerve** supplying the hypothenar muscles.

Of the three **thenar muscles (Fig. 2.49)**, **abductor pollicis brevis** (APB) is the most anterior muscle and it hides the **opponens pollicis** (OP) and party covers the **flexor pollicis brevis** (FPB). To expose the opponens pollicis, lift the APB muscle by introducing an artery forcep or handle of a scalpel behind its lateral border and cut across it. Then separate the **opponens pollicis** (OP) from the **flexor pollicis brevis** (FPB) which is incompletely fused with the medial margin of the OP. Define the margins of FPB and cut across it and reflect the two cut ends one from the other. This exposes the tendon of **flexor pollicis longus** (FPL) and **adductor pollicis** (Figs 2.50 and 2.51) behind the FPL. Note that the recurrent branch of median nerve crosses the superficial surface of the **FPB,** and then disappears deep to the APB muscle. This branch of median nerve helps to locate the correct plane of separation between these muscles.

Hypothenar muscles (Fig. 2.49) are three in number. These are **abductor digiti minimi (ADM), flexor digiti minimi brevis (FDMB)** and **opponens digiti minimi (ODM).** Separate the ADM muscle from the medial margin of FDMB. Between these two muscles, look for the **deep branch of ulnar nerve** and **deep palmar branch of ulnar artery**. Use a probe to elevate the ADM and cut through the middle of the muscle and reflect its parts. Now ODM is exposed.

After the reflection of the cut ends of FDS and FDP tendon distally and reflection of the cut ends of FPB muscles, the **adductor pollicis muscle with its two heads (oblique and transverse)** are properly exposed.

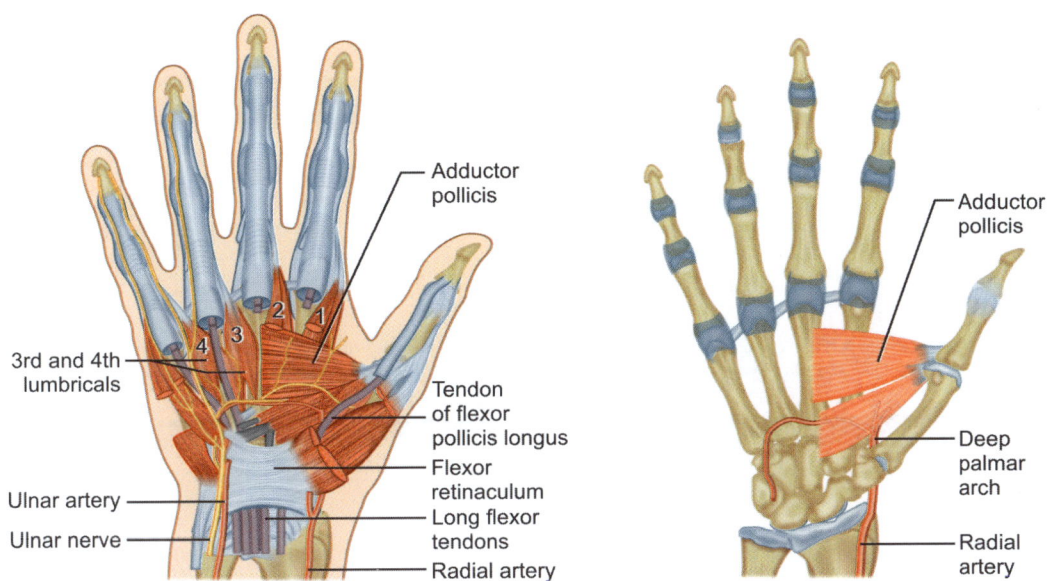

Fig. 2.50: Deep muscles of palm   Fig. 2.51: Adductor pollicis

This also exposes the **deep palmar arterial arch** and **deep branch of ulnar nerve** in the palm. The deep palmar arch courses with the deep branch of the ulnar nerve. The arch is mainly formed by radial artery and partly by a contribution from the deep branch of the ulnar artery. **Radial artery enters the palm between the two heads of first dorsal interosseous muscle.** The deep branch of ulnar nerve accompanies the concavity of the deep palmar arch.

> The **superficial palmar arch and the deep palmar arch** are so named because of their relative position to the long flexor tendons, not to the deep fascia (palmar aponeurosis). Superficial to the flexor tendons (FDS and FDP) is superficial palmar arch and deep to it is the deep palmar arch.

Define the attachments of **adductor pollicis** muscle (Fig. 2.51). Transect the muscle midway between its origin and insertion and follow the branch of the nerve and artery between the two parts of the muscle.

Look for the **four palmar interosseous muscles**. They are all **unipennate** muscle. Do not attempt to dissect these muscles. On the other hand, **dorsal interossei muscles are four** in munber and **bipennate**. They accupy the intervals between the metacarpal bones and well- visualized at the dorsum of the dissected hand. **Palmar interossei** are **AD**ductors **(PAD)** and **Dorsal interossei** muscles are **AB**ductor of the digits **(DAB)**. *Via dorsal digital expansion, lumbricals and interossei muscles (dorsal and palmar) produce flexion at metacarpophalangeal joint and extension of interphalangeal joint.*

> **Lumbricals** are named 1st, 2nd, 3rd and 4th from lateral to medial side. They are so named because of their appearance like the shape of **earthworms. Interossei muscles** are also numbered from lateral to medial side. **Note that** 1st and 2nd lumbricals and four palmar interossei muscles are unipennate whereas 3rd, 4th lumbricals and four dorsal interossei muscles are bipennate.

MUSCLES ON THE PALM

Muscles	Origin	Insertion	Nerve supply	Action
Thenar muscles				
Abductor pollicis brevis	• Flexor retinaculum • Tubercles of scaphoid and trapezium	Base of the proximal phalanx of the thumb	Recurrent branch of **median nerve**	• Abduction of the thumb • Medial rotation of the thumb
Flexor pollicis brevis	• Flexor retinaculum • Trapezoid and capitate	Radial side of the base of the proximal phalanx of the thumb	Recurrent branch of **median nerve**	Flexes the thumb
Opponens pollicis	• Flexor retinaculum • Crest of trapezium	Shaft of the first metacarpal bone	Recurrent branch of **median nerve**	Opposition of the thumb to other digits
Hypothenar muscles				
Abductor digiti minimi	Pisiform bone	Base of the proximal phalanx of little finger	Deep branch of **ulnar nerve**	Abducts the little finger
Flexor digiti minimi brevis	• Flexor retinaculum • Hook of the Hamate	Proximal phalanx of little finger	Deep branch of **ulnar nerve**	Flexion of the proximal phalanx of little finger
Opponens digiti minimi	• Flexor retinaculum • Hook of Hamate	Shaft of 5th metacarpal	Deep branch of **ulnar nerve**	Opposes the little finger to the thumb
Lumbricals				
1st and 2nd lumbricals	Lateral two tendons of flexor digitorum profundus	Lateral sides of the dorsal digital expansion for medial four digits	**Median nerve**	Flexion of meta-carpophalangeal joint and extension of interphalangeal joint
3rd and 4th lumbricals	Medial two tendons of flexor digitoum profundus		**Ulnar nerve**	
Interossei muscles (Figs 2.52 to 2.54)				
Palmar interossei (1st–4th)	From the anterior suface of the 1,2,4,5 metacarpals	Base of the proximal phalanx of corresponding digits and dorsal digital expansion	Deep branch of **ulnar nerve**	Adducts the digits
Dorsal interossei (1st–4th)	Adjacent sides of two metacarpals	Base of the proximal phalanx of 2nd to 4th digits and dorsal digital expansion	Deep branch of **ulnar nerve**	Abducts the digits
Other muscles				
Adductor pollicis	• **Oblique head:** Base of 2nd and 3rd metacarpal, capitate and trapezoid • **Transverse head:** 3rd metacarpal	Base of proximal phalanx of thumb	Deep branch of **ulnar nerve**	Adducts the thumb
Palmaris brevis	Flexor retinaculum and palmar aponeurosis	Dermis of the ulnar side of the hand	Superficial branch of **ulnar nerve**	Tenses the skin and improves the palmar grip

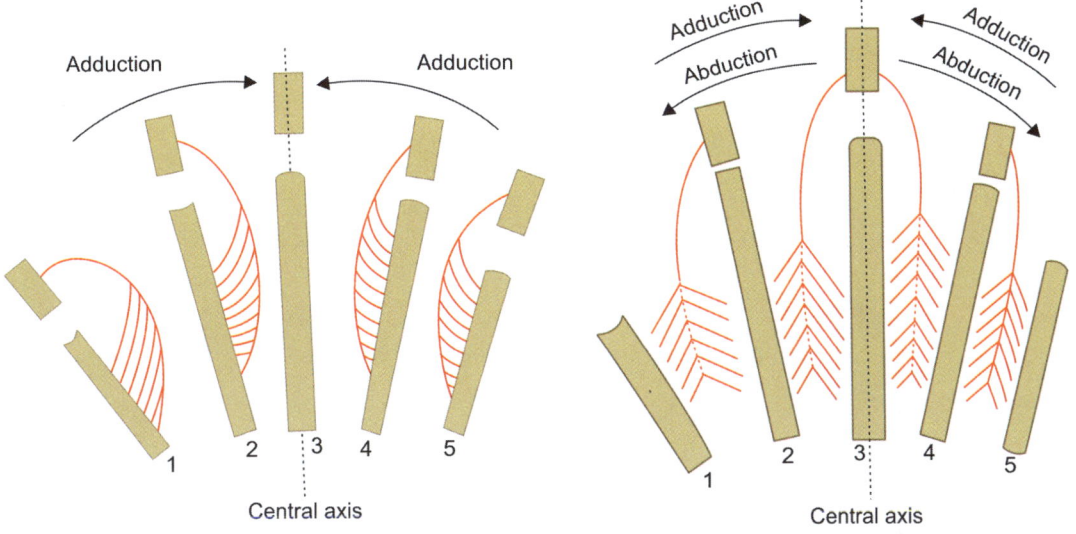

Fig. 2.52: Palmar interossei

Fig. 2.53: Dorsal interossei

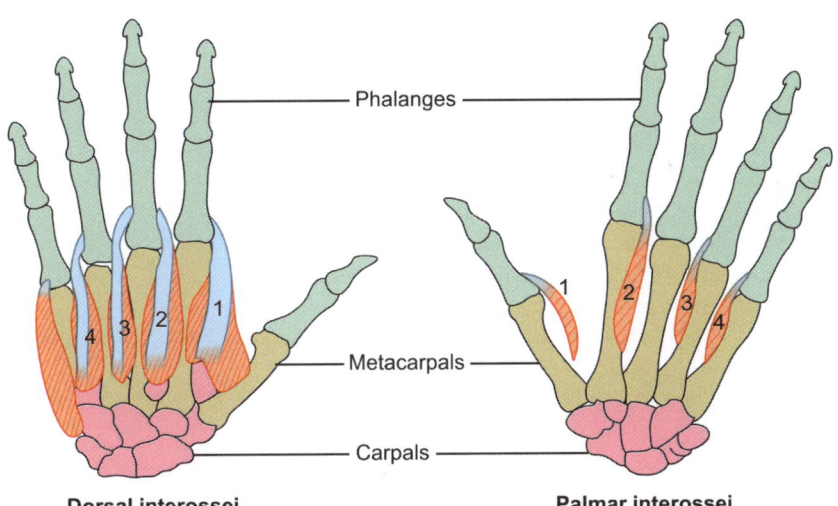

Fig. 2.54: Interossei muscles

Spaces of the Hand

Refer Figure 2.55 and Flowchart 2.1.

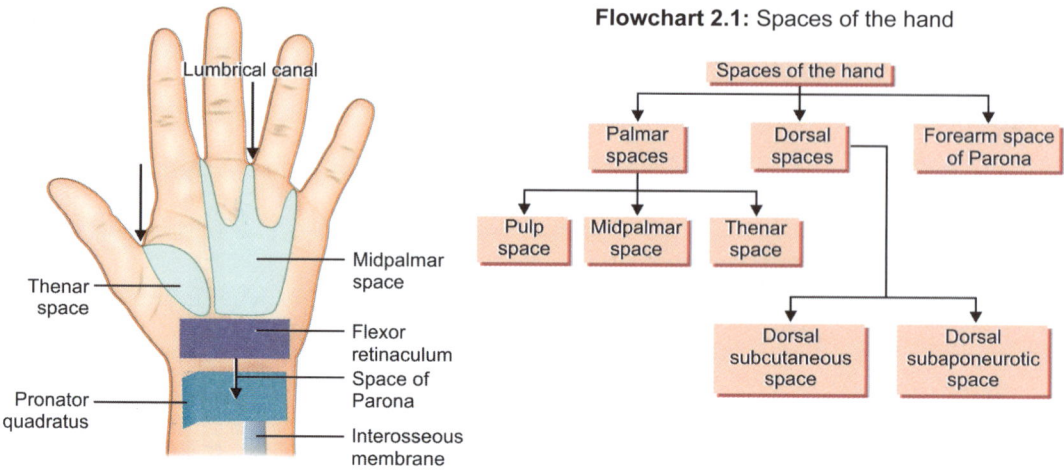

Fig. 2.55: Pulp space

Flowchart 2.1: Spaces of the hand

Summary

1. Incision of skin, superficial fascia and deep fascia as in Figure 2.41.
2. Structures found in superficial fascia:
   i. **Muscle:** Palmaris brevis.
   ii. **Nerves:** Palmar cutaneous branch—median and ulnar nerve.
3. Deep fascia thickened to form three parts:
   i. Flexor reticulum.
   ii. Palmar aponeurosis.
   iii. Fibrous flexor sheaths of the digits.
4. Deep to deep fascia:
   Intrinsic muscles (20, including palmaris brevis)
   - **Thenar muscles:** APB, FPB, OP.
   - **Hypothenar muscles:** ADM, FDM, ODM.
   - 4 lunbicals (1st–4th from lateral to medial).
   - 4 palmar (unipennate) and 4 dorsal (bipennate) muscles.
   - Adductor pollicis (oblique and transverse head) and
   Extrinsic muscles: Tendons of FDS (4), FDP (4) and FPL.
5. Superficial palmar arch—formed mainly by ulnar artery and party by radial artery.
6. Deep palmar arch—formed mainly by radial artery with a contribution from ulnar artery.
7. Out of 20, most of the muscles of hand (15) are supplied by ulnar nerve except thenar muscles and first two lumbricals (5) which are supplied by median nerve.
8. Cutaneous branch of median nerve supplies lateral 3½ of the digits and ulnar nerve supplies medial 1½ of the digits including the digital joints and nail beds and some part of the skin on the dorsal surface of the phalanges.
9. Synovial sheath for flexor pollicis longus is called **radial bursa** and common synovial sheath for four superficialis and four profundus tendons are called **ulnar bursa. These two bursa intercommunicate within the carpal tunnel.**
10. **Carpal tunnel:** A fibro-osseous tunnel between the carpal bones and flexor retinaculum through which passes the median nerve, long flexor tendons (FDS, FDP, FPL, and FCR).

Window Dissections: Upper Limb (Superior Extremity)

PROBABLE QUESTIONS AND ANSWERS

1. What are the features of palmar skin which increase the efficiency of the palmar grip?

Ans.
  i. Immobile palmar skin, because of its firm attachment to the underlying palmar aponeurosis.
  ii. Thick skin.
  iii. Presence of palmar creases.

2. What is water-cushion?

Ans. The dense fibrous bands of the superficial fascia of the skin bind the skin to the palmar aponeurosis and divide the subcutaneous fat into loculi under pressure. This serve as a water cushion to withstand considerable pressure during firm griping.

3. What type of muscle the palmaris brevis is? What is its nerve supply?

Ans. It is a subcutaneous striated muscle, arises from the flexor retinaculum and palmar aponeurosis and inserts into the dermis of the skin of the hand. It is said to be a remnant of **panniculus carnosus**. It is supplied by the superficial branch of the ulnar nerve.

4. Why the thumb is free to move?

Ans. The palmar aponeurosis divides into four slips for medial four fingers. But it does not give any slip to the thumb so that the thumb becomes free to move.

5. What are the structures passing superficial to the flexor retinaculum?

Ans.
  i. Tendon of palmaris longus.
  ii. Palmar cutaneous branch of median nerve.
  iii. Palmar cutaneous branch of ulnar nerve.
  iv. Ulnar vessels.
  v. Ulnar nerve.

6. What are the structures passing deep to flexor reticulum (carpal tunnel)? (Fig. 2.56)

Ans.
  i. The median nerve.
  ii. Tendon of flexor digitorum superficialis et profundus.
  iii. Tendon of flexor pollicis longus.
  iv. Ulnar bursa.
  v. Radial bursa.

7. What is carpal tunnel syndrome?

Ans. The compression of the median nerve in the carpal tunnel by long continued swelling of the synovial sheaths or bony pathology gives rise to the motor and sensory symptoms in the hand which constitute the carpal tunnel syndrome.

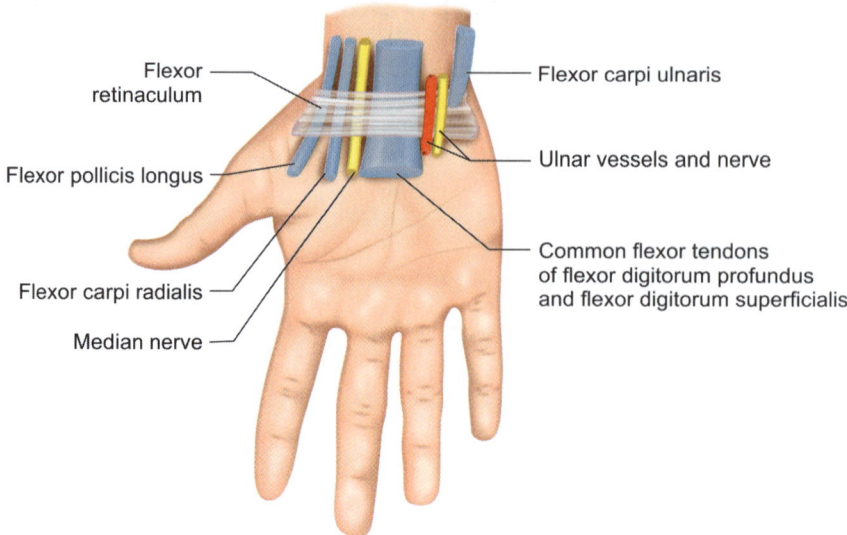

Fig. 2.56: Structures passing deep to flexure retinaculum

8. What are the motor and sensory symptoms of carpal tunnel syndrome?

Ans. **Motor**: Weakness and wasting of the thenar muscles with loss of power of opposition. **Sensory**: Loss of cutaneous sensation of the palmar surface of lateral 3½ fingers.

9. What is the nerve supply of the thenar muscles?

Ans. Thenar muscles (APB, FPB, OP) are supplied by median nerve. **The FPB often receives additional supply from deep branch** of ulnar nerve.

10. What is the axis of adduction and abduction of the fingers?

Ans. The axis passes through the middle finger. Remember that the axis in the foot passes through the second toe.

11. Why does the movement of the thumb take place in different plane?

Ans. This is because of the fact that the first metacarpal (thumb) is rotated medially through 90°.

12. What is the plane of different movement of the thumb?

Ans. The flexion and extension take place in the plane of palm. The abduction and adduction take place at right angles to the plane of the palm.

Window Dissections: Upper Limb (Superior Extremity)

13. What do you mean by opposition of the thumb?

Ans. The medial rotation and adduction of the thumb occur simultaneously and are collectively called the movement of opposition of the thumb.

14. At which joint the opposition movement takes place?

Ans. As the opponens pollicis which is the muscle of opposition movement, inserted into the first metacarpal bone, its action is limited to the **carpometacarpal joint of the thumb.**

15. What is the antagonist of the opponens pollicis muscle?

Ans. The extensor pollicis longus can cause lateral rotation and abduction of the thumb. So, it is the true antagonist to it.

16. What is insertion of opponens digiti minimi?

Ans. Like the opponens pollicis, it is not attached to its phalanges. It is inserted into the shaft of the 5th metacarpal bone.

17. What is superficial transverse ligament of the palm?

Ans. The palmar aponeurosis forms a transverse thickened band opposite the heads of the metacarpal bones—this is known as the superficial transverse palmar ligament.

18. What is deep transverse ligament of the palm?

Ans. Opposite the heads of the medial four metacarpal bones the thin anterior interosseous fascia that covers the front of the interossei muscles is thickened to form the deep transverse ligament of the palm.

19. What are the attachments of flexor retinaculum (Fig. 2.45A)?

Ans. Medially: Pisiform bone and hook of the hamate.
Laterally: Tubercle of scaphoid and the crest of trapezium.

20. Why the median nerve is called workman's nerve or labourer's nerve?

Ans. This nerve supplies most of the large flexor muscles of the forearm and thenar muscles of the hand which are used by a workman or labour more frequently.

21. Why the ulnar nerve is called musician's nerve?

Ans. A musician (violinist) uses most of the intrinsic muscles of the hand. Most of the intrinsic muscles (15 out of 20) are supplied by the ulnar nerve. So, the ulnar nerve is called musician's nerve.

22. What do you mean by the intrinsic muscles of the hand?

Ans. Intrinsic means pertaining exclusively to a part or situated entirely within. Here, the intrinsic muscles of the hand means the origin and insertion of these muscles are within the different bones of the hand. *The muscels are (20 in number):*
  i. Thener muscles (3).
  ii. Hypothenar muscles (4 including palmaris brevis)
  iii. Lumbricals (4)
  iv. Dorsal and palmar interossei muscles (4+4=8)
  v. Adductor pollicis (1).

23. What is the variation of nerve supply of lumbricals and interossei?

Ans. All dorsal and palmar interossei and 3rd and 4th lumbricals are supplied by deep branch of ulnar nerve. 1st and 2nd lumbricals are supplied by the digital branch of the median nerve; But the 2nd lumbrical may be supplied by the deep branch of ulnar nerve; 3rd lumbrical and first dorsal interossei may be supplied by the median nerve.

24. What is radial bursa?

Ans. The flexor synovial sheath of flexor pollicis longus is known as radial bursa.

25. What is ulnar bursa?

Ans. The common synovial sheath of flexor digitorum superficialis et profundus is called ulnar bursa.

26. Do the radial bursa and ulnar bursa intercommunicate with each other?

Ans. In about 50% of cases the two bursa communicate with each other deep to the flexor reticulum.

27. What is compound palmar ganglion?

Ans. The ulnar bursa extends into the forearm up to 2.5 cm above the flexor retinaculum. When the ulnar bursa is infected, the collected pus causes swelling both above and below the wrist with a constriction behind the retinaculum resembling an **'Hour-glass'** appearance called compound palmar ganglion.

28. Which structure divides the origin of adductor pollicis into transverse head and oblique head?

Ans. The deep palmar arch.

29. What is pulp space and whitlow?

Ans. The pulp space is a space between the palmar skin and distal phalanges of all the digits of the hand and lie distal to the fibrous sheaths of the flexor tendons. This space is traversed by the radiating fibrous septa connecting the skin to the periosteum of the distal phalanx and contains subcutaneous fat and blood vessels. Infection of the pulp space is called whitlow (Figs 2.57 and 2.58).

30. What are palmar spaces?

Ans. The fibrous septae from the deeper aspect of the palmar aponeurosis extend backwards to attach with the palmar aspect of the 1st (lateral palmar septum), 3rd (intermediate palmar septum) and 5th (medial palmar septum) metacarpal to divide palm into different potential spaces which are called palmar spaces.

31. What is midpalmar space?

Ans. The potential space between the medial and intermediate palmar septum is called mid-palmar space.

32. What are the contents of midpalmar space?

Ans.
i. Flexor tendons of 3rd, 4th and 5th fingers.
ii. 2nd, 3rd, and 4th lumbricals.
iii. Superficial palmar arch with digital vessels and nerves.

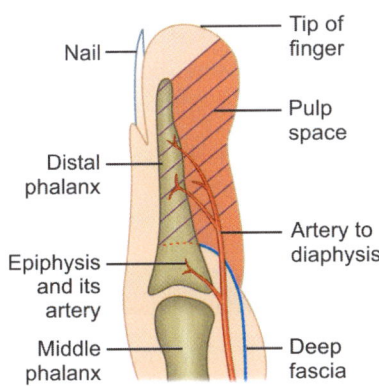

Fig. 2.57: Pulp space

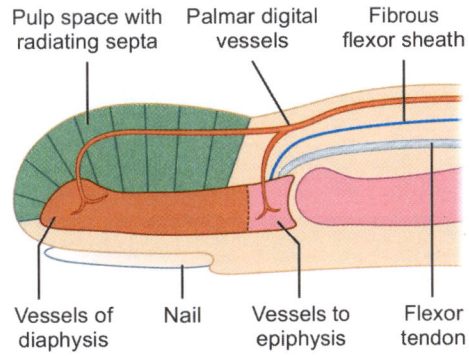

Fig. 2.58: Pulp space with radiating septa

33. What is lumbrical canal?

Ans. The diverticula from the palmar spaces extending to the web of the fingers are called lumbrical canals. Lumbrical canal in the web between 3rd and 4th or 4th and 5th finger is called lumbrical canal of mid palmar space and lumbrical canal in the web between 1st and 2nd finger or sometimes 2nd and 3rd finger is called lumbrical canal of thenar space. **Remember that the 2nd lumbrical canal (between 2nd and 3rd digits) sometimes is connected with the midpalmar space.**

34. Why the palmar cutaneous sensation is spared on the territory of median nerve distribution of the palm in carpal tunnel syndrome?

Ans. The palmar cutaneous branch of the median nerve arises proximal to the flexor retinaculum and passes superficial to it (**not deep to it**) to supply the skin of the palm over the thenar muscles.

35. Why sebaceous cyst never involves the palm?

Ans. The palmar skin is devoid of sebaceous glands. But sweat glands are profuse.

36. Why pisiform bone is a sessamoid bone?

Ans. Pisiform bone ossifies after birth (secondary center of ossification) and develops under the tendon of **flexor carpi ulnaris.**

37. What are the characteristics of a sessamoid bone?

Ans.
 i. It ossifies after birth, i.e. ossifies from secondary center of ossification.
 ii. It develops under a tendon.
 iii. It has no periosteum.
 iv. It has no Haversian system.

38. Name the largest sessamoid bone in the body.

Ans. Patella.

LESSON 9: BACK OF THE ARM

The posterior compartment or back of the arm contains principally **one muscle**—triceps brachi, **one vessel**—profunda brachi vessel and **two nerves**—radial nerve and a part of ulnar nerve after piercing the medial intermuscular septum.

Window Dissections: Upper Limb (Superior Extremity)

STEPS OF DISSECTION

1. **Position of the cadaver:** Place the cadaver in the prone position with the arm rotated medially. The upper extremity is placed on a wooden block at right angle to the body.
2. **Skin incision (Fig. 2.59):**
    a. A transverse incision at the junction of upper ¼th and lower ¾th of the arm on its back (A–B).
    b. Another transverse incision on the back of the elbow joining the two epicondyles of the humerus (C–D).
    c. A median vertical incision joining the midpoint of above two transverse lines (E–F).

   Reflect the skin flaps sideways. Superficial fascia is exposed.
3. **Superficial fascia:** In the superficial fascia few cutaneous nerves are found. These are:
    a. Upper lateral cutaneous nerve of arm (continuation of the posterior division of axillary nerve).
    b. Lower lateral cutaneous nerve of arm (branch of radial nerve).
    c. Posterior cutaneous nerve of the arm (branch of radial nerve).
    d. Posterior cutaneous nerve of the forearm (a branch of radial nerve) which pierces the deep fascia a little above the elbow and supplies mainly the back of the forearm and also lower part of the arm.

   Superficial fascia is reflected in the line of skin incision. Deep fascia is exposed.
4. **Deep fascia:** Using scissors and forceps, open the posterior compartment of arm by making a longitudinal incision through the brachial fascia (deep fascia) and it is reflected like that of skin medially and laterally.

   Deep structures are exposed.
5. Fingers are used to clean and define the borders of the **triceps brachii** muscle. Separate the medially placed **long head** of triceps from the **lateral head** only by using your fingers. Note that the **teres major** muscle crosses the anterior surface of the long head (Fig. 2.60). **Below** the teres major muscle, there is an opening between the long head of triceps and

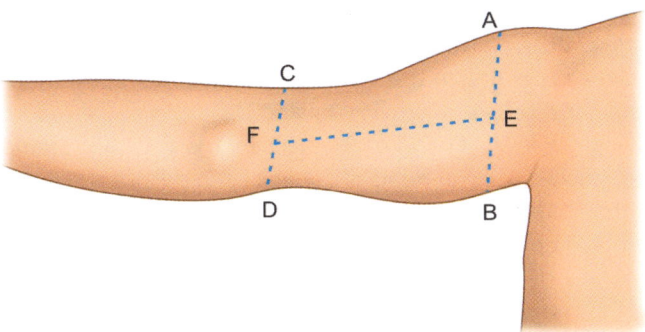

Fig. 2.59: Incision—back of the arm

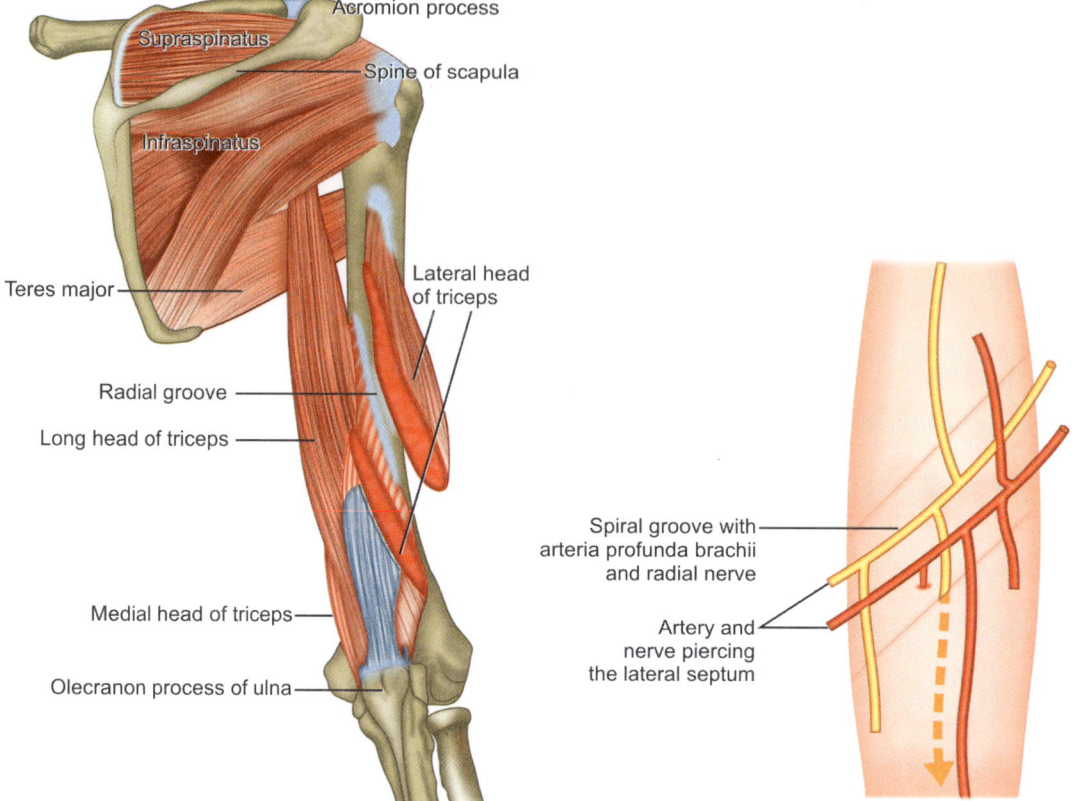

Fig. 2.60: Back of arm (triceps)

Fig. 2.61: Spiral groove and its contents

shaft of the humerus (**lower triangular space**). Widen this opening and identify the **radial nerve** and **arteria profunda brachii**. Observe that the radial nerve and the accompanying vessels lie directly on the posterior surface of the humerus in the **radial (spiral) groove** (Fig. 2.61). For clear exposure of the spiral groove with its contents, transect the lateral head of the triceps and reflect it proximally and distally. Care should be taken not to cut the nerve supplying the medial head of triceps because this twig continues downwards to supply **anconeus** muscle. **Medial head** of triceps arising from the posterior surface of the humerus below the spiral groove joins with the other heads of the triceps.

6. **Use** your fingers to open the connective tissue plane between the **brachioradialis** muscle and **brachialis** muscle. In this plane find the **radial nerve** and trace it proximally and distally. **Radial nerve supplies brachioradialis, extensor carpi radialis longus** and **brachialis** (sensory) in this region. Then it divides into anterior or superficial branch and posterior interosseous (deep) branch at the level of the elbow.

7. **Follow** the **ulnar nerve** with the **superior ulnar collateral** artery in the posterior compartment of the arm. Trace the nerve to the back of the medial epicondyle of the humerus.

Window Dissections: Upper Limb (Superior Extremity)

* All the muscles of the posterior compartment of the arm (**all three heads of triceps** and **anconens**) are supplied by radial nerve.
* All the muscles of the anterior compartment of the arm supplied by musculocutaneous nerve (brachialis partly supplied by radial nerve in addition).
* No muscles are supplied by median or ulnar nerve in the arm.

Summary

1. Incision of skin, superficial fascia and deep fascia as in Figure 2.59.
2. Structures found in superficial fascia: Upper lateral cutaneous nerve of arm (branch of posterior division of axillary nerve) lower lateral cutaneous nerve of arm and posterior cutaneous nerve of arm (both are branches of radial nerve).
3. Deep structures (deep to deep fascia or brachial fascia):
   - 3 heads of triceps (long, lateral medial—all are supplied by radial nerve).
   - Radial nerve and profunda brachii vessels (in radial groove) and **a part of ulnar nerve** after piercing the medial intermuscular septum.

PROBABLE QUESTIONS AND ANSWERS

1. What are the structures found in posterior compartment of arm?

Ans. Three heads of triceps, radial nerve, profunda brachi vessels, a part of ulnar nerve after piercing the medial intermuscular septum, insertion of deltoid muscle.

2. Where does the triceps insert?

Ans. The triceps inserts into the posterior part of upper surface of the olecranon process of ulna.

3. What intervens between its insertion and capsule of elbow joint?

Ans. A bursa intervens between them.

4. What is articularis cubiti or sub-anconeus?

Ans. Few fibers of triceps are inserted into the capsule of the elbow joint. These fibers are called articularis cubitis. It draws up the capsule of the elbow joint during its extension.

5. What are the joints involved in the action of the triceps?

Ans.   i. Elbow joint: It is an strong extensor of the elbow joint.
      ii. Shoulder joint: Only the long head takes part in the movement of shoulder Joint. It supports the humeral head in abducted shoulder.

6. What is the difference between the long head of biceps and triceps in terms of their origin?

Ans.
i. Long head of biceps (from supraglenoid tubercle) is **intracapsular** but **extrasynovial** in its origin.
ii. Long head of triceps is **extracapsular** (from infraglenoid tubercle).

7. Where do you find the spiral groove?

Ans. Behind the shaft of the humerus between the origin of lateral head of the triceps above and medial head of the triceps below.

8. What are the contents of the spiral groove?

Ans. Arteria profunda brachii with its branches and radial nerve with its branches in the groove.

9. What are the branches of arteria profunda brachii in spiral groove?

Ans. The branches are:
i. Muscular branches to deltoid and triceps.
ii. Ascending branch (anastomosing with the descending branch of posterior circumflex humeral artery).
iii. Nutrient artery to the humerus.
iv. Terminal branch:
   a. Radial collateral (anterior descending).
   b. Middle collateral (posterior descending).

10. What are the branches of radial nerve in spiral groove?

Ans.
i. Muscular branches to—medial and lateral head of triceps.
ii. Articular to elbow.
iii. Cutaneous:
   a. Lower lateral cutaneous branch of the arm.
   b. Posterior cutaneous branch of the forearm.

11. What is the nerve supply of medial head of triceps?

Ans. Medial head of triceps is supplied by radial nerve twice—one from the axilla and another from the spiral groove (lateral and long head are supplied by radial nerve once).

12. How does the anconeus muscle get its nerve supply?

Ans. The nerve to medial head of triceps (branch of radial nerve in the spiral groove) supplies the anconeus.

13. What is the root valve of radial nerve?

Ans. C_5, C_6, C_7, C_8, T_1 spinal nerves.

14. From which cord the radial nerve arises?

Ans. It arises from the posterior cord of brachial plexus which is formed by the union of posterior divisions of all three trunks. ***Posterior divisions of the trunks (radial nerve) supply the posterior aspect of upper limb and anterior divisions (median, ulnar and musculocutaneous nerves) supply the anterior aspect of the upper limb.***

LESSON 10: BACK OF THE FOREARM

The back of the forearm or posterior compartment of forearm contains the extensor muscles of the hand and digits. The nerves and vessels of this compartment run in a connective tissue plane that divides the superficial group of extensor muscles from the deep muscles. The floor of the posterior compartment is formed by the two forearm bones and their intervening interosseous membrane. Some **bony landmarks** on the back of the forearm are posterior border of ulna, styloid process and head of radius and ulna, olecranon process of ulna and dorsal tubercle of radius.

STEPS OF DISSECTION

1. **Position of the cadaver:** Place the cadaver in supine position with the forearm pronated elbow extended, arm medially rotated. Otherwise the cadaver is allowed to lie in prone position with pronated forearm, extended elbow and wrist.
2. **Skin incision (Fig. 2.62):**
    a. A transverse incision joining two epicondyles of the humerus (A–B).
    b. Another transverse incision between the styloid process of radius and ulna (C–D).
    c. A third incision given vertically from the mid point of above two incisions (E–F).
   Reflect the skin flap laterally and medially. The superficial fascia is exposed.
3. **Superficial fascia:** In it, few cutaneous nerves are to be cleaned. These are:
    a. Posterior cutaneous nerve of forearm (branch of radial nerve).
    b. Laterally, posterior branch of lateral cutaneous nerve of forearm (continuation of musculo-cutaneous nerve).

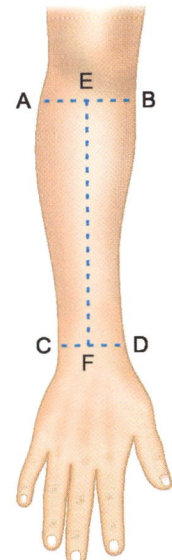

Fig. 2.62: Incision (back of forearm)

c. Medially, posterior branch of medial cutaneous nerve of forearm (branch of medial cord of brachial plexus).

After studying these nerves, the superficial fascia is reflected in the same line of incision. Deep fascia is exposed.

4. **Deep fascia:** It is the antebrachial fascia on the posterior compartment of the forearm. **Note that the antebrachial fascia envelops the whole forearm**. The deep fascia is incised along the line of skin incision but preserve. The **extensor retinaculum** which is the thickened antebrachial fascia is located on the posterior surface of distal forearm.

Separate the fascia from the underlying muscles using a probe or simply by finger dissection. Now detach the deep fascia from its attachments with the radius and ulna. Now the deep structures are exposed. Study them.

5. **Deep structures (Figs 2.63A to C):** The deep structures are the superficial extensor group of muscles, deep extensor group of muscles, vessels and nerves.

The superficial extensor groups of muscles are seven in number and divided into lateral group and medial group. The lateral groups of muscles are **brachioradialis, ECRL and ECRB**. Of these, the **most superficial muscle is the brachioradialis**. Deep and medial to it

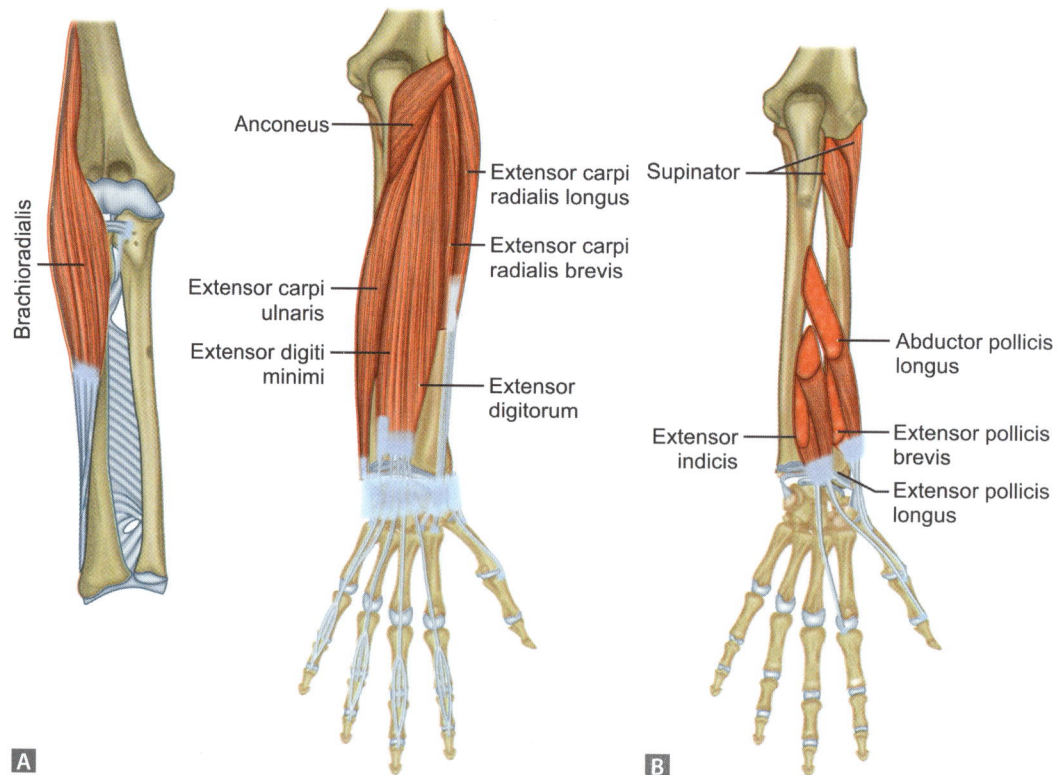

Figs 2.63A and B: (A) Back of forearm—superficial muscles; (B) Back of forearm—deep muscles

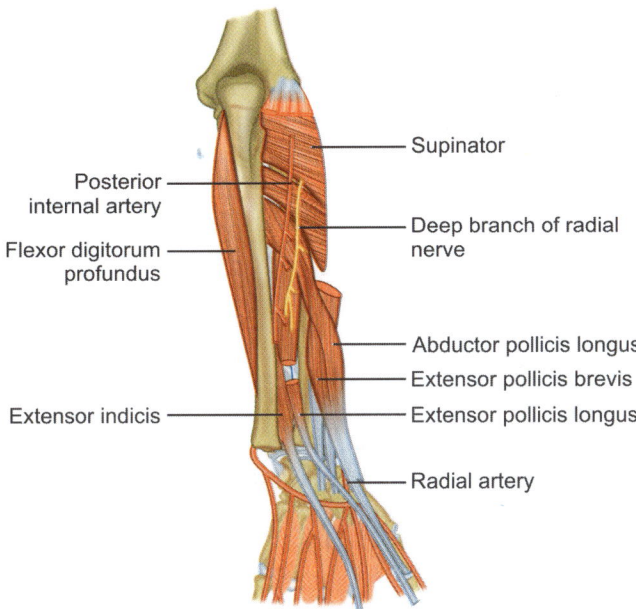

Fig. 2.63C: Posterior compartment of forearm—deep structures

lies ECRL and ECRB. Pull the brachioradialis laterally by fingers or retractors. This exposes **extensor carpi radialis longus** (ECRL) muscle.

Deep to ERCL, there is a relatively samaller muscle **extensor carpi radialis brevis** (ECRB) which is found in contact with the shaft of the radius. This muscle is supplied by the deep branch (posterior interosseous) of radial nerve. Deep to the brevis is the **supinator** muscle which is again supplied by the deep branch of radial nerve before passing between its two heads.

Completely separate these three anterolateral muscles from the **extensor digitorum** (ED) and expose the **supinator** muscle lying deep to them. **Extensor digitorum** is identified by its thick muscle belly ending in a stout tendon and breaking up into four slips for medial four digits. Medial to ED lies a slender muscle belly with stout tendon of **extensor digiti minimi** (EDM). Medial to EDM lies another muscle tendon **extensor carpi ulnaris** (ECU). Separate these superficial muscles from each other starting with the tendons at the back of the wrist and proceed proximally towards their common origin at the lateral epicondyle of the humerus. **Of all these superficial muscles, centrally placed muscles are EDM and ED passing to the fingers. Anconeus** is another superficial muscle at the upper part of the medial aspect of the forearm. This triangular muscle fans out from the lateral epicondyle of the humerus along with the origin of other superficial extensor muscles and inserted into the shaft of the ulna. The upper fibers of this muscle are almost horizontal and lower fibers are vertical. The attachment of ECU to the posterior border of ulna by thick deep fascia should be divided in the proximal third of forearm to demonstrate anconeus. The **deep groups of muscles are five** in number. They are **abductor pollicis longus, extensor pollicis brevis; extensor pollicis longus, extensor indicis and supinator**. They pass only to thumb and index finger except **supinator**.

To expose the deep group of muscles lift the extensor digitorum. **Extensor indicis** (EI) is identified by its narrow muscle belly ending in a slender tendon and found deep to the ED and then passes medial to the tendons of ED. It is the most medial muscle of the deep group. The most lateral deep muscle **abductor pollicis longus** passes obliquely downward and forward, accompanied by a slender muscle medial to it—**extensor pollicis brevis**. The tendons of both these muscles lodge in a groove on the radial side of the distal end of radius. Medial to the above two muscles (APL and EPB), passes another muscle in an oblique course that ends in a long narrow tendon called **Extensor pollicis longus**. Another deep muscle that passes between the two forearm bones and found at the upper part of the forearm deep to the superficial group of muscles is **supinator.**

To expose the **supinator muscle**, completely separate the three superficial antero-lateral muscles from the extensor digitorum. Expose the **posterior interosseous nerve** emerging from the supinator near its distal border. *When the deep branch of the radial nerve emerges from the supinator muscle, its name changes to posterior interosseous nerve*. This nerve provides motor branches to the extensor muscles of the forearm expect **brachioradialis, ECRL and anconeus** which are supplied by radial nerve. To expose the radial nerve at the elbow, pull brachioradialis and ERCL, ERCB laterally. Clean the nerve and note its divisions in superficial and deep branch. The deep branch pierces the supinator. Gently pull on the deep branch to establish its continuity with the posterior interosseous nerve by noticing the movement of the nerve. It is to be mentioned that this nerve ends in a **pseudoganglia** deep to the extensor retinaculum.

Posterior interosseous artery (Fig. 2.64), branch of common interosseous artery passes between the two forearm bones at the upper border of interosseous membrane and below the oblique cord to reach the posterior compartment of forearm from anterior aspect. It is accompanied here by the posterior interosseous nerve. Then it descends downwards and ends by anastomosing with the anterior interosseous branch of common interosseous artery. Note that the **anterior interosseous artery** pierces the interosseous membrane to come to the posterior aspect of forearm about 5–6 cm proximal to the distal end of the radius.

Between these three tendons a triangular depression develops on the radial side of the wrist called **anatomical snuff box** which contains radial artery across the floor.

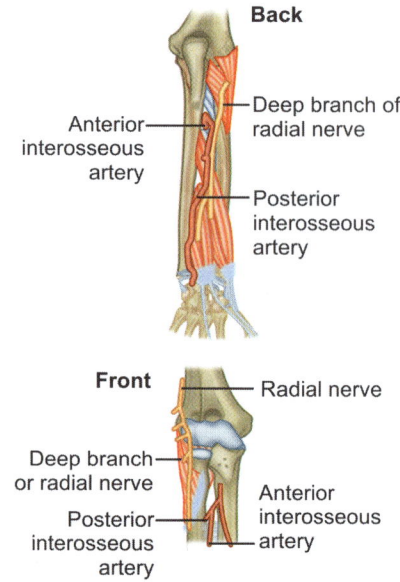

Fig. 2.64: Arteries and nerves—back of forearm

Anatomical Snuffbox (Figs 2.65A and B)

Lateral boundary: Tendons of APL and EPB.
Medial boundary: Tendon of EPL.
Roof: Skin and fascia, crossed by cephalic vein and superficial branch of radial nerve.
Floor: Radial styloid process, scaphoid, trapezium and base of first metacarpal bone.
Content: Radial artery.

Window Dissections: Upper Limb (Superior Extremity)

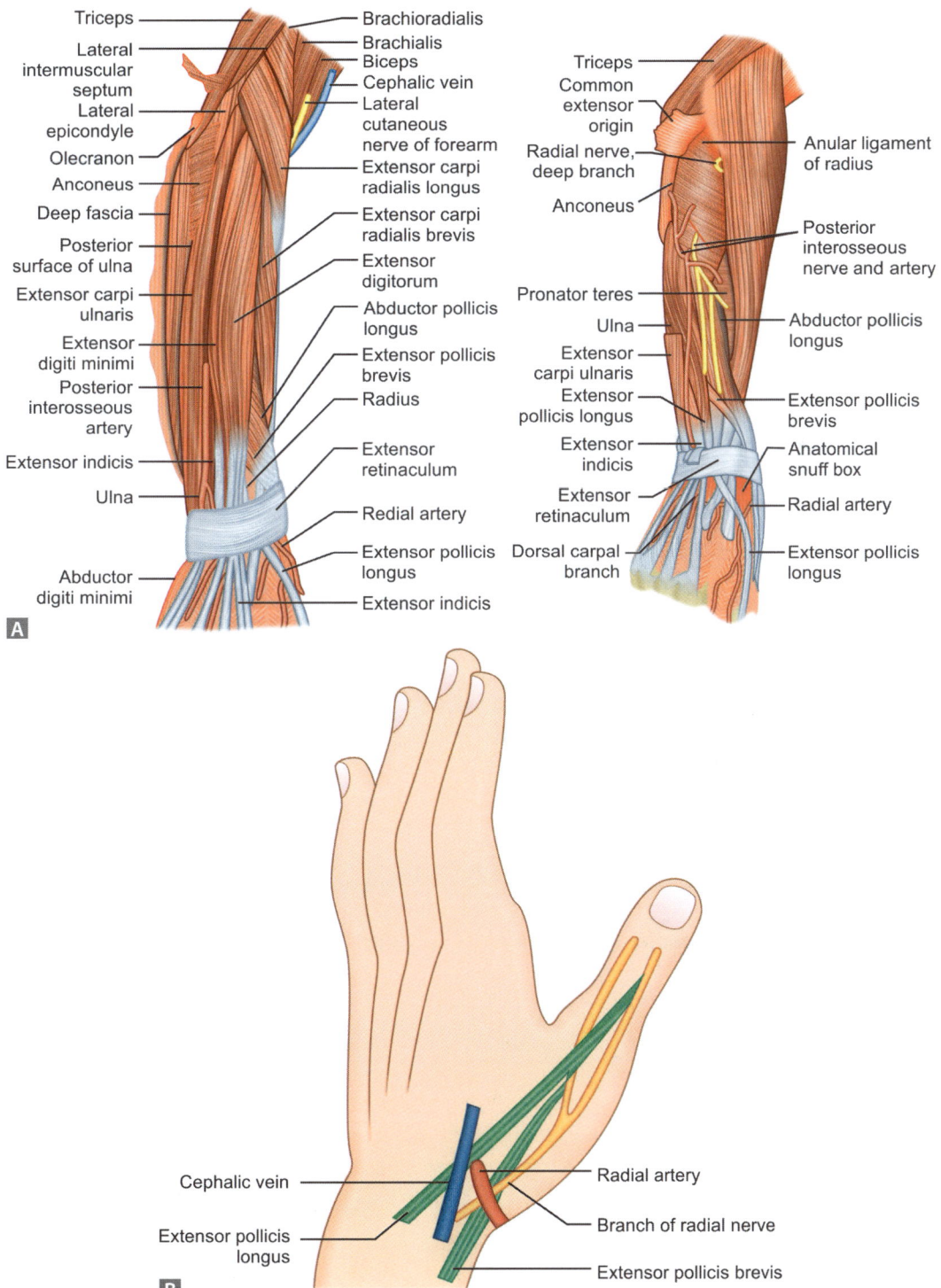

Figs 2.65A and B: Anatomical snuffbox

MUSCLES RELATED TO BACK OF THE FOREARM (TOTAL = TWELVE)

Superficial Muscles (Total = Seven)

Lateral Group (Three)

Muscles	Origin	Insertion	Nerve supply	Action
Brachioradialis	Upper ⅔ of lateral supracondylar ridge	Base of the styloid process of radius	Radial nerve above the elbow	Flexor of the elbow (best in mid-prone position)
ECRL	Lower ⅓ of lateral supracondylar ridge	Dorsal surface of the base of 2nd metacarpal bone	Radial nerve above the elbow	Extension of wrist
ECRB	Lateral epicondyle of the humerus	Dorsal surface of the base of 2nd and 3rd metacarpal bones	Posterior interosseous nerve (deep branch of radial nerve)	Extension of wrist

Superficial Muscles

Posterior Group (Four)

Muscles	Origin	Insertion	Nerve supply	Action
Extensor digitorum	Lateral epicondyle of humerus	Dorsal digital expansion and base of middle and distal phalanx	Posterior interosseous nerve	Extension of wrist, metacarpo-phalangeal and interphalangeal joint
Extensor digiti minimi	Lateral epicondyle of the humerus	Dorsal digital expansion of 5th digit	Posterior interosseous nerve (deep branch of radial nerve)	Extension of little finger
Extensor carpi ulnaris	Lateral epicondyle of humerus	Base of 5th metacarpal bone	Posterior interosseous nerve	Extension and adduction of the wrist
Anconeus	Lateral epicondyle of humerus	Olecranon process of ulna	Radial nerve	Extends elbow

Deep Muscles (Five)

Muscles	Origin	Insertion	Nerve supply	Action
Abductor pollicis longus	Posterior surface of the shaft of radius, ulna and interosseous membrane	Base of the distal phalanx of 1st metacarpal bone	Posterior interosseous nerve	Abduction and extension of the thumb
Extensor pollicis brevis	Posterior surface of the shaft of radius and interosseous membrane	Base of proximal phalanx of thumb	Posterior interosseous nerve	Extends the thumb at 1st carpometa-carpal joint

Contd...

Contd...

Muscles	Origin	Insertion	Nerve supply	Action
Extensor pollicis longus	Posterior surface of the shaft of ulna and adjacent interosseous membrane	Base of the distal phalanx of thumb	Posterior interosseous nerve (continuation of deep branch of radial nerve)	Extends the distal phalanx of the thumb
Extensor indicis	Posterior surface of the shaft of ulna below the extensor pollicis longus and from interosseous membrane	Dorsal digital expansion of the index finger (digit 2)	Posterior interosseous nerve	Extension of index finger
Supinator	Lateral epicondyle and supinator crest of ulna	Shaft of radius distal to radial tuberosity	Deep branch of radial nerve	Supination of forearm

Note that the thumb (EPL, EPB), index (ED, EI) and little finger (ED, EDM) each has two extensor tendons. All the muscles in the extensor compartment of forearm are innervated by radial nerve.

Summary

1. Incision of skin, superficial fascia and deep fascia as in Figure 2.61.
2. Structures found in superficial fascia:
    i. Posterior cutaneous nerve of forearm.
    ii. Posterior branch of medial and lateral cutaneous nerve of forearm.
3. Deep structures (deep-to-deep fascia or antebrachial fascia)
    i. **Muscles: Superficial (7): Lateral group** (3)—brachioradialis, ECRL, ECRB; **Posterior group** (4) ED, EDM, ECU, anconeus.
    ii. **Deep** (5): Supinator, APL, EPB, EPL and EI.
    iii. **Vessels:** Posterior and anterior interosseous artery, both are branches of common interosseous artery (branch of ulnar artery).
    iv. **Nerves:** Posterior interosseous nerve (deep branch of radial nerve after piercing supinator muscle).
4. Most of the superficial muscles take a common origin from the front of the lower part of lateral epicondyle of humerus except brcahioradialis, ECRL and anconeus.
5. All extensor muscles are supplied by the posterior interosseous nerve except brachioradialis, ECRL and anconeus which are directly supplied by the trunk of radial nerve.

PROBABLE QUESTIONS AND ANSWERS

1. **How many muscles are found in the back of the forearm?**

Ans. Total 12 muscles are found. Out of which seven muscles are superficial and five muscles are deep.
Note that all the superficial muscles cross the elbow joint, as they arise from the humerus but none of the deep muscles crosses the elbow joint, as they arise from the two forearm bones and the interosseous membrane.

2. What is the area of origin of the superficial extensor muscles?

Ans. Tip of the lateral epicondyle of the humerus except three [brachioradialis (BR), extensor carpi radialis longus (ECRL) and anconeus].

3. What may be the reason behind the exemption of the origin of these three muscles from the lateral epicondyle?

Ans. The tip of the lateral epicondyle is a very small area. Due to the smaller area, the origin of the multiple muscles is crowded. So the two muscles (BR and ECRL) have shifted upwards on the lower part of lateral supracondylar ridge of the humerus. Anconeous also has shifted backwards to the posterior surface of the lateral epicondyle.

4. How are the compartments formed on the back of the wrist?

Ans. There are six osteofascial compartments on the back of the wrist. The extensor retinaculum, formed by the thickening of the deep fascia on the back of the wrist, sends septa which are attached to the longitudinal ridges on the posterior surface of the lower end of the radius, thus forming six osteofascial compartments through which the structures pass from back of the forearm to the back of the hand.

5. What is tubercle of Lister?

Ans. The posterior surface of the lower end of the radius presents a palpable bony prominence which is called dorsal tubercle of Lister.

6. Name the structures passing medial and lateral to the dorsal tubercle?

Ans. The longitudinal groove lateral to the tubercle lodges the tendons of ECRL and ECRB. The groove just medial to the tubercle is prominent and transmits the tendon of EPL which utilizes the tubercle as a pully before reaching the thumb.

7. What are outcropping muscles?

Ans. In the lower one-fourth of the back of the forearm, the three deep muscles for the thumb (APL, EPB and EPL) emerge obliquely on the surface between three lateral (BR, ECRL, ECRB) and three posterior (ED, EDM, ECU) superficial muscles. These emerging muscles are known as the outcropping muscles.

8. In which position of the forearm, the brachioradialis acts best as a flexor of the elbow joint?

Ans. In the mid-prone position. This is due to the fact that the BR muscle is a preaxial muscle. So, in the preaxial position, i.e. mid-prone position, it acts best as a flexor of the elbow.

Window Dissections: Upper Limb (Superior Extremity)

9. What is the nerve supply of anconeus muscle?

Ans. The nerve to medial head of triceps (branch of radial nerve in the spiral groove) after supplying the medial head, descends down to supply the anconeus. **The anconeus is said to be the part of medial head of triceps muscle.**

10. In which position of the elbow, the biceps brachii is more powerful supinator than supinator muscle?

Ans. In semiflexed position of the elbow.

11. In which position of the elbow, the supinator muscle is the prime mover of supination?

Ans. In extended position of the elbow.

12. What is the nerve supply of the muscle of back of the forearm?

Ans. All the muscles are supplied by the posterior interosseous nerve (deep branch of radial nerve) except BR, ECRL and anconeus which are supplied by the trunk of the radial nerve before division.

13. What is the main arterial supply of the posterior compartment?

Ans. Posterior interosseous artery, branch of common interosseous artery is the main artery of the posterior compartment.

14. How does the posterior interosseous artery come to the posterior compartment of the forearm?

Ans. It is a branch of the common interosseous artery branch of ulnar artery in the anterior compartment. It passes between the upper part of the two forearm bones (radius and ulna) through a gap between the upper border of the interosseous membrane and oblique cord and then comes to the posterior compartment where it anastomoses with the anterior interosseous artery branch of common interosseous artery.

15. What is the effect of injury to the radial nerve in axilla or arm?

Ans. 'Wrist drop' due to the paralysis of the extensor muscles of the forearm.

LESSON 11: DORSUM OF THE HAND

In the dorsum of the hand, the skin is thin and moves freely over the underlying tendon and deep fascia. The bones are superficial. **There are no intrinsic muscles in the dorsum** (dorsal interossei are muscles of the palmar aspect and supplied by ulnar nerve). All the tendons are the continuation of the muscles of the extensor compartment of the forearm. **The cutaneous supply of the dorsum of the hand is contributed by radial, ulnar and median nerves.**

STEPS OF DISSECTION

1. **Position of the cadaver (hand):** Cadaver—prone, wrist extended and fingers abducted.
2. **Skin incision (Fig. 2.66):**
   a. Transverse incision joining the styloid process of radius and ulna (A–B).
   b. 2nd transverse incision at the root of the fingers (2–5) (C–D).
   c. 3rd incision given longitudinally extending from the midpoint of 1st incision (E) to the tip of the middle finger (E–F).
   d. 4th incision starts from the midpoint of 1st incision (E) to the tip of the thumb (G).

   Reflect the flaps of skin laterally and medially. Superficial fascia is exposed.

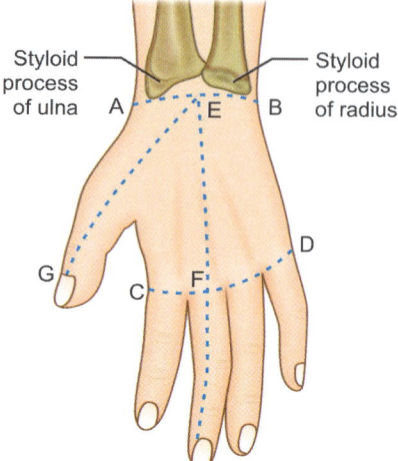

Fig. 2.66: Incision on dorsum of the hand

3. **Superficial fascia:** It presents a **dorsal subcutaneous space**. This space contains the **dorsal venous arch** which is formed by **three dorsal metacarpal veins** which in turn are formed by the union of **dorsal digital veins**. The venous arch continues proximally as **cephalic vein** after joining with the dorsal digital veins of the radial side of index finger and both sides of the thumb on the lateral side and **basilic vein** on the medial side after joining with the dorsal digital vein on the medial side of little finger. The subcutaneous space also contains the dorsal digital nerves derived from **dorsal branch of ulnar nerve and superficial terminal branch of radial nerve.** The superficial radial nerve courses between the ECRB and APL.

   The dorsal digital branch of the radial nerve of the thumb reaches up to the base of distal phalanx and those to index finger extends up to the middle of the middle phalanx. The digital branches to the middle finger and lateral side of ring finger extend upto the middle phalanx, not beyond the distal interphalangeal joint. The dorsal digital branches of the ulnar nerve extend upto the base of distal phalanx of little finger and medial side of the base of the middle phalanx of ring finger (Figs 2.67A and B).

   The **median nerve** supplies lateral 3½ digits including the distal portions of the dorsum of the these digits. **Ulnar nerve** supplies medial 1½ digits both in palmar and dorsal aspect of little and half of ring finger.

   The superficial fascia is reflected sideways. Deep fascia is exposed.

4. **Deep fascia:** It is thin. It is proximally continuous with the extensor retinaculum (thickend antebrachial fascia at the back of the wrist) and at the sides with the palmar fascia. A **sub-aponeurotic space** is present between the deep fascia and metacarpal bones with intervening dorsal interossei muscles. In this space, the deep structures on the dorsum of the hand are found. These are **extensor tendons, dorsal carpal arch** and **their branches.**

   The **dorsal carpal arch** is formed by the union of the dorsal carpal branches of radial and ulnar arteries assisted by anterior and posterior interosseous arteries across the dorsal surface

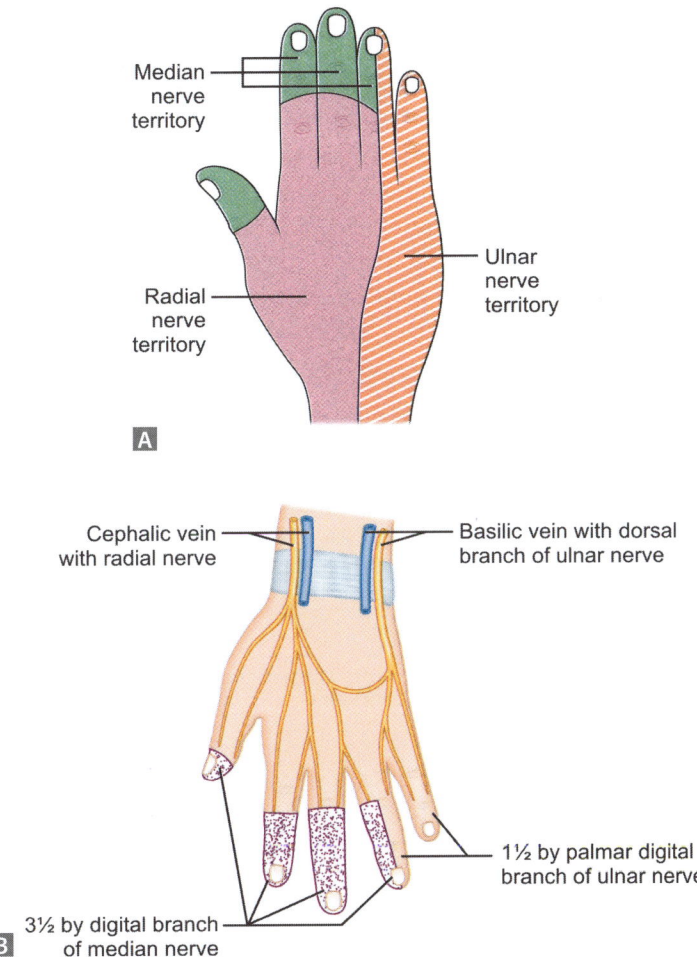

Figs 2.67A and B: (A) Cutaneous innervation of dorsum of hand; (B) Nerves in dorsum of hand

of the carpal bones. This arch gives origin to **three dorsal metacarpal arteries** which in turn divide into **dorsal digital arteries**. Note that **palmar carpal arch** is formed by the palmar carpal branches of radial and ulnar arteries across the lower end of radius and ulna, assisted by the branch from anterior interosseous artery and recurrent branch from deep palmar arch. This arch is **cruciform** in outline.

Extensor tendons are cleaned and identified. Idenfify **the anatomical snuff box** and its boundaries and content (radial artery). **The radial artery** leaves the anterior aspect of radius and lies in the anatomical snuffbox passing deep to the tendons of abductor pollicis longus and extensor pollicis brevis. Then it turns medially into the palm through the proximal end of the first metacarpal space between the two heads of the first dorsal interosseous muscle. Then the artery passes medially deep to the oblique head of adductor pollicis to join the deep branch of ulnar artery to complete the **deep palmar arch.**

The extensor digitorum tendons are tied together by **intertendinous bands** on the dorsum of the hand. These digital tendons of ED also blend with the deep fascia as well. Observe the **dorsal digital expansion (extensor expansion)** of the ED tendons. This expansion is widest at the level of metacarpophalangeal joint. Distal to it, the expansion narrows to the dorsal surface of the proximal interphalangeal joint to reach the middle and distal phalanx. The tendons of lumbricals and interossei muscles attach to the extensor expansion.

It is better to cut through the extensor retinaculum to release the extensor digitorum tendons and retract them medially to note the other extensor tendons on the dorsum of the hand.

The tendons on the dorsum of the hand are properly cleaned and identify them. The tendons lie from **lateral to medial** side are as follows **(Fig. 2.68)**:

> **Note that** there are no intrinsic muscles on the dorsum of the hand. All the tendons (muscles) passing over the dorsum are supplied by radial nerve in the posterior compartment of the forearm.

- Tendons of abductor pollicis longus and extensor pollicis brevis (**first compartment**)
- Tendons of extensor carpi radialis longus and extensor carpi radialis brevis (**2nd compartment**).
- Tendon of extensor pollicis longus (**3rd compartment**).
- Extensor digitorum and extensor indicis (**4th compartment**).
- Extensor digiti minimi (**5th compartment**).
- Extensor carpi ulnaris (**6th compartment**).

Summary

1. Incision of skin, superficial fascia, deep fascia as in Figure 2.66.
2. Structures found in superficial fascia—superficial (cutaneous) branch of radial nerve, cutaneous branch of ulnar nerve, dorsal venous arch and commencement of cephalic vein (laterally), basilic vein (medialy).
3. Structures deep to deep fascia extensor tendons (APL, EPB, ECRL, ECRB, EPL, ED, EI, EDM and ECU); dorsal carpal arch and dorsal metacarpal branches, dorsal digital expansion and anatomical snuffbox with its contents (radial artery) and boundaries (laterally—APL and EPB, medially—EPL, floor—scaphoid, trapezium, base of 1st MC bone).

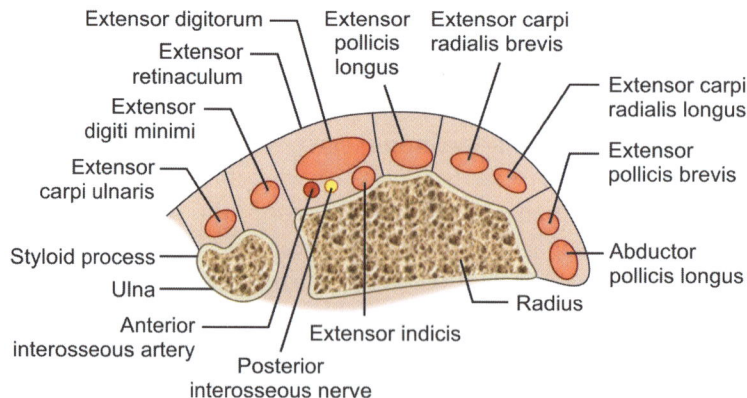

Fig. 2.68: Extensor retinaculum and tendons on dorsum of hand

PROBABLE QUESTIONS AND ANSWERS

1. What are the observations on the dorsum?

Ans.
i. The bones are superficial.
ii. **No intrinsic muscles**; so no motor innervation is required.
iii. The skin in thin and freely moves over the underlying extensor tendons and deep fascia.
iv. All three main nerves (median, ulnar, radial) share the cutaneous innervaton of the dorsum of the skin.
v. Deep fascia in thin.
vi. **Two spaces** are found:
  a. Dorsal subcutaneous space.
  b. Dorsal subaponeurotic space.
vii. **Two arches** are found
  a. Dorsal venous arch in the subcutaneous space.
  b. Dorsal carpal arch in the subaponeurotic space.
viii. **Presence of dorsal digital expansion**.

2. What is anatomical snuffbox?

Ans. Anatomical snuffbox is a space or depression on the radial side of the dorsum of hand bounded laterally by the tendons of APL, EPB and medially by the tendon of EPL.

3. What structures pass over the snuffbox?

Ans. Cephalic vein and superficial (cutaneous) branch of the radial nerve.

4. What is the content of the snuffbox?

Ans. **Radial artery**.

5. What is the chief venous network of the hand?

Ans. Blood from the hand chiefly drains into the dorsal venous network on the back of the hand.

6. What are intercapitular veins?

Ans. The dorsal digital veins communicate with the palmar digital veins by a set of veins which pass between the metacarpal heads. This intercommunicating oblique veins are called intercapitular veins.

7. What is the function of extensor retinaculum?

Ans. It keeps the extensor tendon in position.

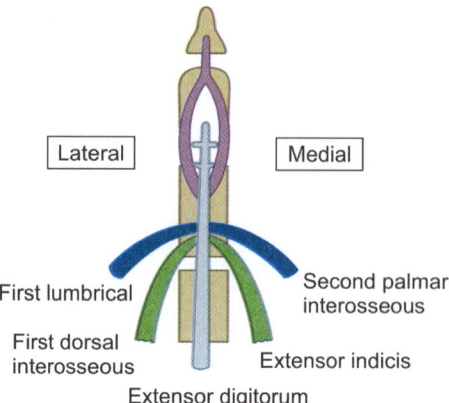

Fig. 2.69: Dorsal digital expansion

8. What is dorsal digital expansion?

Ans. It is an aponeurotic expansion formed by each extensor tendon which covers the dorsum of the metacarpal head and proximal phalanx.

9. What are the other muscles joining this expansion?

Ans. Interossei and lumbrical muscles.

10. What is the shape of the dorsal expansion?

Ans. Triangular, with the base over the metacarpal head and apex is close to the distal end of proximal phalanx.

Dorsal Digital Expansion (Fig. 2.69)

It is triangular expansion of the extensor tendons over the metacarpal head and phalanges of the hand. The base of the expansion forms a hood over the metacarpal head and attached to the deep transverse metacarpal ligament. The apex of the expansion trifurcates close to the distal end of proximal phalanx. The median band in attached to the base of the middle phalanx and the two lateral bands unite distally to be attached with the base of the distal phalanx. The thickened lateral margin receives the insertion of the tendons of lumbricals and interossei muscles and medial margin receives the insertion of interossei muscle only.

CHAPTER 3

Window Dissections: Lower Limb (Inferior Extremity)

- Introduction to Inferior Extremity
- Femoral Triangle
- Adductor Canal
- Anterolateral Compartment of the Leg
- Dorsum of the Foot
- Gluteal Region
- Back of the Thigh
- Popliteal Fossa
- Back of the Leg (Posterior Crural Region)
- Sole of the Foot

LESSON 1: INTRODUCTION TO INFERIOR EXTREMITY

The inferior extremities or lower limbs of a human being are developed from the limb buds which appear as the outgrowths from the trunk at the fourth week of intrauterine life. The lower limb buds (one on each side) appear at the level between T_{12} to S_4 segments of spinal cord which provide innervations to the lower limb.

The lower limb is connected to the trunk and is used for carrying the body weight and for propulsion. The pelvic girdle which is formed by the two hip bones connects the lower limb with the trunk. The two hip bones articulate with each other in front at the symphysis pubis and separated from each other by the sacrum behind.

Axial line is said to be the line of junction of two dermatomes supplied from discontinuous spinal levels. Though adjacent dermatomes overlap, no overlapping occurs across the axial line. **The anterior axial line** (Fig. 3.1) of the lower limb extends from the root of the penis (male) or clitoris (female) → across the front of the scrotum (male)/labia majora (female) → then spirals to the middle of the back of the thigh and leg → almost up to the heel. **The posterior axial line** (Fig. 3.1) starts from the L_4 spine → then across the gluteal region which undergoes a convex curve → lateral side of the back of the thigh and leg → stops above the heel. **It is probable that posterior axial lines do not exist, but evidence of anterior axial line is more convincing** (RJ Last). So, the **preaxial** border means the medial border and the **postaxial** border means the lateral border of the foot. **Due to the medial rotation of the lower limb bud**, the hallux (great toe) becomes medial. The great saphenous vein and the tibia are preaxial in position, whereas the small saphenous vein and fibula are postaxial in position. Due to this rotation of the lower limb bud, the flexor surface of the lower limb becomes posterior and the extensor surface becomes anterior. This explains the nerve supply of the lower limb muscles where the anterior division of the lumbosacral plexus supplies the posterior surface and the posterior division supplies the anterior aspect of the lower limb.

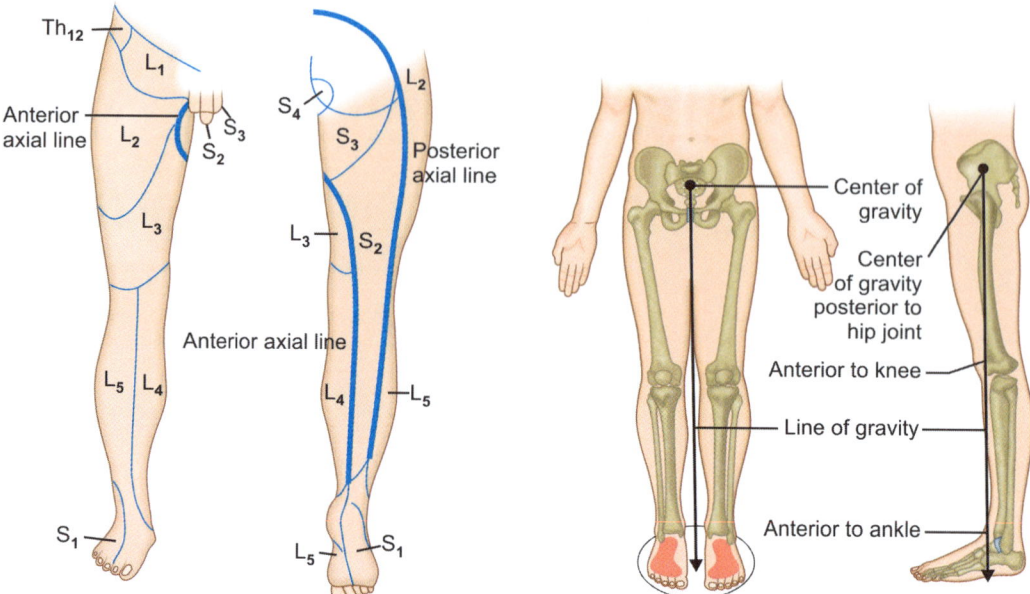

Fig. 3.1: Anterior and posterior axial lines

Fig. 3.2: Center of gravity and line of gravity

On the other hand, it is to be noted that **the upper limb bud rotates laterally** bringing the palm to face anteriorly so that the thumb is placed laterally. The preaxial border of the upper limb is lateral and the postaxial border is medial in position. The cephalic vein and radius become preaxial and the basilic vein and ulna become postaxial. Consequently, the flexor surface becomes anterior and the extensor surface becomes posterior due to this lateral rotation.

In upright posture, the body weight is transmitted through the acetabulum of the hip bone → Femur → Tibia → Foot. **In sitting posture,** the weight is transmitted to the ischial tuberosities. **So, we stand on S_1 and we sit on S_3 spinal nerves** (RJ Last). It is also observed that the line of center of gravity passes behind the hip joint and in front of the knee and ankle joints (Fig. 3.2). So, there is a natural tendency of backward tilting of the pelvis at the hip joint, hyper-extension of the knee joint and forward dislocation of the leg bones at the ankle joints. But these are prevented by the strong ligaments, antagonistic muscles and the bony configuration of the foot (**talus**).

For descriptive purpose, **the lower limb is divided into the following regions:**
  i. **Thigh** which extends from the hip to knee joint is having only one bone called femur (Fig. 3.3). The thigh can be divided into three compartments (Fig. 3.4): (a) Anterior or extensor, (b) Posterior or flexor and (c) Medial or adductor.
 ii. **Gluteal region** which lies above the posterior compartment of the thigh and behind the pelvis and hip.
iii. **Leg** is the region between the knee and foot. It is having two bones called tibia and fibula (Fig. 3.5). The leg can be divided into three compartments (Fig. 3.6): (a) Anterior or extensor, (b) Posterior or flexor and (c) Lateral or peroneal.

Window Dissections: Lower Limb (Inferior Extremity)

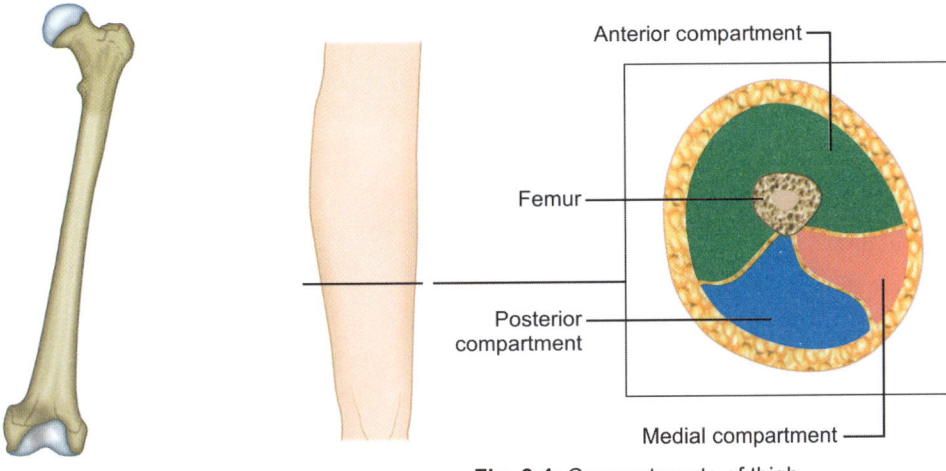

Fig. 3.3: Femur

Fig. 3.4: Compartments of thigh

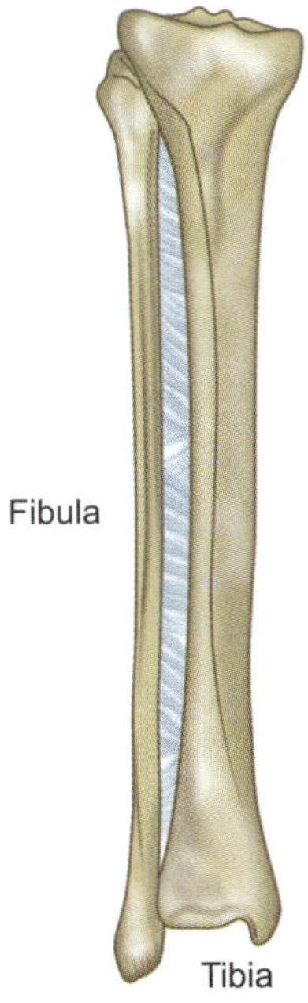

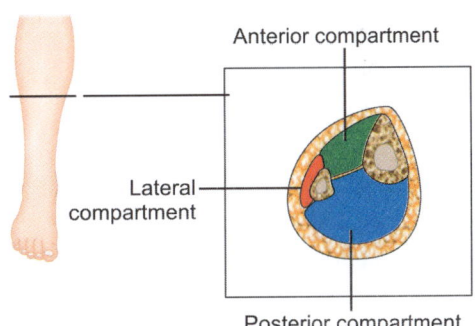

Fig. 3.5: Tibia and fibula

Fig. 3.6: Compartments of leg

iv. **Foot:** Skeleton of the foot consists of the following bones (Fig. 3.7):
   a. **Tarsal bones** (7): Calcaneum, talus, navicular, cuboid and 3 cuneiforms (medial, intermediate and lateral).
   b. **Metatarsal bones** (5): Named as 1st, 2nd, 3rd, 4th and 5th (from medial to lateral)
   c. **Phalanges** (14): 3 for each toe (proximal, middle and distal) except the great toe which has two phalanges (proximal and distal).

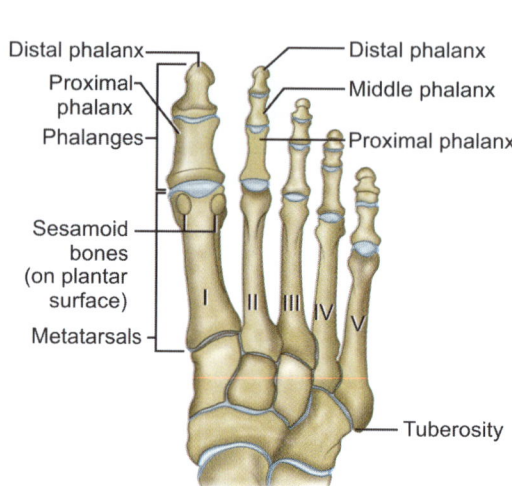

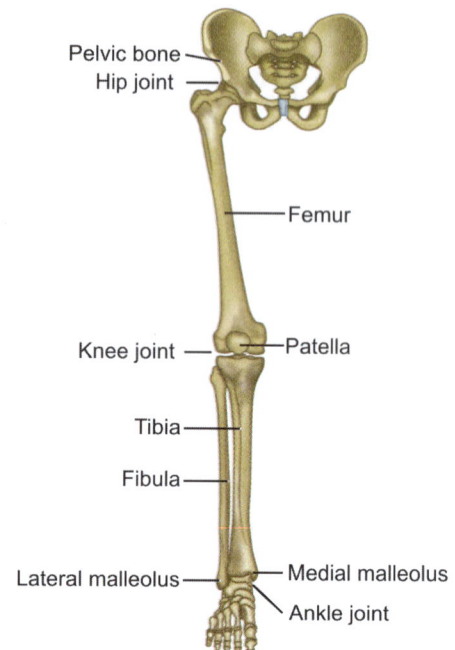

Fig. 3.7: Bones of the foot

Fig. 3.8: Bones and joints of lower limb

The foot has an upper surface called dorsum of the foot and a plantar surface called sole of the foot.

The joints of the lower limb are as follows:
  i. Hip joints
 ii. Knee joints
iii. Ankle joints
 iv. Tibiofibular joint (superior, intermediate and inferior)
  v. Subtalar and midtarsal joints
 vi. Other intertarsal joints
vii. Tarsometatarsal joints (5)
viii. Metatarsophalangeal joints (5)
 ix. Interphalangeal joints (proximal and distal). **Great toe is having only one interphalangeal joint.**

The bones and joints of the lower limb are shown in Figure 3.8.

The deep structures of these above regions are ensheathed by a jacket of common tubular connective tissue membrane called **deep fascia**. This deep fascia only changes its name from region to region such as **fascia lata in thigh, fascia cruris in leg and plantar aponeurosis in the sole of the foot.**

Window Dissections: Lower Limb (Inferior Extremity)

LESSON 2: FEMORAL TRIANGLE

Femoral triangle is a triangular area in front of the upper part of the thigh bounded **superiorly** by the **inguinal ligament, laterally** by the medial border of **sartorius muscle, medially** by the medial border **of adductor longus muscle** and its apex is the meeting point of medial border of sartorius and medial border of adductor longus (Fig. 3.9).

STEPS OF DISSECTION

1. **Position of the cadaver:** Body supine, thigh extended and laterally rotated.
2. **Skin incisions (Fig. 3.10):**
    i. An oblique incision from the anterior superior iliac spine to pubic tubercle (A–B)
    ii. Another transverse incision at the junction of upper 1/3rd and lower 2/3rd of the thigh (C–D)
    iii. A vertical incision from the midpoint of (A–B) to the midpoint of (C–D) which is indicated by the line (E–F)
    Now, reflect the skin flaps laterally and medially.
    Superficial fascia is exposed.
3. **Superficial fascia:** In the superficial fascia, the following structures are identified (Fig. 3.11)
    i. **Blood vessels:**
        a. Great saphenous vein (on the medial aspect).
        b. Superficial external pudendal vessels.
        c. Superficial epigastric vessels.
        d. Superficial circumflex iliac vessels.
    ii. **Cutaneous nerves:**
        a. Lateral femoral cutaneous nerve of thigh.
        b. Intermediate femoral cutaneous nerve of thigh.

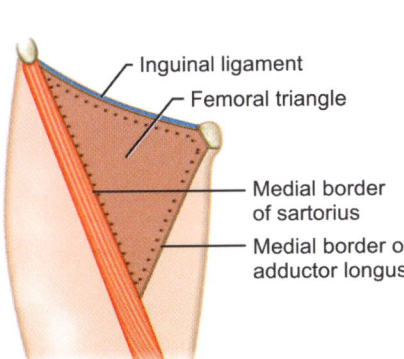

Fig. 3.9: Boundaries of femoral triangle

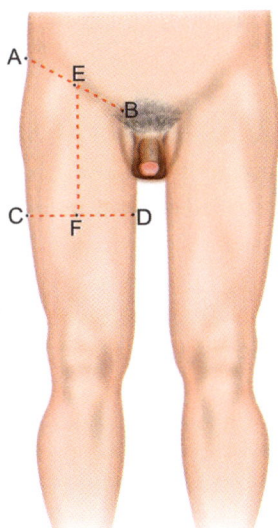

Fig. 3.10: Dissection of femoral triangle (skin incision)

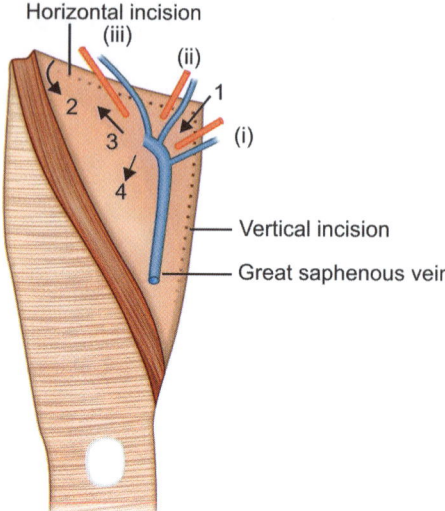

Fig. 3.11: Structures in superficial fascia: (i) Superficial external pudendal vessels; (ii) Superficial epigastric vessels; (iii) Superficial circumflex iliac vessels; (1) Branch of ilioinguinal nerve; (2) Branch of subcostal nerve; (3) Femoral branch of genitofemoral nerve; (4) Branch of medial femoral cutaneous nerve

      c. Medial femoral cutaneous nerve.
      d. Femoral branch of genitofemoral nerve.
      e. A twig from ilioinguinal nerve.
  iii. **Lymphatics:**
      a. Superficial group of inguinal lymph nodes.
      b. Superficial lymph vessels.

Superficial fascia is reflected like skin.

Deep fascia is exposed.

4. **Deep fascia (fascia lata):** There is an oval and twisted gap in the upper and medial part of deep fascia. This opening is called **saphenous opening** which is closed by an areolar membrane called **cribriform fascia**. A number of structures are seen piercing the cribriform fascia. These structures are as follows (Fig. 3.12):
    i. Great (long) saphenous vein.
   ii. Superficial external pudendal artery.
  iii. Superficial epigastric artery.
  iv. Few branches of medial femoral cutaneous nerve.
   v. Few lymph vessels connecting the superficial and deep inguinal lymph nodes.

> **Remember** that the superficial external pudenal **vein** and superficial epigastric **vein** do not pass through the saphenous opening. They accompany their corresponding arteries and drain into the great saphenous vein just before it pierces the cribriform fascia.

Now, insert your index finger into the **saphenous opening** beside the great saphenous vein. Then, you move your finger around the vein to define the margins of the saphenous opening. Trace the great saphenous vein up to its termination into the femoral vein.

Then, again insert your finger into the saphenous opening and push it inferiorly deep to the deep fascia until the fingertip touches the sartorius muscle.

Cut the deep fascia vertically downwards by using scissors from the lower margin of saphenous opening up to the sartorius muscle (Fig. 3.13, incision 1).

A second incision is made through the fascia lata from the upper margin of the saphenous opening. This incision extends laterally below and parallel to the inguinal ligament up to a point just below the anterior superior iliac spine (Fig. 3.13, incision 2).

A third incision is made through the fascia lata from the upper margin of the saphenous opening which extends medially below and parallel to the medial part of inguinal ligament up to a point just below the pubic tubercle (Fig. 3.13, incision 3).

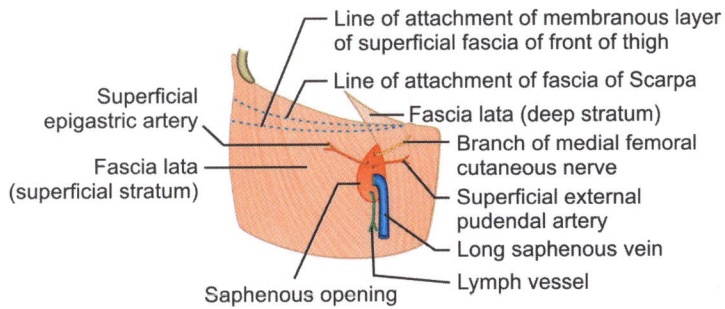

Fig. 3.12: Structures passing through the saphenous opening

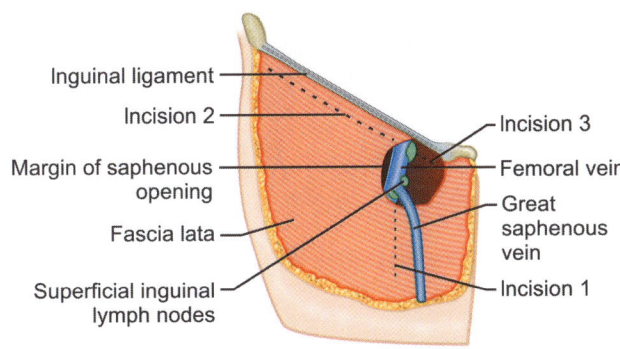

Fig. 3.13: Incisions on deep fascia

Then push your finger through these cuts and move them deep to the fascia lata to separate the deeper structure from the fascia lata.

Then reflect the flaps of fascia lata medially and laterally. Now the roof of the femoral triangle which is formed by the fascia lata (deep fascia of the thigh) is opened.

5. **Identify and clean the boundaries of the femoral triangle**.

   The triangle is bounded **above** (base) by the **inguinal ligament,** laterally by the medial border of **sartorius muscle, medially** by the medial border of **adductor longus muscle** and the **apex** is formed by the junction of these above two muscles.

   The contents of femoral triangle are as follows (Fig. 3.14):
   i. Femoral nerve and its branches.
   ii. Femoral sheath with its contents.
   iii. Terminal part of great saphenous vein.
   iv. Femoral branch of genitofemoral nerve.
   v. Lateral femoral cutaneous nerve (L_2, L_3).

6. **Femoral nerve** is seen to lie on the floor of the femoral triangle in the grove between the **iliacus** and **psoas major** muscle. It lies on the lateral side of the femoral artery and outside the femoral sheath. Clean and trace the femoral nerve inferiorly to expose its branches. It sends a branch to pectineus muscle called **nerve to pectineus** which passes medially behind the femoral sheath before the femoral nerve divides into anterior and posterior branches.

   Identify the **branches of anterior division** of femoral nerve which are as follows:
   i. **Muscular** (1): Nerve to sartorius.
   ii. **Cutaneous**:
      a. Medial femoral cutaneous nerve.
      b. Intermediate femoral cutaneous nerve.

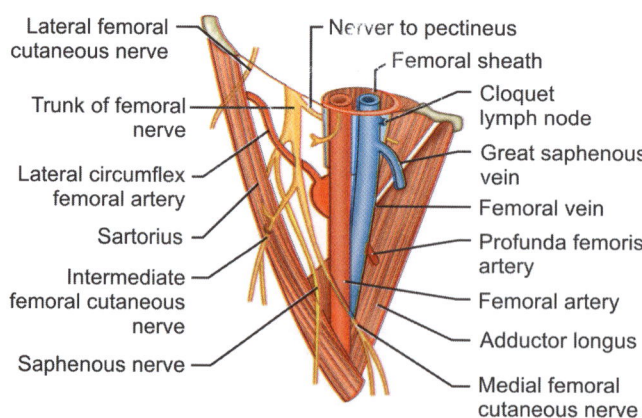

Fig. 3.14: Contents of femoral triangle

Identify the branches of **posterior division of femoral** nerve:
   i. **Muscular (4):**
      a. Nerve to rectus femoris.
      b. Nerve to vastus lateralis.
      c. Nerve to vastus medialis.
      d. Nerve to vastus intermedius.
   ii. **Cutaneous (1): Saphenous nerve.**

> Remember that the lateral circumflex femoral vessels pass between the two divisions of femoral nerve.

7. **Femoral sheath** is identified.

Femoral Sheath (Figs 3.15A and B)

Definition: It is a funnel-shaped fibrous sheath around the proximal part of femoral vessels.
Formation:
   i. Anterior wall by fascia transversalis.
   ii. Posterior wall by fascia iliaca.

Shape: Funnel-shaped.
Lateral wall is vertical but the medial wall is oblique directing laterally and downwards.

Dimension:
   i. Length—about 3–4 cm.
   ii. Breadth—about 2 cm.

Compartments: Three, separated by two vertical septa.
   a. Lateral (**arterial**) compartment containing
      i. Femoral artery.
      ii. Femoral branch of genitofemoral nerve.
   b. Intermediate (**venous**) compartment containing
      i. Femoral vein.
   c. Medial compartment (**femoral canal**) containing
      i. Loose areolar tissue.
      ii. Lymph node and lymph vessels.

Function: It helps in free gliding of the femoral vessels in and out behind the inguinal ligament during the movements of the hip joint.

Changes with age: Femoral sheath is rudimentary in newborn and prolonged below the inguinal ligament after one year.

Structures piercing the sheath:
   a. **In front:**
      i. Superficial external pudendal artery.
      ii. Superficial epigastric artery.
      iii. Superficial circumflex iliac artery.
      iv. Deep external pudendal artery.
   b. **Laterally:** Femoral branch of genitofemoral nerve.
   c. **Medially:** Great saphenous vein.

Applied: Femoral hernia may occur through the femoral ring into the femoral canal.

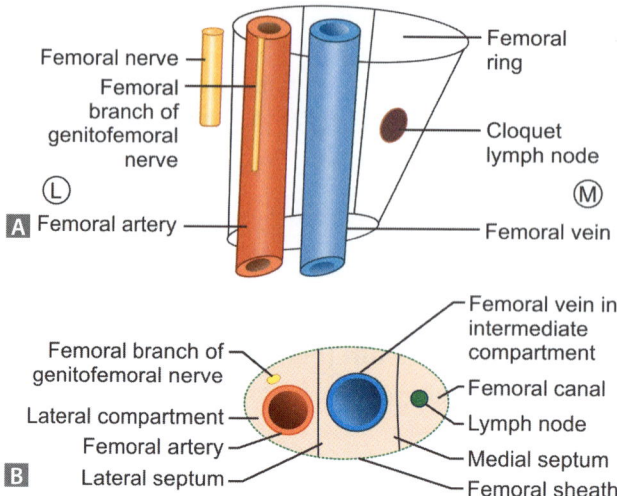

Figs 3.15A and B: Femoral sheath with its contents

The lateral compartment of the femoral sheath contains femoral artery. This artery is cleaned by using blunt dissection. **Three superficial branches** of femoral artery (**superficial external pudendal, superficial epigastric** and **superficial circumflex iliac**) arise just distal to the inguinal ligament. But do not attempt to follow these vessels.

The **largest** branch of femoral artery is the **arteria profunda femoris** which arises from the lateral side of the femoral artery about 3.5 cm below the inguinal ligament. Retract the femoral artery medially and identify the arteria profunda femoris. Very close to the femoral artery the profunda femoris artery gives off lateral and medial circumflex femoral artery. Identify these circumflex branches. The profunda femoris artery leaves the femoral triangle between pectineus and adductor longus muscle. Finally, this artery pierces the adductor magnus muscle as **the fourth perforating artery**. The femoral artery courses distally between the sartorius and adductor longus muscles.

Clean and identify the femoral vein and terminal part of great saphenous vein. Preserve this major vein but remove their tributaries to clear the dissection field.

8. **Floor of the femoral triangle** is cleaned by using blunt dissection. The floor of the triangle is formed by the following muscles from lateral to medial side (Figs 3.16A and B):
    i. Iliacus.
    ii. Psoas major.
    iii. Pectineus.
    iv. Adductor longus.

Window Dissections: Lower Limb (Inferior Extremity)

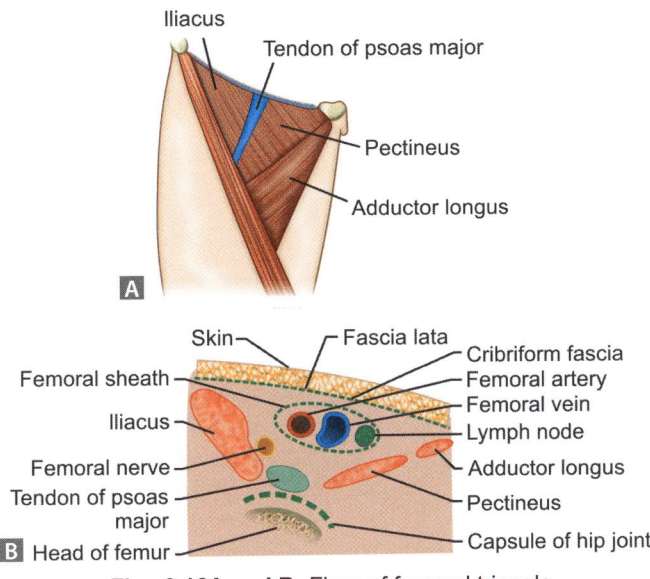

Figs 3.16A and B: Floor of femoral triangle

> **To expose of the obturator nerve the following steps of dissection are to be adopted.**
> i. Adductor longus muscle is divided transversely 2–3 cm below its origin and turn the distal part towards the femur. **Anterior division** of obturator nerve is exposed lying on the adductor brevis.
> ii. Then, adductor brevis is also cut transversely and reflected. **Posterior division** of the obturator nerve is exposed lying on the adductor magnus muscle.

MUSCLES RELATED TO FEMORAL TRIANGLE

Muscles	Origin	Insertion	Nerve supply	Action
Sartorius (Longest strap muscle)	i. Anterior superior iliac spine (ASIS) ii. Notch below ASIS	Upper part of the medial surface of the shaft of tibia	Anterior division of femoral nerve	i. Flexion, abduction and lateral rotation of the hip joint ii. Flexion and medial rotation of the knee in semiflexed position
Iliacus (Triangular muscle)	Upper 2/3rd of iliac fossa	About 2.5 cm below and in front of lesser trochanter of femur	From the trunk of femoral nerve (L_2, L_3)	Flexor of the hip joint

Contd...

Contd...

Muscles	Origin	Insertion	Nerve supply	Action
Psoas major (Fusiform muscle)	Medial part of anterior surface of transverse process and sides of bodies of all lumbar vertebrae (L_1-L_5)	Anterior surface of lesser trochanter of femur	Directly from lumbar plexus (L_2, L_3)	Chief flexor of the hip
Pectineus (Quadrilateral muscle)	i. Pectin pubis ii. Superior ramus of pubis	**Pectineal line** of the femur (line extending from the lesser trochanter to linea aspera)	i. Ventral stratum (lateral part) by the branch from trunk of femoral nerve ii. Dorsal stratum (medial part) by the anterior division of obturator nerve and accessory obturator nerve (if present)	Flexor and adductor of the hip
Adductor longus (Triangular in outline)	Anterior surface of symphysis pubis below the pubic tubercle.	Medial lip of **linea aspera** of the shaft of femur	Anterior division of obturator nerve	Adductor of thigh

Summary

1. Position of the cadaver: Supine, thigh extended and rotated laterally.
2. Skin incision as in Figure 3.10.
3. Superficial fascia and deep fascia are incised and reflected like skin.
4. **Boundaries** of the triangle are identified.
    i. Base (above): Inguinal ligament.
    ii. Laterally: Sartorius (medial border)
    iii. Medially: Adductor longus (medial border)
    iv. Apex: Junction of sartorius and adductor longus.
5. **Contents**:
    i. Femoral nerve and its branches.
    ii. Femoral sheath with its contents (femoral artery, femoral vein, lymphatics).
    iii. Terminal part of great saphenous vein.
    iv. Femoral branch of genitofemoral nerve.
    v. Lateral femoral cutaneous nerve.

PROBABLE QUESTIONS AND ANSWERS

1. **What are the boundaries of femoral triangle?**

Ans. **Laterally:** Medial border of sartorius.
    Medially: Medial border of adductor longus.

Base: Inguinal ligament.
Apex: Meeting point of sartorius and adductor longus.
Roof: Fascia lata (Deep fascia of the thigh)
Floor (From lateral to medial):
  i. Iliacus.
  ii. Psoas major.
  iii. Pectineus.
  iv. Adductor longus.

12. Why the deep fascia of the thigh is called fascia lata?

Ans. Lata (Latin) means broad. The deep fascia envelops the whole of the thigh and is so named because of its wide extent.

13. What are the specialized parts of fascia lata?

Ans.   i. Iliotibial tract.   ii. Gluteal aponeurosis.

14. What is saphenous opening?

Ans. It is an oval opening in the fascia lata. Its center lies about 3 cm below and lateral to the pubic tubercle.

15. What are the dimensions of the saphenous opening?

Ans. i. Length—about 3–4 cm.
ii. Breadth—about 2 cm.

16. What is cribriform fascia?

Ans. It is an areolar membrane which covers the saphenous opening.

17. What is the source of cribriform fascia?

Ans. Regarding its source there are two views:
  i. It is a part of the deep fascia of the thigh.
  ii. It is a part of the fascia of scarpa.

18. What are the structures that pierce the cribriform fascia?

Ans.   i. Great (long) saphenous vein.
  ii. Superficial external pudendal artery.
  iii. Superficial epigastric artery.
  iv. A twig from the medial femoral cutaneous nerve.
  v. Lymph vessels connecting the superficial and deep inguinal lymph nodes.

9. What is iliotibial tract?

Ans.   i. It is the thickened fascia lata along the lateral surface of the thigh.
      ii. It is about 2.5–5 cm wide.

10. What are the muscles that are inserted into the iliotibial tract?

Ans.   i. Gluteus maximus (3/4th of its fibers).
      ii. Tensor fascia lata.

11. How does the femoral vein begin?

Ans. Femoral vein begins as an upward continuation of the popliteal vein at the 5th osseo-aponeurotic opening of adductor magnus.

12. Where does the femoral vein end?

Ans. Femoral vein continues as the external iliac vein behind the inguinal ligament.

13. What are the important tributaries of femoral vein in the femoral triangle?

Ans.   i. Great saphenous vein.
      ii. Profunda femoris vein.

14. What are the important contents of the femoral triangle?

Ans.   i. Femoral artery and its branches.
      ii. Femoral vein and its tributaries.
     iii. Femoral nerve and its branches.
      iv. Lateral femoral cutaneous nerve.
       v. Femoral branch of genitofemoral nerve.
      vi. Deep inguinal lymph nodes.

15. What is the origin of femoral artery?

Ans. Femoral artery is the continuation of the external iliac artery.

16. How does the artery enter the femoral triangle?

Ans. The femoral artery enters the femoral triangle behind the inguinal ligament at the mid-inguinal point.

17. What is mid-inguinal point?

Ans. It is the midpoint of the line drawn between the anterior superior iliac spine and symphysis pubis.

18. What is the midpoint of inguinal ligament?

Ans. It is the midpoint between the attachments of the inguinal ligament (from anterior superior iliac spine to pubic tubercle).

19. What are the branches of the femoral artery?

Ans.
a. Branches in the femoral triangle:
  i. Superficial epigastric.
  ii. Superficial circumflex iliac.
  iii. Superficial external pudendal.
  iv. Deep external pudendal.
  v. Arteria profunda femoris (deep femoral artery).
  vi. Muscular branches.
b. Branch in the adductor canal
  i. Descending genicular artery.

20. What is the last branch of femoral artery?

Ans. Descending genicular artery.

21. What is the largest branch of the femoral artery?

Ans. Arteria profunda femoris.

22. What are the branches of arteria profunda femoris?

Ans.
  i. Lateral circumflex femoral artery.
  ii. Medial circumflex femoral artery.
  iii. Four perforating arteries (1st, 2nd, 3rd and 4th)

23. Where does the arteria profunda femoris arise from the femoral artery?

Ans. It arises from the lateral side of the femoral artery about 3.5 cm below the inguinal ligament.

24. Which of these perforating arteries provides nutrient artery of femur?

Ans. Usually the **second perforating artery** provides nutrient artery of femur. If there are two nutrient arteries, they are derived from first and third perforating arteries.

25. What is the fate of the arteria profunda femoris?

Ans. This artery leaves the femoral triangle between pectineus and adductor longus and finally pierces the adductor magnus as **fourth perforating artery** and anastomoses with superior muscular branch of popliteal artery.

126. What are the branches of lateral circumflex femoral artery?

Ans.
  i. Ascending branch.
  ii. Transverse branch.
  iii. Descending branch.

127. What are the branches of medial circumflex femoral artery?

Ans. i. Ascending branch.
  ii. Transverse branch.

128. What is the relation of spermatic cord with the external pudendal arteries?

Ans. The superficial external pudendal artery passes medially **in front** of the spermatic cord and the deep external pudendal artery passes **behind** the spermatic cord.

129. How does the femoral artery leave the femoral triangle?

Ans. The femoral artery leaves the triangle through its apex beneath the sartorius muscle and appears in the adductor canal.

130. Where does the femoral artery end?

Ans. It transverse through the adductor canal and then passes through the 5th osseo-aponeurotic opening of adductor magnus and appears in the popliteal fossa as **popliteal artery.**

131. What is the extent of femoral artery?

Ans. It extends from the base of the femoral triangle behind the inguinal ligament as a continuation of **external iliac artery** up to the 5th osseo-aponeurotic opening of adductor magnus where it continues as **popliteal artery**.

132. Does the femoral vein possess valves?

Ans. **Yes**. It is provided with valves above the opening of its each tributary.

133. How does the great saphenous vein begin?

Ans. Great saphenous vein begins as a continuation of the medial marginal vein of the foot.

134. Where does the great saphenous vein drain?

Ans. It drains into the femoral vein after passing through the saphenous opening which is about 3 cm below and lateral to the pubic tubercle.

35. What are the tributaries of the great saphenous vein near the saphenous opening.

Ans. Tributaries before piercing the cribriform fascia:
  i. Superficial epigastric vein.
  ii. Superficial circumflex iliac vein.
  iii. Superficial external pudendal vein.

Tributary after piercing the cribriform fascia.
  i. Deep external pudendal vein.

36. How many valves of the great saphenous vein are there?

Ans. About 10–20. Two of them are constant, one just before piercing the cribriform fascia and another at the saphenofemoral junction (**saphenofemoral valve**).

37. What is the relationship between the femoral artery and femoral vein in the femoral triangle?

Ans. The femoral vein lies behind the femoral artery at the apex of the triangle and then it becomes medial to the artery till it becomes external iliac vein behind the inguinal ligament.

38. What is femoral sheath?

Ans. It is a funnel-shaped fascial sheath around the upper part of the femoral vessels.

39. How the femoral sheath is formed?

Ans. Femoral sheath is formed anteriorly by the fascia transversalis and posteriorly by the fascia iliaca.

40. What is the length and breadth of the femoral sheath?

Ans. Length—about 2–4 cm.
Breadth—about 2 cm.

41. How many walls of the sheath are there?

Ans. There are four walls of the sheath. These are:
  i. Another wall (formed by fascia transversalis)
  ii. Posterior wall (formed by fascia iliaca).
  iii. Lateral wall (meeting of above two walls).
  iv. Medial wall (meeting of above two walls).

42. How many compartments of the femoral sheath are there?

Ans. **Three** compartments by two anteroposterior septa.
  i. Lateral (**arterial**) compartment.
  ii. Intermediate (**venous**) compartment.
  iii. Medial (**lymphatic**) compartment or femoral canal.

43. What are the contents of these compartments?

Ans. **Lateral compartment**
  i. Femoral artery.
  ii. Femoral branch of genitofemoral nerve.
Intermediate compartment
  i. Femoral vein.
Medial compartment
  i. One deep inguinal lymph node.
  ii. Lymph vessels.
  iii. Areolar tissue.

44. Name the structures that pierce the different walls of the femoral sheath?

Ans. a. Anterior wall is pierced by:
  i. Superficial external pudendal artery.
  ii. Superficial epigastric artery.
  iii. Superficial circumflex iliac artery.
  iv. Deep external pudendal artery.
 b. Lateral wall is pierced by:
  i. Femoral branch of genitofemoral nerve.
 c. Medial wall is pierced by:
  i. Great saphenous vein.
  ii. Lymph vessels.

45. What is femoral canal?

Ans. The medial compartment of the femoral sheath is called the femoral canal.

46. What is femoral ring?

Ans. The mouth of the femoral canal is known as femoral ring.

47. What is femoral septum?

Ans. It is the condensation of the extraperitoneal fatty tissue which closes the femoral ring.

48. What is femoral fossa?

Ans. It is a peritoneal depression which lies above the femoral septum.

49. What are the boundaries of the femoral ring?

Ans.
  i. **Anteriorly:** Inguinal ligament.
  ii. **Posteriorly:** Pectineus muscle with its fascia.
  iii. **Laterally:** Femoral vein.
  iv. **Medially:** Base of the lacunar ligament.

50. What is the applied importance of the femoral canal?

Ans. **Femoral hernia** may occur through the femoral ring into the femoral canal.

51. What is femoral hernia?

Ans. Protrusion of a part of a viscus out of the abdominal cavity into the femoral canal is called femoral hernia.

52. What is the function of the femoral sheath?

Ans. It allows the femoral vessels to glide freely beneath the inguinal ligament during the movements of the hip joint.

53. Why the femoral sheath is redimentary in newborn?

Ans. This is because of the fetal position of flexion. But gradually the sheath becomes lengthened when the extension of the thigh becomes habitual.

54. Is the femoral nerve a content of the femoral sheath?

Ans. No. It is outside the femoral sheath and lies in the iliopsoas groove in the femoral triangle.

55. Where do you get the femoral nerve in the femoral triangle?

Ans. It lies in the groove between the iliacus and psoas major muscles, outside the femoral sheath and lateral to the femoral artery.

56. What is the root value of the femoral nerve?

Ans. Dorsal divisions of the anterior primary rami of L_2, L_3, L_4 spinal nerves.

57. What are the branches of the femoral nerve from its trunk?

Ans.
  i. Branch to iliacus.
  ii. Nerve to pectineus.
  iii. Vascular branches.

58. How does the nerve to pectineus supply the muscle?

Ans. The nerve to pectineus arises from the medial side of the femoral nerve before its divisions and passes medially behind the femoral sheath to supply the **lateral part of pectineus** from its anterior surface.

59. What is the nerve supply of the medial part of petineus?

Ans. Anterior division of obturator nerve and accessory obturator nerve (if present).

60. What structure intervenes between the two divisions of femoral nerve?

Ans. Lateral circumflex femoral vessels.

61. What are the branches of anterior division of femoral nerve?

Ans.
- a. **Three cutaneous:**
    - i. Lateral intermediate femoral cutaneous.
    - ii. Medial intermediate femoral cutaneous.
    - iii. Medial femoral cutaneous nerve.
- b. **One muscular:**
    - i. Nerve to sartorius.

62. What are the branches from the posterior division of femoral nerve?

Ans.
- a. **Four muscular:**
    - i. Nerve to rectus femoris and hip joint via this nerve.
    - ii. Nerves to three vasti which also supply knee joint.
      (**Articularis genu** is supplied by the nerve to vastus intermedius)
- b. **One cutaneous:**
    - i. Saphenous nerve (**Longest cutaneous nerve of the body**).

63. What do you mean by quadriceps femoris?

Ans. **Four muscles** together are called quadriceps femoris. These are:
- i. Rectus femoris.
- ii. Vastus lateralis.
- iii. Vastus medialis.
- iv. Vastus intermedius.

64. What is the nerve supply of quadriceps femoris?

Ans. Posterior division of femoral nerve.

65. What is articularis genu?

Ans. It is a muscle which arises by two or more heads from the anterior surface of the lower part of the shaft of the femur and inserts into the suprapatellar bursa of the knee joint.

Window Dissections: Lower Limb (Inferior Extremity)

66. What is its nerve supply?

Ans. It is supplied by the nerve to vastus intermedius.

67. What is its action?

Ans. It pulls the suprapatellar bursa upwards during extension of the knee joint and prevents damage to it.

68. What is the nerve supply of sartorius?

Ans. Anterior division of femoral nerve.

69. What is the origin and insertion of sartorius?

Ans. Origins:
  i. Anterior superior iliac spine.
  ii. Notch below the spine.
Insertion: Upper part of the medial surface of the shaft of the tibia.
(**Upright hockey-stick like** insertion)

70. Name the other muscles that are inserted along with the sartorius?

Ans. Insertions of gracilis and semitendinosus behind the insertion of sartorius.

71. What are the nerve supply of these muscles?

Ans. **Gracilis** by anterior division of **obturator nerve**, **semitendinosus** by the tibial part of **sciatic nerve** and **sartorius** by the anterior division of **femoral nerve**.

Note that these three muscles are inserted into the same area of the tibia but arise from three different parts of the hip bone (**Sartorius** from **ilium, semitendinosus** from **ischium** and **gracilis** from **pubis**) and are supplied by three different nerves.

72. What is the other name of sartorius muscle?

Ans. Tailor's muscle [Sartor (Latin) = Tailor]

73. What is the nerve supply of iliacus and psoas major muscles?

Ans. **Iliacus** from the trunk of the femoral nerve (L_2, L_3)
Psoas major directly from the lumbar plexus (L_2, L_3).

74. Why the term iliopsoas is commonly used?

Ans. Because of the facts that:
  i. They are covered in front by a common fascia (fascia iliaca)
  ii. They have common insertion (lesser trochanter of femur)

iii. They possess common functions (flexor of the hip).
iv. Both are supplied by common spinal segments (L_2, L_3).

75. Why the pectineus is called a composite or hybrid muscle?

Ans. Because of its double nerve supply.
Lateral (flexor) part by femoral nerve and medial (adductor) part by obturator nerve.

76. Name the muscles forming the floor of the femoral triangle?

Ans. From lateral to medial:
i. Iliacus.
ii. Psoas major.
iii. Pectineus.
iv. Adductor longus.

77. What is rider's bone?

Ans. A sesamoid bone, developed in the tendinous origin of adductor longus.

78. What is the inguinal ligament?

Ans. It is the condensation of the lower part of the aponeurosis of the external oblique muscle.

79. What is the other name of this ligament?

Ans. Poupart's ligament.

80. What is its average length?

Ans. About 10–14 cm.

81. What are its attachments?

Ans. Laterally: Anterior superior iliac spine.
Medially: Pubic tubercle.

82. What are the different processes/parts of inguinal ligament?

Ans.
i. Lacunar ligament or Gimbernat's ligament (pectineal part of inguinal ligament)
ii. Ligament of Cooper (pectineal ligament)
iii. Reflected part of inguinal ligament.

83. What are the lymph nodes in the femoral triangle?

Ans.
i. **Superficial inguinal lymph nodes** [horizontal group (5–6) and vertical group (4–5), arranged in T-shaped manner].
ii. **Deep inguinal lymph nodes** (1–3).

Window Dissections: Lower Limb (Inferior Extremity)

84. What is Cloquet's lymph node?

Ans. One deep inguinal lymph node, lying in the femoral septum at the femoral ring.

85. Which one is the most constant deep inguinal lymph node?

Ans. The node at the saphenofemoral junction.

86. What is the nerve supply of adductor longus?

Ans. Anterior division of obturator nerve.
(All adductors of thigh are supplied by the obturator nerve).

87. What are the coverings of femoral hernia?

Ans. From within outwards:
   i. Parietal peritoneum.
   ii. Femoral septum.
   iii. Anterior wall of femoral sheath.
   iv. Cribriform fascia.
   v. Superficial fascia.
   vi. Skin.

LESSON 3: ADDUCTOR CANAL

Adductor canal is a musculoaponeurotic tunnel, about 15 cm in length, triangular on cross-section, present in the middle-third of medial side of the thigh, extending from the apex of the femoral triangle to the fifth osseoaponeurotic opening of the adductor magnus, providing the passage for femoral vessels.

STEPS OF DISSECTION

1. **Position of the cadaver:** Body supine, thigh extended and laterally rotated.
2. **Skin incision (Fig. 3.17):**
   i. A transverse incision at the junction of upper third and middle third of the front of the thigh (A–B)
   ii. Another transverse incision at the junction of the middle third and lower third of the front of the thigh (C–D)

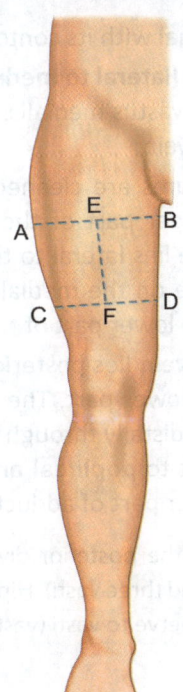

Fig. 3.17: Dissection of adductor canal (skin incision)

iii. A vertical incision from the midpoint of (A–B) to the midpoint of (C–D) which is indicated by the line (E–F)

Reflex the flaps of skin laterally and medially.

Superficial fascia is exposed.

3. **Superficial fascia:** In the superficial fascia, the following structures are found:
    i. Great saphenous vein.
    ii. Medial femoral cutaneous nerve.
    iii. Medial intermediate femoral cutaneous nerve.
    iv. Few lymph vessels.

   Identify these structures and clean them thoroughly. Superficial fascia is cut and reflected like skin. **Deep fascia is exposed**.

4. **Deep fascia:** Cut the deep fascia in the line of skin incision and reflect like that of skin.
5. **Sartorius muscle** is exposed. Use your finger to separate the sartorius muscle from the deep fascia. Retract the sartorius muscle laterally.
6. **Aponeurotic roof** is exposed. This roof is formed by a strong fascia which extends from the anterior border of vastus medialis up to the medial border of adductor longus and adductor magnus muscles. On this roof lies the sub-sartorial plexus of nerve. It means that the sub-sartorial plexus intervenes between the sartorius and aponeurotic roof of adductor canal.
7. Aponeurotic roof is incised longitudinally and retracted to expose the adductor canal with its contents.
8. **Adductor canal** with its contents are exposed (Fig. 3.18)

   Contents are (lateral to medial):
    i. Nerve to vastus medialis.
    ii. Saphenous nerve.
    iii. Femoral vein.
    iv. Femoral artery.

These structures are cleaned thoroughly and identified. The **nerve to vastus medialis** traverses the upper part of the canal and then supplies the vastus medialis muscle. The **saphenous nerve** lies lateral to the femoral artery in the upper part, then crosses the artery superficially to lie on the medial side of the artery in the lower part of the canal. Then the nerve pierces the lower part of the aponeurotic roof and comes out of the canal.

The **femoral vein** lies posterior to the femoral artery at its upper part and then lateral to the artery at its lower part. The **femoral artery** is dissected by using blunt dissection. Then follow the artery distally through the **adductor hiatus** (fifth osseoaponeurotic opening) where its name changes to **popliteal artery**. Find out the **descending genicular branch of femoral artery** in the lower part of adductor canal.

> **Remember** that the posterior division of **femoral nerve** supplies **quadriceps femoris muscle** (rectus femoris and three vasti). Hip joint is supplied via the nerve to rectus femoris and knee joint is supplied via the nerve to vasti (vasti medialis, vasti lateralis and vasti intermedius).

Window Dissections: Lower Limb (Inferior Extremity)

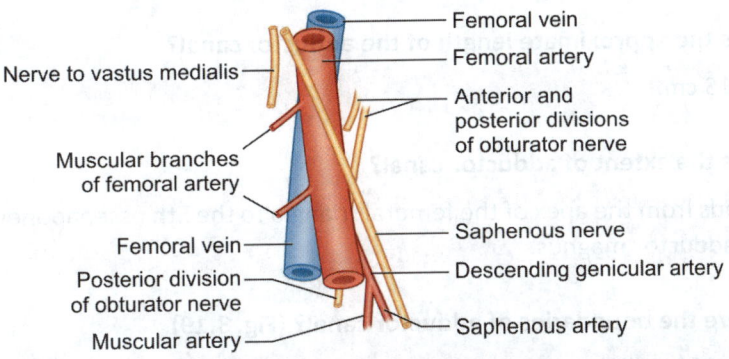

Fig. 3.18: Contents of adductor canal

Summary

1. Position of the body: Supine, thigh extended and laterally rotated.
2. Skin incision as in Figure 3.17.
3. Superficial fascia and deep fascia are reflected in the line of skin incision.
4. Aponeurotic roof is incised longitudinally and retracted.
5. Adductor canal with its contents are exposed:

Contents (lateral to medial):
 i. Nerve to vastus medialis.
 ii. Saphenous nerve.
 iii. Femoral vein.
 iv. Femoral artery.

PROBABLE QUESTIONS AND ANSWERS

1. What is adductor canal?

Ans. Adductor canal is an intermuscular space, triangular on cross-section, situated on the medial side of the middle-third of the thigh.

2. What are the other names for the adductor canal?

Ans. **Sab-sartorial canal** or **Hunter's canal**.

3. Who was Hunter?

Ans. **John Hunter** was an anatomist and surgeon in London. Hunter's operation is popular in vascular surgery for the treatment of popliteal aneurysm by ligating the femoral artery in the adductor canal.

14. What is the approximate length of the adductor canal?

Ans. About 15 cm.

15. What is the extent of adductor canal?

Ans. It extends from the apex of the femoral triangle to the 5th osseoaponeurotic opening of the adductor magnus.

16. What are the boundaries of adductor canal? (Fig. 3.19).

Ans.
a. **Roof** (medial wall):
  i. A strong fascia which extends across the vastus medialis to adductor longus (**aponeurotic roof**).
  ii. Sub-sartorial plexus.
  iii. Sartorius muscle.
b. **Floor** (posterior wall):
  i. Adductor longus in the upper part.
  ii. Adductor magnus in the lower part.
c. **Anterolateral wall**: Vastus medialis.
d. **Apex**: Meeting point of vastus medialis and adductor muscles on the linea aspera of femur.

17. What is sub-sartorial plexus?

Ans. It is a nerve plexus which lies on the aponeurotic roof of the adductor canal under cover of sartorius.

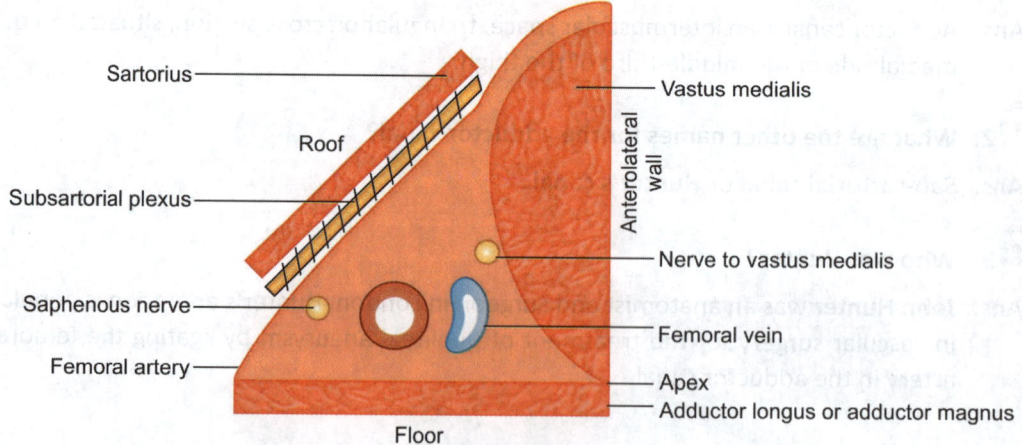

Fig. 3.19: Boundary of adductor canal with its contents

Window Dissections: Lower Limb (Inferior Extremity)

8. How the sub-sartorial plexus is formed?

Ans. It is formed by:
   i. A branch from saphenous nerve.
   ii. A twig from posterior division of medial femoral cutaneous nerve.
   iii. A branch from anterior division of obturator nerve.

9. What are the contents of adductor canal?

Ans.
   i. Femoral vessels (femoral artery and femoral vein).
   ii. Saphenous nerve.
   iii. Nerve to vastus medialis.
   iv. Posterior division of the obturator nerve.
   v. Descending genicular artery (**last branch of femoral artery**).

10. How does the femoral artery leave the adductor canal?

Ans. The femoral artery passes through the 5th osseoaponeurotic opening of adductor magnus and continues as popliteal artery.

11. Name the structures that pass through the 5th osseoaponeurotic opening of adductor magnus?

Ans.
   i. Femoral artery.
   ii. Femoral vein.
   iii. Posterior division of obturator nerve.

12. What are the structures passing through the first to fourth osseoaponeurotic openings of adductor magnus?

Ans. First to fourth perforating arteries, branches of arteria profunda femoris pass through their corresponding osseoaponeurotic openings.

13. What is the relation of the femoral artery with the femoral vein in the adductor canal?

Ans. Femoral vein lies behind the femoral artery in the upper part and posterolateral to the artery in the lower part of the canal.

14. What is the position of the nerve to vastus medialis in relation to femoral artery?

Ans. This nerve lies lateral to the femoral artery in the upper part of the adductor canal.

15. How do the anterior and posterior divisions of obturator nerve end?

Ans. Anterior division ends by supplying the femoral artery and posterior division ends by supplying the knee joint.

16. What are the contents of adductor region?

Ans.
a. **Muscles**:
   i. Pectineus (medial part).
   ii. Adductor longus.
   iii. Gracilis.
   iv. Adductor brevis.
   v. Obturator externus.
   vi. Adductor magnus.
b. **Nerve**: Obturator nerve.
c. **Arteries**:
   i. Arteria profunda femoris.
   ii. Obturator artery.

Remember that all the muscles of the adductor region are supplied by the obturator nerve except lateral part of pectineus and ischial fibers (extensor part) of adductor magnus which are supplied by femoral nerve and tibial part of sciatic nerve respectively.

17. What is adductor minimus?

Ans. The pubic fibers of adductor magnus muscle are sometimes called adductor minimus.

18. How the patellar plexus is formed?

Ans. It is formed by:
   i. Infrapatellar branch of saphenous nerve.
   ii. Anterior division of medial femoral and lateral femoral cutaneous nerves.
   iii. Lateral and medial intermediate femoral cutaneous nerves.

LESSON 4: ANTEROLATERAL COMPARTMENT OF THE LEG

Anterior compartment of the leg (anterior crural compartment) is also called **extensor compartment** of the leg. This compartment is bounded by (Fig. 3.20):
 i. Anteriorly: Deep fascia of the leg (fascia cruris)
 ii. Posteriorly: Interosseous membrane.
 iii. Laterally: Extensor surface of fibula and anterior intermuscular septum.
 iv. Medially: Extensor surface (lateral surface) of the shaft of tibia.

Lateral compartment (peroneal compartment) is bounded by (Fig. 3.20)
 i. Anteriorly: Anterior intermuscular septum.
 ii. Posteriorly: Posterior intermuscular septum.
 iii. Laterally: Deep fascia of the leg.
 iv. Medially: Lateral surface of the shaft of fibula.

Window Dissections: Lower Limb (Inferior Extremity)

Anterior and lateral compartment together are called anterolateral compartment of the leg.

STEPS OF DISSECTION

1. **Position of the cadaver:** Body is supine, knee extended and thigh rotated medially.
2. **Skin incision (Fig. 3.21):**
   i. A transverse incision from tibial tuberosity to head of fibula (A–B)

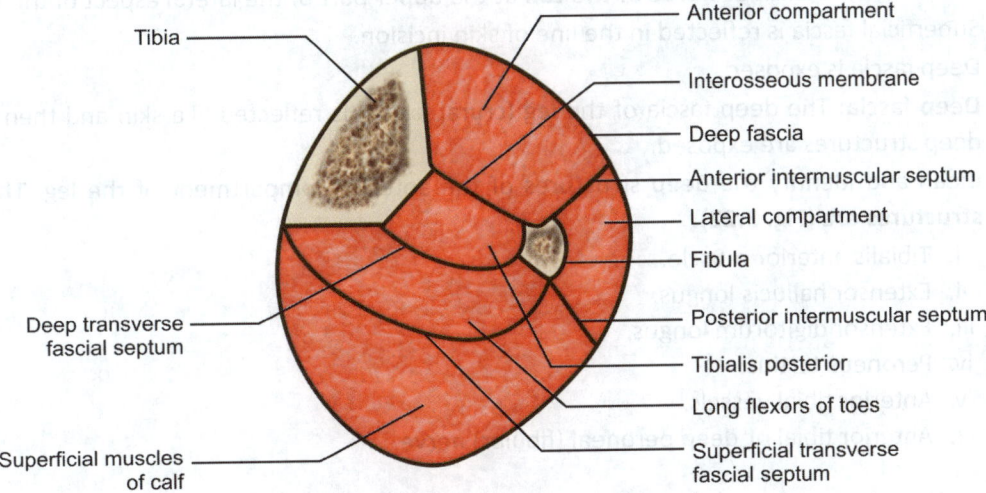

Fig. 3.20: Compartments of leg

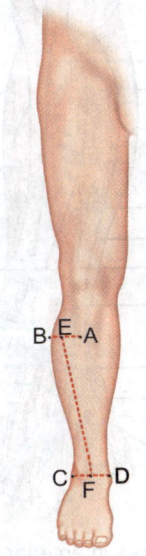

Fig. 3.21: Dissection of anterior crural region (skin incision)

ii. Another transverse incision from medial to lateral malleolus (C–D)
iii. A vertical incision from the midpoint of (A–B) to the midpoint of (C–D) which is indicated by the line (E–F).

Flaps of skin are reflected laterally and medially.

The superficial fascia is exposed.

3. **Superficial fascia:** In the superficial fascia the following structures are found:
   i. **Musculocutaneous nerve.** It pierces the deep fascia at the junction of the upper 2/3rd and lower 1/3rd of the leg and becomes cutaneous.
   ii. **Lateral cutaneous nerve of the calf** at the upper part of the lateral aspect of the leg.

Superficial fascia is reflected in the line of skin incision.

Deep fascia is exposed.

4. **Deep fascia:** The deep fascia of the leg (crural fascia) is reflected like skin and then the deep structures are exposed.
5. Clean and identify the deep structures of the anterior compartment of the leg. **These structures are (Fig. 3.22):**
   i. Tibialis anterior muscle.
   ii. Extensor hallucis longus.
   iii. Extensor digitorum longus.
   iv. Peroneus tertius.
   v. Anterior tibial vessels.
   vi. Anterior tibial or deep peroneal (fibular) nerve.

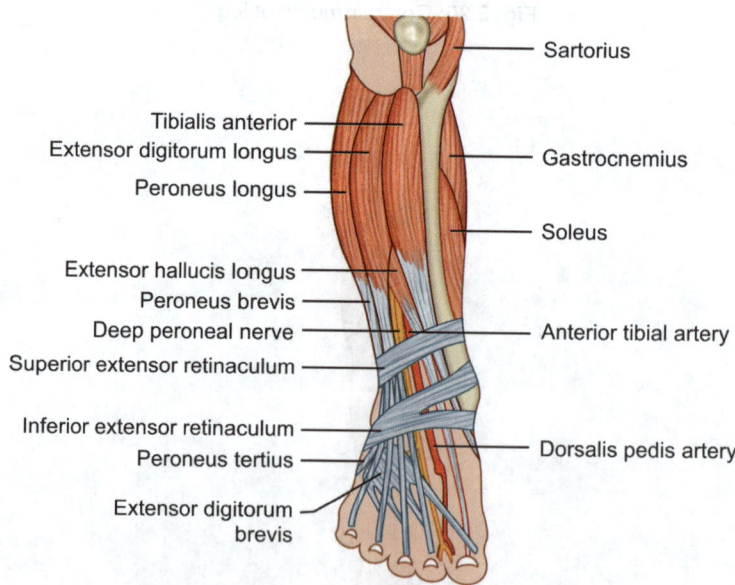

Fig. 3.22: Structures in anterior crural region

The deep structures of the lateral compartment of the leg are (Fig. 3.23):
  i. Peroneus longus.
 ii. Peroneus brevis.
iii. Musculocutaneous (superficial peroneal) nerve.

Tibialis anterior muscle is situated at the medial aspect of the anterior compartment.

The **extensor hallucis longus** lies between tibialis anterior and the extensor digitorum longus. In fact, this muscle is partly hidden by these two muscles in the anterior compartment.

The **extensor digitorum longus** lies on the lateral aspect of the anterior compartment of the leg.

The **peroneus tertius** muscle lies at the lower and lateral part of anterior crural compartment. Its tendon passes downwards deep to the superior extensor retinaculum. In the lateral compartment, the **peroneus longus** lies superficial to the **peroneus brevis muscle.**

Identify the **anterior tibial artery** at the level of superior extensor reticulum. Trace the artery proximally. Separate the extensor digitorum muscle forcibly from the tibialis anterior muscle by using your fingers. Then follow the artery proximally between these two muscle bellies. Now, clean the artery by using a probe. Note that the artery passes over the superior border of the interosseous membrane and lies directly on the anterior surface of the interosseous membrane. As the artery crosses the ankle joint, its name changes to **dorsalis pedis artery**. Remember that the anterior tibial artery is accompanied by **venae comitantes**.

The **anterior tibial nerve (deep peroneal nerve)** is one of the terminal divisions of the **common peroneal** (lateral popliteal) **nerve** on the lateral side of the neck of the fibula. It descends on the interosseous membrane. It intervenes between the tibialis anterior and extensor digitorum longus in the proximal third of the leg but in the distal third of the leg

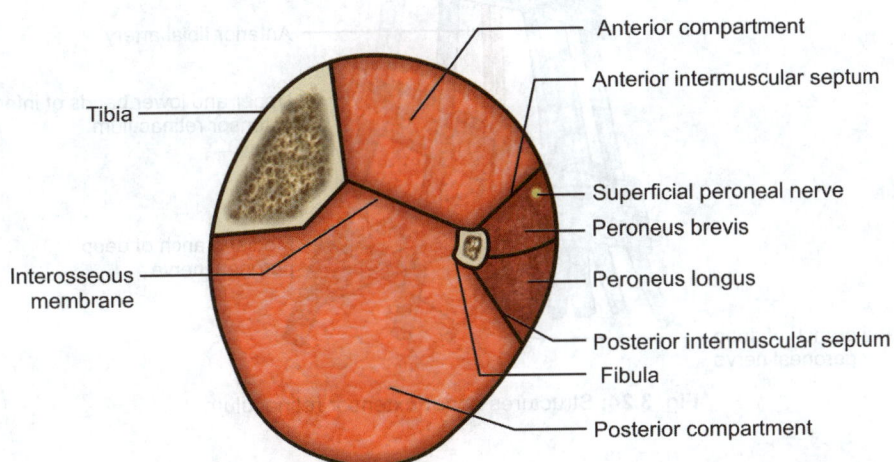

Fig. 3.23: Structures in lateral compartment of leg

the neurovascular bundle (nerve and blood vessels) lies between the tendon of extensor digitorum longus and extensor hallucis longus. Throughout its course the anterior tibial nerve lies lateral to the anterior tibial vessels except in the middle third of the leg where the nerve lies in front of the vessels. This nerve supplies the muscle of the anterior compartment of the leg. It terminates by dividing into lateral and medial branches.

The **musculocutaneous (superficial peroneal)** nerve is another terminal branch of common peroneal nerve on the lateral side of the neck of the fibula between the two heads of peroneus longus. It lies between the peroneus longus and extensor digitorum longus in the lateral crural compartment. It supplies the **peroneus longus** and **peroneus brevis** in the lateral compartment and then becomes cutaneous after piercing the deep fascia at the junction of middle and lower third of the leg.

> **Note the structures** which pass under cover of extensor retinaculum and cross the anterior surface of the ankle joint (Fig. 3.24) (from medial to lateral):
> i. Tibialis anterior tendon.
> ii. Extensor hallucis longus tendon.
> iii. Anterior tibial vessels.
> iv. Anterior tibial (deep peroneal) nerve.
> v. Extensor digitorum longus tendon.
> vi. Peroneus tertius tendon.

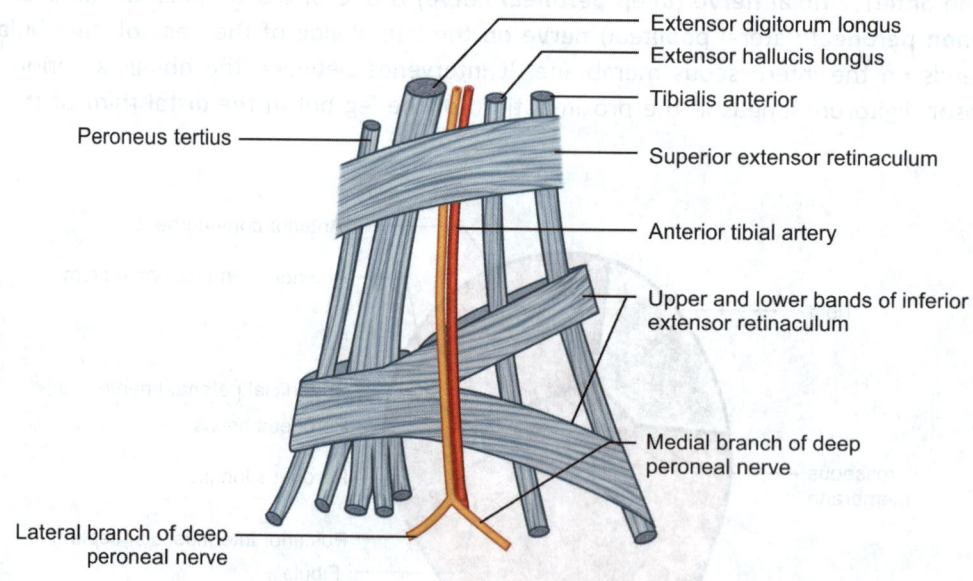

Fig. 3.24: Structures under extensor retinaculum

MUSCLES RELATED TO ANTEROLATERAL COMPARTMENT OF LEG

Muscles	Origins	Insertion	Nerve supply	Actions
1. **Tibialis anterior** (spindle-shaped and multipennate)	i. Upper 2/3rd of the lateral surface of the shaft of the tibia ii. Lateral condyle of the tibia iii. Interosseous membrane	Medial part of under surface of the medial cuneiform and 1st metatarsal bone	i. Deep peroneal nerve ii. Recurrent genicular branch of common peroneal nerve	i. Dorsiflexion of ankle joint ii. Inversion of foot
2. **Extensor hallucis longus** (unipennate)	i. Middle 2/4th of the medial surface of the shaft of the fibula ii. Interosseous membrane	Base of distal phalanx of great toe	Deep peroneal nerve	i. Extension of great toe ii. Dorsiflexion of foot
3. **Extensor digitorum longus**	i. Upper 3/4th of the medial surface of the shaft of the fibula ii. Lateral condyle of tibia iii. Interosseous membrane	i. The medial slip in inserted into the base of middle phalanx ii. Two collateral slips unite and attach to the base of distal phalanx	Deep peroneal nerve	i. Extension of lateral four toes ii. Dorsiflexion of foot
4. **Peroneus tertius**	Lower 1/4th of the medial surface of the shaft of the fibula	Dorsal surface of the base of 5th metatarsal bone	Deep peroneal nerve	Dorsiflexion and eversion of the foot
5. **Peroneus longus** (bipennate muscle)	i. Upper 2/3rd of the peroneal surface of the shaft of fibula ii. lateral condyle of tibia	Lateral part of undersurface of the medial cuneiform and 1st metatarsal bone	Superficial peroneal nerve	i. Eversion of foot ii. Maintains the lateral longitudinal and transverse arches of the foot
6. **Peroneus brevis** (bipennate muscle)	Lower 2/3rd of the peroneal surface of the shaft of fibula	Base of the 5th metatarsal bone	Superficial peroneal nerve	i. Eversion of foot ii. Maintains the lateral longitudinal arch of the foot

> **Summary**
> 1. Position of the body: Supine, knee extended, thigh rotated medially.
> 2. **Skin incision** as in Figure 3.21.
> 3. Superficial fascia and deep fascia are reflected like skin.
> 4. Structures of the **anterior compartment** are exposed and identified. These are:
>   i. Tibialis anterior.
>   ii. Extensor hallucis longus.
>   iii. Extensor digitorum longus.
>   iv. Anterior tibial vessels.
>   v. Anterior tibial nerve (deep peroneal nerve).
>   vi. Peroneus tertius.
> 5. Structures of the lateral compartment are exposed and identified.
>   **These are**:
>   i. Peroneus longus.
>   ii. Peroneus brevis.
>   iii. Superficial peroneal or musculocutaneous nerve.

PROBABLE QUESTIONS AND ANSWERS

1. How many compartments of the leg are there?

Ans. There are three compartments of the leg:
   i. Anterior compartment.
   ii. Lateral compartment.
   iii. Posterior compartment.

2. How these compartments are divided?

Ans. The anterior and posterior intermuscular septa are formed by the extensions of deep fascia of the leg and are attached to the anterior and posterior border of fibula respectively, thus dividing the leg into three compartments.

3. What are the muscles of the anterior crural region?

Ans.
   i. Tibialis anterior.
   ii. Extensor digitorum longus.
   iii. Extensor hallucis longus.
   iv. Peroneus tertius.

4. What is the nerve supply of these muscles?

Ans. All these muscles are supplied by **anterior tibial** or **deep peroneal** nerve.

Window Dissections: Lower Limb (Inferior Extremity)

5. What is the origin of the anterior tibial nerve?

Ans. The anterior tibial nerve arises from the bifurcation of the common peroneal nerve on the lateral side of the neck of the fibula.

6. What is the other branch of common peroneal nerve?

Ans. **Superficial peroneal** or **musculocutaneous** nerve.

7. Where does the superficial peroneal nerve supply?

Ans. It supplies the **muscles** of the lateral compartment of the leg and the **skin** of the lower part of the leg and most of the skin of the dorsum of the foot.

8. What are the muscles of the lateral compartment of the leg?

Ans.   i. Peroneus longus.
        ii. Peroneus brevis.

9. How does the anterior tibial nerve reach the anterior compartment?

Ans. The anterior tibial nerve initially lies in the lateral (peroneal) compartment deep to the peroneus longus. Then it enters the anterior compartment by piercing the anterior intermuscular septum and deep to the fibers of extensor digitorum longus.

10. How does the superficial peroneal nerve passes through the lateral compartment and become cutaneous?

Ans. The superficial peroneal nerve is one of the terminal branches of common peroneal nerve (other terminal branch is anterior tibial or deep peroneal nerve). Initially it lies deep to the peroneus longus muscle and then passes downwards between the peroneus longus and peroneus brevis. **It becomes cutaneous after piercing the deep fascia at the junction of middle and lower third of the leg.**

11. Why the deep peroneal nerve is also called nervous hesitans?

Ans. The deep peroneal nerve lies lateral to the anterior tibial artery in the upper third of the leg. Then the nerve lies anterior to the artery in the middle third of the leg. Again the nerve becomes lateral to the artery in the lower third of the leg. It seems that as if the nerve is hesitating to cross the artery. So the deep peroneal nerve is called the **nervous hesitans**.

12. What is the main artery of the anterior compartment of the leg?

Ans. Anterior tibial artery.

13. What is the origin of anterior tibial artery?

Ans. It is one of the two terminal branches of the popliteal artery at the lower border of popliteus muscle (other terminal branch is posterior tibial artery).

14. How does the anterior tibial artery enter the anterior compartment?

Ans. The artery passes through a gap above the interosseous membrane and appears in the anterior compartment on the medial side of the neck of the fibula.

15. What is venae comitantes of anterior tibial artery?

Ans. While running in the anterior crural region the anterior tibial artery is accompanied by a pair of veins on its either side which is termed as venae comitantes.

16. What are the branches of anterior tibial artery?

Ans.
 i. Anterior tibial recurrent artery.
 ii. Posterior tibial recurrent artery.
 iii. Muscular branches.
 iv. Medial malleolar artery.
 v. Lateral malleolar artery.

17. What is its terminal branch?

Ans. The anterior tibial artery appears on the dorsum of the foot as **arteria dorsalis pedis**.

18. What is retinaculum of the leg?

Ans. The deep fascia of the leg is thickened around the ankle to form bands called retinacula.

19. Why these are so called?

Ans. These are so called because they retain the tendons in place.

20. How many retinacula are found around the ankle?

Ans.
 a. **In front of ankle:**
  i. Superior extensor retinaculum.
  ii. Inferior extensor retinaculum.
 b. **Laterally**:
  i. Superior peroneal retinaculum.
  ii. Inferior peroneal retinaculum.
 c. **Posteromedially**: Flexor retinaculum (**laciniate ligament**).

Window Dissections: Lower Limb (Inferior Extremity)

21. What is the shape and attachments of the inferior extensor retinaculum?

Ans. **Y-shaped** (Fig. 3.24).
Stem is attached to the superior surface of the calcaneum.
Upper band is attached to medial malleolus.
Lower band is attached to plantar aponeurosis.

22. What are the structures passing in front of the ankle and deep to the superior extensor retinaculum?

Ans. From medial to lateral:
  i. **T**ibialis anterior.
  ii. Extensor **H**allucis longus.
  iii. Anterior tibial **V**ein.
  iv. Anterior tibial **A**rtery.
  v. Anterior tibial **V**ein.
  vi. Deep peroneal **N**erve.
  vii. Extensor **D**igitorum longus.
  viii. Peroneus **T**ertius.
Remember the mnemonic: "To Have Vivid Answers Vide N Dey's Text".

23. What is the action of peroneal muscles?

Ans. Peroneal muscles (peroneal longus, peroneus brevis and peroneus tertius) are the **evertors** of the foot.

24. What are the muscles of inversion of the foot?

Ans. **Tibialis anterior** and **tibialis posterior** muscles are the invertors of the foot.

25. What are the joints involved in inversion and eversion of foot?

Ans. The joints of inversion and eversion are:
  a. Subtalar (talocalcaneal) joint.
  b. Midtarsal: (i) Talocalcaneonavicular and (ii) Calcaneocuboid joints.

26. What is the axis of these movements?

Ans. Oblique axis, passing upwards, forwards and medially from the heel to the neck of the talus through the **sinus tarsi**.

Remember that the term **'peroneal'** is now replaced by the term **'fibular'**. For example, common **peroneal** nerve as common **fibular** nerve, deep **peroneal** nerve as deep **fibular** nerve, superficial **peroneal** nerve as superficial fibular nerve, peroneus longus, brevis, tertius as fibularis longus, brevis, tertius respectively.

> Remember that the fibularis tertius muscle is absent in about 5% of specimens (Grant's dissector).

LESSON 5: DORSUM OF THE FOOT

The dorsum of the foot contains little fat in its subcutaneous layer. So, in most of the people the superficial veins and dorsal venous arch are easily seen. The superficial fascia also contains superficial lymphatics and cutaneous nerves.

STEPS OF DISSECTION

1. **Position of the cadaver:** Body supine, ankle plantar-flexed.
2. **Skin incision (Fig. 3.25):**
    i. A transverse incision from medial to lateral malleolus (A–B)
    ii. Another transverse incision along the roots of the toes (C–D)
    iii. A third longitudinal incision from the midpoint of the proximal incision (A–B) to the tip of the 2nd toe (E–F)

    The skin flaps are reflected sideways.
    Superficial fascia is exposed.
3. **Superficial fascia:** In the superficial fascia the following structures are found. Clean and identify them.
    i. Dorsal venous arch and commencement of long and short saphenous veins (Fig. 3.26)
    ii. Cutaneous nerves (Fig. 3.27) such as:

Fig. 3.25: Dissection of dorsum of foot (skin incision)

Window Dissections: Lower Limb (Inferior Extremity)

a. Lateral and medial branches of superficial peroneal nerve.
b. Saphenous nerve along the medial border of the foot.
c. Sural nerve along the lateral border of the foot.
d. Medial terminal branch of deep peroneal nerve.

Superficial fascia is reflected in the line of skin incision.

Deep fascia is exposed.

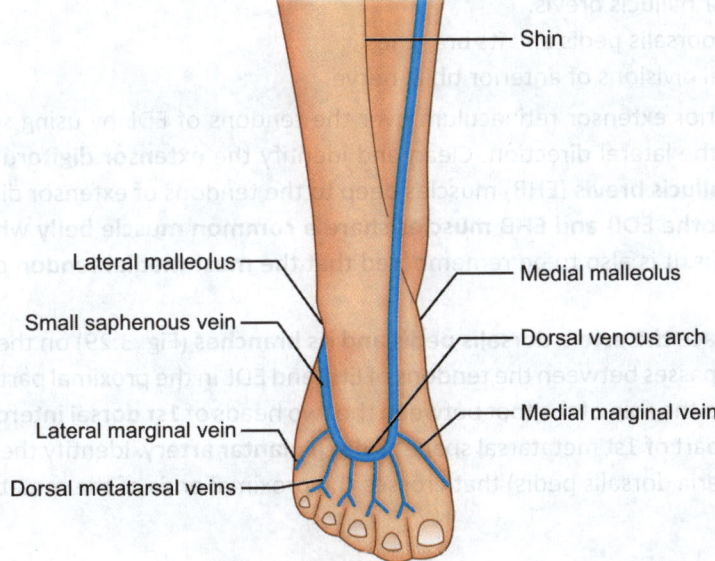

Fig. 3.26: Dorsal venous arch

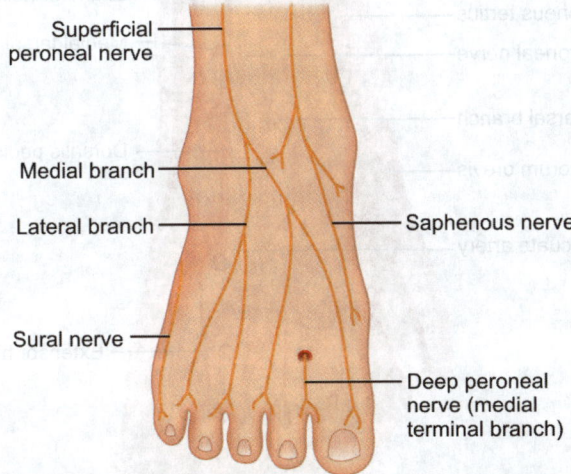

Fig. 3.27: Cutaneous nerves on dorsum of foot

4. **Deep fascia:** Deep fascia is also reflected in the line of skin incision. Now, the structures are exposed deep to the deep fascia. These are (Fig. 3.28):
   i. Tendon of tibialis anterior.
   ii. Tendon of extensor hallucis longus (EHL).
   iii. Tendon of extensor digitorum longus (EDL).
   iv. Tendon of peroneus tertius.
   v. Extensor digitorum brevis.
   vi. Extensor hallucis brevis.
   vii. Arteria dorsalis pedis and its branches.
   viii. Terminal divisions of anterior tibial nerve.

Cut the inferior extensor retinaculum over the tendons of EDL by using scissors. Retract the tendons in the lateral direction. Clean and identify the **extensor digitorum brevis** (EDB) and **extensor hallucis brevis** (EHB) muscles deep to the tendons of extensor digitorum longus (EDL). **Note that the EDB and EHB muscles share a common muscle belly** which is attached to the calcaneous. It is also to be remembered that the **most medial tendon of EDB is called EHB**.

Clean and trace **the arteia dorsalis pedis and its branches** (Fig. 3.29) on the dorsum of the foot. The artery passes between the tendons of EHL and EDL in the proximal part of the dorsum. Distally, it goes to the sole of the foot between the two heads of **1st dorsal interosseous muscle** in the proximal part of 1st metatarsal space as **deep plantar artery.** Identify the **arcuate artery** (a branch of arteria dorsalis pedis) that crosses the proximal ends of the metatarsal bones.

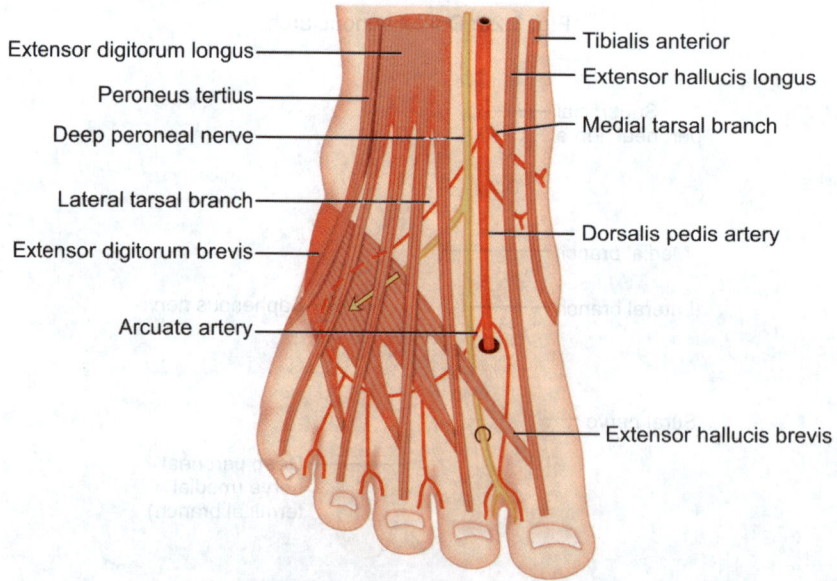

Fig. 3.28: Structures on dorsum of foot

Window Dissections: Lower Limb (Inferior Extremity)

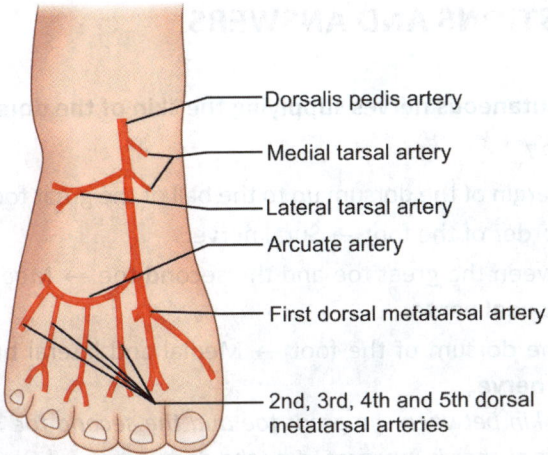

Fig. 3.29: Arteria dorsalis pedis

MUSCLES RELATED TO THE DORSUM OF THE FOOT (INTRINSIC MUSCLES OF THE DORSUM)

Muscles	Origin	Insertion	Nerve supply	Actions
1. **Extensor digitorum brevis (EDB)** 2. **Extensor hallucis brevis (EHB)**	i. Upper surface of calcaneus ii. Inferior extensor retinaculum	i. Medial most tendon of EDB is called EHB which is inserted on the dorsal aspect of the proximal phalanx of the great toe ii. Other three tendons are inserted on the dorsal surface of the bases of middle and terminal phalanx of 2nd, 3rd and 4th toes via dorsal digital expansion	Lateral terminal branch of deep peroneal nerve which is having a **pseudo-ganglia**	Dorsiflexion of medial four toes in dorsiflexed position of the ankle

Summary
1. Position of the body: Supine and ankle plantar-flexed.
2. Skin incision as in Figure 3.25.
3. Superficial fascia is exposed with its superficial veins and cutaneous nerves.
4. Deep fascia is exposed and reflected like skin.
5. **Contents**:
   i. Tendon of tibialis anterior (most medially).
   ii. Tendons of EHL, EDL and PT.
   iii. Intrinsic muscles of dorsum: EDB and EHB.
   iv. Arteria dorsalis pedis and its branches.
   v. Medial and lateral terminal branches of deep peroneal nerve.

PROBABLE QUESTIONS AND ANSWERS

1. What are the cutaneous nerves supplying the skin of the dorsum?

Ans. Refer Figure 3.27
  i. Medial margin of the dorsum up to the ball of the great toe → Saphenous nerve.
  ii. Lateral border of the foot → Sural nerve.
  iii. Cleft between the great toe and the second toe → Medial terminal branch of **deep peroneal nerve**.
  iv. Rest of the dorsum of the foot → Medial and lateral branches of **superficial peroneal nerve**.

Note that the skin between the great toe and the second toe is the only skin on the dorsum of the foot that is innervated by the deep peroneal nerve.

2. What is dorsal venous arch?

Ans. It is a subcutaneous venous arch across the proximal parts of the metatarsal bones and is formed by the union of four **dorsal metatarsal veins** (Fig. 3.26).

3. What is dorsal metatarsal vein?

Ans. It is a subcutaneous vein on the dorsum of the foot and is formed by the union of adjacent **dorsal digital veins** at the web of the toes.

4. How many dorsal digital veins are there?

Ans. There are 10 dorsal digital veins.

5. What is the fate of the dorsal venous arch?

Ans.
  i. Medial part of the dorsal venous arch joins with the dorsal digital vein from the medial side of the great toe and is continued as **medial marginal vein**.
  ii. Lateral part of the dorsal venous arch joins with the dorsal digital vein from the lateral side of the little toe and is continued as **lateral marginal vein**.

6. Do you find great and small saphenous veins on the dorsum?

Ans. **Great saphenous vein** is the continuation of the medial marginal vein of the foot and is found to ascend in front of the medial malleolus. **Small saphenous vein** is the continuation of the lateral marginal vein of the foot and is found to ascend behind the lateral malleolus.

7. Where do the saphenous veins end?

Ans. **Great saphenous vein** ends in the **femoral vein** in the femoral triangle about 3 cm below the inguinal ligament. **Small saphenous vein** ends in the **popliteal vein** 3–7.5 cm above the middle of the knee joint.

8. How the deep fascia is disposed on the dorsum of the foot?

Ans. The deep fascia is **thin** on the dorsum in contrast to the deep fascia in the sole of the foot where the deep fascia is **thickened** called plantar aponeurosis. The deep fascia on the dorsum is continuous with the plantar aponeurosis at the margins of the foot. Proximally, it is thickened to form **Y-shaped inferior extensor retinaculum**.

9. What is the function of the inferior extensor retinaculum?

Ans. It keeps the extensor tendons in position and prevents the tendons from bow-stringing.

10. Name the structures that pass beneath the deep fascia.

Ans.
  a. **Extrinsic muscles** (medial to lateral):
       i. Tibialis anterior.
       ii. Extensor hallucis longus.
       iii. Extensor digitorum longus.
       iv. Peroneus tertius.
     (Peroneous tertius is said to be the 5th tendon of the extensor digitorum longus).
  b. **Intrinsic muscle of the dorsum**: Extensor digitorum brevis.
  c. **Artery**: Arteria dorsalis pedis.
  d. **Nerves**: Terminal branches of deep peroneal nerve.

11. What is extensor hallucis brevis?

Ans. The extensor digitorum brevis ends in four tendons. The medial most tendon is called extensor hallucis brevis.

12. What is the nerve supply of the extensor digitorum brevis?

Ans. Lateral terminal branch of deep peroneal nerve. **Note** that the extrinsic muscles of the dorsum of the foot are supplied by the anterior tibial nerve (deep peroneal nerve) in the anterior compartment of the leg.

13. How does the arteria dorsalis pedis begin?

Ans. It begins as a continuation of the anterior tibial artery distal to the ankle.

14. How does the artery end?

Ans. It enters the sole between the two heads of the first dorsal interosseous muscle and forms the plantar arterial arch by joining with the deep branch of lateral plantar artery.

15. What are the branches of arteria dorsalis pedis?

Ans.
 i. Lateral tarsal artery.
 ii. Medial tarsal artery.
 iii. **First** dorsal metatarsal artery.
 iv. Arcuate artery.

16. What are the origins of other dorsal metatarsal arteries?

Ans. **Second, third** and **fourth** dorsal metatarsal arteries are the branches of the arcuate artery.

17. What is the relation of the dorsalis pedis artery with the deep peroneal nerve on the dorsum?

Ans. Medial terminal branch of the deep peroneal nerve lies lateral to the artery.

18. Clinically, where do you feel the pulsation of dorsalis pedis artery?

Ans. The pulsation of this artery is felt between the tendons of extensor hallucis longus and the first tendon of extensor digitorum longus.

19. Do you expect any anomally of this dorsalis pedis artery?

Ans.
 i. The artery may lie lateral to the deep peroneal nerve.
 ii. The artery may be congenitally absent in about 14% cases.

20. What are the terminal branches of deep peroneal nerve?

Ans.
 i. Lateral terminal branch.
 ii. Medial terminal branch.

21. What are the muscles supplied by these terminal branches?

Ans. Medial branch supplies the first dorsal interosseous muscle. Lateral branch supplies the extensor digitorum brevis and second dorsal interosseous muscle.

Window Dissections: Lower Limb (Inferior Extremity)

22. What is the special observation of the lateral terminal branch of deep peroneal nerve on the dorsum of the foot?

Ans. The lateral terminal branch of the deep peroneal nerve forms a **pseudoganglion** through which it supplies the extensor digitorum brevis muscle.

23. Name some other nerves having pseudoganglion?

Ans.
  i. **Posterior interosseous nerve** on the back of the forearm.
  ii. **Nerve to teres minor.**
  iii. **Median nerve** of the hand.

LESSON 6: GLUTEAL REGION

The **gluteal region** is the most superior part of lower limb. It overlies the side and back of the pelvis, extending from the iliac crest above to the gluteal fold below. **The gluteal fold** is the transverse skin crease of the hip. **Remember that** the gluteal fold does not correspond to the lower border of gluteus maximus, rather the muscle crosses the fold obliquely downwards and laterally. Excessive amount of subcutaneous fat in the lower part of the gluteal region forms a rounded bulge known as **buttock**. The gluteal fold marks the lower **limit of the buttock**.

The **gluteus maximus** is **the largest** muscle of this region and one of the most powerful and bulkiest muscles in men.

STEPS OF DISSECTION

1. **Position of the cadaver:** Body is in prone position and thigh slightly abducted.
2. **Skin incision (Fig. 3.30):**
    i. Make a curved incision along the iliac crest from the anterior superior iliac spine up to the posterior superior iliac spine (A–B)
    ii. 2nd incision starts from the posterior superior iliac spine (B) and carry it downwards and backwards up to the spine of S_2 vertebra in the median plane (B–C)
    iii. Next incision extends from the S_2 vertebral spine (C) up to the tip of the coccyx (C–D)
    iv. Next incision starts from the tip of the coccyx and carry it downwards and laterally with the convexity downwards along the medial aspect of gluteal fold up to the lateral margin of the back of the thigh at the junction of the upper 1/3rd and lower 2/3rd of the thigh (D–E).

   The skin flap is reflected downwards and laterally.

   The superficial fascia and subcutaneous fats are exposed.
3. **Superficial fascia:** The following cutaneous nerves lie in the superficial fascia. **These are:**
    i. Lateral branch of subcostal nerve.
    ii. Lateral branch of iliohypogastric nerve.

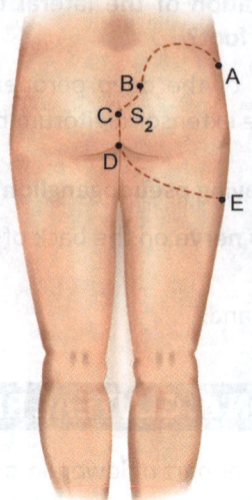

Fig. 3.30: Dissection of gluteal region (skin incision)

    iii. Posterior branch of lateral femoral cutaneous nerve.
    iv. Gluteal branch of posterior femoral cutaneous nerve.
    v. Lateral branches of L_2-L_4 nerves.
    vi. Posterior rami of sacral nerves.

But these cutaneous nerves are difficult to find. The fat and superficial fascia are removed from the gluteal region.

The deep fascia is exposed.

4. **Deep Fascia:** The inferior and superior borders of the **gluteus maximus** muscle should be visible from its proximal attachment to its distal attachment on the iliotibial tract (Fig. 3.31). The fasia lata is relatively this over the superficial surface of the gluteus maximus muscle. But the deep fascia becomes thicker superior to the muscle and forms the **gluteal aponeurosis**. This aponeurosis spans from the superior border of the gluteus maximus to the iliac crest and covers the **gluteus medius** muscle. Along the iliac crest this deep fascia splits twice to enclose the **tensor fascia lata** and **gluteus maximus**. The thickened sheet of deep fascia between these muscles is called **gluteal aponeurosis** (Figs 3.32A and B).

Insert your finger between the upper border of gluteus maximus and the gluteal aponeurosis and separate the muscle from the aponeurosis. Cut the gluteus maximus muscle in an oblique line from superior to its inferior border close to its proximal attachment to the ilium, sacrum and sacrotuberous ligament. Then push your fingers deep to the gluteus maximus muscle to loosen it from the deeper structures. Then reflect the muscle proximally and distally. Inspect and palpate the **inferior gluteal vessels and nerve** deep to the muscle (Fig. 3.33).

Window Dissections: Lower Limb (Inferior Extremity)

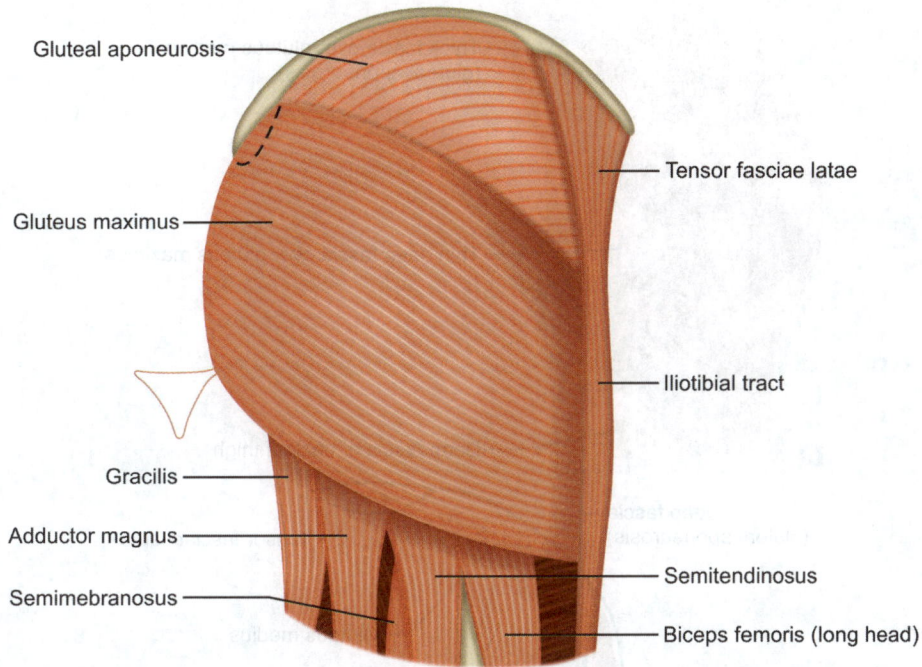

Fig. 3.31: Iliotibial tract and gluteal aponeurosis

Incise the gluteal aponeurosis along the iliac crest by using scalpel. **Remember** that the aponeurosis is firmly attached to the underlying **gluteus medius** muscle. Now, define the inferior border of this muscle and insert your finger along this border to open the interval between the **gluteus medius** and **piriformis muscle**.

Note that the superior border of the piriformis muscle lies adjacent to the inferior border of gluteus medius muscle and then open the interval between these muscles.

Note that the **superior gluteal vessels and nerve** enter the gluteal region between the piriformis and gluteus medius, i.e. above the superior border of piriformis (Fig. 3.33).

Then clean the inferior border of the piriformis by using blunt dissection. Note the following structures to enter the gluteal region passing under the inferior border of piriformis (Fig. 3.34).
  i. Sciatic nerve (**Largest nerve in the body**).
 ii. Posterior cutaneous nerve of thigh.
iii. Inferior gluteal vessels.
iv. Inferior gluteal nerve.
 v. Nerve to obturator internus.
vi. Internal pudendal vessels.
vii. Pudendal nerve.

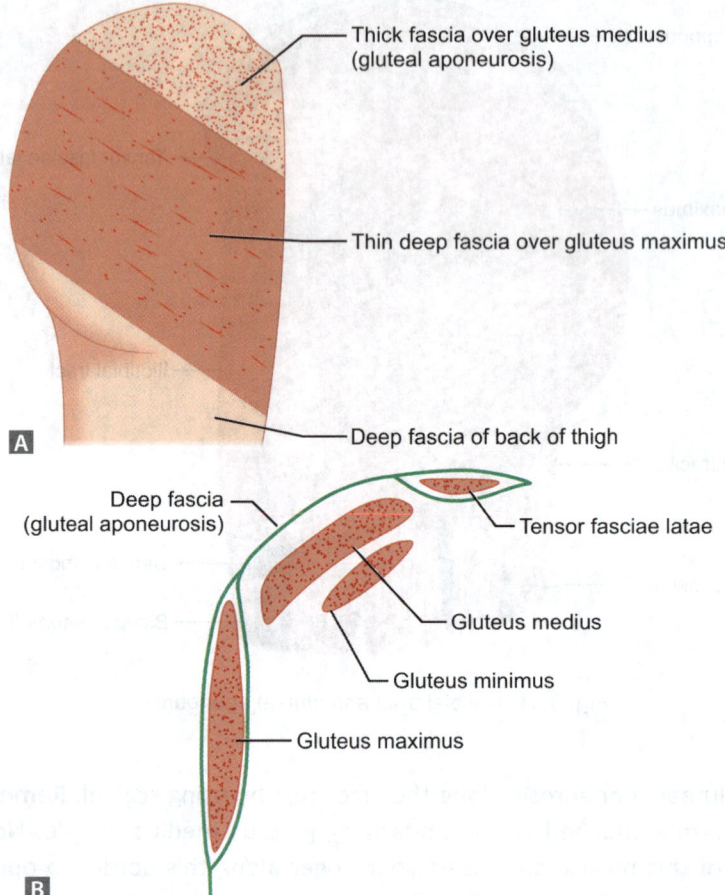

Figs 3.32A and B: Gluteal aponeurosis

The **sciatic nerve** is cleaned and traced into the thigh for 6–7 cm. The **posterior cutaneous nerve of thigh** lies on the medial side of the sciatic nerve. The i**nferior gluteal vessels and nerve** lie on the medial side of the posterior cutaneous nerve of the thigh.

Identify the tendon of **obturator internus** muscle which exits the lesser pelvis by passing through the lesser sciatic foramen. Identify the two **gemelli muscles**. One is above the obturator internus known as **gemellus superior** and another below the internus called **gemellus inferior**. Both gemelli muscles and the tendon of obturator internus are attached on the greater trochanter of the femur.

Identify the **quadratus femoris** muscle below the gemellus inferior. It is **quadrilateral** in shape.

Expose the **gluteus minimus muscle**. To expose this muscle the following steps are to be adopted. Insert your finger superior to the superior gluteal vessels and deep to gluteus

Window Dissections: Lower Limb (Inferior Extremity)

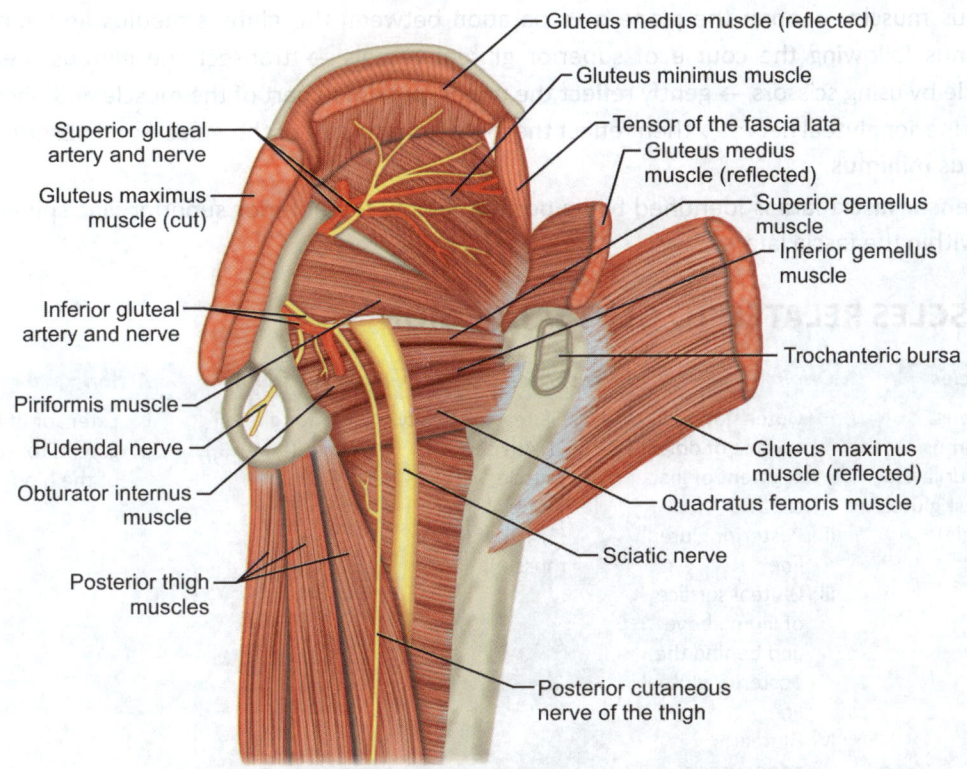

Fig. 3.33: Superior and inferior gluteal vessels and nerves

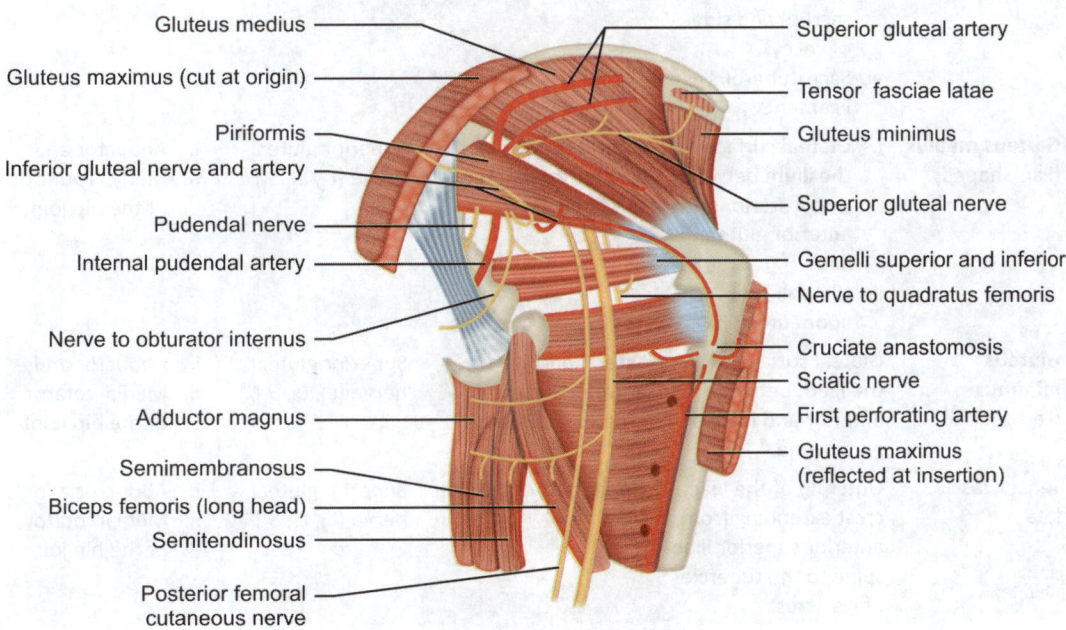

Fig. 3.34: Structures passing below piriformis

medius muscle → open the plane of separation between the gluteus medius and gluteus minimus following the course of superior gluteal vessels → transect the gluteus medius muscle by using scissors → gently reflect the upper (proximal) part of the muscle and observe the superior gluteal nerve → then reflect the lower (distal) part of the muscle and observe the **gluteus minimus**.

Tensor fascia lata is identified by its position below the anterior superior iliac spine and lies within the fascia lata.

MUSCLES RELATED TO GLUTEAL REGION

Muscles	Origin	Insertion	Nerve supply	Action
Gluteus maximus (Quadrilateral, **largest** gluteal muscle)	i. Outer sloping surface of dorsal segment of iliac crest ii. Posterior gluteal line iii. Gluteal surface of ilium above and behind the posterior gluteal line iv. Gluteal aponeurosis v. Dorsal surface of lower part of sacrum and side of coccyx vi. Sacrotuberous ligament	i. Gluteal tuberosity (1/4th of the muscle) ii. Iliotibial tract (3/4th of the muscle)	Inferior gluteal nerve (L_5, S_1, S_2)	i. Extensor of hip ii. Lateral rotator of the hip
Gluteus medius (Fan-shaped)	i. Gluteal surface of the ilium between the posterior and anterior gluteal line ii. Gluteal aponeurosis	Lateral surface of greater trochanter of the femur	Superior gluteal nerve (L_4, L_5, S_1)	i. Abductor and ii. Medial rotator of the hip joint
Gluteus minimus (Fan-shaped)	Gluteal surface of the ilium between anterior and inferior gluteal line	Lateral part of anterior surface of greater trochanter	Superior gluteal nerve (L_4, L_5, S_1)	i. Abductor and ii. Medial rotator of the hip joint
Tensor fascia lata	Outer lip of the iliac crest extending from anterior superior iliac spine to the tubercle of iliac crest	Upper part of iliotibial tract	Superior gluteal nerve (L_4, L_5, S_1)	i. Abductor and ii. Medial rotator of the hip joint

Contd...

Contd...

Muscles	Origin	Insertion	Nerve supply	Action
Piriformis (Key muscle of gluteal region)	i. Pelvic surface of the sacrum between the anterior sacral foramina and the area lateral to the foramina ii. Upper margin of the greater sciatic notch iii. Gluteal surface of the ilium close to the posterior inferior iliac spine	Apex of the greater trochanter of femur	Ventral rami of S_1 and S_2 spinal nerves	i. Abductor in flexed thigh ii. Lateral rotator in extended thigh
Gamellus superior	Posterior surface of ischial spine	Medial surface of the greater trochanter of the femur after blending with upper margin of the tendon of obturator internus	Nerve to obturator internus (L_5, S_1, S_2)	Lateral rotator of the hip joint
Gamellus inferior	i. Lower margin of lesser sciatic notch ii. Upper part of ischial tuberosity	Medial surface of the greater trochanter of the femur after blending with the lower margin of the tendon of of obturator internus	Nerve to quadratus femoris (L_4, L_5, S_1)	Lateral rotator of the hip joint
Obturator internus (Fan-shaped)	i. Inner surface of obturator membrane ii. Upper part of pelvic surface of body of ischium iii. Ischiopubic rami iv. Ilium below the pelvic brim v. Obturator fascia	Medial surface of the greater trochanter of femur above and in front of the trochanteric fossa	Nerve to obturator internus (L_5, S_1, S_2)	Lateral rotator of the hip joint
Obturator externus (Triangular)	i. Outer surface of the bony margins of the obturator membrane ii. Outer surface of the obturator membrane	Trochanteric fossa	Posterior division of obturator nerve (L_2, L_3, L_4)	Lateral rotator of the hip joint
Quadratus femoris (Quadrilateral)	Outer surface of the ischial tuberosity	Quadrate tubercle (Near the middle of the intertrochanteric crest)	Nerve to quadratus femoris (L_4, L_5, S_1)	Lateral rotator of the hip joint

Summary

1. Position of the cadaver: Prone, thigh slightly abducted.
2. Skin incision as in Figure 3.30.
3. Superficial fascia is exposed with its superficial veins and cutaneous nerves.
4. Deep fascia is exposed and reflected.
5. Gluteal aponeurosis is identified.
6. **Contents:**
    A. Gluteus maximus.
    B. Structures under gluteus maximus which include:
      a. **Muscles**
         i. Gluteus medius.
         ii. Gluteus minimus.
         iii. Piriformis.
         iv. Gemellus superior.
         v. Gemellus inferior.
         vi. Obturator internus.
         vii. Quadratus femoris.
         viii. Obturator externus.
         ix. Origins of hamstring muscles.
         x. Pubic fibers of adductor magnus.
      b. **Vessels**
         i. Superior gluteal vessels.
         ii. Inferior gluteal vessels.
         iii. Internal pudendal vessels.
         iv. Trochanteric anastomoses.
         v. Cruciate anastomoses.
      c. **Nerves**
         i. Superior gluteal nerve (L_4, L_5, S_1)
         ii. Inferior gluteal nerve (L_5, S_1, S_2)
         iii. Sciatic nerve (L_4, L_5, S_1, S_2, S_3)
         iv. Posterior cutaneous nerve of thigh (S_1, S_2, S_3)
         v. Pudendal nerve (S_2, S_3, S_4)
         vi. Nerve to obturator internus (L_5, S_1, S_2)
         vii. Nerve to quadratus femoris (L_4, L_5, S_1)

PROBABLE QUESTIONS AND ANSWERS

1. What is the extent of the gluteal region?

Ans.
  i. Above: Iliac crest.
  ii. Below: Gluteal fold.
  iii. Behind: Sacral spines.
  iv. Infront: An imaginary line extending vertically downward from the anterior superior iliac spine to greater trochanter.

Window Dissections: Lower Limb (Inferior Extremity)

2. What is called the gateway of the gluteal region?

Ans. Greater sciatic foramen.

3. What is gluteal aponeurosis?

Ans. It is a thickened sheet of deep fascia which covers the gluteus medius muscle and extends between the posterior border of tensor fascia lata and upper margin of gluteus maximus (Figs 3.31 and 3.32)

4. How the muscles are arranged in gluteal region?

Ans. The muscles of gluteal region are arranged in **three layers**:
  i. **Superficial layer**: Gluteus maximus and tensor fascia lata.
  ii. **Intermediate layer**: Gluteus medius, piriformis, obturator internus, gemelli superior et inferior, quadratus femoris, origins of hamstring muscles.
  iii. **Deep layer**: Gluteus minimus, reflected head of rectus femoris and obturator externus.

Remember that all the muscles of gluteal region are under cover of gluteus maximus except tensor fascia lata.

5. Which one is the largest muscle of the gluteal region?

Ans. Gluteus maximus.

6. Does the lower border of the gluteus maximus correspond with the gluteal fold?

Ans. **No**, the lower border of the gluteus maximus is **oblique** in outline and forms an angle of about 45° with the mid-sagittal plane. But the gluteal fold is a somewhat horizontal skin crease of the hip joint.

7. What are the main origins of gluteus maximus?

Ans. i. Gluteal surface of the ilium above and behind the posterior gluteal line.
  ii. Outer sloping surface of the dorsal segment of the iliac crest.

8. What are the sites of insertion of gluteus maximus?

Ans. i. One fourth of the muscle is inserted into the gluteal tuberosity of the femur.
  ii. Three fourths of the muscle are inserted into the upper part of iliotibial tract.

9. What is the nerve supply of gluteus maximus?

Ans. Inferior gluteal nerve (L_5, S_1, S_2).

10. What is the action of gluteus maximus?

Ans. Extensor of the hip joint.

> **It is to be remembered** that the gluteus maximus acts as an extensor of the hip joint only at the extremes of the hip movements (running, climbing upstairs). But in quiet walking and standing the muscle remains inactive and the hamstring muscles act as extensor of hip.

11. What is the paradoxical action of gluteus maximus?

Ans. Though the gluteus maximus is an extensor muscle of the hip, it regulates the flexion of the hip joint in the act of sitting from standing position by the controlled lengthening of its fibers and acts as the chief anti-gravity muscle.

12. What are the muscles supplied by the superior gluteal nerve?

Ans. Superior gluteal nerve supplies the gluteus medius, gluteus minimus and tensor fascia lata.

13. What is the root value of superior gluteal nerve?

Ans. L_4, L_5, S_1 (Inferior gluteal nerve—L_5, S_1, S_2).

14. What are the vessels and nerves passing through the greater sciatic foramen?

Ans. a. **Structures passing above the piriformis muscle:**
  i. Superior gluteal vessels.
  ii. Superior gluteal nerve.
 b. **Structures passing below the piriformis muscle:**
  i. Sciatic nerve.
  ii. Posterior femoral cutaneous nerve.
  iii. Inferior gluteal vessels.
  iv. Inferior gluteal nerve.
  v. Nerve to quadratus femoris.
  vi. Nerve to obturator internus.
  vii. Internal pudendal vessels.
  viii. Pudendal nerve.

15. What are the structures passing through the lesser sciatic foramen?

Ans. i. Tendon of obturator internus.
 ii. Nerve to obturator internus.
 iii. Internal pudendal vessels.
 iv. Pudendal nerve.

Window Dissections: Lower Limb (Inferior Extremity)

> The piriformis is the **key muscle** of the gluteal region, because it divides the greater sciatic foramen into upper and lower compartments through which a number of structures pass.

16. Name the structures that pass both through the greater and lesser sciatic foramen.

Ans.
  i. Nerve to obturator internus.
  ii. Internal pudendal vessels.
  iii. Pudendal nerve.

17. What are the arrangement of these structures while passing over the ischial spine?

Ans. **From lateral to medial:**
  i. Nerve to obturator internus.
  ii. Internal pudendal vessels.
  iii. Pudendal nerve.

18. What is the site of division of sciatic nerve?

Ans. Sciatic nerve (the **thickest and widest nerve in the body**) usually divides behind the thigh at its junction of upper 2/3rd and lower 1/3rd into **tibial** division and **common peroneal division**. Remember that this division may take place in the pelvic cavity and in such case the common peroneal nerve passes either over the superior border of piriformis or through the piriformis and the tibial nerve emerges below the piriformis.

19. What is the root value of sciatic nerve?

Ans.
  i. **Tibial component:** L_4, L_5, S_1, S_2, S_3
  ii. **Common peroneal component:** L_4, L_5, S_1, S_2.

20. What is arteria comitans nervi ischiadici?

Ans. It is a branch of inferior gluteal artery and supplies the sciatic nerve.

21. What are the nerve supply of gemelli muscles?

Ans. Nerve to obturator internus gives a twig to gemellus superior and nerve to quadratus femoris gives a twig to gemellus inferior.

22. What is the course of the superior and inferior gluteal arteries in the gluteal region?

Ans. The **superior gluteal artery** is a branch of posterior division of internal iliac artery and appears in the gluteal region above the piriformis muscle accompanied by the superior gluteal nerve. Then it divides into superficial and deep branches. The **superficial branch** passes between the gluteus maximus and gluteus medius and the **deep** branch passes between the gluteus medius and gluteus minimus.

The **inferior gluteal artery** is a branch of anterior division of internal iliac artery and appears in the gluteal region through the greater sciatic foramen below the piriformis, accompanied by the inferior gluteal nerve.

Remember that the **internal pudendal artery** is a branch of anterior division of internal iliac artery.

23. What are the arterial anastomosis found deep to gluteus maximus?

Ans. a. **Trochanteric anastomosis** formed by:
  i. Descending branch of superior gluteal artery.
  ii. Ascending branch of medial and lateral circumflex femoral arteries.
  iii. A branch from the inferior gluteal artery.

Trochanteric anastomosis provides the chief source of blood supply to the head of the femur along the retinacular fibres of the capsule of the hip joint.

b. **Cruciate anastomosis** formed by (Fig. 3.35):
  i. Descending branch of inferior gluteal artery.
  ii. Transverse branch of lateral circumflex femoral artery.
  iii. Transverse branch of medial circumflex femoral artery.
  iv. Ascending branch of first perforating artery, a branch of arteria profunda femoris.

24. What is the site of formation of cruciate anastomosis?

Ans. It lies on the posterior surface of the greater trochanter of femur in the intermuscular interval between the quadratus femoris and upper margin of adductor magnus.

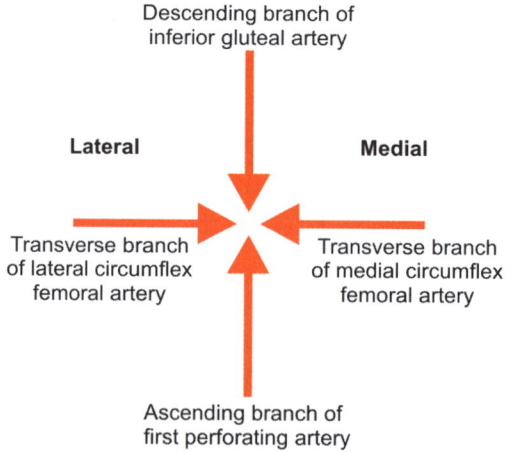

Fig. 3.35: Cruciate anastomosis

Window Dissections: Lower Limb (Inferior Extremity)

25. What is the importance of these anastomosis?

Ans. These anastomoses establish a collateral circulation when the femoral artery is tied above the origin of profunda femoris artery.

> **Remember** that the **superior gluteal nerve** supplies the gluteus medius, gluteus minimus and tensor fascia lata and these muscles are the **abductors** and **medial rotators** of the hip joint. The muscles which are under cover of gluteus maximus are the lateral rotators of the hip.

> The muscles arising from the ilium of the hip bone are supplied by the dorsal division of the limb plexus (lumbar or sacral), whereas the muscles arising from the ischium and pubis are supplied by the ventral division of the limb plexus. This is because of the fact that the ilium of the hip bone represents the dorsal segment and the ischium and pubis represent the ventral segment of the hip bone.

26. Which is called the tricipital tendon?

Ans. The tendon of obturator internus reinforced by gemelli superior and inferior muscles forms the **tricipital tendon**.

27. What is the action of tricipital tendon?

Ans. Lateral rotators of the hip joint.

28. Name the structures lying under the cover of gluteus maximus.

Ans.
a. **Muscles:**
   i. Gluteus medius.
   ii. Gluteus minimus.
   iii. Piriformis.
   iv. Gemellus superior.
   v. Tendon of obturator internus.
   vi. Gamellus inferior.
   vii. Quadratus femoris.
   viii. Origin of hamstring muscles.
b. **Vessels:**
   i. Superior gluteal vessels.
   ii. Inferior gluteal vessels.
   iii. Internal pudendal vessels.
   iv. Trochanteric anastomosis.
   v. Cruciate anastomosis.

c. **Nerves:**
   i. **Sciatic nerve.**
   ii. Superior gluteal nerve.
   iii. Inferior gluteal nerve.
   iv. Posterior cutaneous nerve of thigh.
   v. Nerve to quadratus femoris.
   vi. Pudendal nerve.
   vii. Nerve to obturator internus.
d. **Bursa:**
   i. Ischial bursa (over the ischial tuberosity).
   ii. Trochanteric bursa (on the greater trochanter).
   iii. Gluteofemoral bursa (bursa between gluteus maximus and vastus lateralis).

29. **Name the structures lying under the cover of gluteus medius.**

Ans. i. Gluteus medius.
ii. Superior gluteal nerve.
iii. Deep branch of superior gluteal artery.

30. **Name the structures lying under the cover of gluteus minimus.**

Ans. i. Reflected head of rectus femoris.
ii. Capsule of hip joint.
iii. Bursa on greater trochanter.

LESSON 7: BACK OF THE THIGH

The back of the thigh (posterior compartment of the thigh) is also called the **flexor compartment**. This compartment extends from the lower limit of the gluteal region to the back of the knee. It is separated from the medial (adductor) compartment by the posterior intermuscular septum. The **roof** of this compartment is formed by the deep fascia of the thigh (**fascia lata**) and the **floor** is formed by the **adductor magnus** and **vastus lateralis**.

STEPS OF DISSECTION

1. **Position of the cadaver:** Body prone and the knee extended.
2. **Skin incision (Fig. 3.36):**
   i. A transverse incision below the gluteal fold (A–B)
   ii. Another transverse incision at the junction of upper 2/3rd and lower 1/3rd of the back of the thigh (C–D)

Window Dissections: Lower Limb (Inferior Extremity)

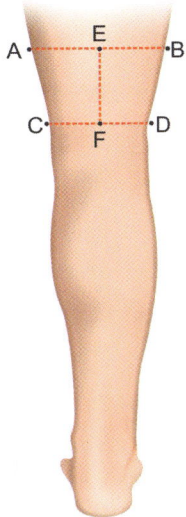

Fig. 3.36: Dissection of back of the thigh (skin incision)

   iii. A vertical incision from the midpoint of the proximal and distal incisions (E–F)
Reflect the skin flaps laterally and medially. The superficial fascia is exposed.
3. **Superficial fascia:** In it, identify the cutaneous branches of posterior cutaneous nerve of thigh. This fascia is reflected in the line of skin incision. The deep fascia is exposed.
4. **Deep fascia:** The deep fascia is incised in the line of skin incision and reflected sideways. The structures in the back of the thigh are exposed. These structures are cleaned by using blunt dissection and studied (Fig. 3.37).

Clean the **sciatic nerve** and follow it inferiorly. Note that the nerve passes deep to the long head of biceps femoris muscle. The nerve gives off unnamed branches to the muscles of posterior compartment of the thigh.

The **long head of biceps femoris** muscle is identified. Clean the muscle and follow it up to its proximal attachment to the ischial tuberosity.

Retract the long head laterally and observe the **short head of biceps femoris** whose proximal attachment is at the lateral lip of linea aspera of femur.

Semitendinosus muscle is identified on the medial side of the back of the thigh by its long, cord like tendon at its distal end.

Separate the semitendinosus muscle from the **semimembranosus** by finger dissection. Retract the semitendinosus medially to expose the semimembranosus which is membranous in the upper half and becomes fleshy in the lower half.

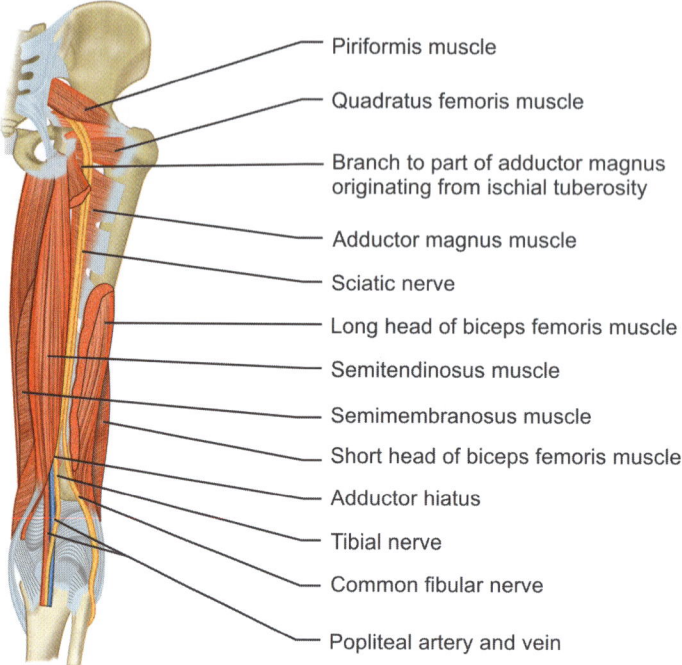

Fig. 3.37: Structures in back of thigh

Hamstring part of adductor magnus muscle (ischial part) is then identified and confirm its proximal attachment to the ischial tuberosity. Note that this muscle forms the deep boundary (floor) of the posterior compartment.

MUSCLES OF THE BACK OF THE THIGH

Muscles	Origins (Fig. 3.38)	Insertions	Nerve supply	Actions
Semitendinosus	Lower and medial part of the quadrilateral area of ischial tuberosity	Upper part of the medial surface of the tibia behind the insertion of sartorius and gracilis	Tibial part of sciatic nerve	i. Extensor of the hip and flexor of the knee (in standing and walking position) ii. Medial rotator of tibia on femur (in semiflexed knee)
Semimembranosus	Upper and lateral part of the quadrilateral area of ischial tuberosity	Posterior surface of the medial condyle of tibia	Same as above	Same as above

Contd...

Contd...

Muscles	Origins (Fig. 3.38)	Insertions	Nerve supply	Actions
Adductor magnus (Hybrid muscle because of its double nerve supply)	i. **Ischial part**: Infero-lateral aspect of ischial tuberosity ii. **Pubic part**: External surface of ischio-pubic ramus	i. **Ischial part**: Adductor tubercle of femur ii. **Pubic part**: In a continuous line along the medial margin of gluteal tuberocity, medial lip of linea aspera and upper part of medial supracondylar line	i. **Ischial part**: Tibial part of sciatic nerve ii. **Pubic part**: Posterior division of obturator nerve	i. Extensor of the thigh (ischial part) ii. Adductor of thigh
Biceps femoris	i. **Long head**: Lower and medial part of quadrilateral area of ischial tuberosity ii. **Short head**: Lower part of lateral lip of linea aspera and upper 2/3rd of lateral supracondylar line of femur	Head of the fubula in front of styloid process	i. **Long head**: By the tibial part of sciatic nerve ii. **Short head**: By the common peroneal part of sciatic nerve	i. Extensor of hip and flexor of knee ii. Lateral rotator of the leg when the foot is off the ground iii. Medial rotator of the thigh when the foot is on the ground

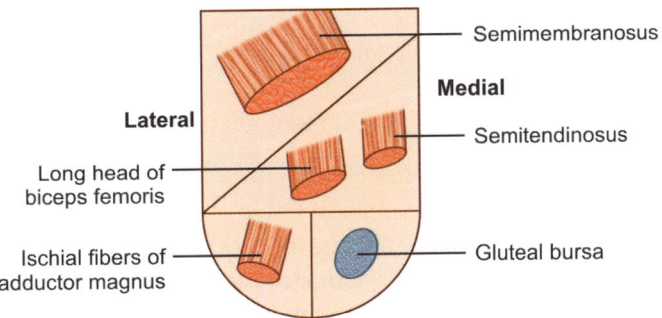

Fig. 3.38: Origins of hamstring muscles

> **Summary**
> 1. Position of the cadaver: Prone, knee extended.
> 2. Skin incision as in Figure 3.36.
> 3. Superficial fascia with cutaneous branches of posterior cutaneous nerve of thigh.
> 4. Deep fascia is reflected in the line of skin incision and **contents are exposed**:
>   i. Hamstring muscles.
>   ii. Short head of biceps femoris.
>   iii. Sciatic nerve.
>   iv. Posterior femoral cutaneous nerve.
>   v. Arterial anastomoses.

PROBABLE QUESTIONS AND ANSWERS

1. What are the muscles of the back of the thigh?

Ans.
   i. Hamstring muscles.
   ii. Short head of biceps femoris.

2. Name the hamstring muscles?

Ans.
   i. Semimembranosus.
   ii. Semitendinosus.
   iii. Long head of biceps femoris.
   iv. Ischial fibers of adductor magnus.

3. What are the characteristic features of hamstring muscles?

Ans. Hamstring muscles fulfill the following common characteristic features.
   i. They arise from ischial tuberosity.
   ii. They are inserted beyond the knee joint (i.e. either tibia or fibula or both bones)
   iii. They are supplied by the tibial component of the sciatic nerve.
   iv. They act as extensors of hip joint and flexors of knee joint.

4. How do the ischial fibers of adductor magnus satisfy the criteria of hamstring group?

Ans. The tendon of ischial fibers of adductor magnus (fibers arising from ischial tuberosity) is inserted into the adductor tubercle of femur. The degenerated part of this muscle forms the medial ligament of knee joint which extends from the adductor tubercle to the tibia. Thus, the ischial fibers of adductor magnus are finally attached to tibia (i.e. beyond the knee joint) as medial ligament of knee joint.

5. What is presemimembranosus?

Ans. In some vertebrates, the ischial part of adductor magnus (hamstring part) is directly inserted to the tibia and is known as the presemimembranosus.

Window Dissections: Lower Limb (Inferior Extremity)

6. What is the degenerated part of ischial part of adductor magnus?

Ans. Medial ligament of the knee joint.

7. What is the morphological remnant of the long head of biceps?

Ans. **Sacrotuberous ligament**.

8. What is adductor minimus?

Ans. The pubic fibers of the adductor magnus are sometimes named as the **adductor minimus**.

9. What are the components (parts) of adductor magnus?

Ans.
    a. **On the basis of origin:**
        i. Ischial part (hamstring part)
        ii. Ischiopubic part (pubic part)
    b. **On the basis of action:**
        i. Adductor component (pubic part)
        ii. Hamstring component.

10. Why the adductor magnus is called a hybrid or composite muscle?

Ans. It is a **composite muscle,** because it is formed by the fusion of the adductor and hamstring part of the muscle. **The adductor** part of the muscle is supplied by the **obturator nerve** and the hamstring part is supplied by the tibial division of sciatic nerve. Because of its double nerve supply, it is a **hybrid muscle**.

11. What is the main nerve of the back of the thigh?

Ans. The sciatic nerve.

12. Where does this nerve bifurcate?

Ans. The sciatic nerve bifurcates into tibial and common peroneal divisions in the back of the thigh at its junction of upper 2/3rd and lower 1/3rd.

13. Why the short head of biceps femoris is not a hamstring group of muscles?

Ans.
    i. It does not arise from the ischial tuberosity (it arises from linea aspera of femur).
    ii. It is supplied by the common peroneal nerve (hamstring muscles are supplied by tibial division of sciatic nerve).

14. What is the nerve supply of hamstring muscles?

Ans. Hamstring muscles are supplied by the tibial division of the sciatic nerve from its medial side.

15. What is the nerve supply of the short head of the biceps?

Ans. Common peroneal nerve from its lateral side.

16. What are the root values of the two divisions of sciatic nerve?

Ans.
  i. **Tibial division:** Ventral divisions of ventral rami of L_4, L_5, S_1, S_2, S_3 spinal nerves.
  ii. **Common peroneal division:** Dorsal divisions of ventral rami of L_4, L_5, S_1, S_2 spinal nerves.

17. Why the biceps femoris is called a locking muscle?

Ans. This is because of the fact that it acts as a medial rotator of the thigh when the leg is on the ground and causes the locking of the knee joint at the end of extension.

18. Which muscle is called an unlocking muscle?

Ans. Popliteus.

LESSON 8: POPLITEAL FOSSA

Popliteal fossa is a diamond-shaped depression at the back of the knee joint. It is bounded (Fig. 3.39):
  i. Above and medially by semitendinosus and semimembranosus.
  ii. Above and laterally by the tendon of biceps femoris.
  iii. Below and medially by medial head of gastrocnemius.
  iv. Below and laterally by lateral head of gastrocnemius.

The **main contents** of the popliteal fossa are:
  i. Popliteal vessels.
  ii. Terminal branches of sciatic nerve: (a) Tibial and (b) Common peroneal nerves.
  iii. Popliteal lymph nodes.

STEPS OF DISSECTION

1. **Position of the cadaver:** Body prone and knee extended
2. **Skin incision (Fig. 3.40):**
     i. A transverse incision at the junction of upper 2/3rd and lower 1/3rd of the back of the thigh (A–B)
     ii. Another transverse incision at the junction of the upper 1/4th and lower 3/4th of the back of the leg (C–D)
     iii. A vertical incision from the midpoints of the above two incisions (E–F).
   Reflect the skin flaps sideways leaving the superficial fascia intact.

Window Dissections: Lower Limb (Inferior Extremity)

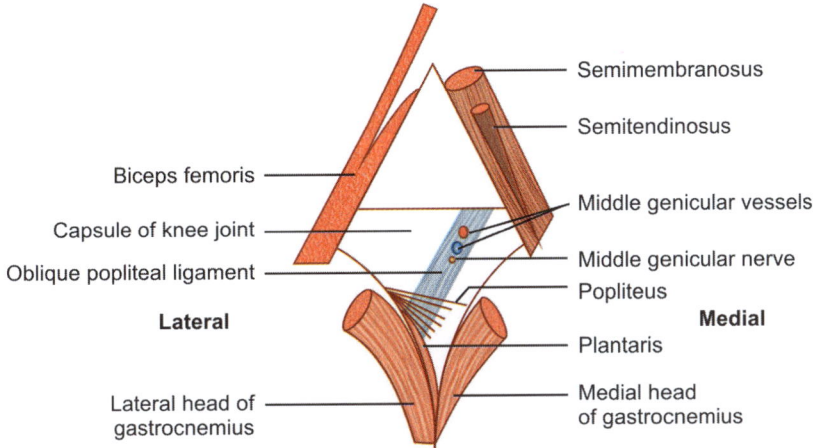

Fig. 3.39: Boundaries of popliteal fossa

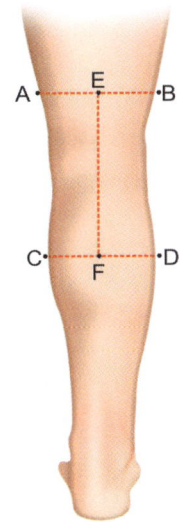

Fig. 3.40: Dissection of popliteal fossa (skin incision)

3. **Superficial fascia:** Strip the superficial fascia from the deep fascia. Look for **posterior cutaneous nerve of thigh** piercing the deep fascia in the upper part of the popliteal fossa and **small saphenous vein** in the lower part of the fossa. In rare situation the **sural nerve** may pierce the deep fascia at this level. Few **lymph vessels** are also available in the superficial fascia.

4. **Deep fascia:** It is also called **popliteal fascia**. It forms the roof of the popliteal fossa.
   Cut through the deep fascia over the **biceps femoris** which forms the upper and lateral boundary of the popliteal fossa. Expose this muscle and its tendon and trace it up to its insertion.

Similarly, cut through the deep fascia over the **semitendinosus** and **semimembranosus** muscles. Follow the tendon of semitendinosus to its insertion on the medial surface of the tibia. Then lift the semimembranosus and trace it downwards. Look for the **bursa** between the semimembranosus and the medial head of gastrocnemius.

Remove the remnants of popliteal fascia from the heads of **gastrocnemius muscles**. Now the lateral and medial heads of gastrocnemius muscles are exposed. Try to search for the **plantaris** muscle and identify it (if it is present). Separate the plantaris from the posteromedial surface of the lateral head of gastrocnemius and care to be taken to avoid injury to the nerve to the lateral head of gastrocnemius which passes between them.

The above mentioned muscles (biceps femoris, semitendinosus, semimembranosus, gastrocnemius) are retracted laterally and medially to observe **the contents of the popliteal fossa** (Fig. 3.41).

At the superior angle of the popliteal fossa, the **sciatic nerve** divides into **tibial** and **common peroneal nerves**. By using fingers, separate the **tibial nerve** from the loose connective tissue that surrounds it and follow the nerve downwards as it courses through the center of the popliteal fossa. Find out its **three articular branches** (superior medial, inferior medial and middle genicular) in the upper part of the fossa. Trace these genicular branches as far as possible. The **cutaneous branch (sural nerve)** lies between the two heads of gastrocnemius. The **muscular branch** to gastrocnemius, plantaris, soleus and popliteus arise near the middle part of the popliteal fossa. Separate the heads of gastrocnemius and trace these muscular branches as far as possible.

Follow the **common peroneal nerve** along the superolateral border of the popliteal fossa by using blunt dissection. Note that the common peroneal nerve lies medial to the tendon of

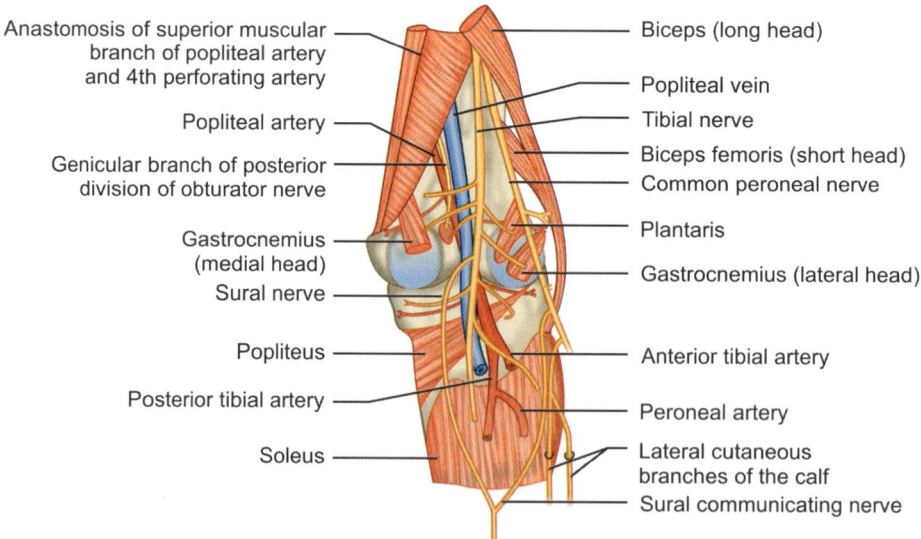

Fig. 3.41: Contents of popliteal fossa

Window Dissections: Lower Limb (Inferior Extremity)

biceps femoris muscle and passes superficial to the lateral head of gastrocnemius. Trace the nerve from the upper angle of the fossa to the back of the head of the fibula. Find its **genicular branches** near the upper part of the fossa.

Insert your index finders between the two bellies of the gastrocnemius muscle and pull them apart for a distance of 5–10 cm to expose the structures that pass from the popliteal fossa into the leg.

The **popliteal vessels** are located deep to the tibial nerve. Lift the upper part of the tibial nerve from the popliteal vessels. **Note** that the popliteal vessels are enclosed by a connective tissue sheath. Open the sheath by scissors and extend the incision above and below.

Use a probe to separate the **popliteal artery** from the popliteal vein. The **popliteal vein** lies on the medial side of the artery in the lower part of the popliteal fossa. It crosses the artery superficially at the back of the knee and then it ascends along the lateral side of the popliteal artery. Note that the vein itself is crossed superficially by the tibial nerve at the middle part of the fossa.

The **popliteal artery** is **the deepest** as seen from the back. It descends downwards and laterally deep to the popliteal vein. Distally, the artery passes deep to the plantaris and gastrocnemius muscle. Gently scrape the fat from the popliteal surface of the femur and find its **genicular branches**.

Find out the structures that form the **floor** of the popliteal fossa (Fig. 3.42). Retract the inferior end of the popliteal vessels and identify the **popliteus muscle** which forms the lower part of the floor of the popliteal fossa. Other structures that form the floor of the fossa are popliteal surface of the femur and capsular ligament and oblique popliteal ligament of the knee joint.

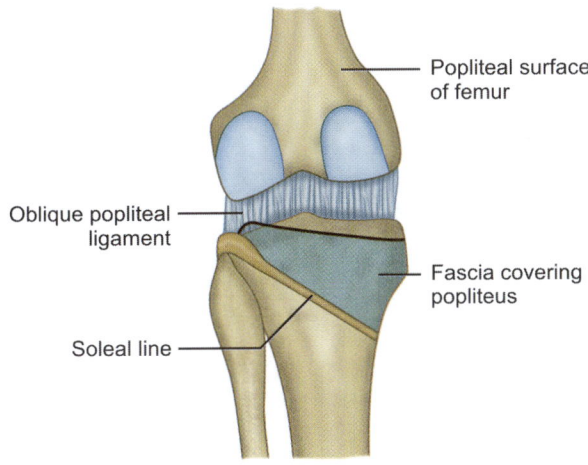

Fig. 3.42: Floor of the popliteal fossa

> ### Summary
> 1. Position of the cadaver: Body prone and knee extended.
> 2. Skin incision as in Figure 3.40.
> 3. Superficial fascia: Structures found
>   i. Posterior cutaneous nerve of thigh.
>   ii. Sural nerve.
>   iii. Short saphenous vein.
>   iv. Some lymph vessels.
> 4. Deep fascia (popliteal fascia) is divided and reflected to expose the contents of the fossa.
> 5. **Contents**:
>   i. Tibial nerve and its branches.
>   ii. Common peroneal nerve and its branches.
>   iii. Popliteal artery and its branches.
>   iv. Popliteal vein and its tributaries.
>   v. Termination of small saphenous vein.
>   vi. Genicular branch of posterior division of obturator nerve.
>   vii. Popliteal lymph nodes.
>   viii. Fat.

PROBABLE QUESTIONS AND ANSWERS

1. What is the boundary of popliteal fossa?

Ans.
　i. **Above and medially**: Semitendinosus and semimembranosus.
　ii. **Above and laterally**: Biceps femoris.
　iii. **Below and medially**: Medial head of gastrocnemius.
　iv. **Below and laterally**: Lateral head of gastrocnemius and plantaris (if present).

2. What forms the roof of the fossa?

Ans. The deep fascia over the popliteal fossa forms the roof.

3. What is popliteal fascia?

Ans. The deep fascia of the popliteal fossa is called popliteal fascia.

4. What are the structures that form the floor of the popliteal fossa?

Ans. The structures (from above downwards) are:
　i. Popliteal surface of the femur.
　ii. Oblique popliteal ligament and the capsule of the knee joint.
　iii. Posterior part of upper end of tibia.
　iv. Fascia covering the popliteus.
Note that the **popliteal fascia** forms the **roof** and the **fascia covering the popliteus muscle** forms the **floor** of the popliteal fossa.

5. What are the expansions from semimembranosus muscle?

Ans.
i. Oblique popliteal ligament.
ii. Fascia covering popliteus muscle.

6. What are the structures that pierce the oblique popliteal ligament?

Ans.
i. Middle genicular vessels.
ii. Middle genicular nerve.
iii. Posterior division of obturator nerve.

7. Which structures pierce the roof of the popliteal fossa?

Ans.
i. Small saphenous vein.
ii. Posterior femoral cutaneous nerve.

8. What are the main contents of the popliteal fossa?

Ans.
i. Popliteal artery and its branches.
ii. Popliteal vein and its tributaries.
iii. Tibial nerve and its branches.
iv. Common peroneal nerve and its branches.
v. Termination of small saphenous vein into the popliteal vein.
vi. Popliteal lymph nodes and fat.

9. Where does the popliteal artery begin and end?

Ans. The popliteal artery begins as a **continuation of the femoral artery** at the 5th osseo-aponeurotic opening of adductor magnus and ends at the lower border of popliteus muscle.

10. How much is the length of the popliteal artery?

Ans. About 20 cm.

11. What are the genicular branches of the popliteal artery?

Ans.
i. Superior medial genicular artery.
ii. Superior lateral genicular artery.
iii. Inferior medial genicular artery.
iv. Inferior lateral genicular artery.
v. Middle genicular artery.

Other branches of the popliteal artery are **cutaneous** branch, **muscular** branch and **terminal** branches.

12. What are the terminal branches of the popliteal artery?

Ans. The popliteal artery terminates at the lower border of the popliteus by dividing into:
  i. Anterior tibial artery.
  ii. Posterior tibial artery.

13. What may be the variations of termination of the popliteal artery?

Ans.
  i. The artery may terminate above the popliteus.
  ii. The artery may terminate by dividing into anterior tibial and peroneal artery.
  iii. The artery may divide into three branches anterior tibial, posterior tibial and peroneal arteries.

14. How the popliteal vein is formed?

Ans. The vein is formed by the union of venae comitantes of the anterior and posterior tibial arteries.

15. Where does the popliteal vein begin and end?

Ans. The vein begins at the lower border of the popliteus and continues as the **femoral vein** after passing through the 5th osseoaponeurotic opening of the adductor magnus.

16. What is the relation of the popliteal vein with the popliteal artery in the popliteal fossa?

Ans. The popliteal vein lies medial to the popliteal artery is the lower part, posterior to the artery in the middle part and posterolateral to the artery in the upper part of the fossa.

17. What are the tributaries of the popliteal vein?

Ans.
  i. Small saphenous vein (at the upper part of the popliteal fossa).
  ii. The veins corresponding to the branches of the popliteal artery.

18. What are the branches of tibial nerve in the popliteal fossa?

Ans.
  a. **Muscular branches to:**
      i. Medial head of gastrocnemius.
      ii. Lateral head of gastrocnemius.
      iii. Plantaris.
      iv. Soleus.
      v. Popliteus.
  b. **Articular branches:**
      i. Superior medial genicular.

ii. Inferior medial genicular.
iii. Middle genicular (it pierces the oblique popliteal ligament).
c. **Cutaneous branches:** Sural nerve.
d. **Vascular branches**.

Remember that all the muscular branches of the tibial nerve arise from its lateral side except the nerve to medial head of gastrocnemius which arises from its medial side.

19. How does the nerve to popliteus supply the popliteus?

Ans. The branch to popliteus passes superficial to the popliteal vessels and then winds round the lower border of popliteus to supply the muscle from its deep surface (Fig. 3.43).

20. What are the other structures supplied by the nerve to popliteus?

Ans.
i. Superior tibiofibular joint.
ii. Interosseous membrane.
iii. Inferior tibiofibular joint.
iv. Tibialis posterior.
v. Medullary branches to the tibia.

21. What is lateral popliteal nerve?

Ans. Common peroneal nerve is also known as lateral popliteal nerve (tibial nerve in the popliteal fossa is known as medial popliteal nerve).

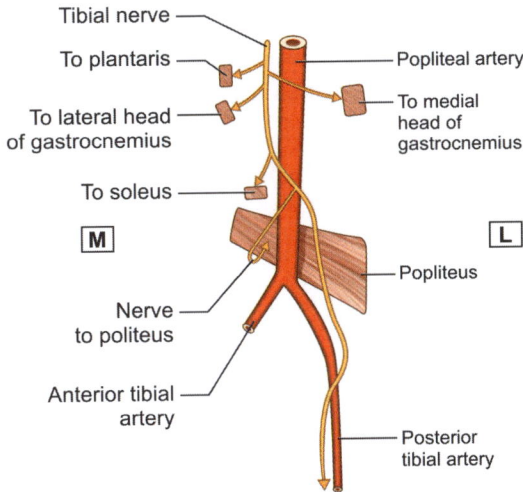

Fig. 3.43: Nerve to popliteus

22. What is the root value of tibial and common peroneal nerves?

Ans. **Tibial nerve:** Ventral divisions of ventral rami of L_4, L_5, S_1, S_2, S_3 spinal nerves.
Common peroneal nerve: Dorsal divisions of ventral rami of L_4, L_5, S_1, S_2 spinal nerves.

23. What are the branches of common peroneal nerve in the popliteal fossa?

Ans.
a. **Cutaneous:**
   i. Sural communicating nerve.
b. **Articular (genicular):**
   i. Superior lateral genicular.
   ii. Inferior lateral genicular.
   iii. Recurrent genicular.

24. What are the structures supplied by the recurrent genicular nerve?

Ans.
i. Capsule of the knee joint.
ii. Superior tibiofibular joint.
iii. Tibialis anterior muscle.

25. Where does the common peroneal nerve terminate?

Ans. It divides into two terminal branches on the lateral side of the neck of the fibula deep to peroneus longus muscle.

26. What are the two terminal branches of this nerve?

Ans.
i. Deep peroneal (anterior tibial)
ii. Superficial peroneal (musculocutaneous nerve).

LESSON 9: BACK OF THE LEG (POSTERIOR CRURAL REGION)

The posterior compartment of the leg is also called the **flexor** compartment. It lies posterior to the interosseous membrane, tibia and fibula (Fig. 3.44). The muscles in this compartment are arranged in **superficial** and **deep groups.** The superficial muscles are powerful and antigravity muscles which are used to raise the heel during walking. So these muscles are quite large in size. The superficial muscles are inserted into the heel, whereas the deep muscles cross the ankle and enter into the sole.

STEPS OF DISSECTION

1. **Position of the cadaver:** Body prone and knee extended.
2. **Skin incision (Fig. 3.45):**
   i. A transverse incision at the junction of the upper 1/4th and lower 3/4th of the back of the leg (A–B).

Window Dissections: Lower Limb (Inferior Extremity)

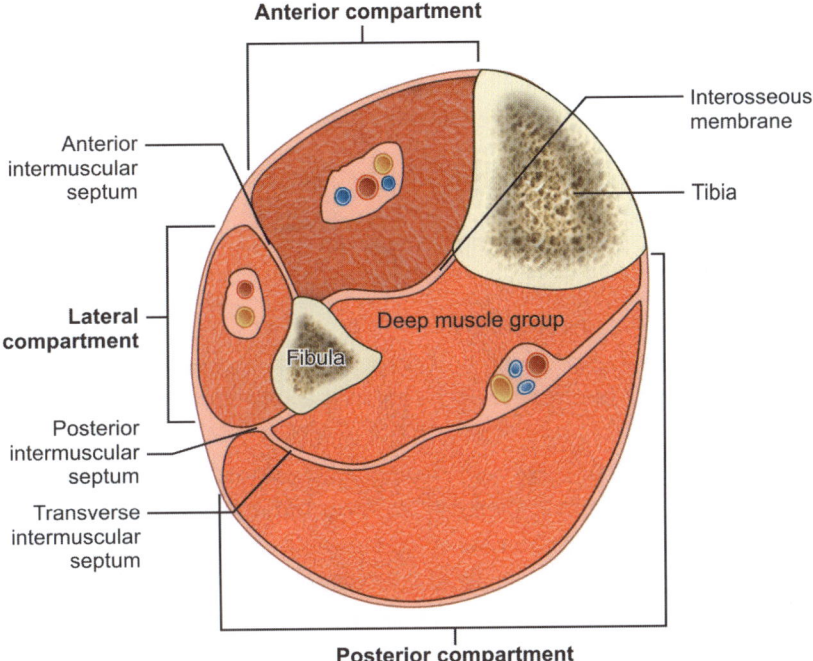

Fig. 3.44: Compartments of the leg

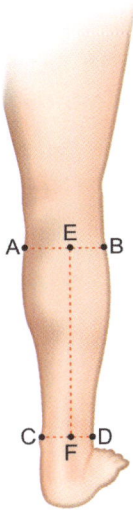

Fig. 3.45: Dissection of posterior crural region (skin incision)

ii. Another transverse incision between the lateral and medial malleoli (C–D).
iii. The third vertical incision extends from the midpoint of proximal incision to the midpoint of distal incision (E–F).

The skin flaps are reflected laterally and medially to expose the superficial fascia.

3. **Superficial fascia:** The structures in the superficial fascia are:
   i. Small saphenous vein.
   ii. Posterior femoral cutaneous nerve.
   iii. Sural nerve.
   iv. Sometimes, sural communicating nerve.

   Note that the **small saphenous vein** enters the back of the leg passing behind the lateral malleolus and lies along the middle of the leg in the superficial fascia. It pierces the deep fascia in the popliteal fossa and opens into the popliteal vein.

   The **sural nerve** (a branch of tibial nerve in the popliteal fossa) lies along the lateral aspect of the small saphenous vein. It pierces the deep fascia of leg and joins with the **sural communicating** branch of common peroneal nerve.

   Reflect the superficial fascia in the line of skin incision. Deep fascia is exposed.

4. **Deep fascia:** The deep fascia is incised vertically and reflected. Now, the posterior compartment of the leg is opened. Define the **flexor retinaculum (laciniate ligament)** which is the condensation of the deep fascia posteroinferior to the medial malleolus. Identify the structures passing deep to the retinaculum.

   After reflecting the deep fascia, identify the **gastrocnemius muscle** which is the most superficial muscle in the posterior compartment of the leg. Now, transect the two heads of gastrocnemius muscles at the point where they join (about 5 cm distal to their origins). Reflect the proximal and distal parts of the muscle (Fig. 3.46).

   Identify the **tendon of plantaris** which passes downwards and medially between the **lateral head of gastrocnemius** and **soleus muscle**. Then the tendon lies on the medial side of the tendon of gastrocnemius muscle and ultimately is attached to the middle part of the back of the calcaneus medial to the insertion of tendocalcaneous. *Remember that the plantaris muscle may be absent.*

   Identify the **soleus** muscle which lies deep to the gastrocnemius muscle. The two heads of gastrocnemius and soleus unite together in the lower part of the back of the leg forming **tendocalcaneous**. The tendocalcaneous is inserted into the middle part of the posterior surface of the calcaneum (Fig. 3.46).

   The **tibial nerve** and **posterior tibial vessels** exit the popliteal fossa after passing deep to the tendinous arch of the soleus and then they course distally in the **transverse intermuscular septum**.

   Transect the **tendocalcaneous** about 5 cm above its insertion into the calcaneum and separate the underlying muscles from the tendon by finger dissection.

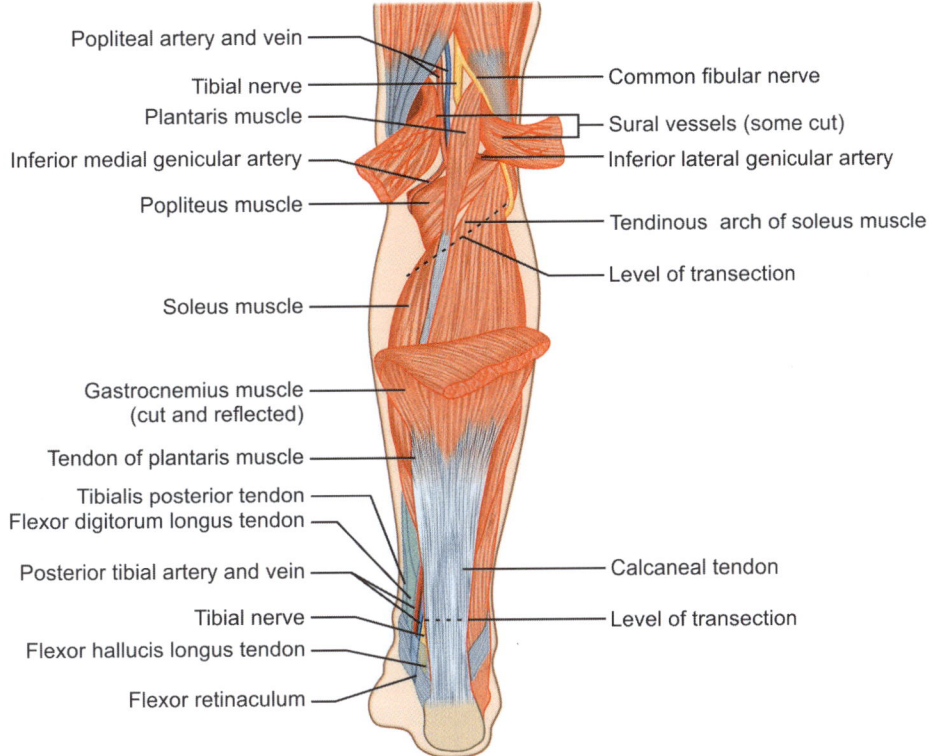

Fig. 3.46: Transection of gastrocnemius muscle

Cut the **soleus** muscle about 2.5 cm inferior to its tendinous arch. Then retract the soleus and distal part of the gastrocnemius muscle laterally to expose **the transverse intermuscular septum** which contains the **tibial nerve** and **posterior tibial vessels**. The posterior tibial artery is accompanied by two veins called **venae comitantes**.

Follow the **posterior tibial artery** proximally and distally. At the inferior border of the popliteus the artery divides into **anterior tibial artery** and **posterior tibial artery**. Distally, the posterior tibial artery lies between the tendons of flexor digitorum longus and flexor hallucis longus deep to the flexor retinaculum. The **peroneal artery** arises from the posterior tibial artery. The **tibial nerve** accompanies the artery (Fig. 3.47).

Incise the transverse intermuscular septum vertically to expose the long flexors of the toes. Here the **flexor hallucis longus lies lateral to the flexor digitorum longus**. Retract the flexor hallucis longus laterally to expose the another intermuscular septum which is then divided to reveal the **tibialis posterior muscle** (Fig. 3.48).

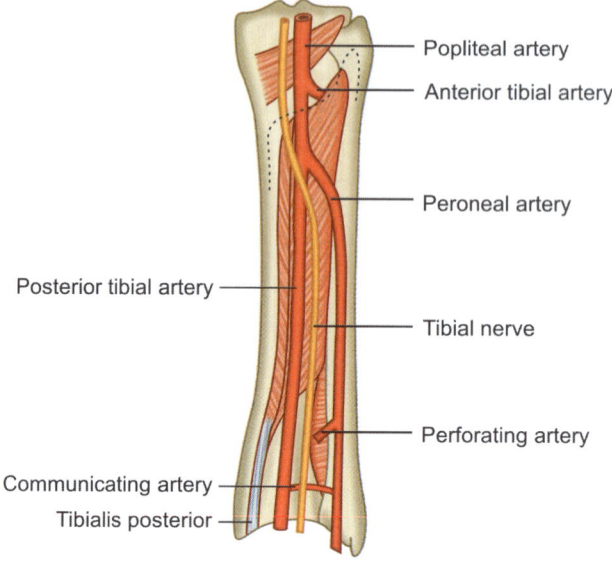

Fig. 3.47: Arteries and nerves in posterior crural region

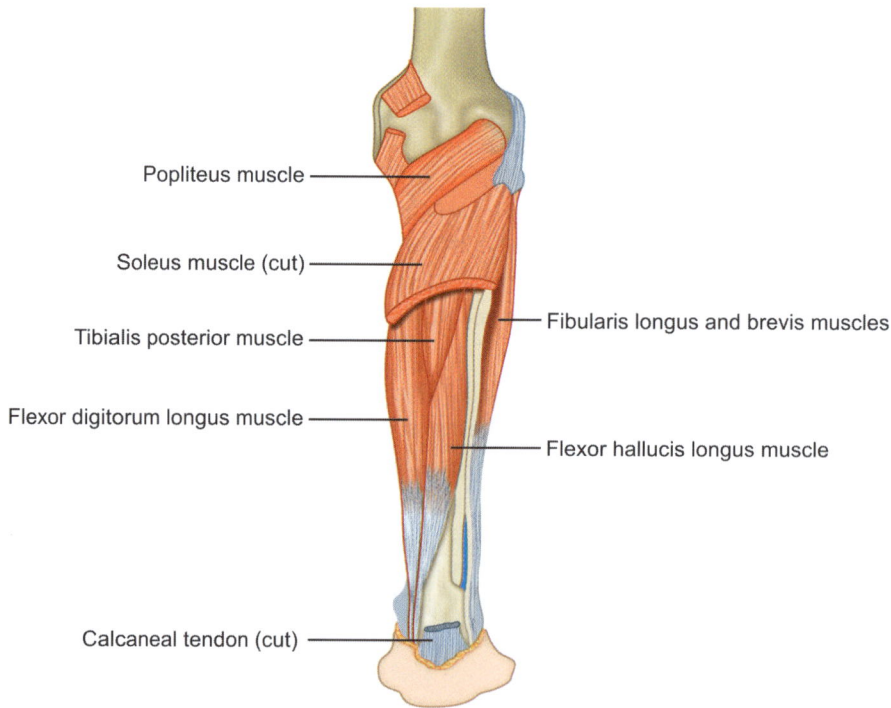

Fig. 3.48: Tibialis posterior

MUSCLES RELATED TO THE BACK OF THE LEG

Muscles	Origins	Insertions	Nerve supply	Actions
Gastrocnemius (Medial head is larger than lateral head)	i. **Medial head**: Upper and posterior part of medial condyle of femur ii. **Lateral head**: Depressed area on the lateral surface of the lateral condyle of femur	Middle 1/3rd of the posterior surface of the calcaneus forming **tendocalcaneus** with the tendon of soleus	Tibial nerve in the popliteal fossa	Plantar flexor of the foot
Soleus (Multipennate muscle)	i. Upper 1/4th of the posterior surface of the shaft and posterior surface of the head and neck of **fibula** ii. Soleal line and middle 1/3rd of the medial border of the **tibia** iii. Tendinous soleal arch	Gastrocnemius and soleus unite to form tendocalcaneous to be inserted into the middle 1/3rd of the posterior surface of the calcaneus	i. Deep surface by the tibial nerve in the leg ii. Superficial surface by the tibial nerve in popliteal fossa	i. Mild plantar flexor of foot ii. Steadying the leg on the foot
Plantaris	i. Lower part of the lateral supracondylar line of femur ii. Oblique popliteal ligament	Posterior surface of the calcaneum medial to the tendocalcaneus	Tibial nerve in the popliteal fossa	Weak plantar flexor of foot
Popliteus	Anterior part of the popliteal groove on the lateral surface of the lateral condyle of femur	Popliteal surface of the tibia above the soleal line	Tibial nerve in the popliteal fossa	Flexor of the knee joint (**Unlocking muscle**)
Flexor digitorum longus (FDL)	Medial part of posterior surface of shaft of tibia below the soleal line	Plantar surface of the base of distal phalanges of 2nd, 3rd, 4th and 5th toes	Tibial nerve in the back of the leg	i. Plantar flexor of lateral four toes ii. Maintains the medial longitudinal arch of foot
Flexor hallucis longus (FHL)	i. Lower 2/3rd of the posterior surface of the shaft of the fibula behind the medial crest ii. Interosseous membrane	Under surface of the distal phalanx of the great toe	Tibial nerve in the back of the leg	i. Flexor of the great toe ii. Maintains the medial longitudinal arch of the foot

Contd...

Contd...

Muscles	Origins	Insertions	Nerve supply	Actions
Tibialis posterior	i. Lateral part of the posterior surface of the middle 1/3rd of tibia below the soleal line ii. Medial part of upper 2/3rd of the posterior surface of the shaft of fibula iii. Upper 2/3rd of interosseous membrane	i. Tuberosity of navicular ii. Undersurface of all tarsal bones except talus iii. Undersurface of the base of 2nd, 3rd, 4th and sometimes 5th metatarsal bones	i. Tibial nerve in the back of the leg ii. Nerve to popliteus	i. Invertor of the foot ii. Plantar flexor of foot iii. Maintains medial longitudinal arch of foot

Special Notes on Popliteus
1. Intracapsular in origin.
2. Its origin is tendinous.
3. Its insertion is fleshy.
4. It receives its nerve supply from its deep surface.

Summary
1. Position of the cadaver: Prone and knee extended.
2. Skin incision as in Figure 3.44.
3. Superficial fascia: Main structures are:
    i. Small saphenous vein.
    ii. Sural nerve.
    iii. Posterior femoral cutaneous nerve.
4. Deep fascia is vertically incised and reflected to expose the contents.
5. Contents:
    a. **Muscles**:
        i. Medial and lateral heads of gastrocnemius.
        ii. Soleus.
        iii. Plantaris.
        iv. Flexor digitorum longus.
        v. Flexor hallucis longus.
        vi. Tibialis posterior.
    b. **Nerve**: Posterior tibial nerve (tibial nerve in the back of the leg).
    c. **Vessels**:
        i. Posterior tibial vessels.
        ii. Peroneal vessels.

Window Dissections: Lower Limb (Inferior Extremity)

PROBABLE QUESTIONS AND ANSWERS

1. What are the boundaries of the posterior compartment of the leg?

Ans.
a. Anteriorly:
   i. Posterior surface of the shaft of the tibia.
   ii. Posterior surface of the shaft of the fibula.
   iii. Interosseous membrane.
b. Posteriorly: Deep fascia of the leg (fascia cruris).
c. Laterally: Posterior intermuscular septum which extends from the deep fascia to the posterior border of the shaft of fibula.

2. How many subdivisions of the posterior compartment are there?

Ans. There are **three** subdivisions of the posterior compartment.
   i. Superficial.
   ii. Intermediate.
   iii. Deep.

3. How they are divided?

Ans. They are divided by the **deep transverse fascia of the leg** and the **fascia covering the tibialis posterior.**

Derivatives of Deep Fascia of the Leg
a. Anterior and posterior intermuscular septum.
b. Deep transverse fascial septum in the back of the leg.
c. Extensor retinaculum: Superior and inferior.
d. Peroneal retinaculum: Superior and inferior.
e. Flexor retinaculum.

---

4. What are the contents of these subcompartments?

Ans. Superficial part:
a. **Muscles:**
   i. Both heads of gastrocnemius.
   ii. Soleus.
   iii. Plantaris.
b. **Nerves:**
   i. Sural nerve (branch of tibial nerve in the popliteal fossa)
   ii. Sural communicating branch of common peroneal nerve.

Intermediate part:
a. **Muscles:**
   i. Flexor digitorum longus.
   ii. Flexor hallucis longus.
b. **Vessels:** Posterior tibial vessels.
c. **Nerve:** Tibial nerve.

Deep part (muscles): Tibialis posterior.

5. What is tendocalcaneus?

Ans. The tendon of both the heads of gastrocnemius fuses with the tendon of soleus to form the **tendocalcaneus**.

6. Where does the tedocalcaneus insert?

Ans. The tendocalcaneus is inserted into the middle third of the posterior surface of the calcaneum.

7. What is triceps surae?

Ans. The two heads of **gastrocnemius** and the **soleus** are collectively named as the **triceps surae**.

8. What is the origin and insertion of plantaris muscle?

Ans.
i. **Origin:** Lower part of lateral supracondylar line.
ii. **Insertion:** Posterior surface of the calcaneum, medial to the insertion of tendocalcaneus.

9. What is the relation of the plantaris with the gastrocnemius?

Ans. The plantaris muscle arises by a short belly and its long tendon passes deep to the lateral head of gastrocnemius and appears between the gastrocnemius and soleus. Then it passes downwards and medially to lie on the medial side of the tendocalcaneus and is inserted into the posterior surface of the calcaneum, medial to the tendocalcaneus.

10. What is the continuation of plantaris?

Ans. **Plantar aponeurosis** of the sole of the foot is the continuation of the plantaris.

Note that the **plantaris** is a vestigial muscle in human beings. In man, it is **accessory to the gastrocnemius** muscle. Sometimes, the plantaris is **absent** in one or both the legs. Rarely, it may be **double**. Its functional importance is negligible.

11. What is fabella?

Ans. **Fabella** is a small sesamoid bone present in the tendon of origin of lateral head of gastrocnemius.

12. What is Brodie's bursa?

Ans. The bursa which lies deep to the medial head of gastrocnemius is known as **Brodie's bursa**. It communicates with the semimembranosus bursa in few occasions and always communicates with the knee joint.

13. What is anserine bursa?

Ans. It is a large bursa which separates the tendons of sartorius, gracilis and semitendinosus from one another at their sites of insertion and from the bony surface of the tibia and from the tibial collateral ligament of the knee joint.

14. What are the detached part of the gastrocnemius and soleus in the sole of the foot?

Ans. The plantar ligament and the flexor digitorum brevis are the detached distal parts of gastrocnemius and soleus respectively.

> **Tendocalcaneus** also known as **tendo-Achilis** is the **thickest** and **strongest** tendon of the body. It is about **15 cm** long and acts as prime mover of plantar flexion of the foot at the ankle joint.

15. Why the muscle soleus is so named?

Ans. It is so named because of its shape like the sole of foot.

16. What is the outline of origin of soleus?

Ans. Horseshoe-shaped in origin (multipennate).

17. What is the nerve supply of soleus?

Ans. Soleus has **double nerve supply.**
  i. **Tibial nerve in the popliteal fossa** (medial popliteal nerve) from its superficial surface.
  ii. **Posterior tibial nerve** (tibial nerve in the leg) from its deep surface.

18. What are the arrangements of soleal fibers and gastrocnemius fibers in the tendo-calcaneus?

Ans. The soleal fibers are mainly inserted on the medial side and the fibers of gastrocnemius are inserted on the lateral side of the tendocalcaneus.

19. Why the soleus is called the peripheral heart?

Ans. The soleus contains a rich venous plexus. This venous plexus communicates with the great saphenous vein (a superficial vein) through the perforating veins. Again, this plexus also communicates with the deep veins of the leg. These veins are provided with valves so that the blood flows only from superficial to deep veins. The contraction of the soleus muscle squeezes the venous blood centripetally and helps in the venous return and acts as **peripheral heart.**

20. Name another muscle which acts as peripheral heart in human body.

Ans. Lateral pterygoid muscle in the head and neck region is also called peripheral heart; because it pumps out the blood of pterygoid venous plexus.

21. Soleus acts as a bottom gear muscle and gastrocnemius acts as a top gear muscle: explain.

Ans. The soleus is more powerful than the gastrocnemius, but the gastrocnemius is faster acting and increases the range of movement. The soleus overcomes the inertia of the body weight and helps to start walking. So, it acts as a bottom gear like that of the bottom gear of a car. On the other hand, the gastrocnemius increases the speed and range of movement like that of the top gear of a car. So, the gastrocnemius is described as top gear muscle.

> *The soleus corresponds with the flexor digitorum superficialis of the forearm from morphological point of view.*

22. What are the muscles deep to the deep transverse fascia of the leg?

Ans. These are three **bipennate** muscles:
  i. Flexor digitorum longus.
  ii. Flexor hallucis longus.
  iii. Tibialis posterior.

23. What is the nerve supply of these muscles?

Ans. **Tibial nerve** in the posterior crural region. Tibialis posterior receives an additional twig from the nerve to popliteus.

> **Remember that** all the muscles of the posterior compartment of the leg are supplied by the posterior tibial nerve, the muscles of the anterior compartment are supplied by the anterior tibial (deep peroneal) nerve and the muscles of lateral compartment are supplied by musculocutaneous (superficial peroneal) nerve.

Window Dissections: Lower Limb (Inferior Extremity)

24. What is the main artery of the posterior crural region?

Ans. The **posterior tibial artery**. It is the larger terminal branch of popliteal artery.

25. What is the other terminal branch of popliteal artery?

Ans. The anterior tibial artery.

26. Where does the popliteal artery terminate?

Ans. The popliteal artery terminates into the anterior and posterior tibial arteries at the lower border of popliteus.

27. How does the anterior tibial artery enter the anterior crural region?

Ans. The anterior tibial artery traverses through an angular gap between the two heads of origin of tibialis posterior and then passes forward above the interosseous membrane to enter the anterior crural region.

Note that the posterior and anterior tibial arteries are accompanied by a pair of **venae comitantes** which establish frequent communications with each other around the arteries.

28. What is venae comitantes?

Ans. An artery accompanied by a pair of veins on its each side is called venae comitantes.

29. Name the branches of posterior tibial artery?

Ans.
  i. Circumflex fibula artery.
  ii. **Peroneal artery.**
  iii. **Nutrient artery to the tibia.**
  iv. Muscular branches.
  v. Communicating branch.
  vi. Calcaneal branch.
  vii. Medial malleolar branch.
  viii. **Two terminal branches:**
    a. Medial plantar artery.
    b. Lateral plantar artery.

30. What are the branches of the peroneal artery?

Ans.
  i. Muscular branch.
  ii. Nutrient artery to the fibula.
  iii. Communicating branch.
  iv. Perforating branch.
  v. Lateral calcaneal artery (terminal branch)

Note that the nutrient artery to the tibia is a branch from the posterior tibial artery and the nutrient artery to the fibula is a branch from peroneal artery.

31. **What is the bony landmark of the peroneal artery?**

    Ans. The artery descends along the **medial crest of fibula**.

32. **What is porta pedis?**

    Ans. It is the channel, situated deep to the flexor retinaculum of the foot through which the structures pass from the posterior crural region to the sole of the foot.

33. **Name the structures that pass through the porta pedis?**

    Ans. **From medial to lateral** (Remember the mnemonic Tom, Dick And Harry):
    Tom → **Tibialis** posterior
    Dick → Flexor **Digitorum** longus.
    And → Posterior tibial **Artery** and tibial nerve.
    Harry → Flexor **Hallucis** longus.

34. **Why the posterior compartment is the bulkiest of three compartments of the leg?**

    Ans. This is because of the powerful **antigravity** superficial muscles (gastrocnemius and soleus) which have to raise the heel during walking.

Remember that the posterior tibial nerve represents both the median and ulnar nerves of the forearm.

LESSON 10: SOLE OF THE FOOT

The skin of the sole is much thicker than the other parts of the body. Moreover, the skin of the sole of the foot is devoid of hairs and sebaceous glands but is provided with numerous sweat glands. The muscles of this region are arranged in **four layers**. All the intrinsic muscles of the sole are supplied by the plantar nerves and the extrinsic muscles are supplied by the nerve of the respective compartment of the leg. Unlike in the hand there is only one plantar arterial arch in the sole.

STEPS OF DISSECTION

1. **Position of the cadaver:** Cadaver is in the **prone** position.
2. **Skin incision (Fig. 3.49):**
    i. A transverse incision along the margin of the heel (A–B)
    ii. Another transverse incision at the roots of the toes (C–D)

Window Dissections: Lower Limb (Inferior Extremity)

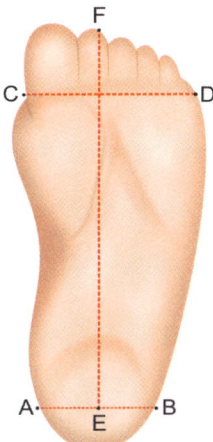

Fig. 3.49: Dissection of sole of the foot (skin incision)

   iii. A longitudinal incision from the midpoint of first incision (A–B) to the tip of the 2nd toe (E–F)

   Reflect the skin flaps medially and laterally. The superficial fascia is exposed.

3. **Superficial fascia:** The superficial fascia is fibrous and dense. It is very thick and dense over the weight-bearing points. It contains cutaneous nerves and vessels. These are the cutaneous branches of the lateral plantar and medial plantar nerves and arteries. Use a scalpel handle to scrape the superficial fascia off the plantar aponeurosis. Deep fascia is exposed.

4. **Deep fascia:** The deep fascia of the sole is called **plantar aponeurosis** (Fig. 3.50). The central part is thick and the lateral and medial parts of the deep fascia are thin. The plantar aponeurosis is approximately 4 mm thick and distally it divides into five bands, one to each toe. Transect the plantar aponeurosis at two sites. One, close to its calcaneal attachment and another in the anterior 1/3rd of the foot. A longitudinal incision is made joining the mid-points of these above two incisions (Fig. 3.51). The plantar aponeurosis is released from the underlying structures and reflect the flaps laterally and medially. The first layer of the sole is exposed.

5. **First layer of the sole (Fig. 3.52):** Clean and identify the muscles, vessels and nerves in this layer. The **abductor hallucis** muscle is located along the medial border of the sole. The **abductor digiti minimi** muscle is located along the lateral border of the sole. The **flexor digitorum brevis** muscle is located between the former two muscles. Trace the tendons of these muscles to their distal attachments.

   The medial plantar vessels and nerve lie between the abductor hallucis brevis and flexor digitorum brevis. The lateral plantar vessels and nerve are found between the flexor digitorum brevis and abductor digiti minimi. **Note that** the lateral plantar vessels and nerve cross the sole from medial to lateral side deep to the proximal end of flexor digitorum

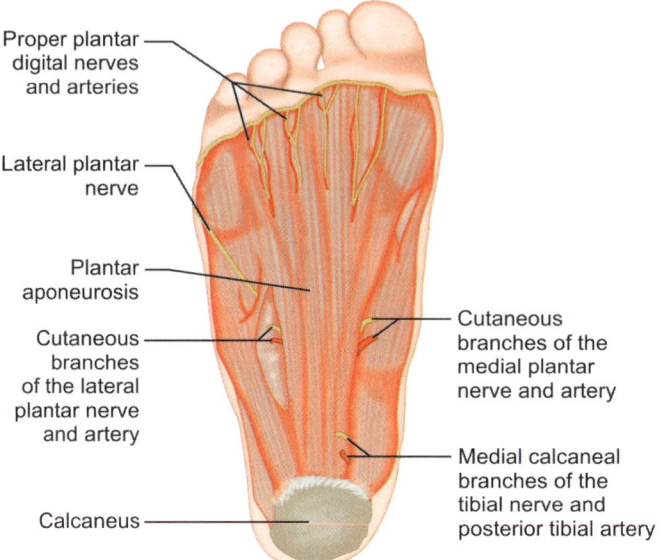

Fig. 3.50: Plantar aponeurosis

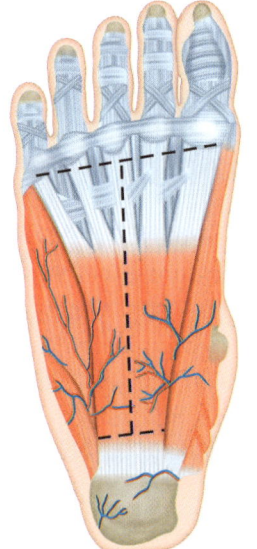

Fig. 3.51: Incisions on plantar aponeurosis

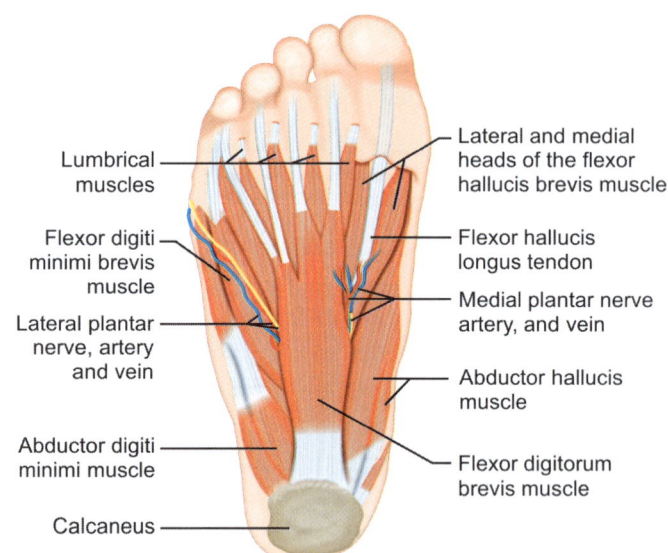

Fig. 3.52: 1st layer of sole of foot

brevis (i.e. between the first and 2nd layer of muscles). Transect the flexor digitorum brevis close to the calcaneus and reflect the muscle distally. Care is to be taken to preserve the underlying lateral plantar vessels and nerve.

6. **2nd layer of the sole (Fig. 3.53):** Retract the abductor hallucis muscle medially and the abductor digiti minimi muscle laterally to display the muscles of the 2nd layer of the sole.

Window Dissections: Lower Limb (Inferior Extremity)

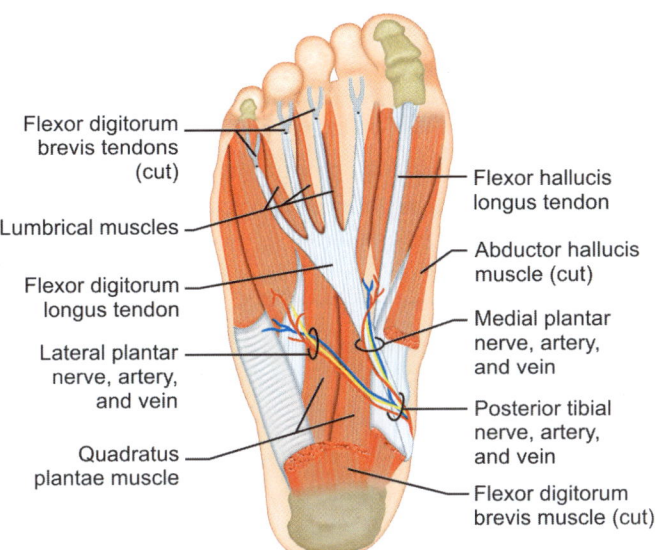

Fig. 3.53: 2nd layer of foot

Alternately, a probe is passed deep to the abductor hallucis and cut the muscle over the probe in its proximal part and detach the abductor digiti minimi muscle from its proximal attachment.

Identify the **flexor digitorum accessorius (quadratus plantae)** muscle deep to the flexor digitorum brevis. Clean and trace the muscle till its insertion into the **tendon of flexor digitorum longus.** The tendon of flexor digitorum longus passes laterally crossing superficial to the tendon of flexor hallucis longus. Then the digitorum tendon is separated into four digital tendons for insertion into the base of the terminal phalanges of the lateral four toes. The **tendon of flexor hallucis longus** inclines medially lying deep to the flexor digitorum longus and finally is inserted into the plantar surface of the base of the distal phalanx of the great toe. Observe the **four lumbrical muscles,** numbered from medial to lateral side. They arise from the adjacent sides of the digital tendons of flexor digitorum longus. The medial and lateral plantar vessels and nerves are also seen in this layer.

Transect the flexor digitorum longus tendon where it is joined by the flexor digitorum accessorius muscle and reflect the tendons distally, along with the lumbrical muscles to reveal the muscles of the third layer of the sole.

7. **3rd layer of the sole (Fig. 3.54):** Identify the **flexor hallucis brevis** muscle with its **two heads** (medial and lateral). A sesamoid bone is found in each of the tendons of its two heads. Identify the **two heads (transverse** and **oblique)** of **adductor hallucis** muscle. Now, separate the flexor hallucis brevis and oblique head of adductor hallucis from their origin and reflect them distally. Reflect the transverse head of adductor hallucis medially. Identify the **flexor digiti minimi brevis muscle** and detach it from its origin and reflect it forwards. Find out the **deep plantar arterial arch** and **deep plantar nerve.** Clean them and preserve them.

8. **4th layer of the sole (Fig. 3.55):** The **deep plantar arch** is formed at the level of the bases of the metatarsal bones between the 3rd and 4th layers of muscles of the sole. The **interossei muscles** are located deep to the plantar arch. The **dorsal interossei** are **four** in number and the **plantar interossei are three** in number. **Note** that the dorsal interossei are **bipennate** (Fig. 3.56A) and plantar interossei are **unipinnate** muscles (Fig. 3.56B). Follow the tendon of **fibularis longus (peroneus longus)** muscle across the deep plantar surface of the foot

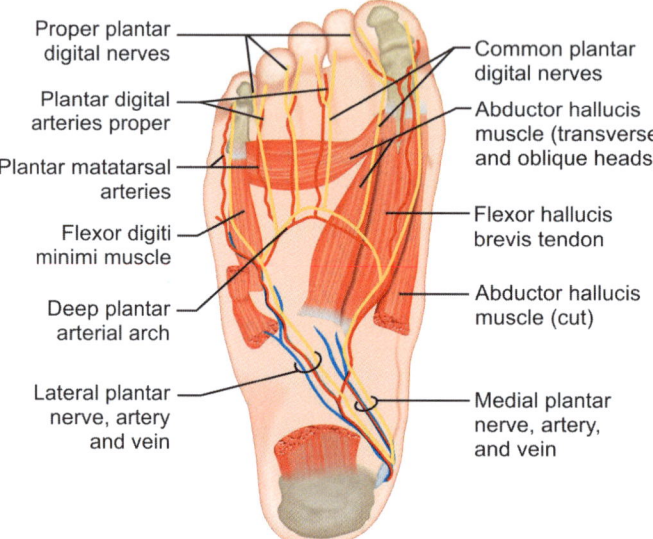

Fig. 3.54: 3rd layer of foot

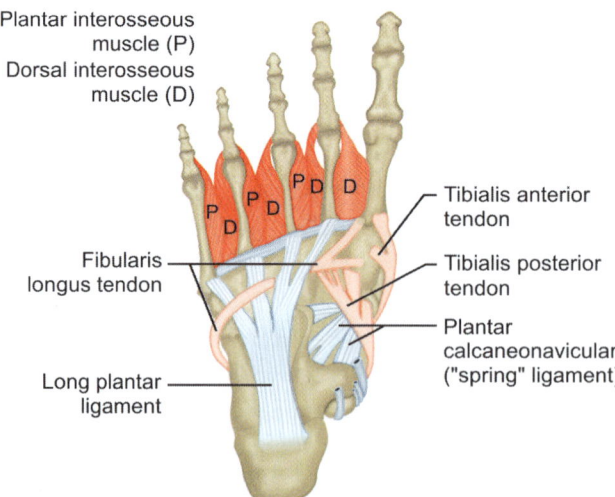

Fig. 3.55: 4th layer of foot

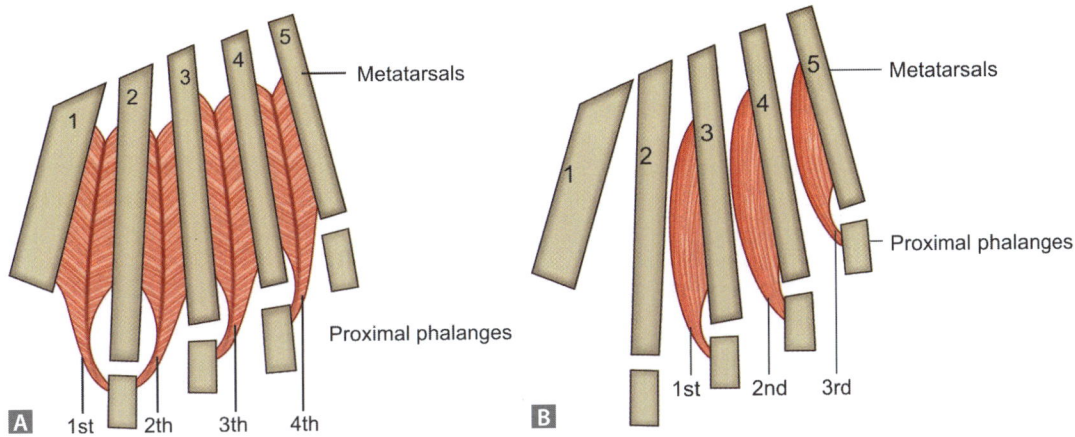

Figs 3.56A and B: (A) Dorsal interossei; (B) Plantar interossei

to the base of the first metatarsal and medial cuneiform bone. Trace the tendon of **tibialis posterior** muscle to its multiple insertions on the tarsal bones. Two important ligaments are found in the deeper aspect of the sole. These are **long plantar ligament** and **plantar calcaneonavicular (spring)** ligament.

Remember that all intrinsic muscles of the sole of the foot are supplied by the lateral plantar nerve except four muscles (abductor hallucis, flexor digitorum brevis, flexor hallucis brevis and first lumbrical) which are supplied by the medial plantar nerve.

Deep plantar arterial arch is formed by the deep branches of **lateral plantar artery** and **arteria dorsalis pedis** between the 3rd and 4th layer of muscles of the sole of the foot.

Summary

1. Position of the cadaver: Prone position.
2. Skin incision as in Figure 3.49.
3. Superficial fascia: In it, cutaneous vessels and nerves are found
4. Deep fascia: It is also called **plantar aponeurosis**. It is having a **central thick part** and covers the flexor digitorum brevis. The **lateral** and **medial** parts are thin and cover abductor digiti minimi and abductor hallucis respectively. Deep fascia is incised and reflected to expose the deeper aspects of the sole.
5. Plantar muscles, lateral and medial plantar vessels and nerves are observed in different layers of the sole.

MUSCLES RELATED TO THE SOLE OF THE FOOT

Layers of the sole	Muscles
First layer (3 muscles)	i. Abductor hallucis ii. Flexor digitorum brevis iii. Abductor digiti minimi

2nd layer (4 muscles)
(2 extrinsic muscles, 2 intrinsic muscles)
	a. **Extrinsic muscles:**
	  i. Flexor hallucis longus
	  ii. Flexor digitorum longus
	b. **Intrinsic muscles:**
	  i. Flexor digitorum accessorius
	  ii. Four lumbricals (1st, 2nd, 3rd, 4th from medial to lateral)

3rd layer (3 muscles)	i. Flexor hallucis brevis ii. Adductor hallucis (two heads: transverse and oblique) iii. Flexor digiti minimi brevis

4th layer
(Extrinsic muscles = 2
Intrinsic muscles =7)
	a. **Extrinsic muscles:**
	  i. Peroneus longus (fibularis longus)
	  ii. Tibialis posterior
	b. **Intrinsic muscles:**
	  i. Four dorsal interossei
	  ii. Three plantar interossei

PROBABLE QUESTIONS AND ANSWERS

1. How does the skin of the sole of the foot differ from the skin of other parts of the body?

Ans. i. The skin of the sole of the foot (and also the skin of the palm) is much thicker than the other parts of the body.
 ii. The skin of the sole and palm is devoid of hairs and sebaceous glands.
 iii. The skin of the sole and palm is provided with numerous sweat glands.

2. Why does the sebaceous cyst never occur in the sole and palm?

Ans. The skin of the sole of the foot and the palm is devoid of sebaceous glands. So, cysts are not formed in these areas.

3. What is plantar aponeurosis?

Ans. The deep fascia in the sole is specialized to form the plantar aponeurosis which is composed of compact bundles of collagen fibers.

4. What are the other specialized forms of the deep fascia of the sole?

Ans.
i. Deep transverse metatarsal ligaments (four in number).
ii. Fibrous flexor sheaths in the toes.

5. What are the superficial transverse metatarsal ligaments?

Ans. The superficial transverse metatarsal ligaments are the thickened bands of superficial fascia stretching across the roots of the toes.

6. What are the parts of the plantar aponeurosis?

Ans. Three parts:
i. Central part (thick).
ii. Lateral part (thin).
iii. Medial part (thin).

> **Morphologically**, the plantar aponeurosis represents the distal part of the plantaris which has become separated from the rest of the muscle because of the enlargement of the heel during evolution.

7. What is the shape of the plantar aponeurosis?

Ans. Triangular. Apex—posteriorly and the broader anterior part splits into five bands, one for each toe (Fig. 3.50).

8. What is plantar fascia?

Ans. The plantar fascia is made up of predominantly longitudinally oriented collagen fibers which is having three distinct structural components: the **medial component**, the **central component (plantar aponeurosis)** and the **lateral component**.

9. How the muscles of the sole are arranged?

Ans. The muscles of the sole (**plantar muscles**) are arranged in **four layers**. They are named as first, second, third and fourth layers from superficial to deep.

10. Where do you find the plantar vessels and nerves?

Ans. The **trunks** of medial and lateral plantar vessels and nerves are found between the first and second layer, whereas the deep branches of the lateral plantar vessels and nerves intervene between the third and fourth layers.

11. Name the muscles of the different layers of the sole?

Ans. Refer to the table above (Muscles related to the sole of the foot).

12. What are the differences between the plantar and palmar aponeurosis?

Plantar aponeurosis	Palmar aponeurosis
1. Weight bearing	1. Not weight bearing
2. Distal part divides into five digital slips	2. Divides into **four** slips
3. Proximal part is attached to the bone (medial tuberosity of calcaneum)	3. Proximal part is attached to the flexor reticulum of the hand
4. **Three muscles of the first** layer of the sole (abductor hallucis, flexor digitorum brevis and abductor digiti minimi) arise from plantar aponeurosis	4. Only one muscle (palmaris brevis) arises from it
5. **Morphologically**, it is the distal part of plantaris	5. It is the distal part of palmaris longus

13. What is quadratus plantae?

Ans. The **flexor digitorum accessorius**, one of the muscles of the second layer of the sole is also known as quadratus plantae.

14. Why the muscle is called flexor digitorum accessorius?

Ans. It is so named because it is accessory to the flexor digitorum longus muscle.

15. What are the origins and insertion of flexor digitorum accessorius?

Ans. **Origins: Fleshly medial head** arises from the medial surface of calcaneum and the **tendinous lateral head** arises from the area in front of the lateral tubercle of the calcaneum.

Insertion: Into the lateral side of the tendon of the flexor digitorum longus.

16. What is its action?

Ans. It converts the diagonal pull of the flexor digitorum longus into a straight pull.

17. How many lumbrical muscles are there?

Ans. There are **four** lumbrical muscles in each foot. They are named as first, second, third and fourth, numbered from medial to lateral side.

18. Why these muscles are so named?

Ans. This is because of their resemblance with the earthworms.

19. From where do these lumbricals arise?

Ans. They arise from the adjacent sides of the tendons of flexor digitorum longus except the first lumbrical which arises only from the medial side of the first tendon of FDL. Therefore, all the lumbricals are **bipennate** in origin, whereas the first lumbrical is **unipennate**.

20. What are the sites of insertions of these lumbricals?

Ans. i. Each lumbrical is inserted into the medial side of the base of the proximal phalanx of the 2nd to 5th toes.
ii. Via **dorsal digital expansion**, these are inserted into the dorsal surface of the base of the middle and distal phalanges of the 2nd to 5th toes.

21. What are the actions of the lumbricals?

Ans. They are the flexors of metatarsophalangeal joints and via dorsal digital expansion, they are extensors of interphalangeal joints.

22. What are the nerve supply of lumbricals?

Ans. **The first lumbrical** is supplied by the **medial plantar nerve** and the **other three lumbricals** are supplied by the **deep branch of lateral plantar nerve**.

> **Note that** the tendon of flexor digitorum longus crosses the tendon of flexor hallucis longus superficially from medial to lateral side in the proximal part of the sole of the foot. The digitorum longus tendon lies medial to the hallucis longus tendon in the back of the leg. In the sole, their relative positions are changed due to the crossing of these tendons.

23. Under which muscle tendon of the foot lies sesamoid bone?

Ans. Two sesamoid bones lie under the tendon of **flexor hallucis brevis**, a third layer muscle.

24. From how many heads the adductor hallucis muscle arises?

Ans. **Two heads:**
i. Oblique head.
ii. Transverse head.

25. How many interossei muscles are there?

Ans. There are **seven interossei** muscles of which four are **dorsal interossei** and three are **plantar interossei**.

26. What are the actions of these interossei?

Ans. Dorsal interossei are **abductors** of the toes and the plantar interossei are **adductors** of the toes. **Axis** of abduction-adduction movement passes through second metatarsal bone and second toe. **Remember** that, in the hand the axis passes through the middle finger. In addition to abduction and adduction, the interossei (both dorsal and plantar) are the **flexors of metatarsophalangeal joint** and via dorsal digital expansion, these are **extensors of the interphalangeal joint**.

27. Do the dorsal interossei abduct all the toes?

Ans. No. Abduction of the great toe and little toe is performed by separate muscles (abductor hallucis and abductor digiti minimi respectively). Dorsal interossei perform abduction only of the middle three toes.

> **Remember that all dorsal interossei muscles** are bipennate in origin, as each of them arises from the adjacent sides of the shafts of five metatarsal bones. **Plantar interossei** muscles arise from the medial side of the 3rd, 4th and 5th metatarsal bones and, therefore, **unipennate** in origin.

28. Why the first and second metatarsal bone do not provide origin to plantar interossei muscles?

Ans. The great toe is provided with a separate adductor muscle (adductor hallucis). So, adduction of the great toe is performed by adductor hallucis, not by plantar interosseous. On the other hand, the axis of the foot passes through the second toe. It means that the second toe is already is in adducted position. So, no plantar interosseous is needed for adduction.

29. What is the nerve supply of interossei muscles?

Ans. All dorsal and plantar interossei muscles are supplied by the **deep branch** of lateral plantar nerve except the interossei occupying the fourth inter-metatarsal space which are supplied by the **superficial branch** of lateral plantar nerve.
In addition, first and second dorsal interossei are supplied by the terminal branches of the **deep peroneal nerve**.

30. What are the interossei occupying the 4th intermetatarsal space?

Ans. 3rd plantar and 4th dorsal interossei muscles.

> **Note that** all interossei muscles and dorsal digital vessels and nerves lie above the deep transverse metatarsal ligament, whereas the lumbricals and plantar digital vessels and nerves lie below that ligament.

31. Name the muscles which maintain both the longitudinal and transverse arches of the foot?

Ans. i. Peroneus longus:
ii. Tibialis posterior.
These two muscle tendons cross each other, the peroneus longus being superficial and the tibialis posterior lies in deep.

32. What are the chief arteries of the sole?

Ans. Lateral and medial plantar arteries, which are the terminal branches of posterior tibial artery.

Window Dissections: Lower Limb (Inferior Extremity)

33. What is plantar arterial arch?

Ans. It is an arterial arch formed by the deep branch of lateral plantar artery and arteria dorsalis pedis.

34. What are the branches of plantar arterial arch?

Ans.
  i. Four plantar metatarsal arteries.
  ii. Three perforating arteries.

35. Where do the arterial arch exist?

Ans. The arch intervenes between the 3rd and 4th layers of muscles of the sole of the foot.

36. What are the chief nerves of the sole?

Ans. Lateral plantar and medial plantar nerves, which are the terminal branches of the tibial nerve.

37. Where does the termination of the tibial nerve occur?

Ans. The termination of the nerve takes place under the cover of the flexor retinaculum.

38. Where do you find the medial plantar vessels and nerve in the sole?

Ans. The medial plantar vessels and nerve lie between the abductor hallucis and the flexor digitorum brevis. Both these muscles are the muscles of first layer of the sole.

39. Where do you find the lateral plantar vessels and nerve?

Ans. The lateral plantar vessels and nerve run obliquely between the first and second layers of the sole.

40. Name the muscles that are supplied by the medial plantar nerve?

Ans. **Four muscles** of the sole are supplied by the medial plantar nerve. These are:
  i. Abductor hallucis (in first layer).
  ii. Flexor digitorum brevis (in first layer).
  iii. First lumbrical (in second layer).
  iv. Flexor hallucis brevis (in third layer).
  All other intrinsic muscles of the sole are supplied by the lateral plantar nerve.

41. What are the branches of the lateral plantar nerve?

Ans.
  i. Superficial branch.
  ii. Deep branch.

Remember that the deep branch of the lateral plantar nerve lies within the concavity of the plantar arterial arch.

42. What is the relative position between the plantar arteries and plantar nerves?

Ans. The plantar arteries (medial and lateral) are closer to the margins of the sole than their corresponding nerves. This is because of the fact that the posterior tibial artery divides into the medial and lateral plantar arteries earlier (higher) than the divisions of the tibial nerve.

*Remember that the **medial plantar** nerve is the larger terminal branch of the tibial nerve and resembles the median nerve of the hand in its distribution. The lateral plantar nerve is the smaller terminal branch of the tibial nerve and resembles the ulnar nerve in the hand in its distribution.*

Muscles of the Lower Limb Having Double Nerve Supply

1. **Pectineus:**
    i. Flexor (lateral) part by femoral nerve.
    ii. Adductor (medial) part by obturator nerve.
2. **Adductor magnus:**
    i. Adductor part by obturator nerve.
    ii. Hamstring (extensor) part by sciatic nerve.
3. **Biceps femoris:**
    i. Long head by tibial part of sciatic nerve.
    ii. Short head by common peroneal part of sciatic nerve.
4. **Soleus:**
    i. Tibial nerve in the popliteal fossa from its superficial surface.
    ii. Tibial nerve in the back of the leg from its deep surface.
5. **Tibialis anterior:**
    i. Anterior tibial (deep peroneal) nerve.
    ii. Recurrent genicular branch of common peroneal nerve.
6. **Tibialis posterior:**
    i. Tibial nerve in the back of the leg.
    ii. Nerve to popliteus.
7. **1st dorsal interosseous:**
    i. Deep branch of lateral plantar nerve.
    ii. Medial branch of deep peroneal nerve.
8. **2nd dorsal interosseous:**
    i. Deep branch of lateral plantar nerve.
    ii. Lateral branch of deep peroneal nerve.

CHAPTER 4

Window Dissections: Abdomen

- Introduction to Abdomen
- Inguinal Canal
- Rectus Sheath
- Exposure of Kidney from Back

LESSON 1: INTRODUCTION TO ABDOMEN

The abdomen is the lower part of the trunk and roughly cylindrical in shape. The **inferior thoracic aperture** forms the superior opening to the abdomen which is closed by a musculo-tendinous partition called **the diaphragm**. So, the abdomen is that part of the trunk which lies below the diaphragm. The abdomen contains a cavity called **abdominal cavity** which is subdivided by the inlet of pelvis into a larger upper part, the **abdomen proper** and a smaller lower part, the lesser pelvis or **pelvic cavity**. The abdominal cavity is lined by an epithelial-like single layer of cells (**mesothelium**) called **peritoneum** which is similar to the pleura and serous pericardium in the thorax. This is why the abdominal cavity is called the **peritoneal cavity** which freely communicates with the **pelvic cavity**. So, infections in one region freely spread to the other. The peritoneal cavity is a closed sac in male but in female the cavity communicates outside through the pelvic ostia of the uterine tubes. So exogenous source of infection in the peritoneal cavity is more common in females than in males. The peritoneum which lines the abdominal wall is called **parietal peritoneum** and the peritoneum which suspends the viscera is called **visceral peritoneum**. Both the layers of peritoneum are continuous to enclose a large potential space called **peritoneal cavity** which contains **peritoneal fluid**. In fact, all the abdominal viscera are intra-abdominal but extraperitoneal. **Retroperitoneal** structures are those which lie between the parietal peritoneum and posterior abdominal wall. These structures include the kidneys, ureters, some parts of small and large intestines.

BOUNDARIES OF THE ABDOMEN

1. **Anterior wall (musculoaponeurotic):** It is formed by:
    i. **Three flat muscles** and their aponeurosis (external oblique, internal oblique and transversus abdominis).
    ii. Rectus abdominis and pyramidalis muscles within the rectus sheath.

2. **Posterior wall (osseomusculofascial)**: It is formed by:
    i. **Bones**: Five lumbar vertebrae and their intervening intervertebral discs in the midline.
    ii. **Muscles**: Lateral to the vertebral column → Quadratus lumborum, psoas major and iliacus.
    iii. **Below iliac crest**: It is bony and formed by the ilium of hip bones.
    iv. **Fasciae**: Fascia iliaca and thoracolumbar fascia.
3. **Lateral walls:**
    i. **Above the iliac crest:** Predominantly **muscular** and formed **by three layers of muscles** → transversus abdominis, internal oblique and external oblique.
    ii. **Below the iliac crest**: Bony and formed by the ilium of hip bones.
4. **Roof (superior wall)**: Under surface of the diaphragm. It forms a partition between the thoracic cavity and abdominal cavity.
5. **Floor (musculofascial)**: It is formed by:
    i. Pelvic diaphragm in the posterior part.
    ii. Urogenital diaphragm in the anterior part.

    It is seen that the abdominal cavity is extensive, its considerable part is overlapped by the thoracic bony cage above and by the bony pelvis below.

CONTENTS

i. Major elements of gastrointestinal system.
ii. Urinary system.
iii. Reproductive system.
iv. Suprarenal glands.
v. Major neurovascular structures.
vi. Lymph nodes and lymphatics.
vii. Muscles and fasciae.

RELATIONSHIP OF THE ABDOMEN TO OTHER REGIONS

1. **Thorax**: The musculotendinous partition between the abdomen and the thorax is called **the diaphragm**. Structures pass between the two regions either through the diaphragm or posterior to the diaphragm.
2. **Pelvis**: The abdominal cavity is continuous with the pelvic cavity at the pelvic inlet. Structures pass between them through the pelvic inlet.
3. **Lower limb:** The abdomen communicates with the thigh of the lower limb through a space between the inguinal ligament and the hip bone (**pelvifemoral space**). Structures pass through this space between these two regions (Fig. 4.1).

    Remember that the abdominal cavity may extend superiorly as high as the fourth intercostal space.

Window Dissections: Abdomen

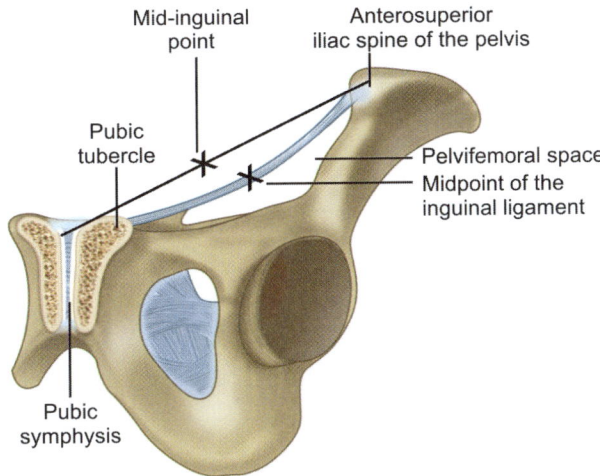

Fig. 4.1: Pelvis

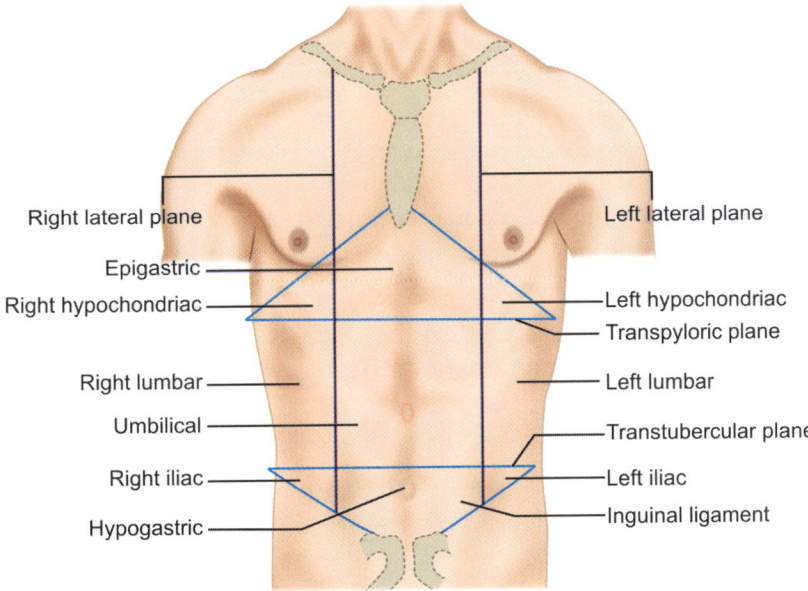

Fig. 4.2: Topographical divisions of abdominal wall

TOPOGRAPHICAL DIVISIONS OF THE ABDOMINAL WALL (FIG. 4.2 AND FLOWCHART 4.1)

The abdomen is divided into **nine** topographical regions by two horizontal planes and two vertical imaginary planes. These divisions are used to describe the location of abdominal organs.

Flowchart 4.1: Topographical regions of abdominal wall

```
                    Nine regions of abdomen
          ┌──────────────────┼──────────────────┐
     Three upper        Three intermediate    Three lower
     ┌────┼────┐                                  │
  Right  Epigastrium  Left                        │
  hypochondrium       hypochondrium               │
          ┌────────────┼────────────┐             │
     Right lumbar   Umbilical    Left lumbar      │
          ┌────────────┼────────────┐
     Right iliac fossa  Hypogastrium  Left iliac fossa
```

Two horizontal planes are:
  i. **Transpyloric plane.**
  ii. **Transtubercular plane.**

Transpyloric plane (at the junction of L_1 and L_2 vertebrae): This plane passes through the tips of both 9th costal cartilages in front and through the lower border of L_1 vertebra behind. This plane roughly corresponds with the midpoint between the xiphisternal joint and the umbilicus. This plane can also be represented by a transverse line at the midpoint between the jugular notch and upper end of the symphysis pubis.

Important Landmarks at Transpyloric Plane

1. Pylorus of stomach.
2. Fundus of gallbladder.
3. Neck of the pancreas.
4. Formation of portal vein.
5. Origin of superior mesenteric artery.
6. Upper part of hilum of right kidney and lower part of hilum of left kidney.
7. Lower end of the spinal cord.

Transtubercular plane (at the junction of L_4 and L_5 vertebrae): The transtubercular line is drawn by joining the **iliac tubercles** of both iliac crests. **Note** that the iliac tubercle lies on the outer lip of the iliac crest of the hip bone, 5 cm away from the anterosuperior iliac spine.

Window Dissections: Abdomen

> **Important Landmarks at Transtubercular Plane**
> 1. Junction of L_4 and L_5 vertebrae.
> 2. Bifurcation of abdominal aorta is about 1.25 cm above this plane.
> 3. Formation of inferior vena cava lies about 1.25 cm below this plane.
> 4. Position of ileocecal orifice is at the junction of transtubercular and right lateral plane.
> 5. Base of the vermiform appendix lies about 2 cm below the ileocecal orifice.

Remember that the **transcristal plane** (supracristal plane) is represented by a line joining the highest point of both iliac crests of hip bones and passes through the L_5 spine behind.

Two vertical plans are:
i. **Right lateral plane**
ii. **Left lateral plane**.

Each lateral plane is represented by a vertical line drawn from the **midpoint of the clavicle** extending inferiorly to **midinguinal point** [midway between the anterosuperior iliac spine (ASIS) and the symphysis pubis]. These lateral lines are also called **midclavicular** or **mammary lines**.

LESSON 2: INGUINAL CANAL

Inguinal canal is an about 4 cm long oblique musculoaponeurotic canal in the lower part of the anterior abdominal wall, situated about ½ inch above and parallel to the medial half of the inguinal ligament, extending from the deep inguinal ring to the superficial inguinal ring, formed embryologically by the descent of the gubernaculum of the testis or ovary.

STEPS OF DISSECTION

1. **Position of the cadaver**: Body supine, thigh extended and laterally rotated.
2. **Skin incisions (Fig. 4.3):**
   i. Make a transverse incision from the anterior superior iliac spine up to the midline of the anterior abdominal wall (A–B).
   ii. Another vertical incision from the point B up to the symphysis pubis (B–C).
   Reflect the triangular flap of skin downwards and laterally. Superficial fascia is exposed.
3. **Superficial fascia:** The superficial fascia of this region consists of two layers.
   a. Outer fatty layer of superficial fascia (**fascia of Camper**).
   b. Inner membranous layer of superficial fascia (**fascia of Scarpa**).
   The fatty layer is reflected in the line of skin incision. The membranous layer is exposed. On it, the following structures lie (Fig. 4.4):
   i. Superficial epigastric vessels.
   ii. Superficial external pudendal vessels.
   iii. Superficial circumflex iliac vessels.
   iv. Terminal ends of the iliohypogastric and ilioinguinal nerves.

Clean and observe these structures. The membranous layer is now reflected in the line of skin incision. The aponeurosis of the external oblique muscle is exposed.

4. **External oblique aponeurosis (Figs 4.5A and B):** A triangular aperture is found in the aponeurosis of the external oblique muscle immediately superolateral to the pubic tubercle. This aperture is the **superficial inguinal ring**. The lower border of the external

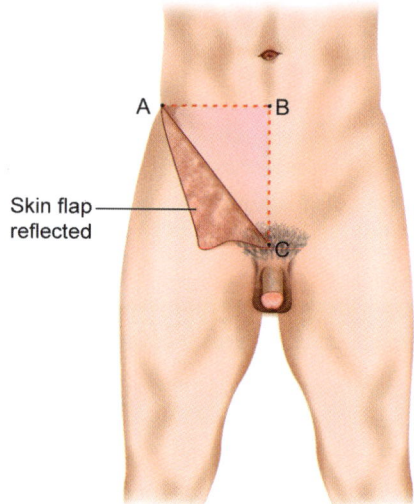

Fig. 4.3: Dissection of inguinal canal

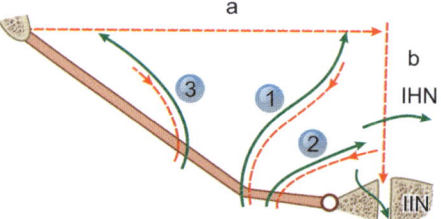

1 = Superficial epigastric vessels
2 = Superficial external pudendal vessels
3 = Superficial circumflex iliac vessels
IHN = Iliohypogastric nerve
IIN = Ilioinguinal nerve

Fig. 4.4: Structures on membranous layer of superficial fascia

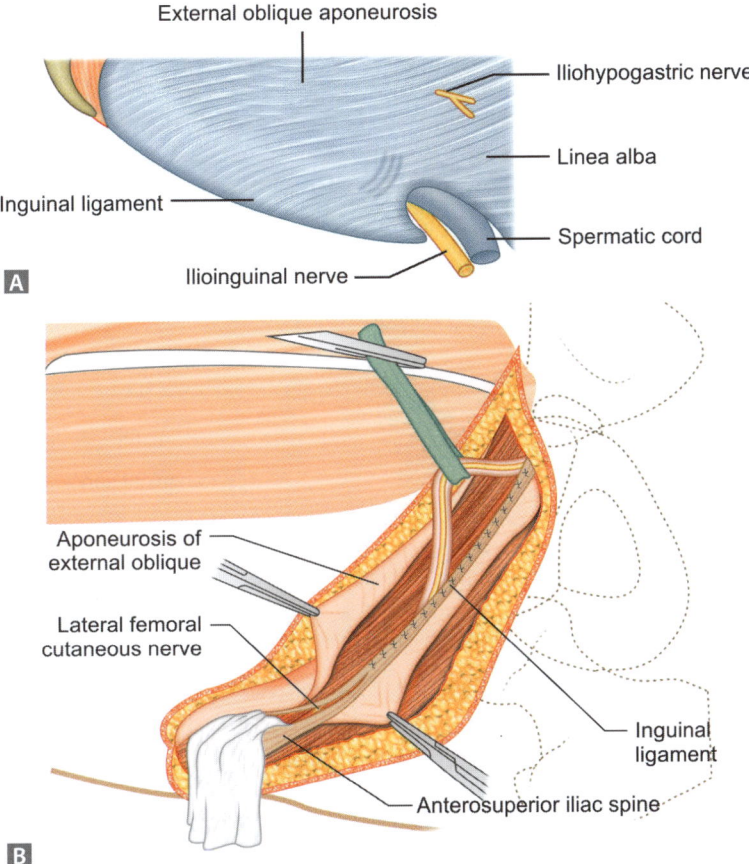

Figs 4.5A and B: External oblique aponeurosis

oblique aponeurosis is thickened to form the **inguinal ligament**. Note the structures which are coming out through the superficial inguinal ring. These are:
  i. **Spermatic cord** (in male) or **round ligament of uterus** (in female).
  ii. **Ilioinguinal nerve.**

Remember that the ilioinguinal nerve passes only through the superficial inguinal ring, not through the deep inguinal ring.

Now, cut the aponeurosis of the external oblique muscle by a transverse incision starting from the anterosuperior iliac spine up to the lateral border of the rectus abdominis. Then make a vertical incision along the lateral border of the rectus abdominis starting from the medial end of the previous incision up to the symphysis pubis.

Reflect the flap of external oblique aponeurosis downwards and laterally onto the thigh.

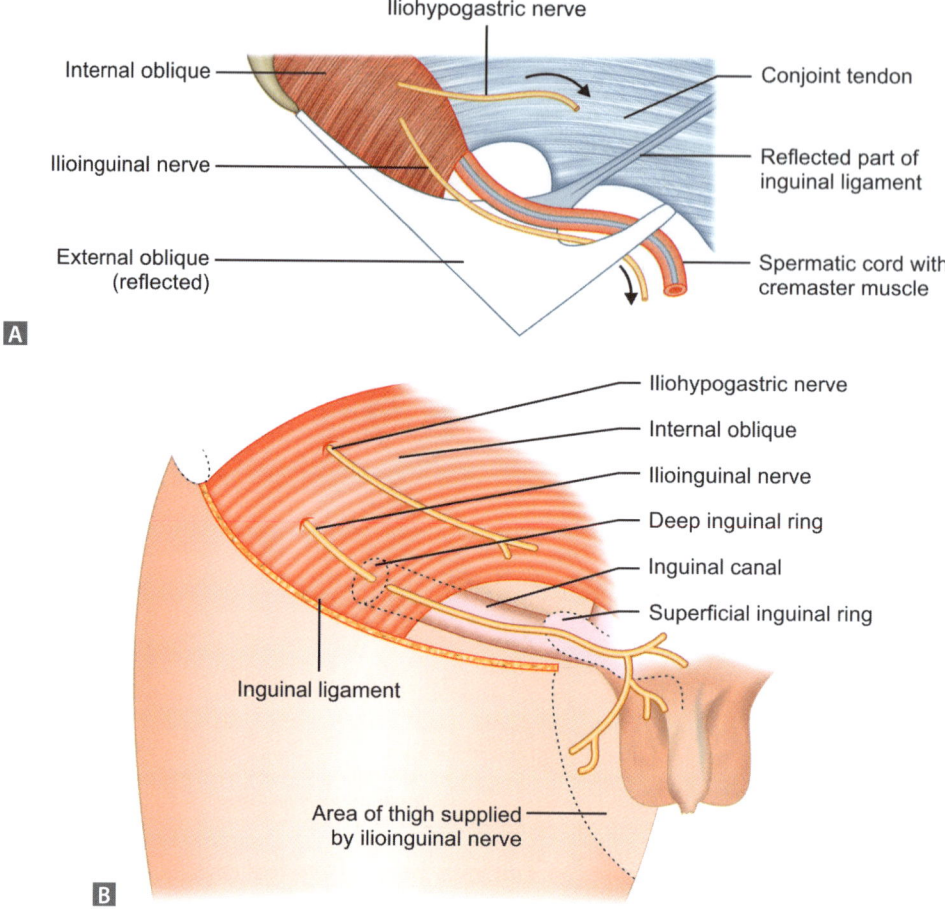

Figs 4.6A and B: (A) Internal oblique muscle is exposed; (B) Internal oblique muscle and ilioinguinal nerve

Most of the part of the inguinal canal is exposed. Only the lateral part of the canal is blocked by the internal oblique muscle (Figs 4.6A and B)

5. Now, cut the fibers of the internal oblique muscle from the lateral part of the inguinal ligament and reflect it medially up to the conjoint tendon. The inguinal canal with its contents is exposed (Fig. 4.7).

The contents of the inguinal canal are:
  i. Spermatic cord in male or round ligament of uterus in female.
  ii. Ilioinguinal nerve.

Identify the **deep inguinal ring** in the fascia transversalis which is situated 1.25 cm above the midinguinal point. The **inferior epigastric artery** lies medial to the deep inguinal ring.

Window Dissections: Abdomen

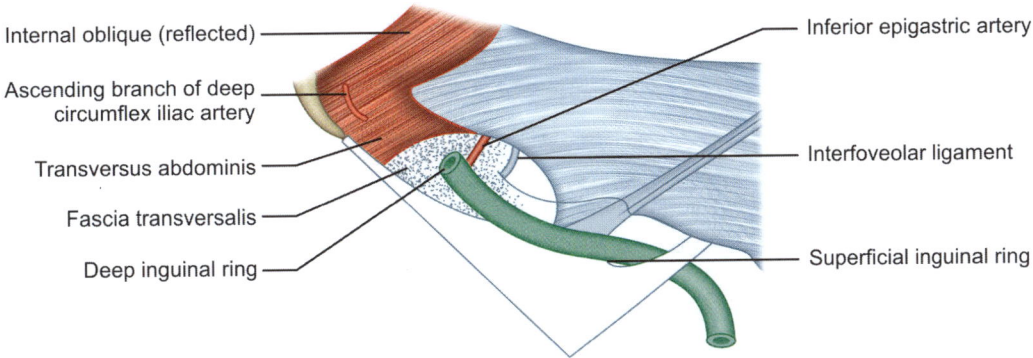

Fig. 4.7: Inguinal canal

Summary

1. **Position of the cadaver**: Supine with thigh extended and laterally rotated.
2. **Skin incision** as in Figure 4.3.
3. **Superficial fascia with** its two layers (fascia of Camper and fascia of Scarpa) is exposed. The structures between these two layers are:
    i. Superficial epigastric vessels.
    ii. Superficial external pudendal vessels.
    iii. Superficial circumflex iliac vessels.
    iv. Terminal ends of iliohypogastric and ilioinguinal nerves.
4. External oblique aponeurosis is exposed after reflecting superficial fascia.
5. External oblique aponeurosis is cut and reflected downwards and laterally.
6. Internal oblique fibers are cut and reflected medially.
7. **Inguinal canal** with its contents is exposed.
   Contents:
    i. Spermatic cord (male) or round ligament of uterus (female).
    ii. Ilioinguinal nerve.

PROBABLE QUESTIONS AND ANSWERS

1. **Define inguinal canal.**

Ans. Inguinal canal is an about 4 cm long oblique musculoaponeurotic canal in the lower part of the anterior abdominal wall, situated about ½ inch above and parallel to the medial half of the inguinal ligament, extending from the deep inguinal ring to the superficial inguinal ring, formed embryologically by the descent of the gubernaculum of the testis or ovary.

12. What are the boundaries of inguinal canal?

Ans. Boundaries:
  a. **Anterior wall**:
      i. Skin.
      ii. Superficial fascia.
      iii. External oblique aponeurosis.
      iv. Fleshy fibers of internal oblique in its lateral one-third.
  b. **Posterior wall:**
      1. **Lateral to the obliterated umbilical artery:**
          i. Fascia transversalis.
          ii. Extraperitoneal fatty tissue.
          iii. Parietal peritoneum.
      2. **Medial to the obliterated umbilical artery:**
          i. Reflected part of inguinal ligament.
          ii. Conjoint tendon.
          iii. Fascia transversalis.
          iv. Extraperitoneal fatty tissue.
          v. Parietal peritoneum.
  c. **Floor:**
      i. Grooved upper surface of the inguinal ligament.
      ii. Upper concave surface of the lacunar ligament at the medial end.
  d. **Roof:** Arched fibers of the internal oblique and transversus abdominis muscle.
  e. **Inlet:** Deep inguinal ring.
  f. **Outlet**: Superficial inguinal ring.

13. What is deep inguinal ring?

Ans. It is an **oval** gap in the fascia transversalis (Fig. 4.8).

14. Is the deep inguinal ring an opening in the fascia transversalis?

Ans. No. It represents the mouth of the pouch of the internal spermatic fascia which is dragged down by the gubernaculum of the testis or ovary.

15. What is the location of the deep inguinal ring?

Ans. It is located about 1.25 cm above the midinguinal point.

16. What is midinguinal point?

Ans. It is the midpoint between the anterior superior iliac spine and **symphysis pubis**. It must not be confused with the middle point of the inguinal ligament which is the midpoint between the anterior superior iliac spine and **pubic tubercle** (sites of attachments of inguinal ligament).

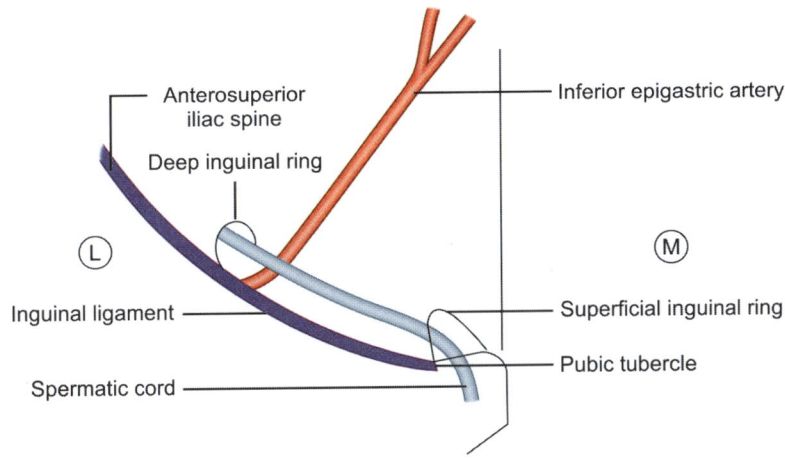

Fig. 4.8: Deep and superficial inguinal ring

7. What is the relation of the deep ring with the inferior epigastric artery?

Ans. The deep ring lies lateral to the inferior epigastric artery (this artery is the guide to identify the deep inguinal ring).

8. What structures pass through the deep ring?

Ans. Spermatic cord in male or round ligament of uterus in female.

9. What is superficial inguinal ring?

Ans. It is an oblique **triangular** gap in the aponeurosis of external oblique muscle (Fig. 4.8)

10. What is the size of the ring?

Ans. Apex to base: About 2.5 cm.
Base: About 1.25 cm.

11. What is the location of the ring?

Ans. It is located about 1 cm above and lateral to the pubic tubercle.

12. What are the boundaries of the superficial ring?

Ans. i. **Base** is formed by the pubic crest.
  ii. **Apex** is formed by the convergence of the two crura.
  iii. **Laterally** by the inferior crus of external oblique aponeurosis.
  iv. **Medially** by the superior crus of the external oblique aponeurosis.
Both the crura are interconnected by the **intercrural fibers** of the external oblique.

13. What structures pass through the superficial ring?

Ans.
i. Spermatic cord or round ligament of uterus.
ii. Ilioinguinal nerve.
(Ilioinguinal nerve does not pass through the deep inguinal ring)

14. What are the contents of the inguinal canal?

Ans.
i. Spermatic cord in males or round ligament of uterus in females.
ii. Ilioinguinal nerve: This nerve is the partial content of the canal, because it enters the canal by piercing the internal oblique muscle and leaves the canal through the superficial inguinal ring.

15. What is the root value of ilioinguinal nerve.

Ans. Ventral ramus of L_1 nerve.

16. What are the distributions of this nerve?

Ans. Ilioinguinal nerve is a **mixed nerve**. While passing between the internal oblique and transversus abdominis muscles, it supplies both of them (**motor supply**). It becomes **sensory** after piercing the internal oblique muscle and then supplies the skin of the upper and medial part of the thigh, root of the penis and anterior 1/3rd of the scrotum (in males) or mons pubis and labia majora (in females).

17. What are the abnormal contents of the inguinal canal?

Ans.
i. Undescended testis (in males)
ii. More descended ovary (in females)
iii. Accessory spleen.
iv. Accessory suprarenal cortical tissue.
v. Remnants of processus vaginalis.

18. What are the constituents of spermatic cord?

Ans.
i. Vas deferens.
ii. Artery to vas (a branch of superior or inferior vesical artery)
iii. Artery to cremaster (a branch of inferior epigastric artery).
iv. Testicular artery (a lateral branch of abdominal aorta).
v. Pampiniform plexus (veins of the testis and epididymis form pampiniform plexus).
vi. Testicular lymph vessels (drain into lateral aortic group of lymph nodes).
vii. Testicular sympathetic plexus.
viii. Nerve to cremaster (genital branch of genitofemoral nerve).
ix. Extraperitoneal fatty tissue.

Window Dissections: Abdomen

19. What are the abnormal constituents of the spermatic cord?

Ans.
i. Accessory spleen.
ii. Accessory suprarenal cortical tissue.
iii. Remnants of processus vaginalis.

20. What are the coverings of the spermatic cord?

Ans. From within outwards (Fig. 4.9):
i. Internal spermatic fascia, derived from fascia transversalis.
ii. Cremasteric fascia, derived from internal oblique and transversus abdominis muscles.
iii. External spermatic fascia, derived from external oblique aponeurosis.

21. What is the length of the spermatic cord?

Ans. About 7 cm.

22. What is the extent of the spermatic cord?

Ans. It extends from the upper pole of the testis to the deep inguinal ring.

23. What is vas deferens?

Ans. Vas deferens is the continuation of the tail of the epididymis at the lower pole of testis.

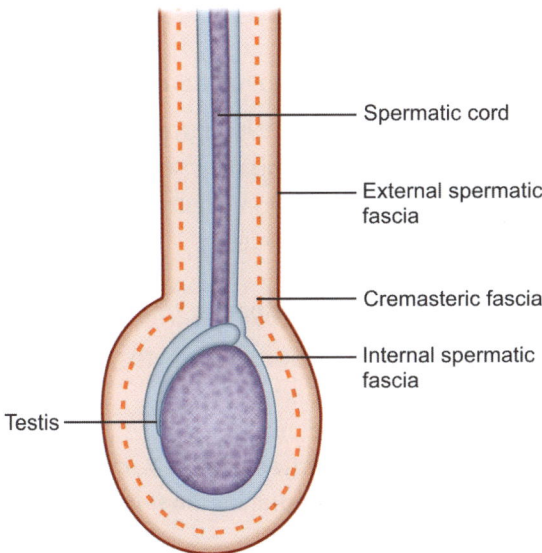

Fig. 4.9: Coverings of spermatic cord

24. What is its length?
Ans. About 45 cm.

25. What is the fate of vas deferens?
Ans. The vas deferens joins with the duct of seminal vesicle to form ejaculatory duct which opens at the colliculus seminalis, one on each side of the prostatic utricle.

26. What is round ligament of uterus?
Ans. It is a fibromuscular band and about 10–12 cm long. Developmentally, it is a remnant of the **distal part of the gubernaculum of ovary.**

27. What is the fate of the proximal part of the gubernaculum of ovary?
Ans. It forms the ligament of ovary.

28. How does the inguinal canal develop?
Ans. The inguinal canal is developed from the differentiation of the muscles of the anterior abdominal wall and represents the passage of gubernaculum of testis or ovary during the descent of the respective sex gland.
Remember that the inguinal canal is formed before the descent of the testes.

29. What is inguinal hernia?
Ans. Abnormal protrusion of abdominal contents into the inguinal canal is known as inguinal hernia.

30. What are the different varieties of inguinal hernia?
Ans.
i. **Indirect hernia (oblique hernia):** When the herniation occurs through the deep inguinal ring, lateral to inferior epigastric artery.
ii. **Direct hernia:** When the herniation occurs through the **Hesselbach's triangle**, medial to the inferior epigastric artery.

31. What are the boundaries of the Hesselbach's triangle?
Ans.
i. **Medially (Fig. 4.10):** Lateral border of rectus abdominis muscle.
ii. **Laterally:** Inferior epigastric artery.
iii. **Base (below):** Inguinal ligament.

The triangle is divided by the obliterated umbilical artery into a medial part or **supravesical fossa** and a lateral part or **medial inguinal fossa**.

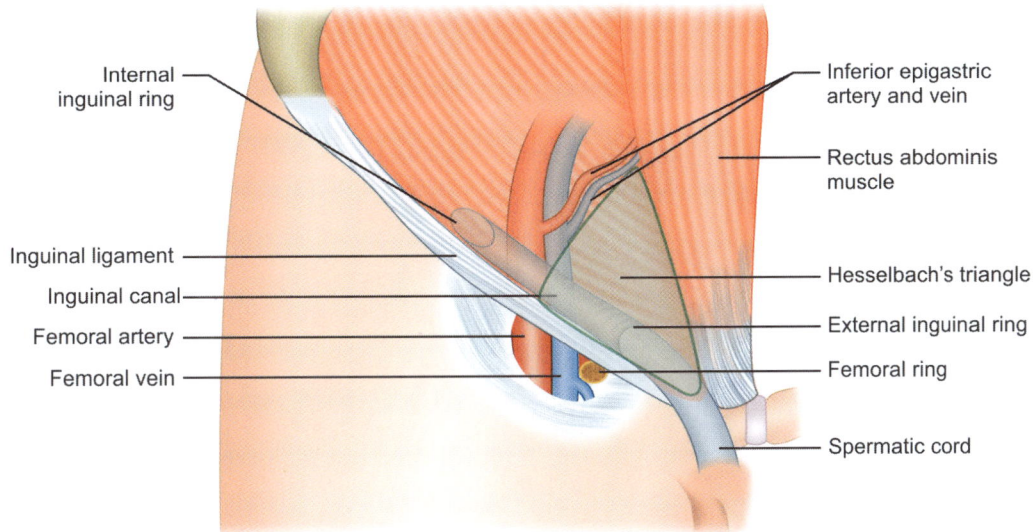

Fig. 4.10: Boundaries of Hesselbach

LESSON 3: RECTUS SHEATH

Rectus sheath is an aponeurotic envelope enclosing the rectus abdominis muscle on either side of the linea alba.

STEPS OF DISSECTION

1. **Position of the cadaver:** Body supine.
2. **Skin incisions (Fig. 4.11):**
    i. An oblique incision from the anterosuperior iliac spine to the symphysis pubis (A–B)
    ii. Another transverse incision from the xiphisternal junction laterally for 3 inches (C–D)
    iii. A vertical incision from the midpoints of the above two incisions (E–F).
    Reflect the flaps of skin laterally and medially. The superficial fascia is exposed.
3. **Superficial fascia:** The superficial fascia consists of two layers below the line joining the anterosuperior iliac spines of both sides.
    Between these two layers lie the following structures:
    i. Superficial external pudendal vessels.
    ii. Superficial epigastric vessels.
    iii. Superficial circumflex iliac vessels.
    iv. Anterior cutaneous branches of lower 5 or 6 intercostal nerves.

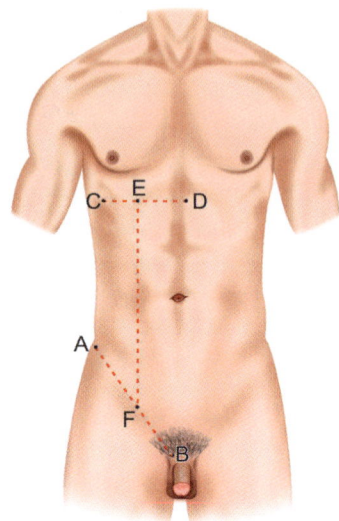

Fig. 4.11: Dissection of rectus sheath

Above the line joining the anterosuperior iliac spines of both sides the superficial fascia consists of single layer. The superficial fascia is reflected in the line of skin incision. The anterior layer of the rectus sheath is exposed.

4. **Anterior layer of rectus sheath:** The anterior layer of the rectus sheath stretches from the linea alba to the lateral border of the rectus abdominis muscle and from the costal arch to symphysis pubis.

   Now, cut the anterior layer of the rectus sheath by a vertical incision along the middle of the muscle from above downwards. Then cut the attachments of the rectus abdominis muscle to its **tendinous intersections** (3–4) and reflect the anterior layer of the sheath sideways. The rectus abdominis muscle is exposed.

5. **Rectus abdominis muscle and the posterior layer of rectus sheath (Fig. 4.12):** Rectus **abdominis muscle** is identified and cleaned. Then divide the muscle transversely at its middle and reflect the parts superiorly and inferiorly. The posterior wall of the rectus sheath is now exposed. The **superior and inferior epigastric vessels** lie in front of the posterior layer of the rectus sheath. Identify these vessels and trace them. Before transection, the muscle is lifted and the lower five (7–11) intercostal nerves and the subcostal nerve are identified. These nerves are seen to enter the rectus sheath by piercing the posterior lamella of internal oblique in the posterior wall of the rectus sheath. Then these nerves supply the rectus abdominis muscle segmentally and become cutaneous leaving through the anterior wall of the sheath. The rectus muscle is retracted from medial to lateral side in order to avoid injury to these nerves, as these nerves enter the substance of the muscle from its lateral border.

   Define the lower extent of the posterior wall of the rectus sheath and identify the **arcuate line** which is the free margin of the posterior layer of the rectus sheath. This line is concave

Window Dissections: Abdomen

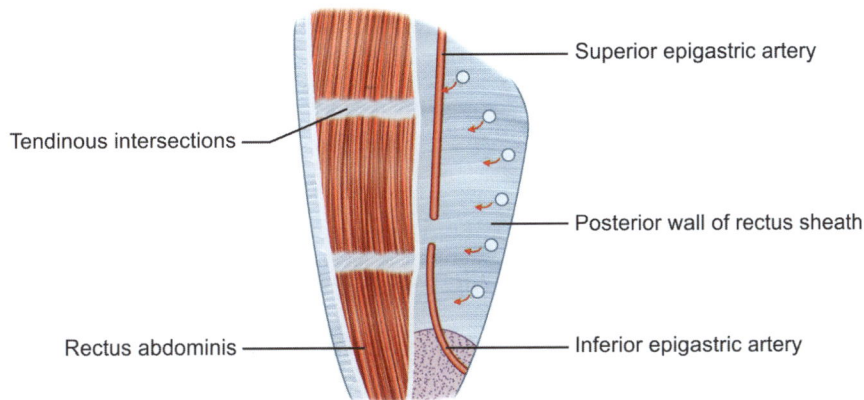

Fig. 4.12: Rectus abdominis with tendinous intersections and posterior wall of rectus sheath

downwards and the **inferior epigastric artery** is seen to enter the sheath in front of the arcuate line.

Remember that he anterior wall of the rectus sheath is attached with the rectus abdominis muscle by 3–4 tendinous intersections whereas the posterior wall is entirely free from the posterior surface of the muscle.

Summary

1. **Position of the body**: Supine.
2. **Skin incisions** as in Figure 4.11.
3. **Superficial fascia**: In it, the following structures are found.
    i. Superficial external pudendal vessels.
    ii. Superficial epigastric vessels.
    iii. Superficial circumflex iliac vessels.
    iv. Anterior cutaneous branches of lower 5–6 intercostal nerves.
4. Anterior layer of rectus sheath with **tendinous intersections** (3–4)
5. Posterior layer of rectus sheath and its **arcuate line**.
6. Contents of the rectus sheath between these two layers:
    i. **Muscles**:
        a. Rectus abdominis.
        b. Pyramidalis (if present).
    ii. **Vessels**:
        a. Superior epigastric vessels.
        b. Inferior epigastric vessels.
    iii. **Nerves**:
        a. Lower five intercostal nerves.
        b. Subcostal nerve.

PROBABLE QUESTIONS AND ANSWERS

1. What is rectus sheath?

Ans. It is an aponeurotic sheath enveloping the rectus abdominis muscle on each side of the linea alba.

2. How is the sheath formed?

Ans. The sheath is formed by the aponeurosis of the external oblique, internal oblique and transversus abdominis muscles. But the formation of the sheath is not uniform throughout its whole extent. So, its formation is described at different levels in the following way.

A. **Above the costal margin (Figs 4.13 and 4.18):**
   a. **Anterior wall**: Solely by the external oblique aponeurosis (internal oblique and transversus abdominis do not extend above the lower costal margin).
   b. **Posterior wall**: **Deficient** above the costal margin. The rectus muscle lies directly on the 5th, 6th and 7th costal cartilages.

B. **From the costal margin to midway between the xiphoid process and umbilicus (Figs 4.14 and 4.18):**
   a. **Anterior wall**:
      i. Aponeurosis of external oblique.
      ii. Anterior lamella of the aponeurosis of internal oblique.
   b. **Posterior wall**:
      i. Posterior lamella of the aponeurosis of internal oblique.
      ii. **Musculoaponeurotic** layer of transversus abdominis.

C. **From the lower end of the previous level up to midway between umbilicus and symphysis pubis (Figs 4.15 and 4.18):**
   a. **Anterior wall**:
      i. Aponeurosis of external oblique.
      ii. Anterior lamella of the aponeurosis of internal oblique.
   b. **Posterior wall**:
      i. Posterior lamella of the aponeurosis of internal oblique.
      ii. **Aponeurosis** of transversus abdominis.

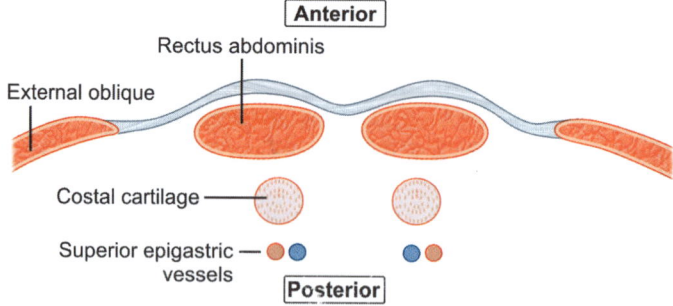

Fig. 4.13: Formation of rectus sheath above the costal margin

Window Dissections: Abdomen

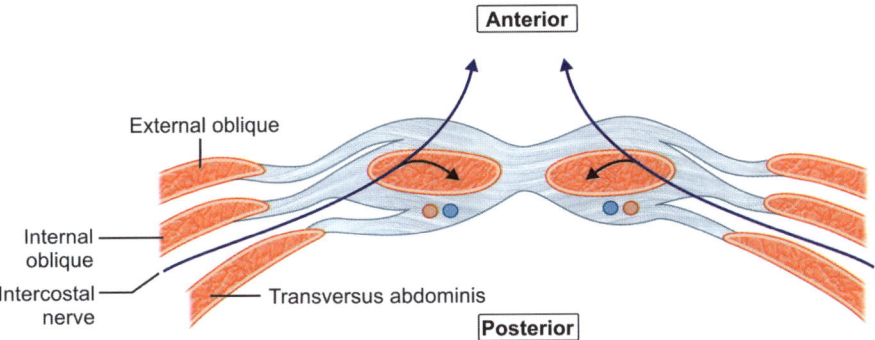

Fig. 4.14: From the costal margin to midway between the xiphoid process and umbilicus

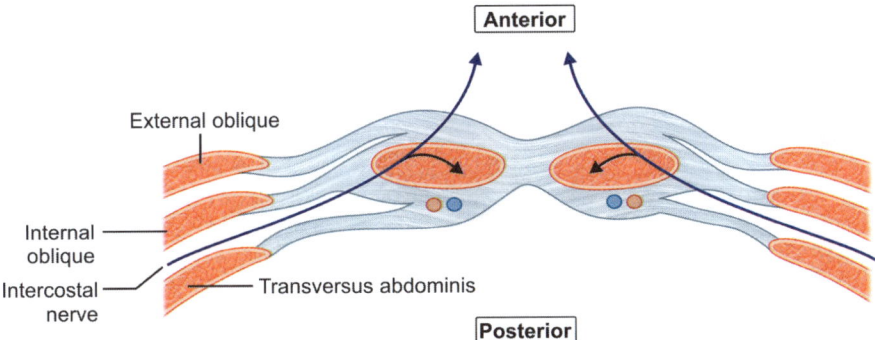

Fig. 4.15: From the lower end of the previous level up to midway between umbilicus and symphysis pubis

Note that at this level the transversus abdominis contributes its aponeurotic part only to form the posterior wall whereas at the previous level, it was partly muscular and partly aponeurotic (musculoaponeurotic). So, at this level the walls of the sheath are completely aponeurotic; hence called **true rectus sheath**.

D. **From the lower end of the previous level up to the symphysis pubis (Fig. 4.16):**
   a. **Anterior wall:**
      i. Aponeurosis of external oblique.
      ii. Aponeurosis of internal oblique.
      iii. Aponeurosis of transversus abdominis.
      (Note that all three aponeurosis of the three flat muscles of abdomen pass in front of the rectus abdominis).
   b. **Posterior wall: Deficient** at this level. Here the rectus abdominis muscle lies directly on the fascia transversalis.

3. What is iliopubic tract?

Ans. The posterior wall of rectus sheath is deficient below the arcuate line. Here, the fascia transversalis is thickened and the rectus muscle lies directly on this fascia tarnsversalis. This thickened part of fascia transversalis below the arcuate line is called **iliopubic tract**.

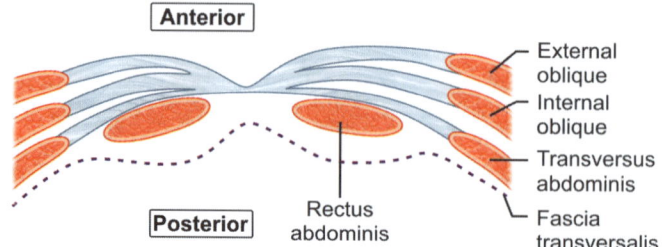

Fig. 4.16: From the lower end of the previous level up to the symphysis pubis

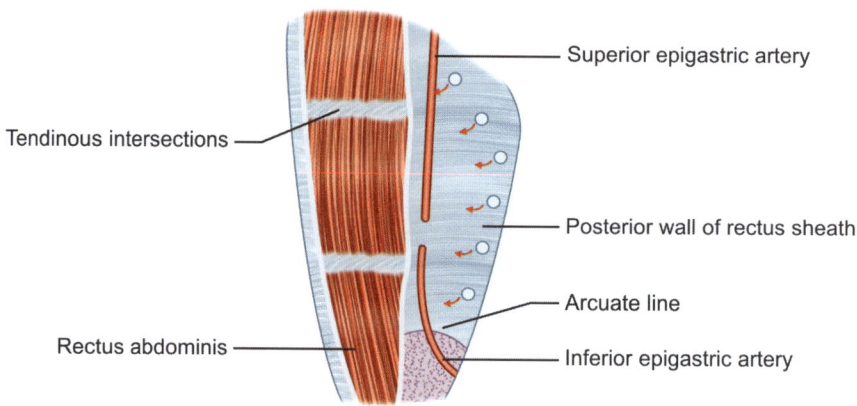

Fig. 4.17: Posterior wall of rectus sheath and arcuate line

4. What is arcuate line?

Ans. At the level of midway between the umbilicus and symphysis pubis, the posterior wall of rectus sheath ends in a free concave margin called the **arcuate line (Fig. 4.17).**

5. What are the other terms used for arcuate line?

Ans. The arcuate line is also termed as **linea semicircularis** or **semicircular line of Douglas.**

6. What important structure is related to the arcuate line?

Ans. The **inferior epigastric artery**, a branch of external iliac artery enters the rectus sheath in front of the arcuate line.

7. Is the rectus muscle free from the walls of the sheath?

Ans. The posterior wall of the sheath is entirely free from the rectus muscle. But the anterior wall of the sheath is attached to the anterior surface of the rectus abdominis muscle by means of its **tendinous intersections.**

Window Dissections: Abdomen

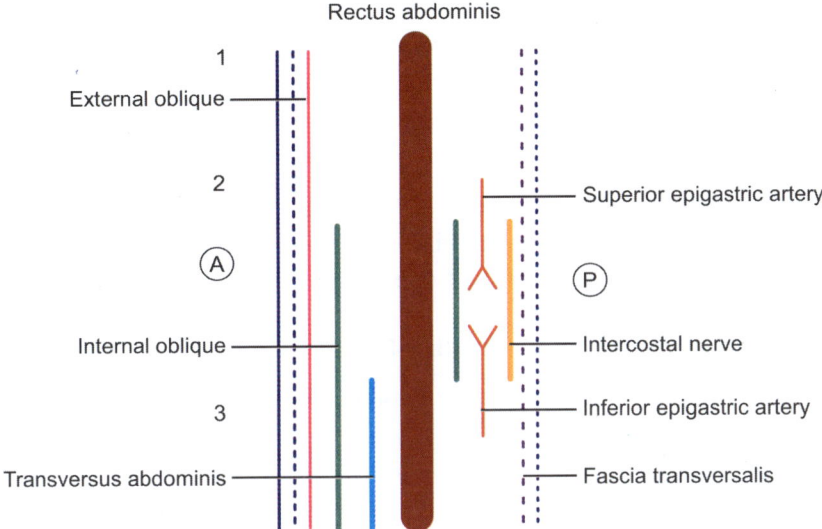

Fig. 4.18: Formation of rectus sheath (sagittal view)

8. What are the tendinous intersections?

Ans. These are the fibrous bands across the muscle occurring horizontally in a zig-zag manner.

9. How many intersections are there?

Ans. These are usually three in number.
   i. One, opposite the tip of xiphoid process.
   ii. One, at the level of umbilicus.
   iii. One, at the level midway between the xiphoid process and umbilicus.
Sometimes a 4th intersection may be present a little below the umbilicus.

10. What do these intersections represent?

Ans. These intersections represent intersegmental tissues between the different segments of the rectus muscle. It means that the rectus muscle is segmental in origin and each segment has its own nerve supply and arterial supply.

It is said that these intersections divide the long rectus abdominis muscle into a number of shorter segments to give strength to the muscle.

11. How the medial margin of the rectus sheath is formed?

Ans. The medial margin of the sheath is formed by **linea alba**.

12. What is linea alba?

Ans. Linea alba is linear whitish structure formed by the interlacement of the aponeurotic fibers of the three flat muscles of abdomen, extending from the tip of the xiphoid process to the upper margin of the symphysis pubis.

13. How the lateral margin of the sheath is formed?

Ans. The lateral margin is formed by **linea semilunaris**.

14. What is linea semilunaris?

Ans. It is a curved fibrous line extending from the tip of the 9th costal cartilage to the pubic tubercle.

15. What is the importance of the lateral margin of the sheath?

Ans. Anterior rami of lower five intercostal nerves and subcostal nerve enter the sheath after passing across the lateral margin. Therefore, a surgical incision along this margin is not preferred because of the chance of injury of these nerves. For this reason, the rectus abdominis muscle is retracted laterally to avoid undue stretching of these thoracic nerves during the operative procedures.

16. What are the contents of the rectus sheath?

Ans.   1. **Muscles**:
   i. Rectus abdominis.
   ii. Pyramidalis (sometimes absent on one or both sides).
2. **Arteries**:
   i. **Superior epigastric artery**, one of the terminal branches of internal thoracic artery. Other terminal branch of internal thoracic artery is **musculophrenic artery**.
   ii. **Inferior epigastric artery**, a branch of the external iliac artery. Other branch of external iliac artery is deep circumflex iliac artery.
3. **Veins**:
   i. Superior epigastric vein draining into internal iliac vein.
   ii. Inferior epigastric vein draining into external iliac vein.
4. **Nerves**:
   i. Lower five (7th to 11th) intercostal nerves.
   ii. Subcostal nerve.

17. What are the attachments of the rectus abdominis muscle?

Ans. **Origins:**
  i. **Lateral head:** From pubic crest and pubic tubercle.
  ii. **Medial head:** From the anterior surface of symphysis pubis.
Insertions:
  i. Anterior surface of xiphoid process.
  ii. 5th, 6th and 7th costal cartilages.

18. What is the nerve supply of this muscle?

Ans. It is supplied by the lower five intercostal nerves and subcostal nerve.

19. What are the origins and insertions of pyramidalis?

Ans. **Origins:** Symphysis pubis and pubic crest.
Insertion: Linea alba.

20. What is the nerve supply and action of pyramidalis?

Ans.
  i. Nerve supply—subcostal nerve.
  ii. Action—mild tensor of linea alba.

21. How does the superior epigastric artery enter the rectus sheath?

Ans. This artery enters the sheath through a triangular gap between the xiphoidal origin and the 7th costal origin of the diaphragm.

22. What is name of this triangular gap?

Ans. **Space of Larrey. Remember** that when the 7th costal origin of the diaphragm is absent, the space of Larrey becomes larger and then it is known as **foramen of Morgagni.**

23. How do the intercostal nerves enter the rectus sheath and what is the fate of these nerves?

Ans. The intercostal nerves run between the aponeurosis of the internal oblique and the transversus abdominis muscles. Then they pierce the posterior lamella of the internal oblique aponeurosis to enter the rectus abdominis muscle through its lateral border and supply this muscle. Thereafter, these nerves come out as anterior cutaneous nerves by piercing the anterior wall of the rectus sheath.

24. Does the subcostal nerve supply the rectus abdominis muscle?

Ans. Yes, the subcostal nerve supplies the rectus abdominis muscle. In addition, it sends a twig to supply the pyramidalis (if present).

25. What is the function of the rectus sheath?

Ans. The rectus sheath acts as a retinaculum and prevents the long rectus muscle from bow stringing.

LESSON 4: EXPOSURE OF KIDNEY FROM BACK

The kidneys are a pair of excretory organs. Each kidney is situated in the posterior abdominal wall behind the peritoneum. The long axis of the kidney is directed downwards and laterally so that the upper pole is nearer to the vertebral column than the lower pole. The hilum of the kidney lies about 5 cm away from the midline and occupies the medial border. Each kidney is situated mostly in the lumbar region, through the right kidney is slightly lower than the left kidney due to the presence of large right lobe of the liver. The left kidney is more towards the midline and is longer and narrower than the right one. Each kidney lies within the **Morris parallelogram** on the back.

STEPS OF DISSECTION

1. **Position of the body:** Prone.
2. **Skin incision (Fig. 4.19):**
    i. A transverse incision starting 2.5 cm (1") away from the T_{11} spine for 7.5cm (3") (A–B).
    ii. Another transverse incision in the same way like the previous incision at the level of L_3 spine (C–D).

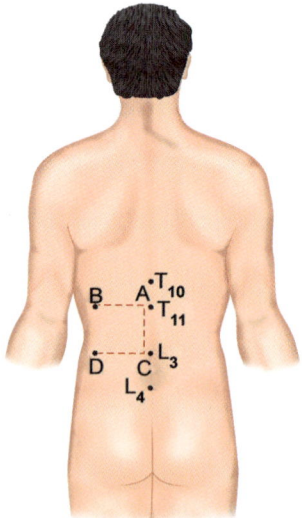

Fig. 4.19: Incision for kidney from back

iii. A vertical incision (A–C) from the medial end of the proximal incision (A–B) to the medial end of the distal incision (C–D).

Reflect the flap of skin laterally. The superficial fascia is exposed.

3. **Superficial fascia:** Cutaneous branches of the posterior primary rami of the lumbar nerves are found in the superficial fascia. Then the fascia is cut and reflected in the line of skin incision. The posterior layer of thoracolumbar fascia is exposed.
4. **Posterior layer of thoracolumbar fascia (Fig. 4.20):** In this layer of fascia, the fibers of latissimus dorsi and serratus posterior inferior muscles are available. The posterior layer of thoracolumbar fascia is cut in the line of skin incision and reflected laterally. The erector spinae (sacrospinalis) muscle is exposed.
5. **Erector spinae:** This muscle can be removed for convenience. Otherwise, retract the muscle medially as far as possible towards the vertebral column. The middle layer of thoracolumbar fascia is exposed.
6. **Middle layer of thoracolumbar fascia (Fig. 4.20):** Cut the attachments of this fascial layer from the tips of the transverse processes of the lumbar vertebrae and reflect it laterally. The quadratus lumborum muscle is exposed.
7. **Quadratus lumborum:** This muscle is retracted medially as far as possible to expose the anterior layer of thoracolumbar fascia.

   The **anterior layer** looks glistening and is cut in the same line of skin incision and reflected laterally. The paranephric pad of fat is exposed.

Note that the posterior, middle and anterior layer of thoracolumbar fascia are cut in the line of skin incision and reflected **laterally** but the exposed muscles (**erector spinae** and **quadratus lumborum**) are retracted **medially**.

8. **Paranephric pad of fat:** It consists of a variable amount of fat lying outside the renal fascia. Renal fascia is a fibroareolar sheath which surrounds the kidneys. Paranephric fat is more

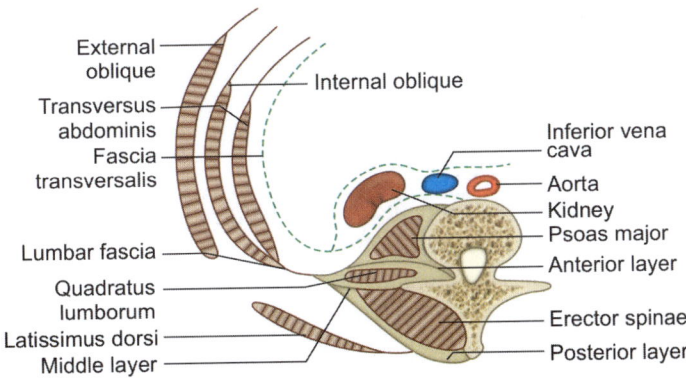

Fig. 4.20: Anterior, middle and posterior layer of thoracolumbar fascia

abundant on the posterior aspect and towards the lower pole of the kidneys. This fatty layer is cleaned carefully to preserve the following structures which come in relation to the posterior surface of the kidney. These structures are as follows from above downwards:
  i. Subcostal vessels.
  ii. Subcostal nerves.
  iii. Iliohypogastric nerve.
  iv. Ilioinguinal nerve.
9. **Now the kidney is exposed from the back**.

Summary
1. **Position of the body:** Prone.
2. **Skin incision** as in Figure 4.19.
3. The following structures are reflected one by one:
     i. Posterior layer of thoracolumbar fascia with latissimus dorsi and serratus posterior inferior.
     ii. Erector spinae.
     iii. Middle layer of thoracolumbar fascia.
     iv. Quadratus lumborum.
     v. Anterior layer of thoracolumbar fascia.
4. Paranephric pad of fat.
5. Structures on the back of the kidney from above downwards:
     i. Subcostal vessels and nerves.
     ii. Iliohypogastric nerve.
     iii. Ilioinguinal nerve.
6. Posterior aspect of kidney is exposed.

PROBABLE QUESTIONS AND ANSWERS

1. What is Morris parallelogram?

Ans. It is an area on the back within which kidney lies.

2. How do you draw a Morris parallelogram?

Ans. It is drawn as follows (Fig. 4.21):
  i. Point A—2.5 cm away from the T_{11} spine.
  ii. Point B—7.5 cm away from the point A.
  iii. Point C—2.5 cm away from L_3 spine.
  iv. Point D—7.5 cm away from the point C.

Now, join these points to draw Morris parallelogram.

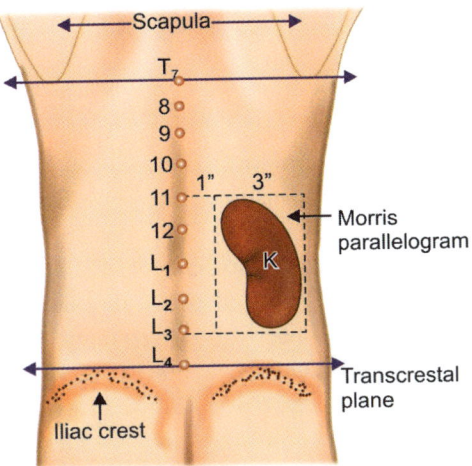

Fig. 4.21: Morris parallelogram

3. Is the level of parallelogram equal on both the sides?

Ans. No. The above parallelogram represents the right sided parallelogram. It is about 1.25 cm above on the left side. This is due to the fact that the right kidney occupies a lower position than the left kidney due to the presence of large right lobe of liver.

4. What are the layers do you come across to expose the kidney from back?

Ans. Skin → Superficial fascia → Posterior layer of thoracolumbar fascia (lumbar fascia) → Erector spinae → Middle layer of lumbar fascia → Quadratus lumborum muscle → Anterior layer of lumbar fascia → Paranephric pad of fat kidney.

5. How many layers of thoracolumbar fascia are there?

Ans. Three layers.
  i. Anterior
  ii. Middle
  iii. Posterior.

6. What is the fascial disposition of these layers?

Ans. A. **Anterior layer of thoracolumbar fascia:**
  i. **Above:** It is thickened to form the **lateral arcuate ligament** (lateral lumbocostal arch) which is attached medially to the tip of transverse process of L_1 vertebra, and laterally to the lower border of the 12th rib.
  ii. **Below:** It is thickened to form **iliolumbar ligament** which is attached to the posterior part of the iliac crest.

iii. **Medially**: Attached to a vertical ridge on the anterior surface of the transverse process of lumbar vertebra.
iv. **Laterally**: It blends with the middle layer of thoracolumbar fascia at the lateral border of quadratus lumborum.

B. **Middle layer of thoracolumbar fascia:**
i. **Above**: It is thickened to form the **lumbocostal ligament** which extends from the lower border of 12th rib to the transverse process of L_1 vertebra.
ii. **Below**: Attached to the posterior part of the iliac crest.
iii. **Medially**: Attached to the tips of the transverse processes of lumbar vertebrae and to the intertransversus muscles.
iv. **Laterally**: It blends with the anterior layer of the thoracolumbar fascia at the lateral border of the quadratus fascia at the lateral border of the erector spinae.

C. **Posterior layer of thoracolumbar fascia:**
i. **Above**: In the thorax, it is attached medially to the spines of the thoracic vertebrae and laterally to the angles of the ribs. In the neck, it blends with the investing layer of deep cervical fascia.
ii. **Below**: Attaches to the dorsal surface of the sacrum and dorsal segment of the iliac crest.
iii. **Medially**: Attached to the spines and supraspinous ligaments of the lumbar vertebrae.
iv. **Laterally**: It blends with the middle layer of thoracolumbar fascia.

17. What are the muscles enclosed between the layers of lumbar fascia?

Ans. The **quadratus lumborum** is enclosed between the anterior and middle layer and the **erector spinae** muscle between the middle layer and posterior layer of the thoracolumbar fascia.

18. What are the muscles attached to the different layers of lumbar fascia?

Ans.
a. From the anterior layer: Few fibers of the diaphragm.
b. From the fusion of the anterior and middle layers of fascia
   i. Transversus abdominis
   ii. Internal oblique.
c. From the posterior layer:
   i. Latissimus dorsi.
   ii. Serratus posterior superior.

19. Name the structures that pass behind the lateral arcuate ligament?

Ans.
i. Subcostal vessels.
ii. Subcostal nerve.

10. What is the medial arcuate ligament?

Ans. It is the thickened part of the psoas fascia which is attached medially to the body of the L_1 and L_2 vertebrae, and laterally to the tip of transverse process of the L_1 vertebra.

11. What are the structures that pass behind the medial arcuate ligament?

Ans.
i. Sympathetic trunk.
ii. Least splanchnic nerve.

12. Name the structures which run between the anterior layer of thoracolumbar fascia (i.e. behind the anterior layer) and quadratus lumborum.

Ans. The structures from above downwards are:
i. Subcostal vessels and subcostal nerve.
ii. Iliohypogastric nerve.
iii. Ilioinguinal nerve.
iv. Fourth lumbar artery.

CHAPTER 5

Window Dissections: Thorax

- Introduction to Thorax
- Dissection of Intercostal Space (Upper Intercostal Spaces)

LESSON 1: INTRODUCTION TO THORAX

The thorax is the part of the body between the neck above and abdomen below. Its skeletal framework is called **thoracic cage**.

THORACIC CAGE

i. **Thoracic cage** (also called **rib cage**) (Fig. 5.1).
ii. It is an **elastic, osseocartilagenous** structure enclosing a conical cavity called **thoracic cavity.**

Boundaries

a. **In front:**
  i. Sternum.
  ii. Anterior ends of ribs and their costal cartilages.
b. **Behind:**
  i. Bodies of 12 thoracic vertebrae and intervertebral discs between them.
  ii. Posterior parts of 12 pairs of ribs up to their posterior angles.
c. **On each side**: 12 pairs of ribs with their costal cartilages, separated by 11 intercostal spaces containing intercostal muscles, vessels and nerves.
d. **Above: Thoracic inlet** (upper aperture of thorax).
e. **Below: Thoracic outlet** (lower aperture of thorax), closed by the diaphragm.

INLET OF THORAX (FIG. 5.2)

Features

i. It is the upper aperture of the thorax through which many important structures like trachea, esophagus, apex of the lung with its apical pleura, big vessels, nerves, etc. pass.

Window Dissections: Thorax

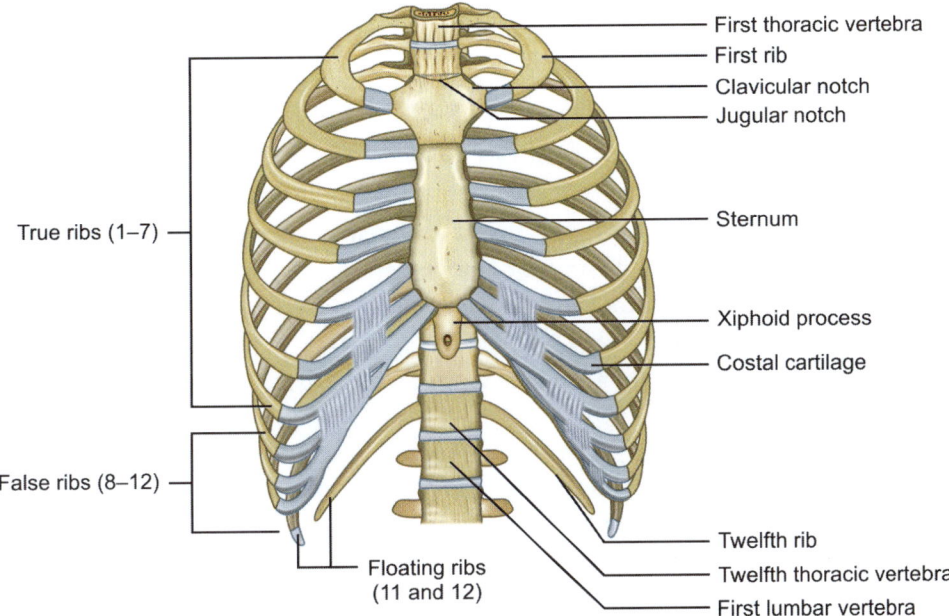

Fig. 5.1: Thoracic cage

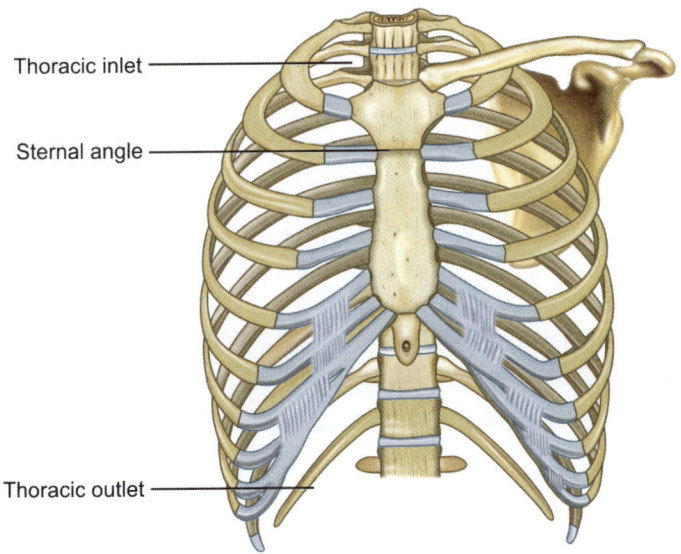

Fig. 5.2: Thoracic inlet

ii. **Shape**: Kidney-shaped (reniform), body of T_1 vertebra forming the hilum.
iii. **Measurements:**
   a. Anteroposterior diameter—about 5 cm.
   b. Transverse diameter—about 10 cm.

iv. **Direction**: Oblique slopes downwards and forwards at an angle of 45° with the horizontal plane. Anterior part is about 1½ inch below the posterior part.
v. **Boundaries**:
   a. **In front**: Upper border of manubrium sterni.
   b. **Behind**: Upper border of the body of T_1 vertebra.
   c. **On each side**: Inner border of the first rib with its cartilage.
vi. **Character**: Inlet is made more rigid, because the **first sternocostal** joint is a **primary cartilaginous joint** providing less range of movement.
   Note that all other sternocostal joints are **synovial** joints with greater range of mobility).
vii. **Structures passing through**:
   a. **Viscera**: Esophagus, trachea, apex of lungs of both sides, suprapleural membranes of both sides.
   b. **Vessels**:
      – **Arteries** (brachiocephalic trunk, left common carotid artery, left subclavian artery, internal thoracic arteries, etc.).
      – **Veins** (brachiocephalic veins, 1st posterior intercostal veins etc.).
   c. **Nerves**: Vagus, phrenic, sympathetic trunks, 1st thoracic nerve.
   d. **Muscles**: Sternothyroid, sternohyoid, longus coli.

OUTLET OF THORAX (FIG. 5.2)

Features

i. It is the lower aperture of the thorax, closed by the diaphragm.
ii. **Boundaries**:
   a. **In front**: Xiphoid process of the sternum.
   b. **Behind**: Lower border of the body of T_{12} vertebra.
   c. **At sides**: Costal cartilages of 7th to 10th ribs and lower two ribs (11th and 12th ribs).

LESSON 2: DISSECTION OF INTERCOSTAL SPACE (UPPER INTERCOSTAL SPACES)

STEPS OF DISSECTION

1. **Position of cadaver**: Supine.
2. **Skin incision (Fig. 5.3)**:
   a. A horizontal incision extending from the lateral margin of the sternum, about 3–4 inches, along the upper border of a space (A–B).
   b. Another horizontal incision, 4–5 spaces below, along its lower border (C–D).
   c. A vertical incision along the sternal margin from the point B to D.
   Reflect the skin with fascia laterally.

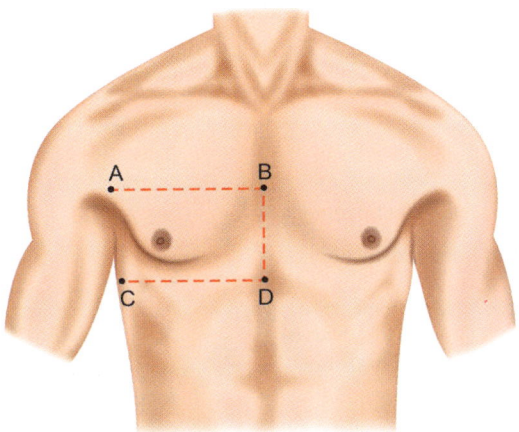

Fig. 5.3: Incision of intercostal space

3. Clean the area by removing the remains of the muscles of the upper limb.
4. Intercostal spaces are exposed.
5. Clean the middle 3–4 spaces and expose **external intercostal muscle** with its anterior intercostal membrane.
6. Using a scalpel, carefully cut through the external intercostal muscle and its membrane from the upper border of the space in 2–3 spaces and reflect them downwards to expose the underlying **internal intercostal muscle**. **Note** that the fibers of internal intercostal muscle course at a right angle to the fibers of external intercostal muscle (Fig. 5.4).
7. Divide the internal intercostal muscle close to the lower border of the rib in 2–3 intercostal spaces to expose **intercostal vein, artery and nerve**. **Note** that intercostal vein, artery and nerve lie in the costal groove along the lower border of a rib between the internal intercostal muscle and innermost intercostal (**intercostalis intimus**) muscle (Fig. 5.5).
8. Trace these structures with their branches carefully.
   Lower border of the rib can be removed by using a bone forceps for better exposure of the vessels and nerve.

PROBABLE QUESTIONS AND ANSWERS

1. What do you mean by an intercostal space?

Ans. An intercostal space is defined as a space between the two consecutive ribs with their cartilages.

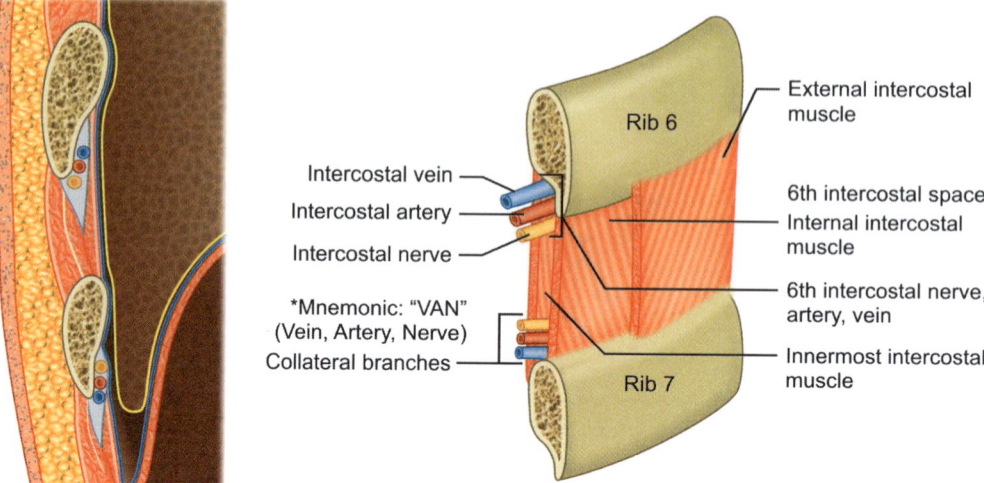

Fig. 5.4: Intercostal space and its contents

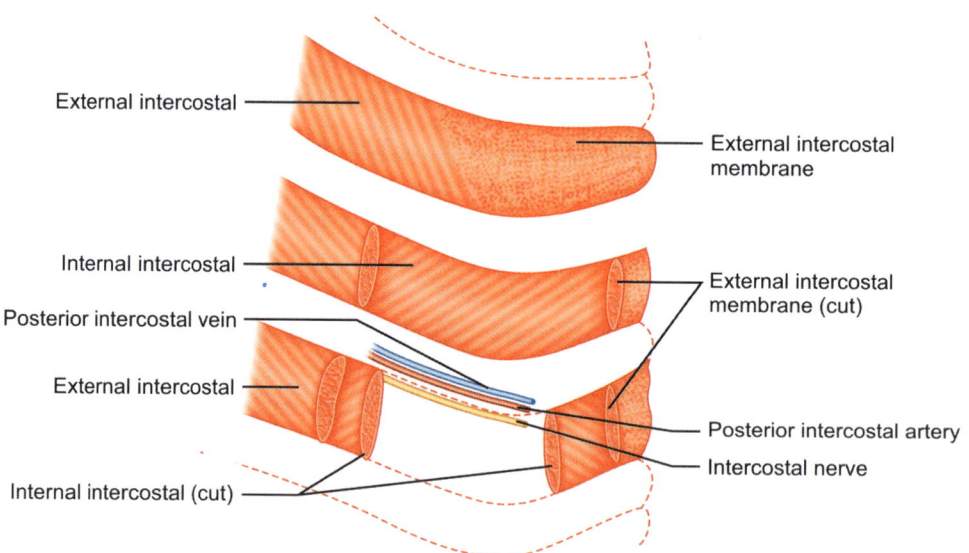

Fig. 5.5: Intercostal muscles and intercostal vein, artery and nerve

2. How many spaces are there?

Ans. Eleven on each side. 10th and 11th spaces are open in front, because 11th and 12th ribs are floating ribs.

3. What are the types of intercostal spaces?

Ans.
a. Typical intercostal spaces (3rd to 6th spaces) those intervening between typical ribs and traversed by intercostal vessels and typical intercostal nerves.
b. Atypical spaces (1st, 2nd and 7th to 11th spaces).

4. What are the boundaries of a typical intercostal space?

Ans.
a. **Above:** Lower border of upper rib and its cartilage.
b. **Below:** Upper border of lower rib with its cartilage.
c. **In front:** Lateral border of the sternum in between two costal facets.
d. **Behind:** Lateral surface of the body of the corresponding thoracic vertebra.

5. What are the contents of a space?

Ans.
A. **Muscles** (from outside inwards):
   a. External intercostal.
   b. Internal intercostal.
   c. Intercostalis intimus.
B. **Intercostal arteries.**
C. **Intercostal veins.**
D. **Intercostal nerves (11 pairs):** T_1-T_{11} nerves.

6. How many sets of intercostal arteries are there?

Ans. Two sets.
   i. Anterior set.
   ii. Posterior set.

7. What is the origin of anterior intercostal arteries?

Ans. **In the upper six spaces** (1st to 6th space), anterior intercostal arteries are the branches of internal thoracic artery.
   i. **In the 7th–9th spaces**, anterior intercostal arteries are the branches of musculophrenic artery, branch of internal thoracic artery.
   ii. **In 10th and 11th space**—no artery, as they are open in front.

8. What is the origin of posterior intercostal arteries?

Ans.
   i. **In upper two spaces**, the arteries are the branches of superior intercostal artery, branch of costocervical trunk.
   ii. **In lower nine spaces** (3rd to 11th), the arteries are the branches of descending thoracic aorta.

9. Where do the anterior intercostal veins drain?

Ans.
i. **From 1st to 6th spaces**, drains into internal thoracic vein.
ii. **From 7th to 9th spaces**, drains into musculophrenic vein.

10. Where do the posterior intercostal veins drain?

Ans.

Summary of drainage of posterior intercostal veins		
Veins	On right side	On left side
1st posterior intercostal vein	Right brachiocephalic vein	Left brachiocephalic vein
Superior intercostal vein (formed by 2nd, 3rd and 4th posterior intercostal veins)	Arch of azygos vein	Left brachiocephalic vein
5th to 8th posterior intercostal veins	Vertical part of azygos vein	Accessory hemiazygos vein
9th to 11th posterior intercostal veins	Vertical part of azygos vein	Hemiazygos vein

11. What do you mean by an intercostal nerve?

Ans. An intercostal nerve is the ventral ramus of a thoracic nerve.

12. What is a typical intercostal nerve?

Ans. Ventral rami of **3rd to 6th thoracic spinal nerves** are confined only to the thoracic wall, hence called typical intercostal nerves.
T_1 and T_2 nerves are distributed to both thorax and upper limb.
T_7 to T_{11} are distributed to both thorax and abdomen.

13. What are the branches of intercostal nerve?

Ans. **Branches**:
i. **Ganglionic branches** to corresponding sympathetic ganglia.
ii. **Muscular branches** to intercostal muscles.
iii. **Collateral branch**.
iv. **Lateral cutaneous branch** which again divides into:
   a. Anterior branch.
   b. Posterior branch.
v. **Anterior cutaneous branch** which divides into:
   a. Medial branch.
   b. Lateral branch.
vi. Branches to costal pleura.
vii. Branches to joints of the ribs.

14. What do you mean by the neurovascular plane?

Ans. Neurovascular plane: It is the interval or plane between the intercostalis internus and intercostalis intimus through which the intercostal vessels and nerves run.

15. What are the muscles of intercostal space?

Ans.

Muscles	Origin	Insertion
External intercostal	Lower border of the upper rib	Outer lip of the upper border of the lower rib
Internal intercostal	Floor of the costal groove of the upper rib	Intermediate part of the upper border of lower rib
Transversus thoracis	Upper lip of the costal groove of the upper rib	Inner lip of the upper border of the lower rib
Intercostalis intimus	Inner surface of the rib above, near the angle	Inner surface of the 2nd or 3rd rib below
Subcostalis (absent in a typical space)	Posterior surface of the lower ⅓rd of the body and posterior surface of the xiphoid process of sternum	Inner surface of the 2nd or 3rd rib below
Sternocostalis	Posterior surfaces of the 5th, 6th and 7th costal cartilages	2nd to 6th costal cartilages

CHAPTER 6

Window Dissections: Head and Neck

- Introduction to Head and Neck
- Face
- Anterior Triangles of Neck
- Posterior Triangles of Neck

LESSON 1: INTRODUCTION TO HEAD AND NECK

The **head** is the upper most part of the body in anatomical position of a person. The **neck** connects the head with the rest of the body. Though the head and neck are considered as a single region, there is an anatomical demarcation between the head and neck. The neck is demarcated from the head by a line starting from the lower border of the body of the mandible in the midline → along the lower border of the mandible up to its angle → mastoid process → superior nuchal line of the occipital bone → external occipital protuberance. From this line of demarcation it is obvious that the neck is higher or longer in the posterior aspect than its anterior side. Similarly, the neck is demarcated from the thorax (i.e. the lower limit of the neck) by an imaginary line extending anteriorly from the upper border of the manubrium sterni in the midline → along the clavicle → acromion process of the scapula → up to the spine of seventh cervical vertebra (vertebral prominence). The lower limit of the neck also called the base of the neck corresponds to the plane of **superior thoracic aperture** (thoracic inlet).

Though the head and neck region is relatively small, it contains very important organs and structures which include a number of **endocrine glands** (pituitary, pineal gland, thyroid gland and parathyroid glands), **salivary glands** (parotid glands, submandibular glands, sublingual glands), **special sense organs** (eyes for vision, ears for hearing and equilibrium, nose for olfaction and tongue for taste sensation), **parts of the digestive tract** (pharynx and cervical part of esophagus), parts of other structures like teeth, tonsils, palate, etc.

The respiratory tract (larynx and upper part of trachea), **nerves** (twelve pairs of cranial nerves and their branches, eight pairs of cervical nerves and their plexuses, sympathetic chain with 3 cervical sympathetic ganglia), **large blood vessels** (carotid arterial system and vertebral arterial system, jugular venous system), huge number of **lymph nodes and lymphatic vessels**

and last but not the least is the **brain** with meninges in the cranial cavity and **spiral cord** in the cervical part of vertebral canal are also important structures of the head and neck region.

Besides the above mentioned organs and structures, it is necessary to understand the skeletal basis of the head and neck with their muscular attachments. The skeleton of the head and neck is made up of skull, hyoid bone and seven cervical vertebrae. The skull consists of 22 bones excluding the ossicles of the middle ear. The skull can be subdivided into two parts:

1. **Calvaria** or **brain–box** which surrounds the cranial cavity and encloses the brain. The bones which form the calvaria are parietal (2), temporal (2), frontal (1), occipital (1), ethmoid (1) and sphenoid (1) (**Total of 8 bones** excluding 3 pairs of middle ear ossicles—malleus, incus and stapes).
2. **Facial skeleton** is formed by 6 paired bones (maxilla, zygomatic, lacrimal, palatine, nasal and inferior nasal concha) and 2 unpaired bones (mandible and vomer) **(Total of 14 bones)**.

The bones of the skull are interconnected mostly by sutures (sutural variety of fibrous joint). Only 3 pairs of synovial joints are found in the head. The largest synovial joint in the head is the **temporomandibular joint**. Other two pairs of synovial joints are between the three ossicles of the middle ear. The cervical vertebrae in the cervical region are interconnected by all three types of joints. In the following chapters, dissections of face and triangles of the neck will be discussed.

LESSON 2: FACE

EXTENT OF FACE

Above: Hairline of the scalp.
Below: Chin and base of the mandible.
On each side: Auricle.

The facial skin is rich in vascularity, sebaceous and sweat glands, very elastic and thick. Greater part of the facial skin is lax, more so in the eyelids. The fixity of the skin to the underlying cartilages of the nose and ear is responsible for acute pain in boils over these regions. **Note that the forehead is common to both the face and the scalp.**

STEPS OF DISSECTION

1. **Position of the cadaver:** Supine, face turned to the opposite side and the neck is slightly extended.
2. **Skin incision (Figs 6.1A and B):**
   Vertical incision: The incision starts from the root of the nose (A), up to the center of the philtrum of the upper lip (B), then encircling the margin of the upper lip, angle of the mouth (C) and margin of the lower lip up to the center of the lower lip (D) then vertically downwards up to the midpoint of the chin (E).

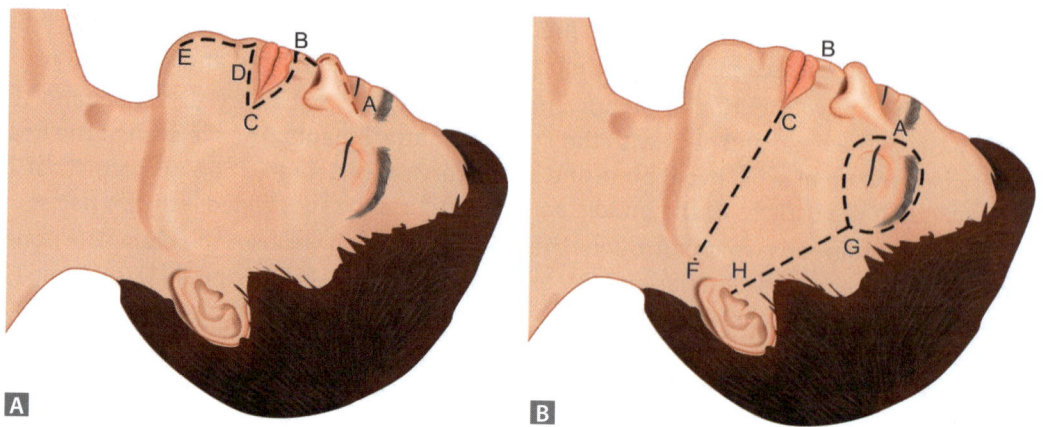

Figs 6.1A and B: Skin incision for face: (A) Vertical incision; (B) Horizontal incision

Horizontal incision:
  i. One incision extends from the angle of the mouth to the posterior border of the mandible (C–F).
  ii. Another incision starts from the bridge of the nose then around the upper and lower margin of the orbit meets at the lateral angle of the eye (G) then continued up to the tragus or upper end of the auricle (H).

Reflect the lower flap downwards and laterally to the lower border of the mandible, middle flap towards the ear and the upper flap upwards. During this procedure one must be very careful not to damage the structures (vessels, nerves, muscles) which are immediately deep to the skin. For this reason the scalpel is to be applied more towards the skin.
Superficial fascia is exposed.

3. **Superficial fascia:** It is present all over the face expect in the eyelids. Subcutaneous fat is abundant in the cheeks (buccal pad of fat) but absent is the eyelids. So, **eyelids are devoid of fat and superficial fascia.** After removal of the skin, using blunt dissection and scraping technique isolate the underlying structures. The following structures are found in the superficial fascia.
    A. **Muscles:** The facial muscles are collectively known as the **muscles of facial expression.** These muscles are attached to the dermis of the skin and represent morphologically the remnants of **panniculus carnosus.**
       The muscles to be identified are (Figs 6.2A and B):
         i. **Orbicularis oculi**—around the orbit.
         ii. **Corrugator supercilii**—found obliquely from the medial end of superciliary arch to the middle of supraorbital margin.

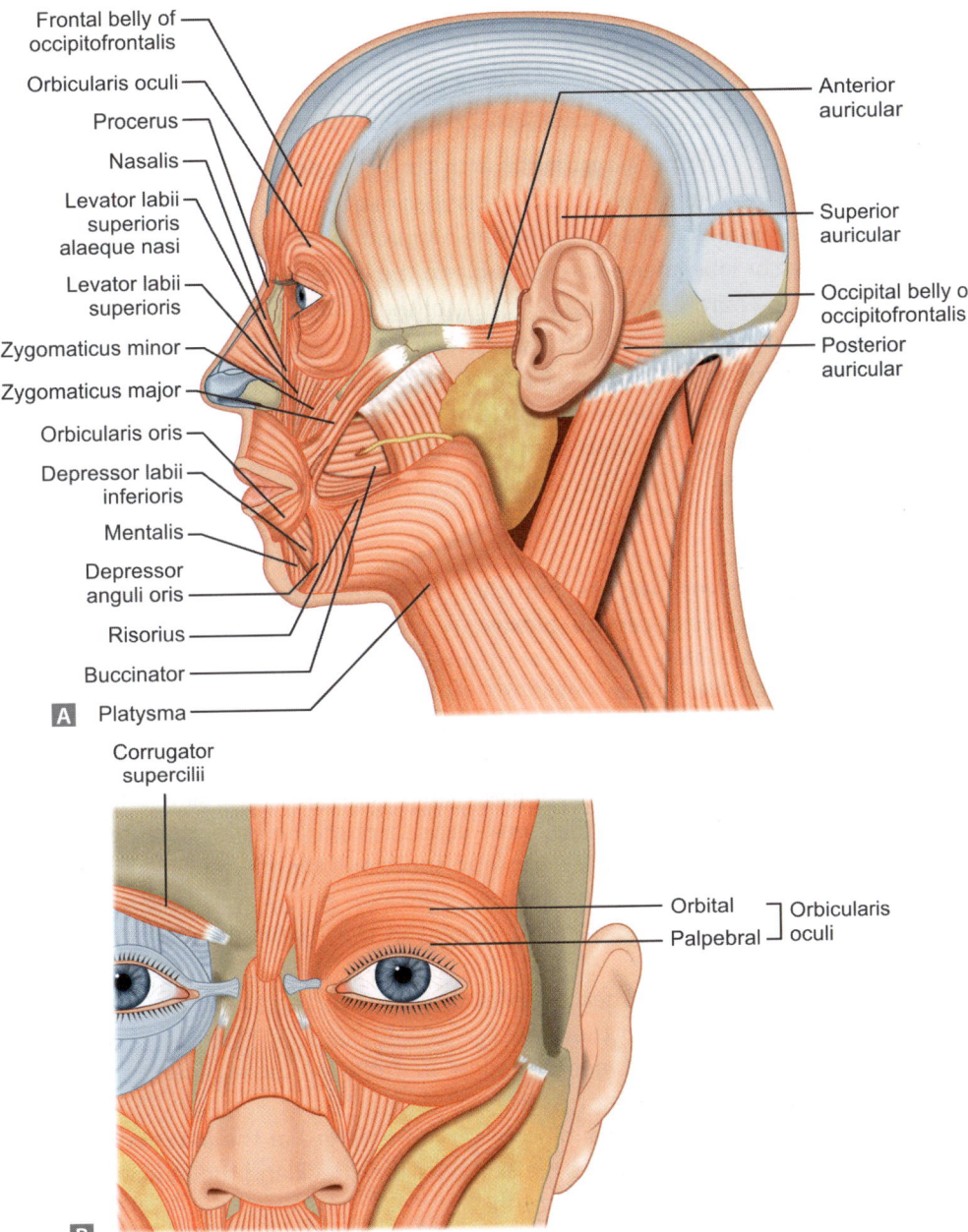

Figs 6.2A and B: (A) Muscles of face; (B) Orbicularis oculi muscle

iii. Procerus—between the medial ends of eyebrows.
iv. Levator labii superioris alaque nasi—over the dorsum of the nose up to the ala of the nose and upper lip.

v. Levator labii superioris—found lateral to the alaque nasi muscle running vertically up to upper lip.
vi. Levator anguli oris—from the maxilla to the angle of the mouth.
vii. Zygomaticus major—zygomatic arch to the angle of the mouth.
viii. Zygomaticus minor—zygomatic arch to the upper lip close to the angle of the mouth and found medial to zygomaticus major.
ix. Depressor anguli oris—oblique line of mandible to the angle of the mouth.
x. Depressor labii inferioris—mandible to the lower lip.
xi. Risorius—parotid fascia to the angle of the mouth.
xii. Mentalis—incisive fossa to the skin of the chin.
xiii. **Buccinator** (deep to buccal pad of fat)—from external alveolar margin of the maxilla and mandible to the angle of the mouth.
xiv. **Orbicularis oris**—around the oral fissure.
xv. Frontal belly of occipitofrontalis—from epicranial aponeurosis to the skin of forehead and eyebrows.

B. **Nerves:**
  a. **Sensory nerves: All are branches from the trigeminal nerve except great auricular nerve of cervical plexus.**
   i. Supratrochlear nerves—at the medial part of supraorbital margin.
   ii. Supraorbital nerve—exits the skull through the supraorbital foramen which can be located on the cadaver's skull by separating the corrugator supercilii muscle and orbicularis oculi muscle.
   iii. Infratrochlear nerve—supplies skin of the eyelids and side of the nose.
   iv. Infraorbital nerve—this nerve exits through the infraorbital foramen which is located deep to levator labii superioris muscle and if necessary this is exposed by cutting the muscle.
   v. Zygomaticotemporal nerve—found in the temple behind the supraorbital nerve.
   vi. Zygomaticofacial nerve—exits through the zygomaticofacial foramen supplies the skin over the cheek.
   vii. Auriculotemporal nerve—found in front of the auricle and behind the superficial temporal vessels.
   viii. Buccal branch of mandibular nerve—emerges at the anterior border of the masseter and supplies the skin over the cheek and the mucous membranes of the vestibule of the mouth.
   ix. Mental nerve—exits through the mental foramen of the mandible and it is located by separating the fibers of mentalis muscle from the depressor anguli oris muscle.
   x External nasal nerve—supplies tip and ala of the nose.
   xi. **Great auricular nerve (C_2, C_3)—branch of cervical plexus. It supplies the skin overlying the angle of the mandible.**

b. **Motor nerves:** The motor supply of the face is derived from the **facial nerve**. It emerges from the parotid gland as five terminal branches and diverges to supply the various facial muscles. These branches of the facial nerve are named according to the region they innervate. These are (Fig. 6.3):
   i. Temporal branch.
   ii. Zygomatic branch.
   iii. Buccal branch.
   iv. Marginal mandibular branch.
   v. Cervical branch (appears in the neck to supply platysma).

These branches are identified by their white thread like appearance. Follow these branches through the parotid gland towards the **stylomastoid foramen** through which the facial nerve comes out from the cranium. For proper exposure of the nerve, it may be necessary to remove the parotid gland piece by piece. All the five terminal branches are dissected parallel to the course of the nerves and studied.

C. **Arteries:** The chief artery of the face is **facial artery** (Fig. 6.4). It is a branch of the external carotid artery is the neck. It enters the face by winding around the base of the mandible at the anteroinferior angle of the masseter. Then it courses **tortuously** upwards and forwards to a point about 1.25 cm lateral to be angle of the mouth. It then ascends by the side of the nose up to the medial angle of the eye and ends by anastomosing with the dorsal nasal branch of ophthalmic artery. The facial artery runs in the face between the superficial muscles (risorius, zygomaticus major et minor) and deep muscles (buccinators, levator anguli oris) of the face.

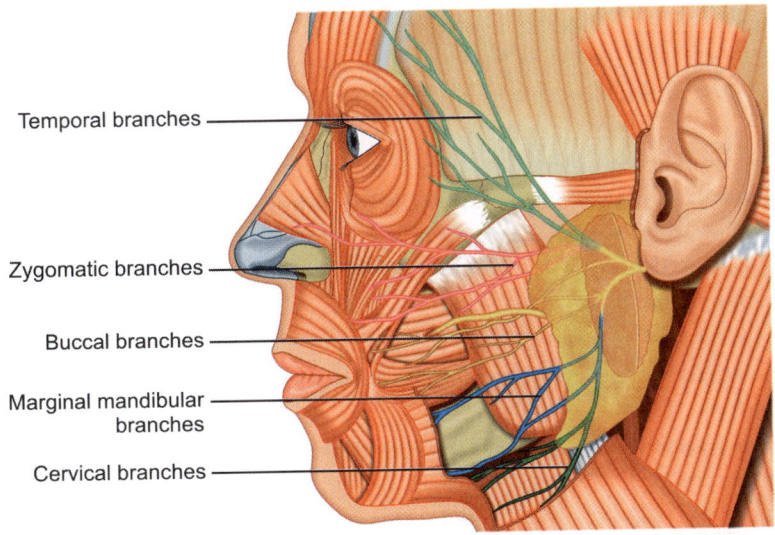

Fig. 6.3: Motor nerves in the face

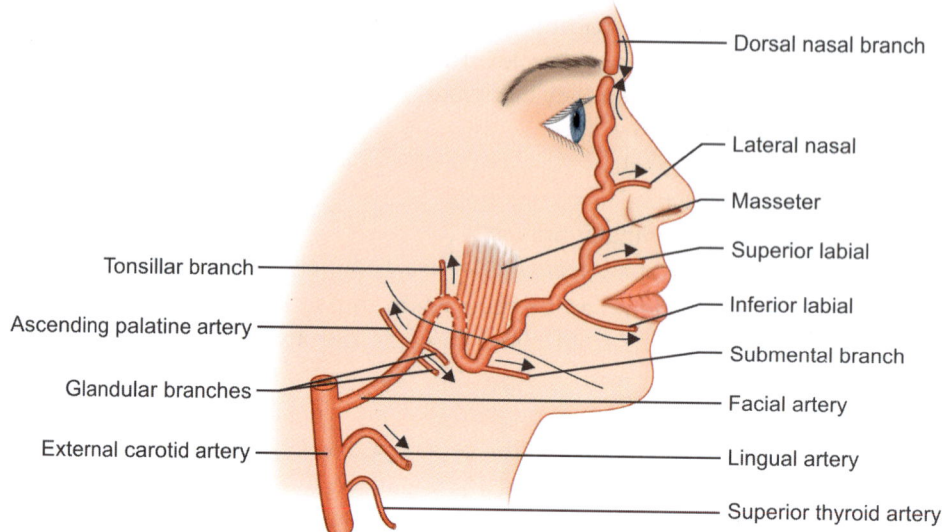

Fig. 6.4: Facial artery and its branches

Superficial temporal artery is a terminal branch of the external carotid artery. It is located anterior to the ear in front of the auriculotemporal nerve and behind the temporal branch of facial nerve.

Transverse facial artery is a branch of superficial temporal artery. It emerges from the parotid gland and then it passes forwards over the masseter between the zygomatic arch above and parotid duct below. **Supratrochlear** and **supraorbital vessels** accompany the cutaneous nerves.

D. **Veins:** Facial vein begins at the medial angle of the eye as the angular vein and runs downwards and backwards behind the facial artery with a straighter and superficial course. At the side of the nose facial vein lies on the surface of levator labii superioris alaque nasi muscle and then it passes deep to zygomaticus major. It forms the common facial vein after joining with the anterior division of the retromandibular vein and ultimately drains into the internal jugular vein.

E. **Parotid Duct:** It emerges from the parotid gland through its anterior border and passes forwards over the masseter between the upper and lower buccal branches of facial nerve, 2–3 cm below the zygomatic arch. It then pierces the **buccophayngeal fascia** and **buccinator muscle** to open is the vestibule of the mouth. It is 5 cm in length and 3 mm in width.

F. **Accessory parotid gland:** It may be found between the upper buccal branch of facial nerve and the parotid duct.

4. **Deep fascia:** Deep fascia is conspicuously absent is the face. Only the **masseter** and **parotid gland** are covered by the investing layer of the deep cervical fascia which is considered to be the deep fascia of the face and called **parotidomasseteric fascia**.

MUSCLES RELATED TO FACE

The muscles of the face are subcutaneous in position and represent remnants of **panniculus carnosus**. All the muscles of the face (except masseter) are supplied by facial nerve as they are derived from the mesoderm of second branchial arch. Remember that facial nerve is the nerve of second branchial arch. Though there are many muscles in the face, only few of them are discussed here.

Muscles	Origin	Insertion	Nerve supply	Action
Buccinator	i. Outer surface of the alveolar process of maxilla and mandible opposite the 3rd molar tooth ii. Pterygomandibular raphe	Angle of the mouth: i. Upper fibers pass along the upper lip ii. Lower fibers pass along the lower lip iii. Intermediate fibers decussate with each other	Lower buccal branch of facial nerve	i. Flattens the cheek against the gum and teeth and helps in mastication ii. Helps in sucking, blowing and whistling iii. Acts as a sphincter of parotid duct

Orbicularis oculi	Three parts:
	1. **Orbital part:**
	  i. Medial palpebral ligament
	  ii. Frontal and maxillary bone around the orbit
	2. **Palpebral part:** Medial palpebral ligament
	3. **Lacrimal part:**
	  i. Crest of lacrimal bone
	  ii. Lacrimal fascia
		Orbital part:
		i. Upper fibers blend with frontalis and corrugator supercilli muscle
		ii. Subcutaneous tissue of the eye brow (depressor supercilli)
		Palpebral part: Lateral palpebral raphe
		Lacrimal part:
		i. Tarsi of both eye lids
		ii. Lateral palpebral raphe
			Temporal branch of facial nerve	i. Closing of the eye lids
				ii. Lacrimal part dilates the lacrimal sac
				iii. Palpebral part closes the eyelids gently in sleep
				iv. When entire muscle contracts the skin of the forehead, temple and check are drawn towards the medial angle of the eye. This is a permanent feature in old age and it is called as **crow's feet**

Frontal belly of occipitofrontalis	No bony attachment. It arises from the skin and subcutaneous tissue of the eye brow and the root of the nose	Epicranial aponeurosis (**galea aponeurotica**)	Temporal branch of facial nerve	i. Elevation of eye brow ii. Transverse wrinkling of forehead iii. Moves entire scalp by alternate contraction of frontalis and occipitalis

MUSCLES OF MASTICATION

Muscles	Origin	Insertion	Nerve supply	Action
Masseter	Lower border and deep surface of the zygomatic arch	Lateral surface of the ramus of the mandible	Masseteric branch of anterior division of mandibular nerve	i. **Elevator** of the mandible ii. **Protraction** by superficial fibers iii. **Retraction** by deep fibers
Temporalis (Fan-shaped)	i. Temporal fossa below the inferior temporal line ii. Temporal fascia	i. Tip, medial surface and anterior border of coronoid process of mandible ii. Anterior border of ramus of mandible	Deep temporal branch of anterior division of mandibular nerve	i. **Elevator** of the mandible ii. **Retraction** by its posterior fibers iii. **Side-to-side** movements by temporalis of both sides
Lateral pterygoid	i. **Upper head:** Infratemporal surface and infratemporal crest of greater wing of sphenoid ii. **Lower head:** Lateral surface of lateral pterygoid plate of sphenoid	i. Front of the neck of the mandible ii. Capsule and articular disc of temporo-mandibular joint	Anterior division of mandibular nerve	i. **Depression** of mandible ii. **Protrusion** of mandible by action of lateral and medial pterygoid muscles of both sides iii. **Side-to-side chewing movement** by alternate action of both pterygoids of one side with that of the other side
Medial pterygoid	i. **Deep head:** Medial surface of lateral pterygoid plate ii. **Superficial head:** Tuberosity of maxilla and pyramidal process of palatine bone	Medial surface of ramus and angle of the mandible below and behind the mandibular foramen and mylohyoid groove	Branch from trunk of mandibular nerve	i. **Elevation** of mandible ii. **Protrusion** of mandible by combined action with lateral pterygoid iii. Side-to-side chewing movement by alternate action of both pterygoids

Summary

1. Position of the cadaver—supine, head turned to the opposite side.
2. Skin incision as in Figure 6.1.
3. **Superficial fascia**—structures found in it → sensory (trigeminal) and motor (facial) nerves, facial artery and vein, parotid duct, accessory parotid gland and muscles of facial expression.
4. Deep fascia—found only over masseter and parotid gland (**parotidomasseteric fascia**).

PROBABLE QUESTIONS AND ANSWERS

1. What is the extent of the face?

Ans. The face is limited above by the hairline of the scalp, below by the chin and base of the mandible and on each side by the auricle.

2. Why the forehead is common to both face and scalp?

Ans. The scalp is limited anteriorly by the supraorbital margin and the face is limited above by the hairline of the scalp. So, the forehead is common to both face and scalp.

3. What are the special characteristics of the skin of face?

Ans.
  i. Facial skin highly vascular → so wound healing is rapid.
  ii. Sebaceous glands are more in facial skin and scalp → so sebaceous cysts are usually found on face and scalp.
  iii. Facial skin is thick and elastic because of the attachment of facial muscles into it → so facial wounds tend to gape and bleed profusely.
  iv. Skin is lax over the greater part of the face → so edema spreads rapidly in the face.

4. Which part of the face is devoid of superficial fascia and subcutaneous fat?

Ans. Eyelids.

5. Why deep fascia is absent in the face?

Ans. Absence of deep fascia allows facial expression.

6. What are the other regions of the body where deep fascia is absent?

Ans. Deep fascia is absent in front of the trunks (abdomen and thorax).

7. Why do the muscles of the face morphologically represent panniculus carnosus?

Ans. This is because of the fact that the facial muscles are subcutaneous in position and all of them are inserted into the skin.

8. Name other panniculus carnosus group of muscles?

Ans.
  i. Dartos muscle in the scrotum.
  ii. Palmaris brevis in the hand.

9. What is the sensory supply of the face?

Ans. The sole sensory supply of the face is the cutaneous branches of the trigeminal nerve except an area of skin overlying the angle of mandible which is supplied by the greater auricular nerve (C_2, C_3) from the cervical plexus.

10. How many cutaneous branches of trigeminal nerve are there in the face?

Ans. There are eleven sensory branches are there in each half of the face. These are:
  a. From the **ophthalmic division (5)**:
     i. Lacrimal.
     ii. Supraorbital.
     iii. Supratrochlear.
     iv. Infratrochlear.
     v. External nasal.
  b. From the **maxillary division (3)**:
     i. Infraorbital.
     ii. Zygomaticofacial.
     iii. Zygomaticotemporal.
  c. From the **mandibular division (3)**:
     i. Auriculotemporal
     ii. Buccal.
     iii. Mental.

11. What is the motor supply of the face?

Ans. All the muscles of the face are supplied by the facial nerve except masseter which is supplied by the mandibular nerve.

12. What is the embryological reason of supplying the facial nerve to all the facial muscles?

Ans. Because, all the facial muscles are derived from the mesoderm of second branchial arch and the facial nerve is the nerve of that arch.

Note that the sensory supply of the face is trigeminal nerve except over the angle of the mandible (great auricular nerve) and the motor supply is the facial nerve except masseter (mandibular division of trigeminal nerve).

13. How does the facial nerve appear in the face?

Ans. The facial nerve appears in the face radiating through the anterior border of the parotid gland as its five terminal branches.

14. **What are the terminal branches of facial nerve and their areas of supply?**

Ans. Refer Figure 6.5:
   i. **Temporal branch:** Orbicularis oculi, frontalis, corrugator supercilli, auricularis anterior et superior.
   ii. **Zygomatic branch:** Orbicularis oculi (lower part).
   iii. **Buccal branch:**
      a. Superficial branches to procerus.
      b. Deep branches to zygomaticus major et minor, levator labii superioris, levator labii superioris alaque nasi, levator anguli oris, dilator naris, compressor naris, buccinator, orbicularis oris.
   iv. **Marginal mandibular branches:** Rhisorius, depressor anguli oris, depressor labii inferioris, mentalis.
   v. **Cervical branches:** Platysma (**in the neck**)

Note that mental nerve, branch of inferior alveolar nerve is a sensory nerve. It comes out through the mental foramen of the mandible and supplies the skin overlying the mentalis muscle. Mentalis muscle is supplied by facial nerve.

Muscles of upper lip are supplied by buccal branch and the muscles of lower lip are supplied by marginal mandibular branch of facial nerve.

Remember that buccal branch of the facial nerve (**motor**) supplies the buccinator muscle, but the buccal branch of the anterior division of mandibular nerve (**sensory**) supplies the skin overlying the buccinator muscle.

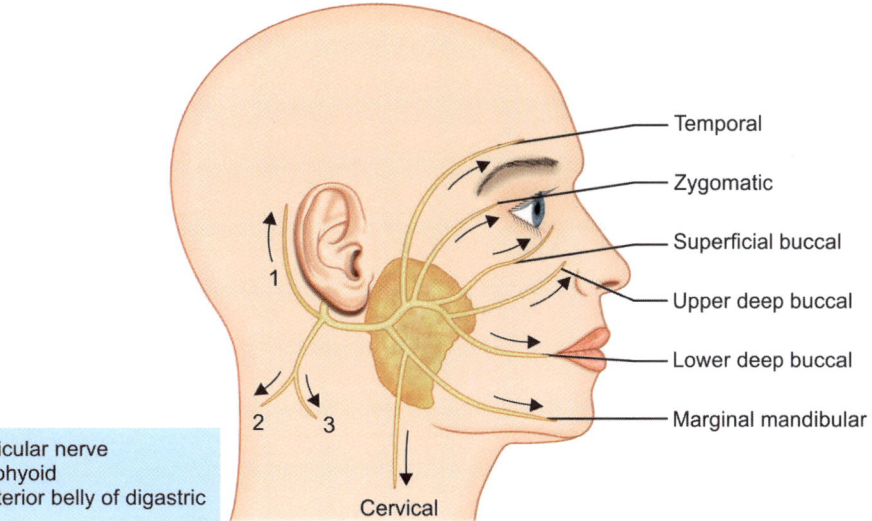

1 = Posterior auricular nerve
2 = Nerve to stylohyoid
3 = Nerve to posterior belly of digastric

Fig. 6.5: Terminal branches of facial nerve

Q15. How does the facial nerve come out of the cranial cavity?

Ans. It comes out through the **stylomastoid foramen** of the temporal bone.

Q16. What is the area of sensory distribution of trigeminal nerve in the face?

Ans. The three divisions of trigeminal nerve supply the face (Fig. 6.6). These three cranial dermatomes meet at the angles of the mouth and eye. Of these three dermatomes, the area supplied by **the maxillary division is least extensive**.
The first division (**ophthalmic division**) of trigeminal nerve supplies:
  i. Upper eyelid.
  ii. Dorsum and upper part of the side of the nose.
  iii. Forehead.
  iv. Scalp.
The second division (**maxillary division**) supplies:
  i. Lower eye lid.
  ii. Side of the lower part of the nose.
  iii. On and above the prominence of the cheek and upper lip.
The third division (**mandibular division**) supplies:
  i. Anterior part of lateral surface of external ear.
  ii. Skin of the cheek.
  iii. Skin in front of the ear extending onto the scalp.
  iv. Lower lip.
  v. Chin.

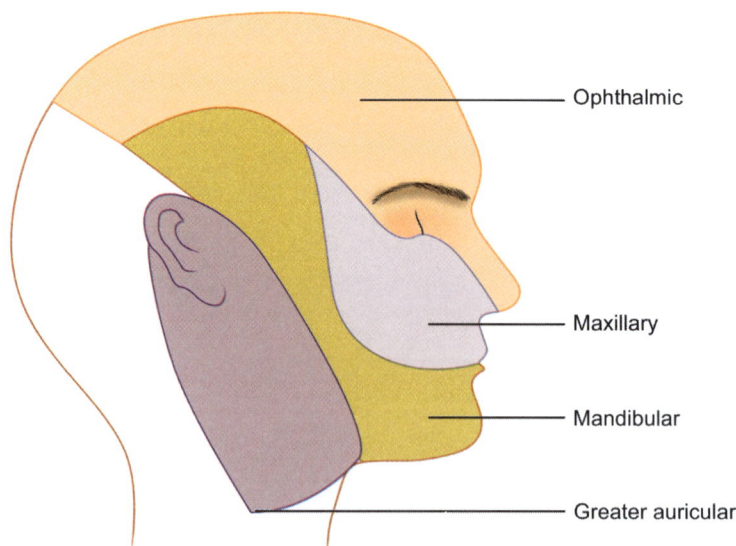

Fig. 6.6: Distribution of trigeminal nerve in the face

Window Dissections: Head and Neck

17. Why the muscles of the face are known as the muscle of facial expression?

Ans.
  i. Absence of deep fascia in the face allows facial expression by the contraction of the facial muscles.
  ii. The facial muscles are attached into the skin. So during contraction of these muscles they produce wrinkles or dimples on the skin and thus showing various kinds of facial expression.

18. What is modiolus?

Ans. It is a palpable thickening felt towards the lateral angle of the month opposite the upper second premolar tooth due to the chiasmatic decussation of the intermediate fibres of buccinator muscle and interlacement of other muscles around the oral fissure. The upper (maxillary) fibers and the lower (mandibular) fibers pass straight into the corresponding lips.

19. Which muscle is known as blowing muscles?

Ans. Buccinator, because it forcibly expels the air between the lips from the inflated vestibule.

20. Which muscle is known as accessory muscle of mastication?

Ans. **Buccinator**, because it prevents accumulation of food in the vestibule of the mouth.

21. What is Bugler's muscle?

Ans. Buccinator muscle (A Halim).

22. What are the muscles of mastication and their nerve supply?

Ans. Four muscles:
  i. Temporalis.
  ii. Lateral pterygoid.
  iii. Medial pterygoid.
  iv. Masseter.
  These four muscles are supplied by mandibular nerve.

23. Name the structures that pierce the buccinator muscle?

Ans.
  i. Parotid duct (opposite the 3rd upper molar tooth).
  ii. Buccal branch of mandibular nerve.
  iii. 4–5 molar mucous glands.

24. Which muscle is called laughing muscle?

Ans. Zygomaticus major—by drawing the angle of the mouth upwards and laterally.

25. What is the source of Rhisorius muscle?

Ans. It is regarded as a part of platysma. It arises from the parotid fascia and inserted into skin at the angle of mouth.

26. What is risus sardonicus?

Ans. Due to contraction of the muscle rhisorius, it pulls the angle of the mouth upwards and laterally producing the sardonic grin. It is seen in patients suffering from tetanus.

27. Which muscle surrounds the palpebral fissure and how many parts of this muscle are there?

Ans. **Orbicularis oculi.** It has three parts:
 i. Orbital part.
 ii. Palpebral part.
 iii. Lacrimal part.

Remember that the action of frontalis part of occipitofrontalis is antagonistic to the orbital part of orbicularis oculi.

28. What are the different parts of orbicularis oris?

Ans. It is a composite muscle around the oral fissure consisting of intrinsic part (muscle fibers which extended from the mucous membrane of the lips to the skin) and extrinsic part (which are arranged in three layers). The extrinsic muscles are:
 a. Superficial group of muscles:
  i. Levator labii superioris.
  ii. Levator labii superioris alaque nasi.
  iii. Levator anguli oris.
  iv. Zygomaticus major et minor.
  v. Depressor anguli oris.
  vi. Depressor labii inferioris.
  vii. Mentalis.
 b. Intermediate group of muscle: Buccinator (deep to buccal pad of fat).
 c. Deep group of muscles:
  i. Incisivus labii superioris.
  ii. Incisivus labii inferioris.

29. What are the arteries supplying the face.

Ans.
i. **Facial artery (principal artery of the face).**
ii. Superficial temporal artery.
iii. Transverse facial artery.
iv. Supraorbital and supratrochlear arteries.
v. Dorsal nasal artery.

30. What is the origin of facial artery?

Ans. It arises from the external carotid artery in the neck.

31. What the branches of facial artery in the face?

Ans.
i. Superior labial.
ii. Inferior labial.
iii. Lateral nasal.

32. Why the facial artery is tortuous?

Ans. The tortuosity of the artery prevents its wall from being unduly stretched during the movements of the pharynx in deglutition and during the movements of the lips, cheeks and mandible. This artery is vulnerable to stretching as it lies away from the axis of temporomandibular joint, i.e. the line joining the two mandibular foramen.

33. What are the branches of facial artery in the neck?

Ans.
i. Tonsillar.
ii. Glandular.
iii. Ascending palatine.
iv. Submental.

34. What is linguofacial artery?

Ans. When the facial artery arises as a common trunk with the lingual artery, it is called linguofacial artery.

35. Name other tortuous arteries in the body.

Ans.
i. Nutrient artery of the long bones.
ii. Splenic artery.
iii. Uterine artery (in female).
iv. Posteroinferior cerebellar artery.

36. What is the fate of facial artery in the face?

Ans. It ends by anastomosing with the dorsal nasal branch of ophthalmic artery at the medial angle of the eye.

37. How does facial artery establish free communication between external and internal carotid artery?

Ans. Facial artery is the branch of external carotid artery. Ophthalmic artery is a branch of internal carotid artery. So, by anastomosing with the dorsal nasal branch of ophthalmic artery (internal carotid), facial artery (internal carotid) establishes free communication between them.

38. What are the veins draining the face?

Ans.
i. Facial vein.
ii. Retromandibular vein.

39. What is the largest vein of the face?

Ans. The facial vein.

40. How the facial vein is formed?

Ans. The union of supratrochlear and supraorbital vein forms the angular vein at the medial angle of the eye. The angular vein continues as the facial vein.

41. What is common facial vein?

Ans. It is formed by the union of anterior division of retromandibular vein and facial vein below the angle of the mandible (Fig. 6.7).

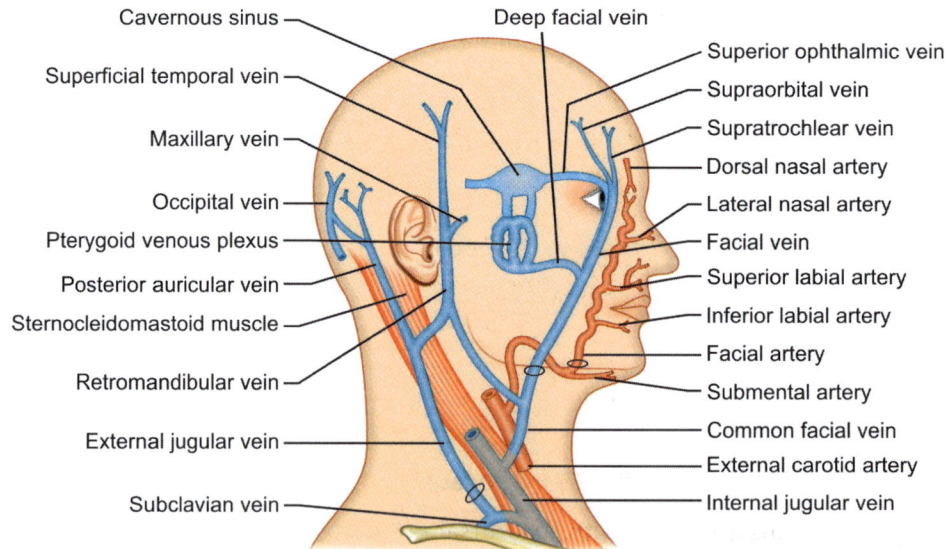

Fig. 6.7: Formation of common facial vein and external jugular vein

Window Dissections: Head and Neck

42. What is retromandibular vein?

Ans. It is a vein formed within the substance of the parotid gland behind the ramus of the mandible by the union of superficial temporal vein and maxillary vein.

43. What is the fate of posterior division of retromandibular vein?

Ans. It joins with the posterior auricular vein to form external jugular vein which drains into the subclavian vein about 4 cm above the clavicle (Refer to Flowchart 6.1 and Fig. 6.8).

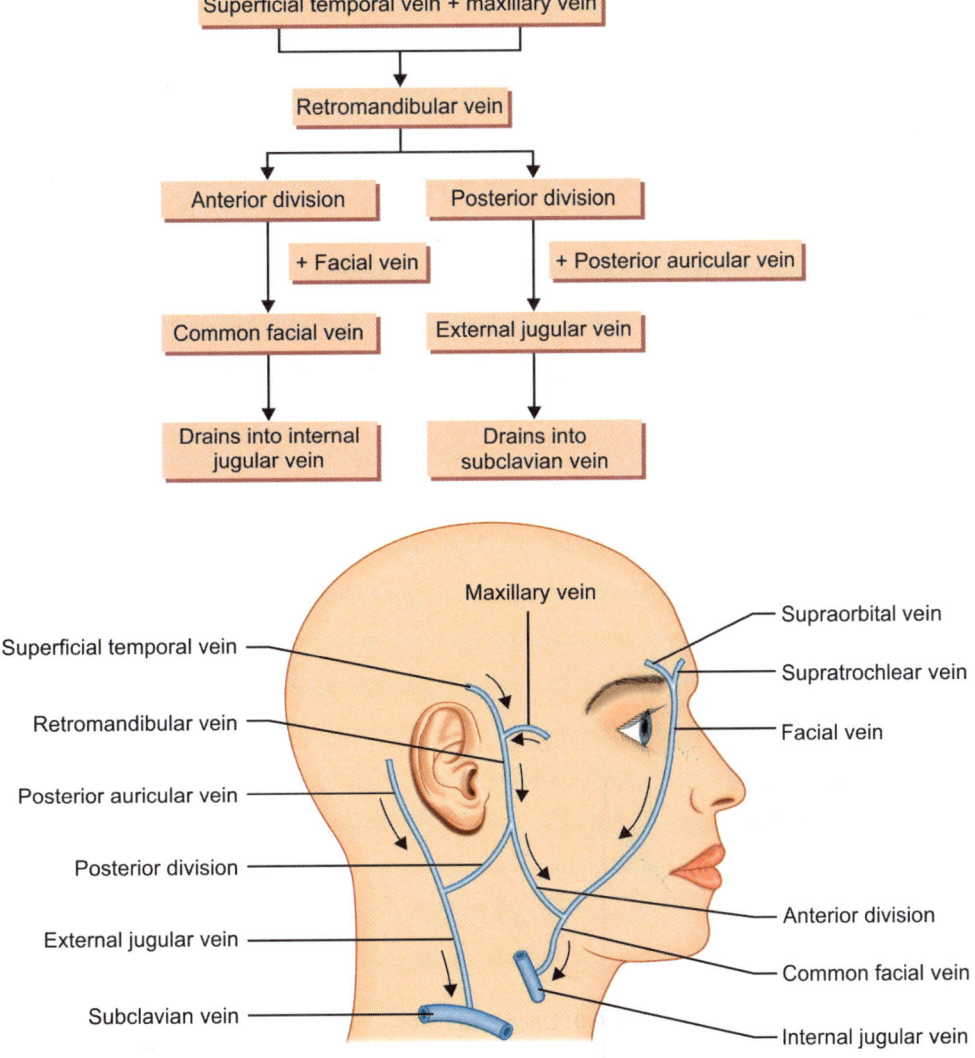

Flowchart 6.1: Formation of retromandibular vein and its fate

Fig. 6.8: Veins of the face

44. Is the facial vein tortuous?

Ans. No. It is straight, because it lies close to the axis of temporomandibular joint.

45. How do you identify facial artery and facial vein in the dissected face?

Ans.
  i. Facial artery is tortuous, but facial vein is straight.
  ii. Facial vein runs downwards and backwards behind the facial artery.

46. Why does the infection from the face spread rapidly?

Ans. The facial vein has no valves. So the septic emboli spread easily during the movements of facial muscles.

47. How does the facial vein communicate with the cavernous sinus?

Ans.
  i. Via superior ophthalmic vein.
  ii. Via deep facial vein → Pterygoid venous plexus → Emissory vein → Cavernous sinus.

48. Which area of the face is dangerous area and why?

Ans. The upper lip and the adjoining nose lying between the angular vein and deep facial vein is the dangerous area of the face, because infection from this area passes to the cavernous sinus (Fig. 6.9).

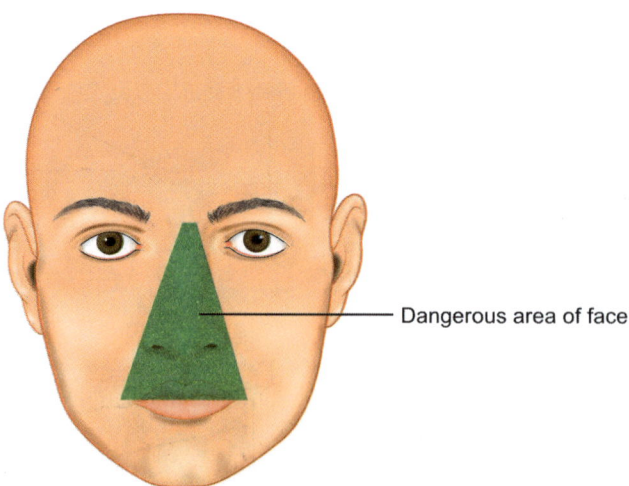

Fig. 6.9: Dangerous area of face

149. What is the lymphatic drainage of the face?

Ans.
i. From the upper part: Preauricular or superficial parotid lymph node.
ii. From the intermediate area: Submandibular lymph node.
iii. From the lower area: Submental lymph node.

150. What is the site of emergence of the parotid duct?

Ans. At the anterior border of the parotid gland.

151. What is the other name of this duct?

Ans. **Stenson's duct.**

152. What is its length and width?

Ans. Length—5 cm, width—3 mm.

153. What are the layers of the duct?

Ans.
i. Outer fibroelastic coat.
ii. Inner mucous membrane behind by simple cubical epithelium.

154. Where does the parotid duct open?

Ans. The parotid duct opens in the vestibule of the mouth opposite the crown of upper second molar tooth.

155. How many bents it takes before its opening in the vestibule?

Ans. Three bents (Fig. 6.10):
i. At the anterior border of the masseter and pierces the buccopharyngeal fascia and buccinator muscle.
ii. At the site of entrance into the submucous coat.
iii. At its opening through the mucous membrane in the vestibule of the mouth.

156. What is parotidomasseteric fascia?

Ans. The investing layer of deep cervical fascia splits to enclose the parotid gland. The superficial layer of this splitted fascia continues as parotidomasseteric fascia which covers the external surface of the parotid gland and masseter muscle. This is considered as the **deep fascia of the face.**

157. Do you know any other deep fascia of the face?

Ans. Buccopharyngeal fascia.

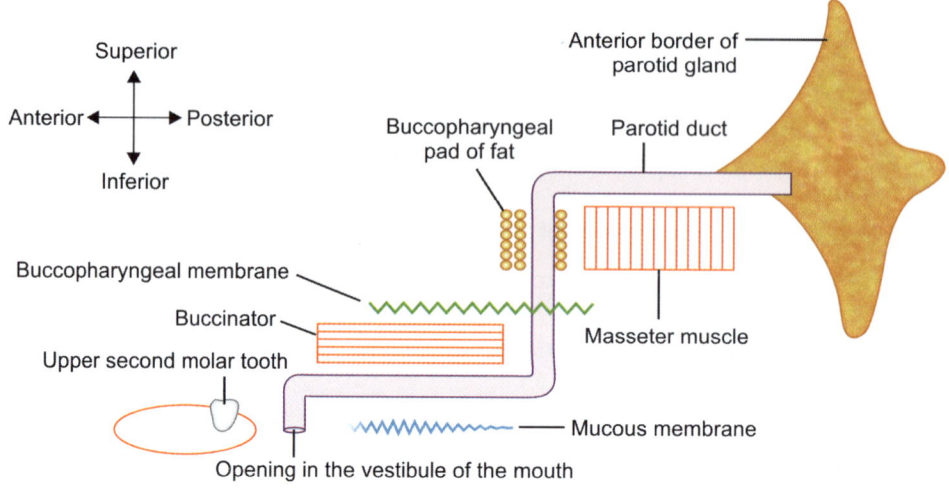

Fig. 6.10: Bents of parotid duct

LESSON 3: ANTERIOR TRIANGLES OF NECK

For descriptive purposes, the neck region is divided into many small and large triangles—**anterior triangles, posterior triangles, suboccipital triangles, scalenovertebral triangle**, etc. The anterior triangle of the neck is a large area which lies on either side of the midline of the neck. So there are two anterior triangles. Each triangle is bounded as follows (Fig. 6.11):

- **Medially:** Midline of the neck which extends from the symphysis menti to the suprasternal notch of the sternum.
- **Laterally:** Anterior border of sternocleidomastoid muscle.
- **Above:** Lower border of the body of the mandible up to the angle of the mandible and then a line extending from the angle of the mandible to the tip of the mastoid process of the temporal bone.
- **Apex:** Suprasternal notch of the manubrium sterni.

Each anterior triangle again is subdivided by digastric and omohyoid muscles into four smaller triangles for descriptive purposes. These are **carotid triangle, submental triangle, digastric triangle and muscular triangle**. Each of these triangles is having the boundaries as follows (Fig. 6.11):

1. **Submental triangle:**
   Base: Body of the hyoid bone.
   Apex: Chin (symphysis menti)
   Each side: Anterior belly of digastric muscle of the corresponding side.
   Floor: Mylohyoid muscle.

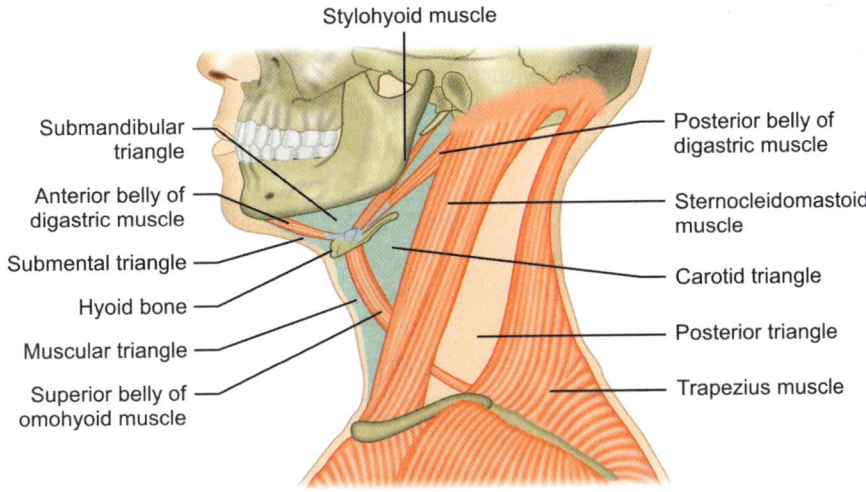

Fig. 6.11: Subdivisions of anterior triangle of neck

2. **Digastric triangle:**
   Above: Lower border of the body of the mandible and a line extending from the angle of the mandible to the tip of mastoid process of temporal bone.
   Anteroinferiorly: Anterior belly of digastric muscle.
   Posteroinferiorly: Posterior belly of digastric muscle.
   Floor: Mylohyoid and hyoglossus muscle.
3. **Carotid triangle:**
   Posteriorly: Anterior border of sternocleidomastoid muscle.
   Above and in front: Posterior belly of digastric and stylohyoid muscle.
   Below and in front: Superior belly of omohyoid muscle.
   Floor: Middle constrictor and inferior constrictor muscles of pharynx, thyrohyoid and hyoglossus muscles.
4. **Muscular triangle:**
   Above and behind: Superior belly of omohyoid muscle.
   Below and behind: Anterior border of sternocleidomastoid muscle.
   Medially: Midline of the neck extending from the hyoid bone to the suprasternal notch of the manubrium sterni.
   Floor: Sternohyoid and sternothyroid muscles.

The **roof** of all these four triangles are formed by the skin, superficial fascia with platysma and deep fascia of the neck from outside inwards.

During dissection of the anterior triangle of the neck we will observe and study the boundaries and contents of these triangles.

STEPS OF DISSECTION

1. **Position of the cadaver:** Supine with neck extended and the head is turned slightly to the opposite side. A wooden block is applied behind the shoulder blades to maintain the extended position of the neck.
2. **Incision of the skin (Fig. 6.12):**
   a. **Vertical incision:** It starts from the middle of the chin → along the midline of the neck up to the suprasternal notch (A–B).
   b. **Transverse incision:** It starts from the middle of the chin → along the base of the mandible → from there up to the tip of the mastoid process of the temporal bone to reach the attachment of the sternocleidomastoid muscle (A–C).

   Reflect the skin flap downwards and laterally up to the anterior border of sternocleidomastoid muscle. During the incision and reflection of the skin care to be taken to avoid cutting the skin deeply so that the fibers of platysma in the posterosuperior part are not torn.
3. **Superficial fascia with platysma is exposed.**

   The superficial fascia contains a variable amount of fat. In this fascial layer the following structures are to be dissected and identified.
   a. **Platysma (Fig. 6.13):** It is a thin sheet of quadrilateral, striated subcutaneous muscle (**panniculus carnosus**) extending from the second rib to face. Deep to this muscular sheet lies the superficial veins and cutaneous nerves of the neck. The platysma is then reflected upwards onto the mandible.
   b. **Transverse cervical nerve** (branch of cervical plexus): It crosses the superficial surface of the sternocleidomastoid muscle and passes forward towards the anterior triangle of the neck. Its upper branch communicates with the cervical branch of facial nerve.

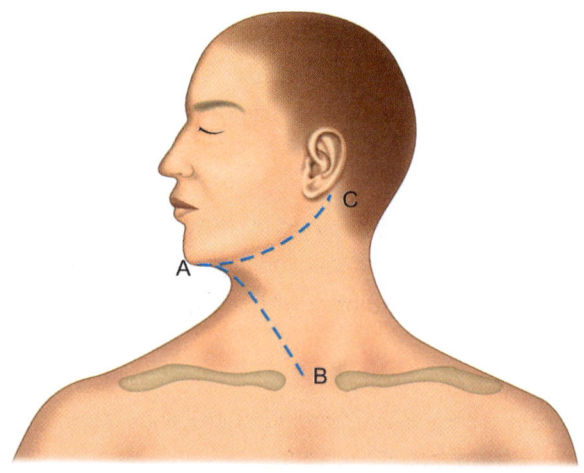

Fig. 6.12: Incision for anterior triangle of neck

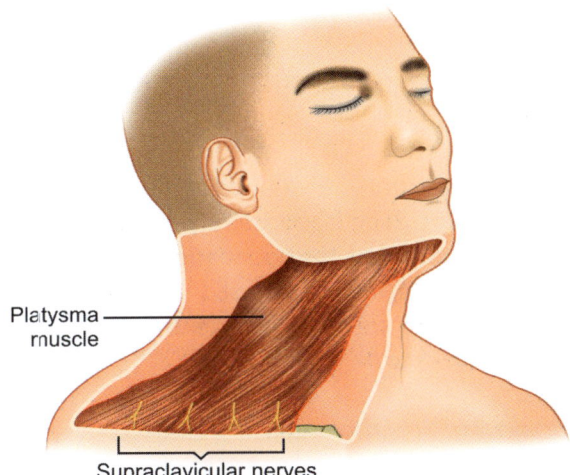

Fig. 6.13: Platysma

    c. **Cervical branch of facial nerve:** It leaves the lower border or apex of the parotid gland and appears in the anterior triangle of the neck.

    d. **Anterior jugular vein:** It lies near the anterior median line of the neck. It begins below the chin and runs downwards to pierce the investing layer of deep cervical fascia about 2 cm above the sternum.

> **Note** that the anterior jugular veins of both sides are united in the midline of the neck to form a single trunk which is known as **median cervical vein.**

Superficial fascia is then cut and reflected like that of skin. The deep fascia is exposed.

4. **Deep fascia:** The investing layer of deep cervical fascia is cut and the borders of the anterior triangle are identified.

The borders include:
    i. Midline of the neck.
    ii. Anterior border of sternocleidomastoid muscle.
    iii. Base of the mandible and a line drawn from the angle of the mandible to the tip of mastoid process.

5. Identify and palpate the important structures along the anterior midline of the anterior triangle from above downwards. These are:
    i. Hyoid bone.
    ii. Thyroid cartilage.
    iii. Cricoid cartilage.

iv. Thyroid gland.
v. Trachea.
6. Use forceps to reveal and indentify the boundaries of the four smaller triangles into which the anterior triangle is subdivided. The boundaries and contents of each triangle are cleaned and identified one by one.
7. **Submental triangle:** It is better to hyperextend the neck for proper visualization of the submental triangle. Remove the deep fascia from the anterior bellies of digastric muscles and the area between them. Now clean the boundaries and contents of submental triangle. The mylohyoid muscles of both sides are exposed forming the floor of the triangle. The following contents of this triangle are clean and identified (Fig. 6.14):
   i. Submental lymph nodes—3 to 4.
   ii. Submental vessels.
   iii. Commencement of anterior jugular vein.
   iv. Mylohyoid vessels and nerves.
   v. Superficial part of submandibular salivary gland.
8. **Diagastric triangle (submandibular triangle) (Fig. 6.15):** The submandibular region means the area which lies between the body of the mandible above and the hyoid bone below. The superficial part of the submandibular region includes the submental triangle and the digastric triangle. The deep part of the region exposes the deep structures which include deep part of the submandibular gland and the structures on hyoglossus muscle. Here, we will dissect the superficial and deep structure of the digastric triangle.
9. Hyperextend the neck for proper exposure and visualization of the structures of this triangle.
10. Identify and clean the borders of the digastric triangle, i.e. the lower border of the mandible and both the bellies (anterior and posterior) of the digastric muscle.
11. **Submandibular gland** almost fills this triangle. The gland forms a U-shape hooking round the free posterior border of the mylohyoid muscle. Thus the gland is having superficial and deep parts in respect to the superficial and deep to the mylohyoid muscle.

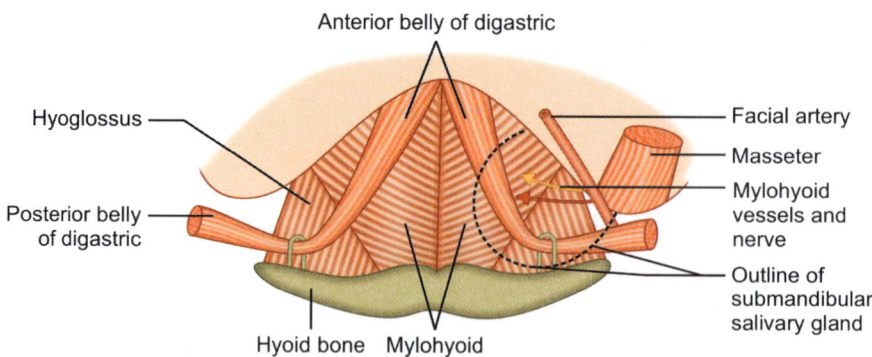

Fig. 6.14: Submental triangle

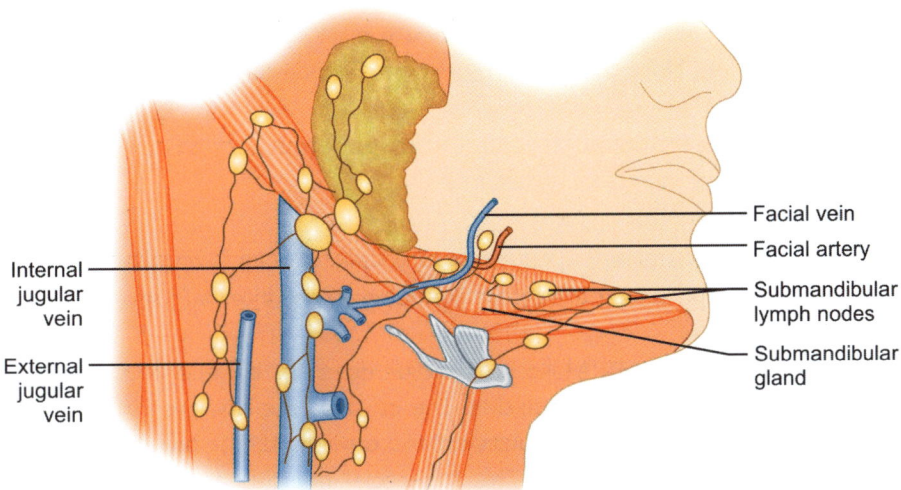

Fig. 6.15: Digastric (submandibular triangle)

12. Very small **submandibular lymph nodes** are found to lie on the surface of the submandibular gland.
13. The **facial artery** lies deep to the stylohyoid and posterior belly of digastric muscle and it usually takes a course deep to the submandibular gland. The artery curves round the lower border of the mandible and gives off the submental branch and then pierces the deep fascia of the neck. **Note that the stylohyoid muscle** lies anterior to the posterior belly of digastric muscle and the intermediate tendon of digastric muscle splits the stylohyoid near the greater cornu of the hyoid bone.
14. The **facial vein** joins the anterior branch of retromandibular vein to form the common facial vein and drains into internal jugular vein.
15. Remove the superficial part of the submandibular gland to expose the deep structures of this triangle. Otherwise, turn the gland posteriorly and expose the mylohyoid muscle.
16. To exposure the mylohyoid nerve, pull the anterior belly of the digastric muscle and it is found that this nerve lies on the inferior surface of the mylohyoid muscle near the mandible.
17. The **hypoglossal nerve** is found on a deeper plane. This nerve passes posterior and deep to the intermediate tendon of digastric muscle and then disappears between the mylohyoid and hyoglossus muscle. Try to reveal the course of this nerve on the hyoglossus turning the submandibular gland anteriorly.
18. Identify the **lingual nerve** which crosses the hyoglossus muscle at a higher level.
19. The **submandibular ganglion** is suspended from the lower margin of the lingual nerve and is found to lie on the hyoglossus muscle.

20. The **submandibular duct** passes forwards from the deep part of the gland over the hyoglossus and opens in the floor of the mouth on a sublingual papilla on each side of the frenulum linguae. **Note that the lingual nerve winds round the lower border of the submandibular duct.**
21. Besides the above mentioned structures, the posterior part of the digastric triangle contains the following important structures:
    i. Lower part of the parotid gland.
    ii. External carotid artery with its posterior auricular branch (sometimes).
    iii. Carotid sheath containing the internal carotid artery, internal jugular vein and vagus nerve.
22. **Muscular triangle (Fig. 6.16):** Muscular triangle does not contain important structures. It contains the **infrahyoid strap muscles**. The muscles are arranged in two planes. The muscles of superficial plane are **sternohyoid** and **omohyoid**. The muscles of deeper plane are **sternothyroid** and **thyrohyoid**. The superficial muscles are longer than the deep muscles.
23. Retract the sternal head of sternocleidomastoid muscle laterally to expose the anterior jugular vein. Find out the intermediate tendon of omohyoid muscle which lies deep to the sternocleidomastoid. Lift the superior belly of omohyoid muscle to expose its nerve. The origin of sternocleidomastoid muscle may be cut through and reflected upwards to expose the nerve supply of the infrahyoid muscles.
24. Clean the fascia from the infrahyoid muscles. Identify the omohyoid and sternohyoid muscles. The latter is medial to the former. Define the sternal attachment of sternohyoid

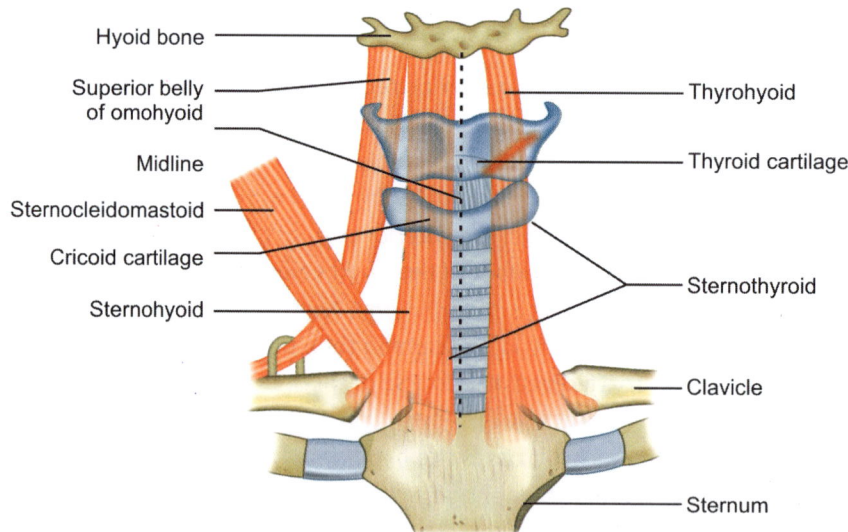

Fig. 6.16: Muscular triangle

muscle by passing the handle of the scalpel downwards between muscular attachment and the sternum. Divide the muscle as low down as possible and reflect it upwards to expose the deep muscles.

25. Deep muscles are **sternothyroid** and **thyrohyoid**. Thyrohyoid is the upward continuation of sternohyoid muscle and both are attached to the oblique line of thyroid cartilage.
26. Beneath these four infrahyoid muscles lie thyroid gland parathyroid glands, larynx, trachea, esophagus, prelaryngeal lymph nodes, pretracheal and paratracheal lymph nodes, etc.
27. **Carotid triangle (Fig. 6.17):** Identify and clean the triangular area between the posterior belly of digastric and stylohyoid anterosuperiorly, superior belly of omohyoid anteroinferiorly and the anterior border of upper 2/3rd of sternocleidomastoid muscle posteriorly.
28. Retract the sternomastoid muscle laterally to expose the **carotid sheath** which contains the neurovascular bundle. The neurovascular bundle includes common carotid artery, internal carotid artery, internal jugular vein and the vagus nerve. Remove the fat and fascia from this area to expose these structures.

    Note the relative position of these important structures. The **common and internal carotid arteries** lie medial to the **internal jugular vein**. The **external carotid artery** lies anteromedial to the **internal carotid artery. The vagus nerve** lies behind and between the artery and the vein (Fig. 6.18).
29. Gently clean and preserve the **descendens hypoglossi** (superior root of ansa cervicalis**),** and **descendens cervicalis** (inferior root of ansa cervicalis) in relation to the anterior or superficial layer of the carotid sheath.

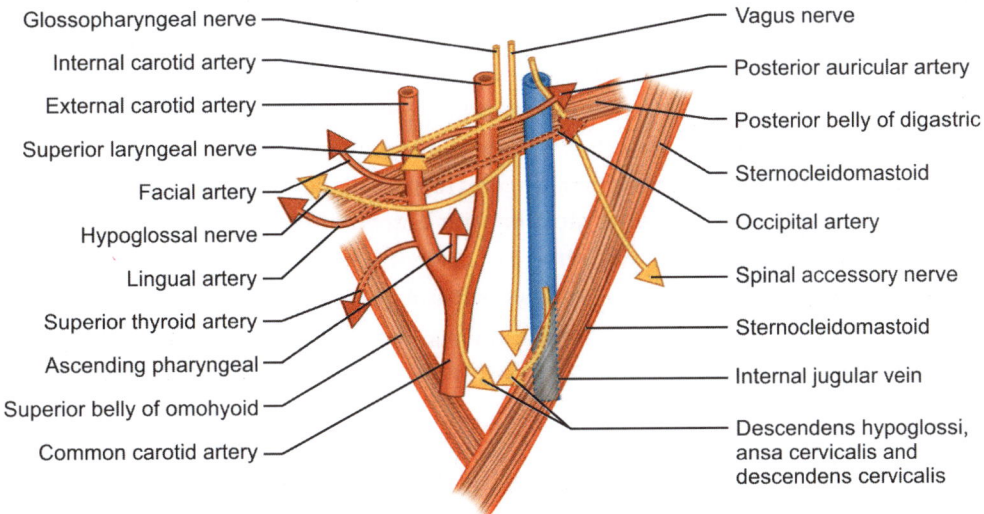

Fig. 6.17: Carotid triangle (left side)

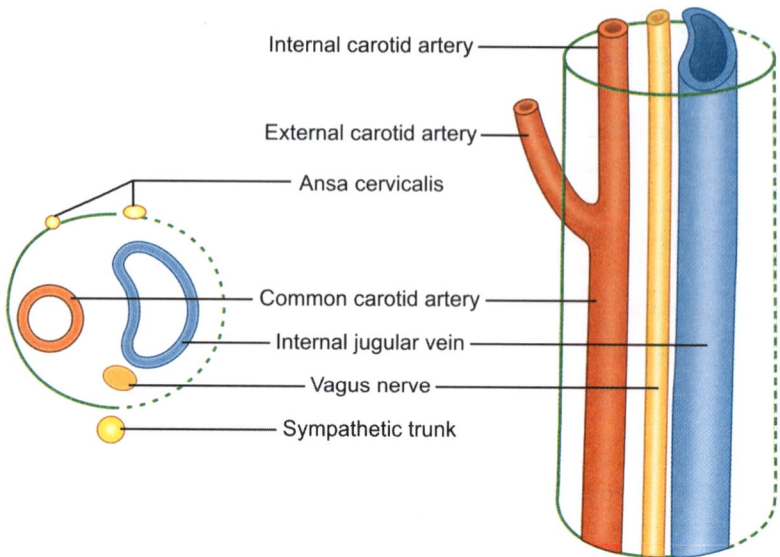

Fig. 6.18: Carotid sheath

30. Cut open the anterior layer of the carotid sheath without destroying the ansa which lies embedded in it. Now the carotid artery, internal jugular vein are properly exposed. Find out the **vagus nerve** which lies in between and behind the internal carotid artery and the internal jugular vein.

> **Note that** the descendens hypoglossi (C_1) lies superficial to internal and common carotid arteries. The descendens cervicalis (C_2, C_3) lies along the lateral margin of the internal jugular vein and they unite to form ansa (C_1, C_2, C_3) below the midpoint of the neck.

31. Find out the cervical part of the **sympathetic trunk** which runs vertically downwards posteromedial to the carotid sheath. To expose this nerve trunk, retract the carotid sheath laterally.
32. Expose the **external carotid artery (Fig. 6.18)** anteromedial to the internal carotid artery. Then dissect and clean the branches of external carotid artery in the neck. The branches are—superior thyroid, ascending pharyngeal, facial, lingual, occipital, posterior auricular. The **superior thyroid artery is the most inferior branch** of the external carotid artery.
33. The **lingual artery** arises from the external carotid artery below the facial artery and then it forms a loop which is crossed by the hypoglossal nerve (Fig. 6.19).
34. Facial and lingual veins drain into the internal jugular vein in the upper part of the triangle and the superior thyroid vein drains into the internal jugular vein in the lower part of the carotid triangle.
35. **Hypoglossal nerve** runs parallel to the inferior border of the posterior belly of digastric muscle. Push that muscle upwards to identify the hypoglossal nerve. This nerve emerges between the internal jugular vein laterally and the common carotid artery medially and then crosses anterior to the common carotid artery and external carotid artery. Then this nerve passes deep to the posterior belly of digastric muscle to enter into the digastric triangle.

36. A part of the **spinal accessory nerve** is found to cross the upper angle of the carotid triangle either superficial or deep to the internal jugular vein and then disappears beneath the sternocleidomastoid muscle.
37. Cut and reflect the omohyoid and sternohyoid muscles. This allows proper exposure of the **thyrohyoid membrane** extending from the upper border of the thyroid cartilage to the hyoid bone. The **superior laryngeal nerve** arises from the vagus nerve and passes deep to the carotid artery (Fig. 6.20). Then it divides into internal and external laryngeal branches. The **internal laryngeal nerve** pierces the thyrohyoid membrane to reach the larynx and the **external laryngeal nerve** supplies the cricothyroid muscle. The external laryngeal nerve lies deep to the superior thyroid artery.

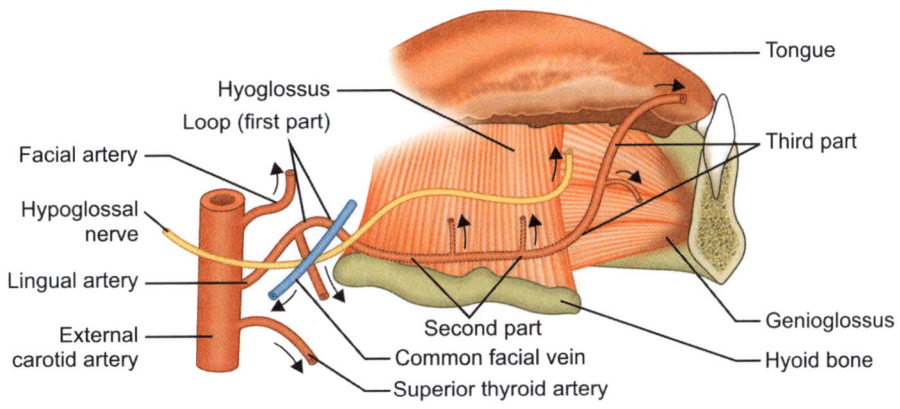

Fig. 6.19: Lingual artery

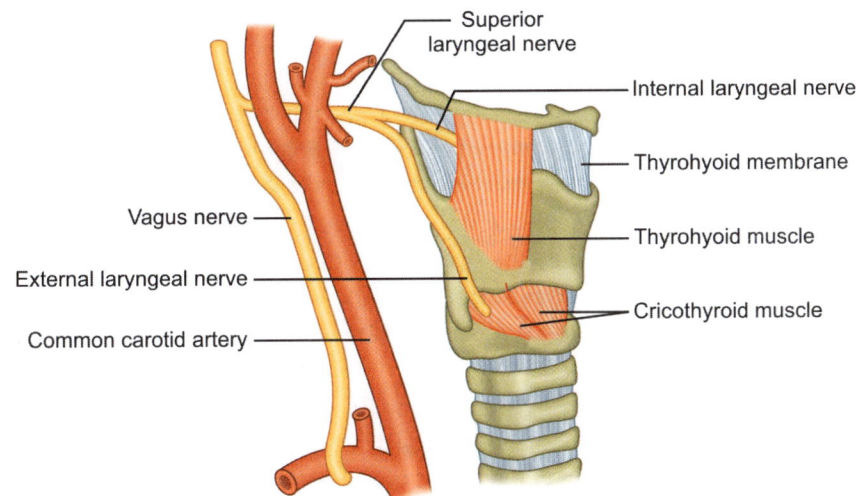

Fig. 6.20: Superior laryngeal nerve

38. **Deep cervical lymph nodes** (jugulodigastric group) are found along the carotid sheath, near the internal jugular vein.
39. **Carotid body** is a small, oval, reddish-brown structure found behind the bifurcation of common carotid artery.

> Note that the common carotid artery bifurcates at the level of the upper border of the lamina of thyroid cartilage. It corresponds to the disc between C3 and C4 vertebrae.

40. Find out the floor of the carotid triangle which is formed by four muscles. These are:
    i. Thyrohyoid.
    ii. Hyoglossus.
    iii. Middle constrictor of pharynx.
    iv. Inferior constrictor of pharynx.

MUSCLES RELATED TO ANTERIOR TRIANGLE OF NECK

Suprahyoid Muscles (Four Pairs) (Fig. 6.21)

Muscles	Origin	Insertion	Nerve supply	Action
Digastric	i. Posterior belly from mastoid notch of temporal bone ii. Anterior belly from the digastric fossa on the lower border of mandible close to symphysis menti	Intermediate tendon which is connected to the junction of greater cornu and body of hyoid by means of an inverted 'U' shaped sling of deep cervical fascia	**Posterior belly:** Facial nerve **Anterior belly:** Mylohyoid branch of inferior alveolar nerve branch of posterior division mandibular nerve	i. Depresses and retracts the chin in opening of the mouth ii. Help in deglutition by pulling the hyoid bone upwards
Stylohyoid	Posterior surface of the styloid process of temporal bone	Junction of body and greater cornu of hyoid bone	Facial nerve	Elongates the floor of the mouth by pulling the hyoid bone upward land backward
Mylohyoid (Diaphragma oris)	Mylohyoid line on the inner surface of the mandible	Posterior one-fourth Body of hyoid bone. Anterior three-fourth median fibrous raphe extending from symphysis menti to hyoid bone	Mylohyoid branch of inferior alveolar nerve branch of posterior division of mandibular nerve	i. Elevates the floor of the mouth and tongue ii. Assists in depression of the mandible
Geniohyoid	Inferior genial tubercle on the posterior aspect of symphysis menti of mandible	Body of hyoid bone	Fibers of C_1, conveyed by hypoglossal nerve	Shortens the floor of the mouth by drawing the hyoid bone upward and forward

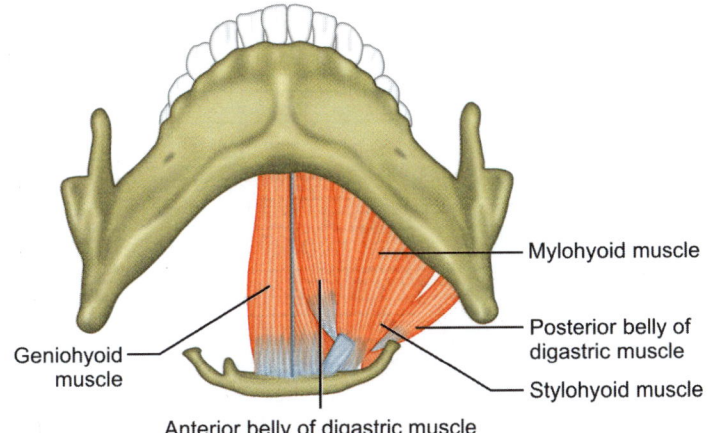

Fig. 6.21: Suprahyoid muscles

Infrahyoid Muscles (Fig. 6.22)

Muscles	Origin	Insertion	Nerve supply	Action
Sternohyoid	i. Posterior surface of manubrium sterni ii. Posterior aspect of capsule of sternoclavicular joint and adjoining clavicle	Lower border of body of hyoid bone	Ansa cervicalis (C_1, C_2, C_3)	
Sternothyroid	Posterior surface of manubrium sterni	Oblique line of thyroid cartilage		
Thyrohyoid	Oblique line of thyroid cartilage	Lower border, body and greater cornu of hyoid bone	C_1 fibers conveyed by hypoglossal nerve	i. Stabilizes the hyoid bone ii. Depressors of larynx, hyoid bone and floor of the mouth
Omohyoid (arises by inferior belly and inserted through superior belly)	i. Inferior belly: From the upper border of scapula close to suprascapular notch ii. Suprascapular ligament	Superior belly: Inserted at the lower border of the body of hyoid bone Intermediate tendon: Passes deep to the sternocleidomastoid muscle where the both bellies meet	Superior belly by fibres of C_1 (descendens hypoglossi) Inferior belly by ansa cervicalis (C_1, C_2, C_3)	

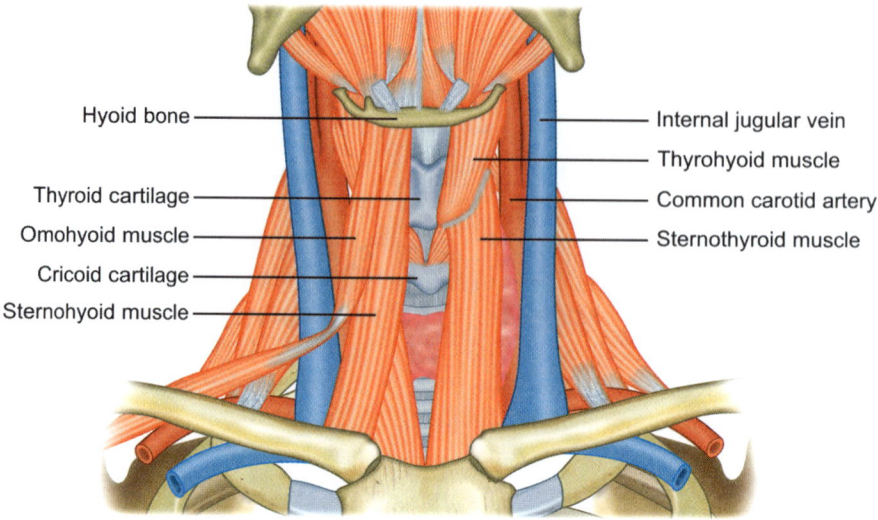

Fig. 6.22: Infrahyoid muscles

Summary

1. Position of the cadaver—supine with extended neck.
2. Incision of skin as in Figure 6.12.
3. Superficial fascia with platysma. Structures found in this layer:
    i. Platysma.
    ii. Transverse cervical nerve.
    iii. Cervical branch of facial nerve.
    iv. Anterior jugular vein.
    v. External jugular vein in the upper part.
4. Deep fascia (investing layer).
5. Anterior triangle of the neck with its boundary, subdivisions and contents are exposed.
6. Boundaries of different triangles are discussed in the beginning of this chapter.
7. **It contents of:**
    a. **Submental triangle (refer Fig. 6.14):**
        i. Submental lymph nodes (3–4)
        ii. Submental vein.
        iii. Commencement of anterior jugular vein.
        iv. Mylohyoid vessels ad nerve.
        v. Superficial part of submandibular gland.
    b. **Digastric triangle (refer Fig. 6.15):**
        i. Submandibular gland.
        ii. Submandibular lymph nodes.
        iii. Facial artery and vein.

            iv. Mylohyoid nerve.
            v. Hypoglossal nerve.
            vi. Lingual nerve.
            vii. Submandibular ganglion.
            viii. Submandibular duct.
            ix. Lower part of parotid gland.
            x. External carotid artery.
    c. **Muscular triangle (refer Fig. 6.16):**
            i. Superficial muscles—**sternohyoid** and **omohyoid.**
            ii. Deep muscles—**sternothyroid** and **thyrohyoid.**
    d. **Carotid triangle (refer Fig. 6.17):**
            i. Carotid sheath and its contents (internal jugular vein, common and internal carotid artery and vagus nerve).
            ii. Ansa cervicalis.
            iii. Sympathetic trunk.
            iv. External carotid artery and its branches in the neck.
            v. Facial artery and vein.
            vi. Lingual artery and vein.
            vii. Hypoglossal and spinal accessory nerve.
            viii. Superior laryngeal nerve
            ix. Deep cervical lymph nodes
            x. Carotid body.

PROBABLE QUESTIONS AND ANSWERS

1. What are the triangles of the neck?

Ans.
  i. **Anteriorly:** Anterior triangle (one on each side).
  ii. **Anterolaterally:** Posterior triangle (one on each side).
  iii. **Posteriorly:** Suboccipital triangle (one on each side).

2. What structure forms demarcation between anterior and posterior triangle?

Ans. Sternocleidomastoid muscle.

3. What is the boundary of anterior triangle?

Ans.
  i. **Above:** Lower border of the body of the mandible up to its angle and then a line extending from the angle up to the tip of mastoid process.
  ii. **Medially:** Anterior median line extending from the symphysis menti to suprasternal notch.
  iii. **Laterally:** Anterior border of sternocleidomastoid muscle.
  iv. **Apex:** Suprasternal notch.

4. What are the subdivisions of the anterior triangle?

Ans. Four subdivisions:
  i. Submental triangle.
  ii. Digastric triangle.
  iii. Carotid triangle.
  iv. Muscular triangle.

5. What structures divide the anterior triangle into four smaller triangles?

Ans. Digastric and omohyoid muscles.

6. What are the bellies of digastric muscle?

Ans. Two bellies:
  i. Anterior belly (shorter).
  ii. Posterior belly (longer).

7. Which muscle accompanies the posterior belly?

Ans. Stylohyoid muscle.

8. What is intermediate tendon of digastric muscle?

Ans. The intermediate tendon is the area where both the digastric muscles meet. The tendon is connected to the junction of body and greater cornu of hyoid bone by means of an inverted 'U' shaped facial sling of deep cervical fascia. This sling anchors the tendon of digastric muscle to the hyoid bone.

9. What is the gap through which the intermediate tendon passes?

Ans. The stylohyoid muscle at its insertion splits to form a triangular gap through which the tendon passes.

10. Why the two bellies of digastric muscles are supplied by two different nerves?

Ans. Anterior belly is supplied by the mylohyoid branch of the inferior alveolar nerve (branch of posterior division of mandibular nerve) because it is developed from the myotome of first branchial arch (**mandibular nerve is the nerve of first branchial arch**). Posterior belly is supplied by the facial nerve because it is developed from the myotome of second branchial arch (**facial nerve is the nerve of second branchial arch**).

11. Which muscle is known as the key muscle of the neck?

Ans. The posterior belly of digastric is known as the **key muscle of the neck**, because three great vessels (internal and external carotid arteries, internal jugular vein) and three cranial nerves (10th, 11th, 12th) pass under cover of the posterior belly of digastric muscle (Fig. 6.23).

Window Dissections: Head and Neck

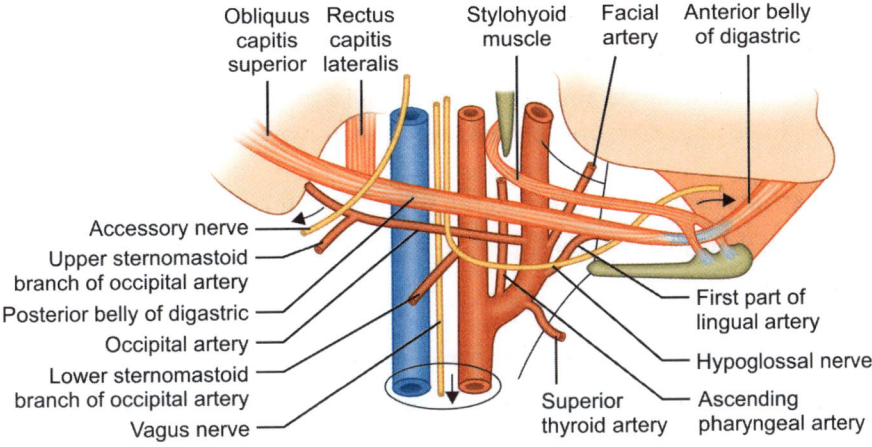

Fig. 6.23: Vessels and nerves under posterior belly of digastric

12. Where do you find the platysma muscle?

Ans. Platysma is found in the superficial fascia of the front of the neck.

13. Where do you find the superficial veins and cutaneous nerves of the neck?

Ans. The superficial vein and cutaneous nerves of the neck lie deep to the platysma.

14. What is the feature of platysma?

Ans. Platysma is a thin, quadrilateral subcutaneous striated muscle.

15. What is its origin, insertion and nerve supply?

Ans. The **platysma** arises from the fascia covering the anterior part of deltoid and pectoralis major muscles up to the 2nd rib and runs upwards and medially in the neck to be inserted into the lower border of the body of the mandible and to the angle of the mouth as **risorius. The nerve supply of platysma is cervical branch of facial nerve, because platysma is developed from the myotome of second branchial arch.**

16. What is the morphological basis of platysma?

Ans. Morphologically, platysma represents a remnant of **panniculus carnosus** and is developed from the myotome of second branchial arch.

Note that *platysma crosses over two bones—clavicle and mandible which are first and second bones to ossify in human body.*
Subptatysmal bones are mostly ossified in membrane.

17. What are the layers of deep cervical fascia?

Ans. The deep cervical fascia also called **fascia colli** consists of three layers from outside inwards **investing layer**, **pretracheal layer** and **prevertebral layer**.

18. Which structure forms the roof of the triangles of the neck?

Ans. The roof of both anterior and posterior triangles of the neck is formed by the investing layer of deep cervical fascia.

19. Name the structures piercing the investing layer?

Ans. The superficial veins and cutaneous nerves pierce the investing layer.

20. What are the derivatives of the investing layer?

Ans.
  i. **Parotidomassteric fascia** covering the outer surface of the masseter and parotid gland in the face.
  ii. **Stylomandibular ligament** thickened fascia extending between the styloid process of the temporal bone and angle of the mandible.

21. What are muscles enclosed by the investing layer?

Ans.
  i. Sternocleidomastoid muscle.
  ii. Trapezius muscle.

22. What are the spaces enclosed by the investing layer of the deep fascia?

Ans.
  i. Supraclavicular space.
  ii. Suprasternal space of burns.

Suprasternal Space (of Burn) (Fig. 6.24)

It is a triangular space above the manubrium sterni and formed due to the splitting of the investing layer of deep cervical fascia into anterior (superficial) and posterior (deep) layers which are attached to the anterior and posterior borders of the suprasternal notch respectively. At sides, the two layers are fused.

Contents:
  i. Sternal head of sternocleoidomastoid.
  ii. Interclavicular ligament.
  iii. Jugular venous arch connecting two anterior jugular veins.
  iv. Loose areolar tissue.
  v. A lymph node (occasionally).

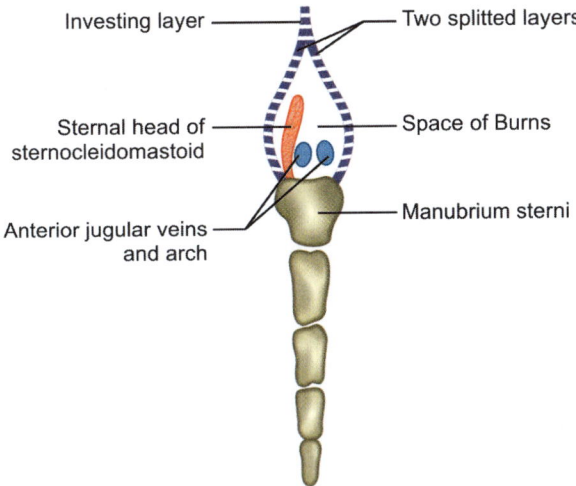

Fig. 6.24: Suprasternal space of Burn

23. What are the different laminae of deep cervical fascia?

Ans.
  i. Investing layer.
  ii. Pretracheal layer.
  iii. Prevertebral layer.
  iv. Carotid sheath.
  v. Buccopharyngeal fascia.
  vi. Pharyngobasilar fascia.
  vii. Temporal fascia, etc.

24. Why the skin of the anterior triangle of the neck is freely movable?

Ans. The superficial fascia in this region contains a variable amount of fat and it is loosely connected to the skin and deep fascia. So the skin is freely movable.

25. What are the boundaries and contents of the four subdivisions of the anterior triangle of the neck?

Ans. Boundaries and contents were discussed earlier in this chapter.

26. What is the motor supply of the sternocleidomastoid muscle?

Ans. Spinal accessory nerve.

27. Name another muscle which is supplied by spinal accessory?

Ans. Trapezius.

28. What is carotid sheath?

Ans. The carotid sheath is tubular investment of deep cervical fascia around the carotid vessels and internal jugular vein.

29. What is the extent of the sheath?

Ans. From the base of the skull to the arch of the aorta.

30. How the carotid sheath is formed?

Ans. Anterior wall of the sheath is formed by the pretracheal layer and posterior wall is formed by the prevertebral layer of deep cervical fascia.

31. What are the contents of the sheath (refer Fig. 6.23)?

Ans.
  i. Common and internal carotid artery on the medial side.
  ii. Internal jugular vein on the lateral side.
  iii. Vagus nerve is in between and behind the artery and vein.

Note that the sheath is thick over the arteries and thin over the vein in order to allow expansion of the vein during increased venous return.

32.. What important structure is present within the anterior wall of the carotid sheath?

Ans. Ansa cervicalis.

33. What structure is present behind the sheath?

Ans. Sympathetic trunk.

34. How many cervical sympathetic ganglia are there?

Ans. Three:
  i. Superior cervical ganglia (C_1-C_4)
  ii. Middle cervical ganglia (C_5-C_6)
  iii. Inferior cervical ganglia (C_7-C_8)

35. Name the structures piercing the carotid sheath?

Ans.
  i. External carotid artery.
  ii. Tributaries of internal jugular vein.
  iii. Glossopharyngeal, cervical branches of vagus, accessory and hypoglossal nerve.

36. What is ansa cervicalis?

Ans. A loop formed by the ventral rami of the 1st, 2nd and 3rd cervical nerves on the anterior wall of the carotid sheath (Fig. 6.25).

37. What are the roots of the ansa cervicalis?

Ans.
  i. Superior root (descendens hypoglossi) carries the fibers from ventral rami of C_1 nerve.
  ii. Inferior root (descendens cervicalis) carries the fibers from ventral rami of C_2 and C_3 nerves.

These two roots join in front of the carotid sheath to form **ansa cervicalis**.

38. What are the branches of the ansa cervicalis?

Ans. The branches from the ansa (loop) supply three muscles:
  i. Sternohyoid.
  ii. Sternothyroid.
  iii. Inferior belly of omohyoid.

39. What is the nerve supply of superior belly of omohyoid?

Ans. Superior root or descendens hypoglossi (C_1) supplies the superior belly of omohyoid.

40. Name others muscles supplied by the C_1 nerve?

Ans.
  i. Geniohyoid.
  ii. Thyrohyoid muscle.

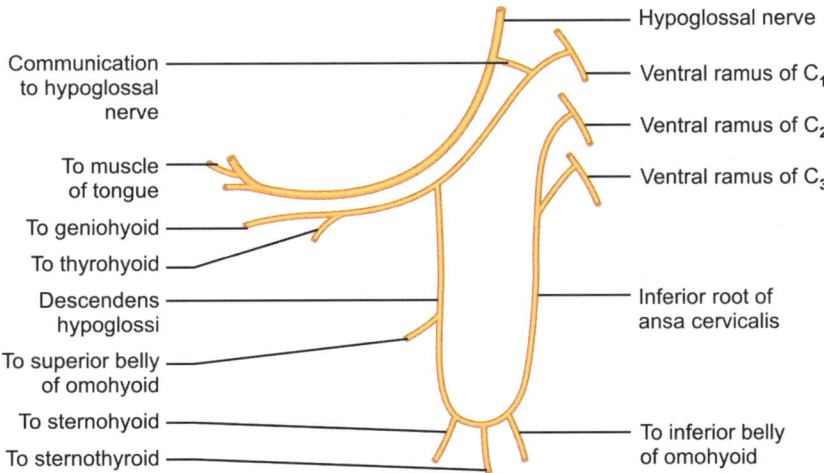

Fig. 6.25: Ansa cervicalis and its branches

41. Name suprahyoid muscles and their nerve supply?

Ans.
i. Anterior and posterior bellies of digastric muscle (mylohyoid branch of inferior alveolar nerve and facial nerves respectively).
ii. Stylohyoid (facial nerve).
iii. Mylohyoid (mylohyoid branch of inferior alveolar nerve).
iv. Geniohyoid (C_1 fibers conveyed by hypoglossal nerve).

42. Name the infrahyoid muscles and their nerve supply?

Ans.
i. Sternohyoid (ansa cervicalis).
ii. Sternothyroid (ansa cervicalis).
iii. Omohyoid (superior belly by descendens hypoglossi and inferior belly by ansa cervicalis).
iv. Thyrohyoid (fibres of C_1 nerve through the hypoglossal nerve).

43. Which of these muscles in known as diaphragma oris?

Ans. **Mylohyoid**. The mylohyoid muscles of both sides form a gutter shaped floor of the mouth and support the weight of the tongue. Hence, it is known as diaphragma oris.

44. What are the branches of external carotid artery in the neck?

Ans. Refer Figure 6.26:
i. Ascending pharyngeal.
ii. Superior thyroid.
iii. Facial.
iv. Lingual
v. Posterior auricular.
vi. Occipital.

45. What are the terminal branches of external carotid artery?

Ans. Superficial temporal artery and maxillary artery (within the parotid gland).

46. What is the first and smallest branch of external carotid artery?

Ans. Superior thyroid artery.

47. Name the branches of facial artery in the neck?

Ans. Refer Figure 6.26:
i. Ascending palatine.
ii. Tonsillar.
iii. Submental.
iv. Glandular (submandibular gland).

Window Dissections: Head and Neck

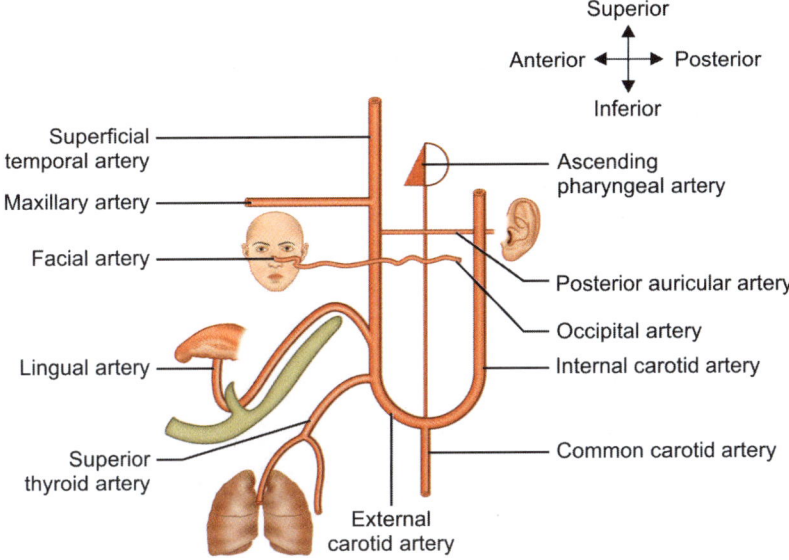

Fig. 6.26: Branches of external carotid artery

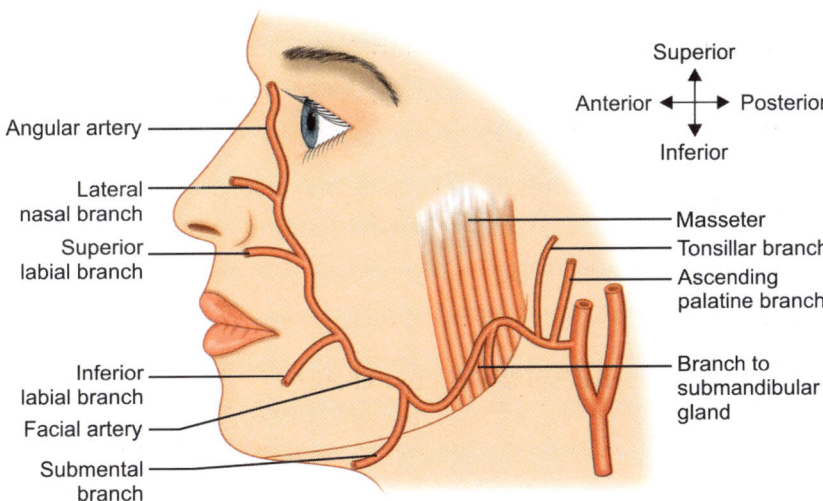

Fig. 6.27: Branches of facial artery in the neck and face

Remember that common carotid artery and internal carotid artery provide no branch in the neck.

48. What is the origin of common carotid artery?

Ans.   i. On right side: It is one of the terminal branches of brachiocephalic trunk (other one is right subclavian).
   ii. On left side: It arises directly from the arch of aorta.

49. What is the level of bifurcation of common carotid artery?

Ans. Upper border of the lamina of thyroid cartilage. Vertebral level → Disc between C_3 and C_4 vertebrae.

50. What is carotid sinus?

Ans. At the bifurcation, the common carotid artery forms a fusiform dilatation which is called carotid sinus. It acts as a baroreceptor for controlling intracranial blood pressure.

51. What is carotid body?

Ans. It is a small, oval-shaped neurovascular structure situated close to the posterior wall of the carotid sinus. It acts as a chemoreceptor monitoring oxygen tension within the artery.

52. What is the area of drainage of internal jugular vein?

Ans. Brain, superficial part of face and neck.

53. Where from the internal jugular vein starts?

Ans. Internal jugular vein starts as a continuation of sigmoid sinus in the posterior compartment of jugular foramen.

54. How many bulbs (dilatation) the internal jugular vein presents?

Ans. It presents two bulbs:
  i. Superior bulb (at its beginning)
  ii. Inferior bulb (close to its termination).

55. Where does the internal jugular vein terminate?

Ans. It unites with the subclavian vein behind the medial end of clavicle to form brachio-cephalic vein.

56. Name the tributaries of internal jugular vein?

Ans.
  i. Inferior petrosal sinus.
  ii. Pharyngeal veins.
  iii. Common facial veins.
  iv. Lingual vein.
  v. Superior and middle thyroid vein.
  vi. Thoracic duct (left side), right lymphatic duct (right side).

57. Name the first and last tributary of internal jugular vein?

Ans.
i. First tributary: Inferior petrosal sinus.
ii. Last tributary: Thoracic duct (left side), right lymphatic duct (right side).

58. Name the structures passing between the external and internal carotid artery?

Ans.
i. Stylopharyngeus muscle.
ii. Glossopharyngeal nerve.
iii. Pharyngeal branch of vagus nerve.
iv. Tip of styloid process of temporal bone.
v. A part of parotid gland.

59. How anterior jugular vein is formed?

Ans. It is formed by the confluence of several superficial veins in the submental region and runs downward near the anterior median line of the neck.

60. Where does it drain?

Ans. Anterior jugular vein passes laterally deep to the sternomastoid muscle and drains into the external jugular vein close to its termination.

61. Where does the external jugular vein terminate?

Ans. External jugular veins is formed by the union of posterior auricular vein and posterior retromandibular vein. It drains into the subclavian vein.

62. What is jugular venous arch?

Ans. The two anterior jugular veins are connected by a transverse branch in the suprasternal **space of Burn**. This is how the jugular venous arch is formed.

63. What is median cervical vein?

Ans. When both the anterior jugular veins are united along the midline of the neck to form a single trunk, it is termed as median cervical vein.

64. What are the branches of vagus nerve in the neck?

Ans.
i. Pharyngeal branch.
ii. Superior laryngeal nerve.
iii. Branch to carotid sinus and carotid body.
iv. Superior cervical cardiac branch.
v. Inferior cervical cardiac branch.
vi. Right recurrent laryngeal nerve.
(**Note** that **left recurrent** laryngeal nerve arises from the vagus nerve in the **thorax**).

65. Name the structures found between mylohyoid and hyoglossus muscle (structures on hyoglossus).

Ans.
 i. Lingual nerve.
 ii. Submandibular ganglion.
 iii. Deep part of submandibular gland and its duct.
 iv. Hypoglossal nerve and vena comitantes hypoglossi.

66. How the submandibular gland is divided into superficial and deep part?

Ans. The superficial part (larger part) is superficial to the mylohyoid muscle and the smaller deep part is deep to the mylohyoid muscle. Both the parts are continuous around the posterior border of mylohyoid muscle.

67. Name the structures (important landmarks) in the midline of the neck?

Ans. From above downwards, these are:
 i. Symphysis menti.
 ii. Hyoid bone.
 iii. Thyrohyoid membrane.
 iv. Thyroid cartilage and thyroid angle.
 v. Cricoid cartilage.
 vi. Isthmus of thyroid gland.
 vii. Tracheal rings.
 viii. Suprasternal notch.

68. What is 'potato tumor'?

Ans.
 i. It is a rare feature of the neck and produced by the enlargement of the carotid body.
 ii. It moves transversely and shows transmitted pulsation.

69. Why the infrahyoid muscles are called 'strap muscles'?

Ans. The infrahyoid muscles (sternothyroid, sternohyoid, origin of thyrohyoid and superior belly of omohyoid) strap the thyroid in position. Hence, the name.

LESSON 4: POSTERIOR TRIANGLES OF NECK

The posterior triangle of the neck is a triangular space behind the posterior border of the sternocleidomastoid muscle situated on each side of the neck. Though the triangle is described as posterior triangle, it is situated on the anterolateral aspect of the neck (not posterior aspect of the neck). It is so named because of its relative position to the sternocleidomastoid muscle. It is bounded as follows (Fig. 6.28):

Window Dissections: Head and Neck

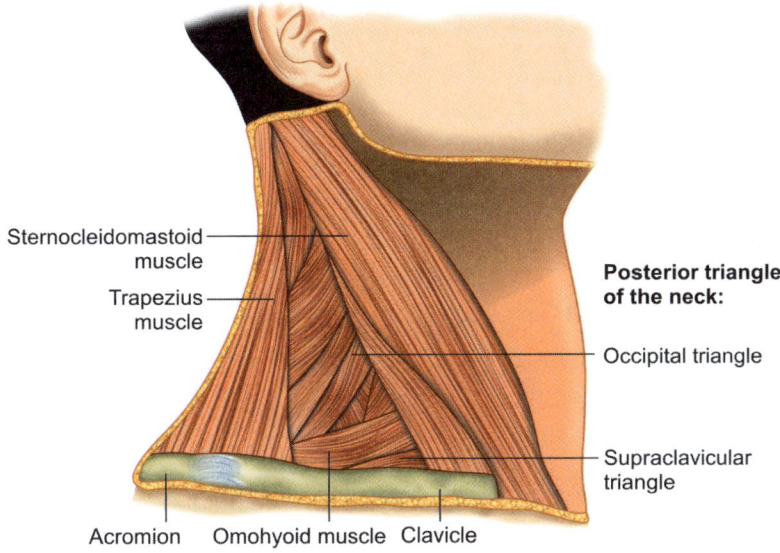

Fig. 6.28: Boundary of posterior triangle of neck

i. **Anteriorly**: Posterior border of sternocleidomastoid muscle.
ii. **Posteriorly**: Anterior border of trapezius muscle.
iii. **Base**: Middle 1/3rd of the clavicle.
iv. **Apex:** Meeting point of the above mentioned two muscles on the superior nuchal line of occipital bone.
v. **Roof**: Skin, superficial fascia with platysma and investing layer of deep cervical fascia.
vi. **Floor**: Prevertebral fascia covering the deep muscles of the neck.

The posterior triangle of the neck is again divided into a upper larger **occipital triangle** and lower smaller **subclavian** or **supraclavicular triangle** by the inferior belly of omohyoid muscle which passes obliquely through the posterior triangle about 2.5 cm above the clavicle.

STEPS OF DISSECTION

1. **Position of the cadaver:** Supine and the head is turned to the opposite side.
2. **Incision on the skin (Fig. 6.29):**
   i. **1st incision (A–B)**: From the mastoid process of the temporal bone (point A) downwards and obliquely along the middle of the sternocleidomastoid up to the sternal end of the clavicle (point B).
   ii. **2nd incision (B–C)**: From point B—a transverse incision along the clavicle up to its acromial end (point C). Skin flap is reflected downwards and backwards up to the anterior border of the trapezius muscle. During the incision on the skin, do not cut into the superficial fascia to avoid damage to the cutaneous nerves and external jugular vein.

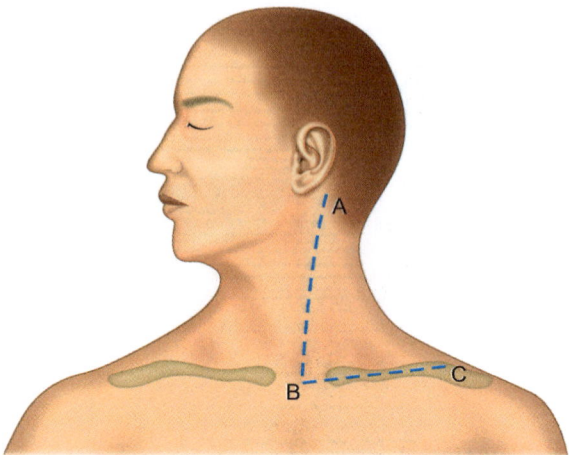

Fig. 6.29: Incision for posterior triangle of neck

3. **Superficial fascia is exposed:** In the superficial fascia the following structures are noted.
    i. **Platysma:** It forms the lower and anterior part of the roof of the posterior triangle. Beneath the platysma lies the supraclavicular nerve and external jugular vein. The supraclavicular nerve descends beneath the platysma and emerges through the platysma muscle 1–2 cm above the clavicle. Using forceps, separate the supraclavicular nerve (ventral rami of C_3, C_4) from the clavicular end of the platysma. Then reflect the platysma from the clavicle upwards and forwards towards the face superficial to the supraclavicular nerve and the external jugular vein. Do not cut the platysma from the mandible.
    ii. **External jugular vein:** It lies beneath the platysma. Trace the vein upwards across the sternomastoid muscle till it joins the posterior auricular vein. Clean and trace it downwards till it pierces the investing layer of deep cervical fascia about 2.5 cm above the clavicle.
    iii. Lesser occipital, great auricular and transverse cervical cutaneous nerves pierce the deep fascia and appear in the superficial fascia to supply the skin of the front and side of the neck.

   Then superficial fascia is cut and reflected to expose the deep fascia. Deep fascia is exposed.
4. **Deep fascia:** Deep fascia is cut carefully along the posterior border of sternomastoid muscle and reflect it downwards and backwards towards the trapezius muscle. The boundaries of the posterior triangle is cleaned and identified. The inferior belly of omohyoid which crosses the posterior triangle obliquely and divides the triangle into upper occipital and lower supraclavicular triangle is cleaned and indentified. Now the contents of the above two triangles are dissected, cleaned and identified separately.
5. **Occipital triangle:** The boundary of this triangle is demarcated anteriorly by the posterior **border of sternocleidomastoid,** posteriorly by the anterior border of **trapezius** and below by the inferior belly of **omohyoid muscle.**

Window Dissections: Head and Neck

The floor of this triangle is formed by the following muscles from above downwards:
a. Semispinalis capitis (vertically disposed fibers).
b. Splenius capitis.
c. Levator scapulae.
d. Scalenus posterior.
e. Scalenus medius.

Except semispinalis capitis, all other muscles of the floor slope downwards and backwards. These muscles are carpeted by the prevertebral layer of deep cervical fascia.

The structures of this triangle are cleaned and identified one by one.

i. **Spinal accessory nerve (Fig. 6.30):** It emerges a little above the middle of the posterior border of sternocleidomastoid and appears in the triangle and then runs on the **levator scapulae** muscle separated from it by prevertebral fascia. The course of the nerve is downwards and backwards. Then the nerve passes deep to the anterior border of **trapezius** about 5 cm above the clavicle where it forms **subtrapezoid plexus** with C3 and C4 nerves and supply **trapezius** muscle.

   Note that the spinal accessory nerve is hooked below by the lesser occipital nerve at the middle of the posterior border of sternocleidomastoid.

ii. **C_3 and C_4 nerves:** They pass a little below and parallel to the spinal accessory nerve under the cover of prevertebral fascia and supply the **trapezius** (forming subtrapezoid plexus with the spinal accessory nerve passing deep to the anterior border of trapezius) and **levator scapulae.**

iii. **Dorsal scapular nerve (nerve to rhomboids):** This slender nerve carrying the fibers from C_5 root passes backwards and downwards. It pierces the scalenus medius muscle and passes deep to the levator scapulae to supply rhomboideus muscle.

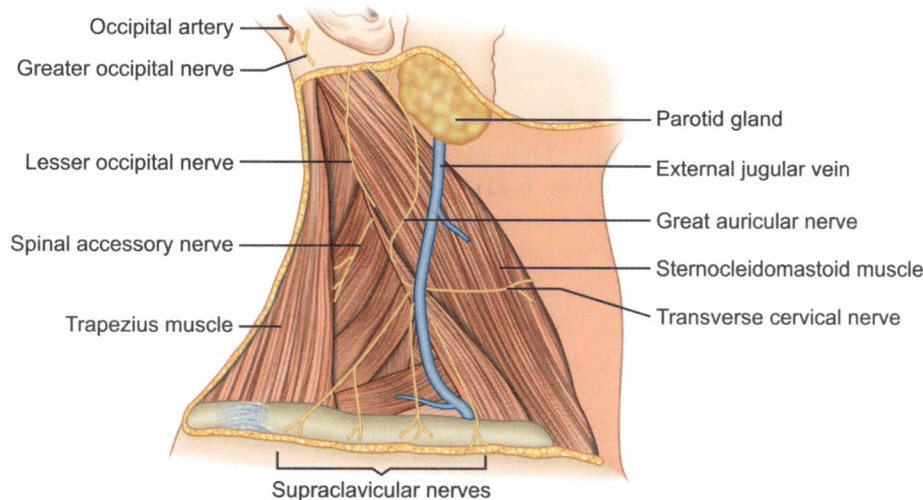

Fig. 6.30: Spinal accessory nerve in posterior triangle of neck

iv. **Great auricular nerve (C_2, C_3):** This cutaneous branch of cervical plexus emerges at the middle of the posterior border of sternomastoid muscle along with other three cutaneous branches of cervical plexus. Then it ascends on the superolateral surface of the sternocleidomastoid muscle posterior and parallel to the upper part of the external jugular vein.

> Four cutaneous branches of cervical plexus are (a) great auricular (C_2, C_3), (b) lesser occipital (C_2), transverse cervical (C_2, C_3) and (d) supraclavicular (C_3, C_4). They emerge at the middle of the posterior border of sternocleidomastoid muscle and runs in the posterior triangle of the neck and then pierce the investing layer of deep cervical fascia to supply their respective areas.

v. **Lesser occipital nerve (C_2):** It hooks the spinal accessory nerve at the middle of the posterior border of sternocleidomastoid and runs upwards along the posterior border of the sternomastoid muscle.

vi. **Transverse cervical nerve (C_2, C_3):** It traverses forwards over the superficial surface of the sternomastoid muscle and then divides into ascending and descending branches in the anterior triangle of the neck along the anterior border of the sternomastoid muscle.

vii. **Supraclavicular nerve (C_3, C_4):** It emerges as a common trunk and runs downwards under cover of deep cervical fascia. Then it divides into three diverging branches which pierce the deep cervical fascia and platysma 1–2 cm above the clavicle.
   a. The lateral branches pass obliquely over the trapezius.
   b. The intermediate branch passes over or through the clavicle.
   c. The **medial branch** passes superficial to the external jugular vein and sternocleidomastoid muscle.

viii. **Upper trunk of brachial plexus:** Remove the fascia from the inferior belly of omohyoid muscle and turn the muscle forwards to expose the upper trunk of brachial plexus which peeps between the scalenus medius and inferior belly of omohyoid.

ix. **Suprascapular nerve (C_5, C_6):** It is a branch from the upper trunk of brachial plexus. It runs posteroinferiorly immediately above the brachial plexus under cover of inferior belly of omohyoid muscle.

x. **Long thoracic nerve (C_5, C_6, C_7):** The C_5 and C_6 roots of this nerve pierce the scalenus medius muscle and the C7 root joins the nerve at a lower level. The thoracic nerve of bell passes downwards behind the brachial plexus and enters into the axilla through its apex.

xi. **Transverse cervical artery (Fig. 6.31):** It passes at the upper border of the inferior belly of omohyoid muscle and runs posterolaterally over the scalenus medius muscle in the occipital triangle. In fact all three scalene muscles (scalenus anterior, scalenus medius and scalenus posterior) are crossed superficially by transverse cervical artery.

xii. Inferior and some superficial cervical lymph nodes.

6. **Subclavian or supraclavicular triangle (Fig. 6.32):** It is bounded as follows:
   Above: Inferior belly of omohyoid muscle.
   Below: Middle third of the clavicle.
   Infront: Lower part of the posterior border of sternocleidomastoid muscle.
   Roof: Investing layer of deep cervical fascia and superficial fascia with platysma.
   Floor: First rib and first digitations of serratus anterior muscle, a part of scalenus medius.

The above-mentioned boundaries are defined. The contents of this triangle are dissected and identified one by one.
  i. **Third part of subclavian artery:** It begins about a fingers breadth above the clavicle deep to the posterior border of lower part of sternocleidomastoid muscle. It then descends downwards and laterally from the lateral border of the scalenus anterior muscle to lie in the subclavian groove of the first rib anterior to the lower trunk of brachial plexus. It ends at the outer border of the first rib behind the middle of the clavicle and continued as the first part of axillary artery.
  ii. **Subclavian vein:** It is a continuation of the axillary vein at the outer border of the first rib. It usually lies behind the clavicle. So, depress the clavicle to expose the subclavian vein. It passes medially, anteroinferior to the third part of the subclavian artery and the

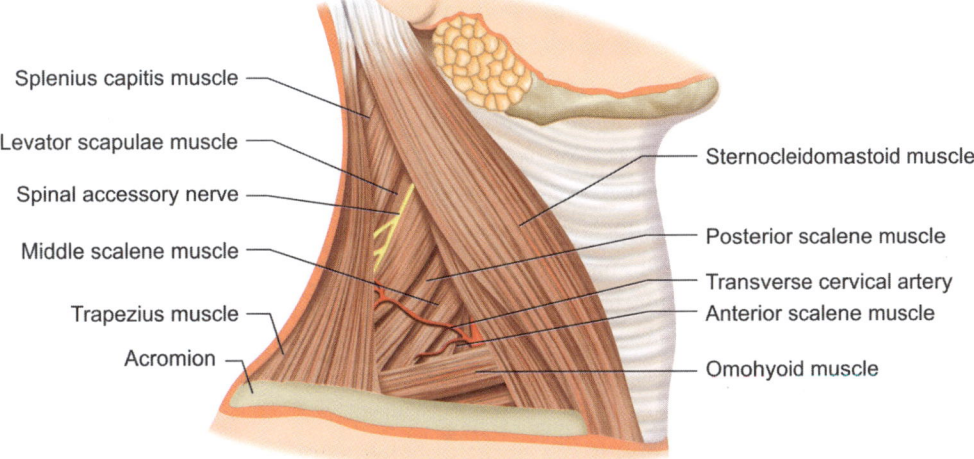

Fig. 6.31: Transverse cervical artery in posterior triangle

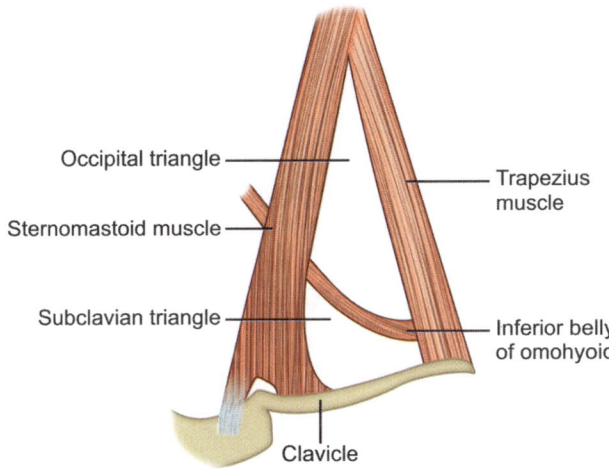

Fig. 6.32: Supraclavicular triangle or subclavian triangle

scalenus anterior muscle intervene between the subclavian artery and the subclavian vein. **The only tributary of the subclavian vein is the external jugular vein** which joins the subclavian vein at the lateral border of the scalenus anterior muscle behind the clavicle. Note that the external jugular vein emerges posterior to the ramus of the mandible and crosses superficial to the fascial roof (investing layer of deep cervical fascia) of the posterior triangle at the posterior border of sternocleidomastoid, 2–3 cm above the clavicle.

iii. **Trunks of brachial plexus:** The trunks of brachial plexus emerge downwards and laterally between the scalenus medius and scalenus anterior muscle. The brachial plexus lies deep to the scalenus anterior muscle. To expose the scalenus anterior muscle cut through the clavicular attachment of the sternocleidomastoid muscle and reflect it medially. Then remove the underlying fatty tissue to expose the muscle in front of the brachial plexus. Though the upper and middle trunk of brachial plexus are situated above and lateral to the 3rd part of subclavian artery, the lower trunk lies behind that artery.

iv. **Suprascapular nerve and nerve to subclavius:** Both are the branches from the upper trunk of brachial plexus. The nerve to subclavius (C_5, C_6) descends in front of the brachial plexus and subclavian vessels and then passes behind the clavicle to reach the deep surface of the subclavius muscle.

v. **Phrenic nerve (C_3, C_4, C_5):** This nerve is located on the anterior surface of the scalenus anterior muscle but deep to the sternocleidomastoid muscle. This structure is seen after retracting the sternocleidomastoid muscle medially.

vi. **Long thoracic nerve (C_5, C_6, C_7):** This nerve passes behind the brachial plexus and 3rd part of subclavian artery.

vii. **Transverse cervical artery:** It is a branch of thyrocervical trunk. It crosses over the scalenus anterior and scalenus medius muscle and runs laterally to the anterior border of the levator scapulae muscle. This artery is crossed superficially by the inferior belly of omohyoid muscle.

viii. **Suprascapular artery:** This artery is a branch of thyrocervical trunk. It passes inferolaterally behind the clavicle and in front of the scalenus anterior and subclavian artery. Its corresponding vein ends in the external jugular vein.

ix. **Supraclavicular lymph nodes:** These are found between the fascial roof and fascial carpet of the triangle in relation to the inferior belly of omohyoid muscle. These nodes are located superficial to the brachial plexus and subclavian vessels.

MUSCLES RELATED TO POSTERIOR TRIANGLE OF NECK

Muscles related to posterior triangle of the neck are—**platysma**, inferior belly of omohyoid, semispinalis capitis (occasional) splenius capitis, levator scapulae, scalenus medius, scalenus posterior, first digitation of serratus anterior, **sternocleidomastoid** and **trapezius**. Only few of them are described here.

Window Dissections: Head and Neck

Muscles	Origin	Insertion	Nerve supply	Action
Platysma (Morphologically—panniculus carnosus)	Fascia covering the pectoralis major muscle up to the 2nd rib and fascia covering the anterior part of deltoid muscle	**Anterior fibers:** Lower border of the body of mandible **Posterior fibers:** Skin at the angle of the mouth through the **risorius**	Cervical branch of facial nerve	i. **Depresses** mandible by its anterior fibers ii. Pulls the angle of the mouth downwards through its attachment of the posterior fibers iii. Reduces concavity of the side of the neck and facilitates venous return by releasing pressure from the underlying veins
Sternocleidomastoid	i. **Sternal head:** Upper part of the anterior surface of manubrium sterni. ii. **Clavicular head:** Anterior surface and upper border of the medial 1/3rd of the clavicle	i. Later surface of mastoid process of temporal bone from its tip to superior border ii. Lateral half of superior nuchal line of squamous part of temporal bone	i. Motor supply by spinal accessory nerve ii. Proprioceptive fibers by ventral rami of C_2 and C_3 nerves	i. **Acting from one side:** a. Pulls the head towards the same shoulder b. Turns the chin to the opposite side ii. **Acting bilaterally:** a. Flex the neck b. Act as accessory muscles of inspiration by elevating the thorax when both the muscles act from above
Trapezius	i. Medial 1/3rd of the superior nuchal line and external occipital protuberance of occipital bone ii. Ligamentum nuchae extending from external occipital protuberance to C_7 spine iii. Spines of C_7 and all thoracic (T_1-T_{12}) vertebrae and supraspinous ligaments between these spines	i. Posterior border of lateral 1/3rd of the clavicle ii. Medial margin of the acromial process of scapula iii. Upper lip of the crest of the spine of scapula	Sub-trapezoid plexus which is formed by spinal accessory nerve and ventral rami of C_3 and C_4 nerves	i. Elevates the shoulder along with levator scapulae ii. Raises the hand above the head acting with the serratus anterior iii. Fix the scapula iv. Maintains the level of shoulder

Summary

1. Position of the cadaver: Supine with head turned to opposite side.
2. Skin incision as in Figure 6.29.
3. Superficial fascia with platysma; external jugular vein and cutaneous nerves of the cervical plexus.
4. Deep fascia (investing layer of deep cervical fascia).
5. **Occipital triangle** (above the inferior belly of omohyoid).

    Contents (Fig. 6.33):
    i. Spinal accessory nerve.
    ii. C_3 and C_4 nerves.
    iii. Dorsal scapular nerve (C_2, C_3)
    iv. Great auricular nerve (C_2).
    v. Lesser occipital nerve.
    vi. Transverse cervical nerve (C_2, C_3).
    vii. Supraclavicular nerve (C_3, C_4).
    viii. Upper trunk of brachial plexus.
    ix. Suprascapular nerve (C_5, C_6).
    x. Long thoracic nerve (C_5, C_6, C_7).
    xi. Transverse cervical artery.
    xii. Occipital artery.
    xiii. Cervical lymph nodes.

    Floor (above downwards) (Fig. 6.34):
    i. Semispinalis capitis (vertical fibers)
    ii. Splenius capitis.
    iii. Levator scapulae (fibers slope downwards and backwards)
    iv. Scalenus posterior.
    v. Scalenus medius.

6. **Supraclavicular** (subclavian) triangle (below the inferior belly of omohyoid and above the clavicle).

    Contents:
    i. Third part of subclavian artery.
    ii. Subclavian vein and external jugular vein.
    iii. Trunks of brachial plexus.
    iv. Suprascapular nerve and nerve to subclavius (C_5, C_6)
    v. Phrenic nerve (C_3, C_4, C_5)
    vi. Longthoracic nerve (C_5, C_6, C_7).
    vii. Transverse cervical artery.
    viii. Suprascapular artery.
    ix. Supraclavicular lymph nodes.

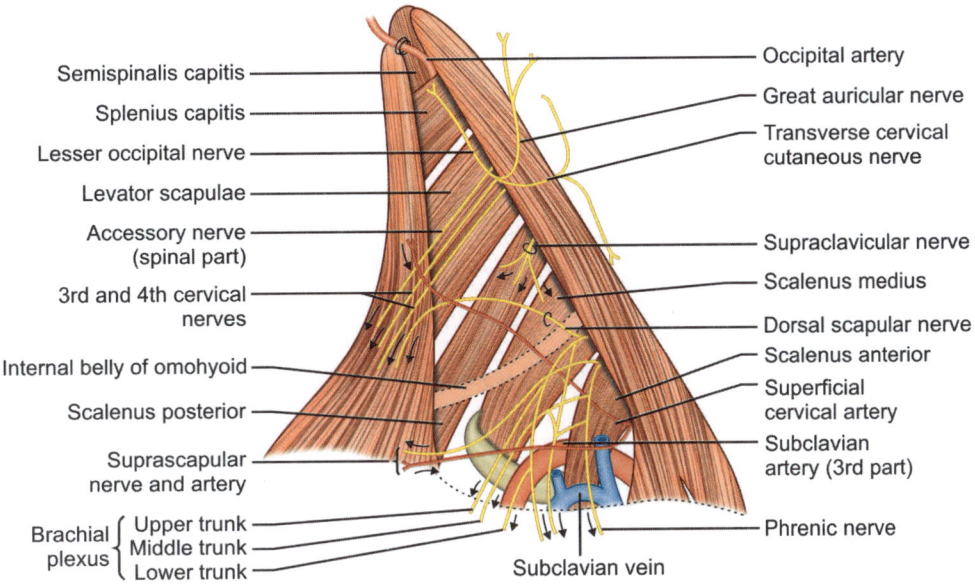

Fig. 6.33: Contents of posterior triangle

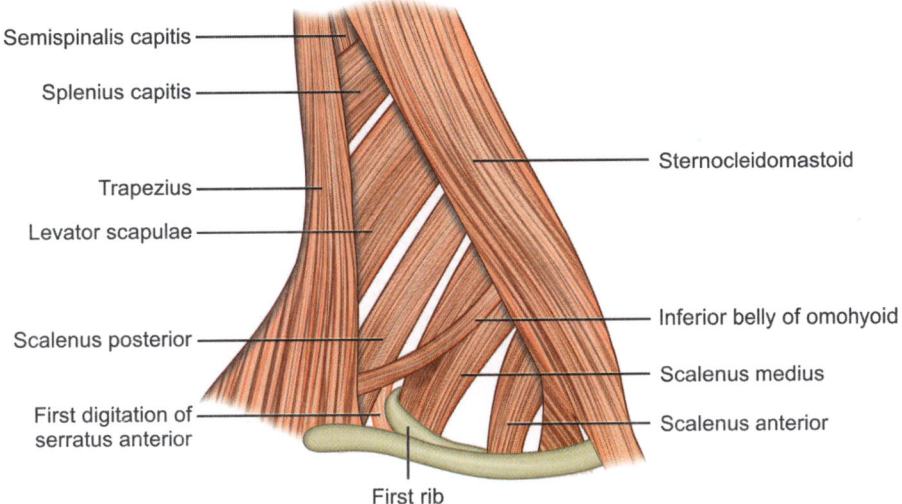

Fig. 6.34: Floor of posterior triangle

PROBABLE QUESTIONS AND ANSWERS

1. **What are the boundaries, subdivisions and contents of the posterior triangle?**

   Ans. Described earlier in this chapter.

2. What are the muscles forming the floor of the triangle (occipital triangle)?

Ans. Described earlier in this chapter?

3. What structures divides the posterior triangle in upper occipital and lower supraclavicular triangles?

Ans. The inferior belly of omohyoid muscle which crosses the posterior triangle obliquely upward and forward about 2.5 cm above the clavicle subdivides the posterior triangle into two parts.

4. How the inferior belly of omohyoid is anchored to the clavicle?

Ans. It is anchored to the clavicle by an inverted sling of omohyoid fascia.

5. What is omohyoid fascia?

Ans. Regarding its source there are two views:
   i. Posterior layer of clavipectoral fascia.
   ii. Posterior lamella of the investing layer of deep cervical fascia.

6. Do you know any other subdivisions of posterior triangle?

Ans. The posterior triangle may be divided by the spinal accessory nerve which lies on the levator scapulae and runs downwards and backwards to disappear beneath the trapezius muscle into an upper **carefree triangle** (containing no important structures) and a lower **careful triangle** which contains most of the important structures like trunks of brachial plexus, subclavian vessels, etc.

7. Where do you find the platysma muscle?

Ans. It is found in the superficial fascia of both anterior and posterior triangle of the neck.

8. What is the nerve supply of platysma?

Ans. Cervical branch of the facial nerve.

9. What are the superficial veins of the neck?

Ans. External jugular veins in the posterior triangle and anterior jugular vein in the anterior triangle of the neck.

10. How is the external jugular vein formed?

Ans. It is formed by the union of posterior auricular vein and posterior division of retromandibular vein.

11. Where does it begin and end?

Ans. External jugular vein begins within the lower part of the parotid gland or just below the angle of the mandible. It ends in the subclavian vein (in supraclavicular triangle) about 2.5 cm above the clavicle by piercing the investing layer of deep cervical fascia.

12. Is there any valve of external jugular vein?

Ans. Yes, there are two pairs of valves. First pair is at its termination into the subclavian vein and the second pair is present about 4 cm above the clavicle. **But these valves do not prevent regurgitation of blood.**

13. What is the clinical importance of external jugular vein?

Ans. The right external jugular vein serves as a **venous monometer,** i.e. it is used to assess the venous pressure. The right atrial pressure is reflected in the right external jugular vein. In right heart failure, there is increased venous pressure causing engorgement of jugular vein.

14. Why this vein is used to assess venous pressure?

Ans. This vein is straight and the valves do not prevent regurgitation of blood form the right atrium.

15. What is the special feature of this vein?

Ans. The wall of the vein is adherent to the deep fascia through which it passes. So the vein does not collapse if the vein is cut at this point and it causes sucking of air from the atmosphere due to negative intrathoracic pressure resulting in air embolism.

16. What are the tributaries of external jugular vein?

Ans.
 i. Transverse cervical vein.
 ii. Suprascapular vein.
 iii. Posterior external jugular vein.
 iv. Anterior jugular vein.
 v. Oblique jugular vein.

17. What is oblique jugular vein?

Ans. It is a communicating vein running obliquely across the middle 1/3rd of the sternocleidomastoid muscle connecting the external jugular vein with the internal jugular vein.

18. Name the cutaneous nerves of the posterior triangle?

Ans.
  i. Lesser occipital nerve (C_2).
  ii. Great auricular nerve (C_2, C_3).
  iii. Transverse cervical nerve (C_2, C_3).
  iv. Supraclavicular nerve (C_3, C_4).

19. How do these nerves appear in the posterior triangle?

Ans. These nerves radiate from the middle of the posterior border of sternocleidomastoid muscle and then piercing the investing layer of deep cervical fascia become cutaneous and found in the superficial fascia.

20. What is the relation of spinal accessory nerve with the lesser occipital nerve near the posterior border of sternocleidomastoid?

Ans. The spinal accessory nerve is hooked by the lesser occipital nerve (feature of identification).

21. What important structure accompanies great auricular nerve?

Ans. The upper half of external jugular vein accompanies the great auricular nerve.

22. Why the clavicle is considered as a dermal bone?

Ans. Sometimes, the intermediate branch of supraclavicular nerve (cutaneous nerve) pierces the clavicle. So it is considered as a dermal bone.

23. What are the muscles supplied by spinal accessory nerve?

Ans. **Two muscles**:
  i. Sternocleidomastoid.
  ii. Trapezius.

24. What is subtrapezoid plexus?

Ans. About 5 cm above the clavicle, the spinal accessory nerve forms a plexus with C_3 and C_4 nerves deep to the trapezius called subtrapezoid plexus and supplies trapezius muscle.

25. What is interscalene triangle?

Ans. It is a triangular area bounded by:
  i. **Laterally:** Medial border of scalanus medius.
  ii. **Medially:** Lateral border of scalenus anterior.
  iii. **Below:** Upper border of first rib.

Window Dissections: Head and Neck

26. What are the contents of the interscalene triangle?

Ans. i. Brachial plexus.
 ii. Subclavian artery.

27. Which part of the subclavian artery is seen in the posterior triangle?

Ans. 3rd part.

28. How the subclavian artery is divided into three parts?

Ans. By scalenus anterior muscle:
 i. 1st part: From its origin up to the medial border of scalenus anterior muscle.
 ii. 2nd part: Behind the muscle.
 iii. 3rd part: From the lateral border of the muscle to the outer border of first rib.

29. What is the branch from 3rd part of subclavian artery?

Ans. Dorsal scapular artery.

30. What are the other branches of subclavian artery?

Ans. i. Verterbral artery.
 ii. Internal thoracic artery.
 iii. Thyrocervical trunk.
 iv. Costocervical trunk.

31. What is the origin of suprascapular and transverse cervical artery?

Ans. Thyrocervical trunk (which is the branch of subclavian artery).

32. Where do you find the phrenic nerve?

Ans. It lies in front of the scalenus anterior muscle in the supraclavicular triangle and it is exposed after the clavicular head of sternocleidomastoid muscle is retracted.

33. Which important structure is related to the superficial cervical lymph nodes?

Ans. Spinal accessory nerve.

34. What is wry neck or torticollis?

Ans. This is deformity is which the neck is bent to one side and the chin points to the other side. This deformity results from the spasm or contracture of the muscles (sternocleidomastoid and trapezius) supplied by the spinal accessory nerve.

35. What is supraclavicular space?

Ans. The investing layer of deep cervical fascia splits to enclose a space in the supraclavicular triangle called supraclavicular space.

36. What are the contents of this space?

Ans.
  i. Terminal part of external jugular vein.
  ii. Anterior jugular vein at its termination into the external jugular vein.
  iii. Few lymph nodes.

37. What is scalenovertebral triangle?

Ans. It is a triangle bounded by:
  i. **Medially:** Longus cervicis on C_6 and C_7 verterbral bodies.
  ii. **Laterally:** Medial border of scalenus anterior.
  iii. **Apex:** Tranverse process of C_6 vertebra.
  iv. **Base:** 1st part of subclavian artery.

38. What are the contents of this triangle?

Ans.
  i. Carotid sheath with its contents.
  ii. Inferior thyroid artery and cervical part of thoracic duct.
  iii. Vertebral vein.
  iv. 1st part of vertebral artery.
  v. Inferior cervical ganglia.

39. Which structure is used as a guide to identify middle cervical ganglia?

Ans. Inferior thyroid artery.

40. Name the structures on the anterior surface of scalenus anterior?

Ans.
  i. Ascending cervical branch of inferior thyroid artery.
  ii. Cervical part of thoracic duct.
  iii. Transverse cervical artery.
  iv. Suprascapular artery.
  v. Subclavian vein.
  vi. Internal jugular vein.
  vii. Terminal part at anterior jugular vein opening into the external jugular vein.

Window Dissections: Head and Neck

REVIEW OF SKIN INCISIONS FOR DISSECTIONS AT A GLANCE

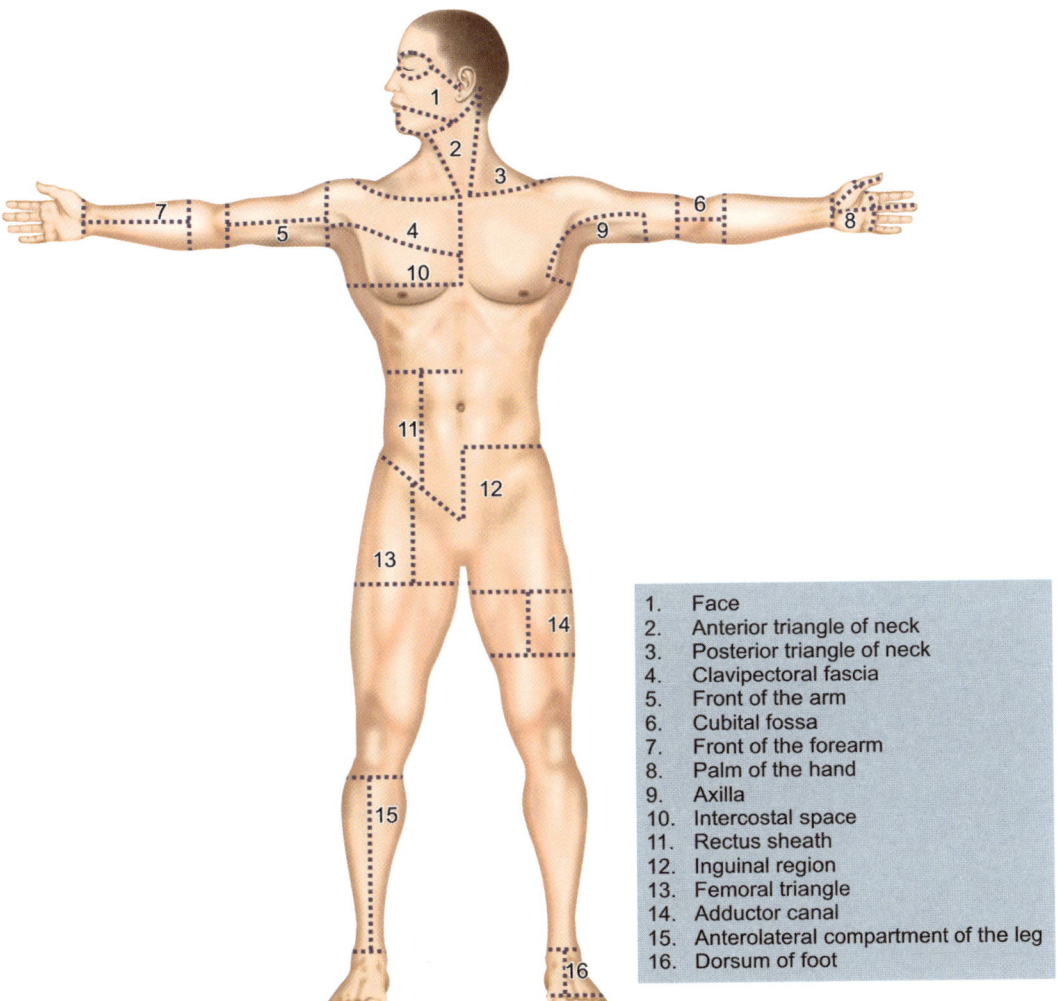

1. Face
2. Anterior triangle of neck
3. Posterior triangle of neck
4. Clavipectoral fascia
5. Front of the arm
6. Cubital fossa
7. Front of the forearm
8. Palm of the hand
9. Axilla
10. Intercostal space
11. Rectus sheath
12. Inguinal region
13. Femoral triangle
14. Adductor canal
15. Anterolateral compartment of the leg
16. Dorsum of foot

Skin Incisions on Front of the Body

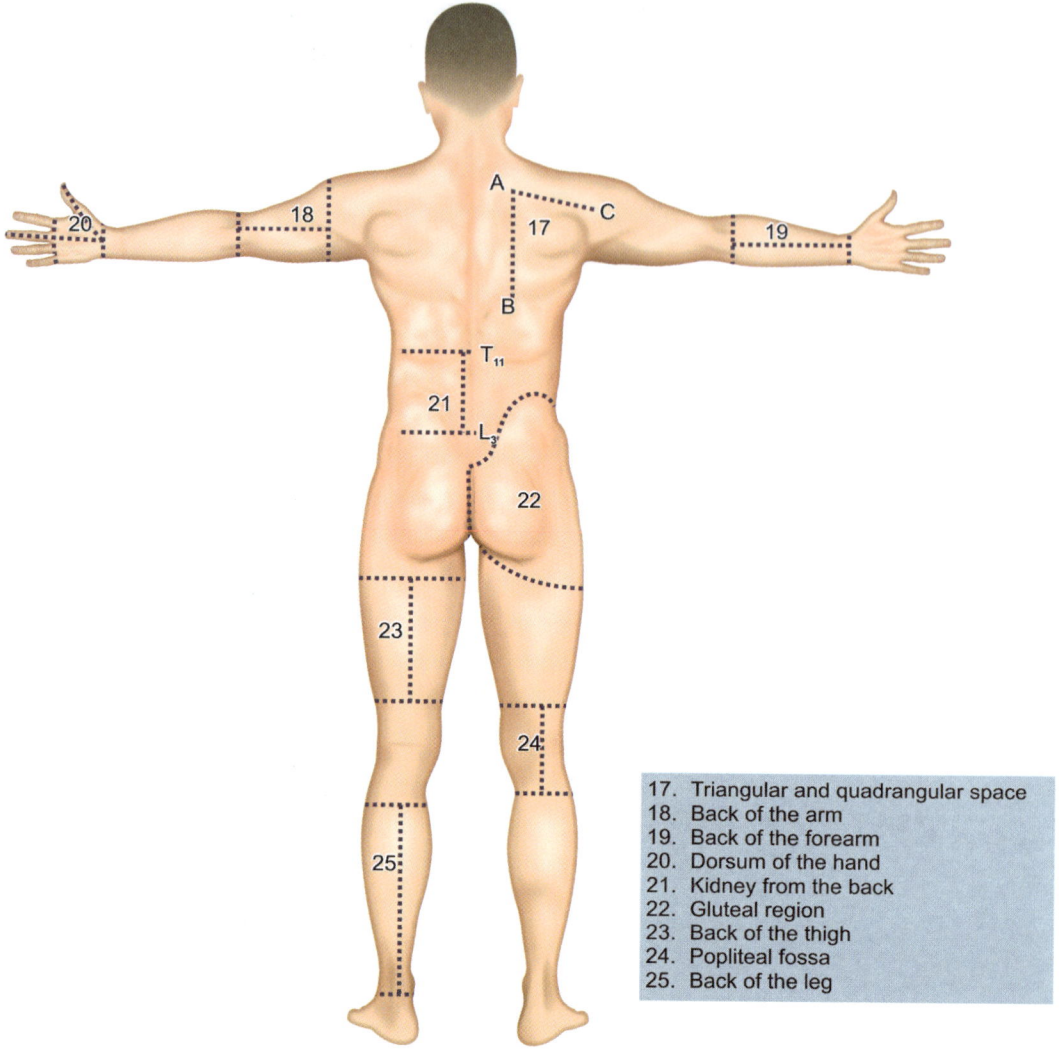

17. Triangular and quadrangular space
18. Back of the arm
19. Back of the forearm
20. Dorsum of the hand
21. Kidney from the back
22. Gluteal region
23. Back of the thigh
24. Popliteal fossa
25. Back of the leg

Skin Incisions on Back of the Body

SECTION 2

Surface Anatomy

Section Outline

- Surface Anatomy: Upper Limb (Superior Extremity)
- Surface Anatomy: Lower Limb (Inferior Extremity)
- Surface Anatomy: Abdomen
- Surface Anatomy: Thorax
- Surface Anatomy: Head and Neck

CHAPTER 7
Surface Anatomy: Upper Limb (Superior Extremity)

- Points
  1. Head of Radius
  2. Head of Ulna
  3. Styloid Process of Radius
  4. Styloid Process of Ulna
  5. Pisiform Bone
  6. Tip of Coracoid Process
  7. Acromial Angle
  8. Hook of Hamate
  9. Bifurcation of Brachial Artery
  10. Beginning of Brachial Artery, Radial Nerve, Median Nerve and Ulnar Nerve
- Lines
  1. Radial Nerve in the Back of the Arm
  2. Ulnar Nerve in Forearm
  3. Axillary Artery
  4. Brachial Artery
  5. Radial Artery in the Forearm
  6. Ulnar Artery in Forearm
  7. Superficial Palmar Arch
  8. Flexor Retinaculum

LESSON 1: POINTS

1. HEAD OF RADIUS (FIG. 7.1)

Lateral epicondyle is felt in the upper part of a depression on the lateral side of the olecranon process of ulna in an extended elbow. Head of the radius is palpable below the lateral epicondyle at the bottom of the depression in the upper part of the back of the extended and pronated forearm. The posterior surface of the radial head can be easily felt during supination and pronation of the forearm in an extended elbow. A point is given on this area.

PROBABLE QUESTIONS AND ANSWERS

1. What is the general feature of radial head?

   Ans. It has a cup-shaped (concave) upper surface and a peripheral margin.

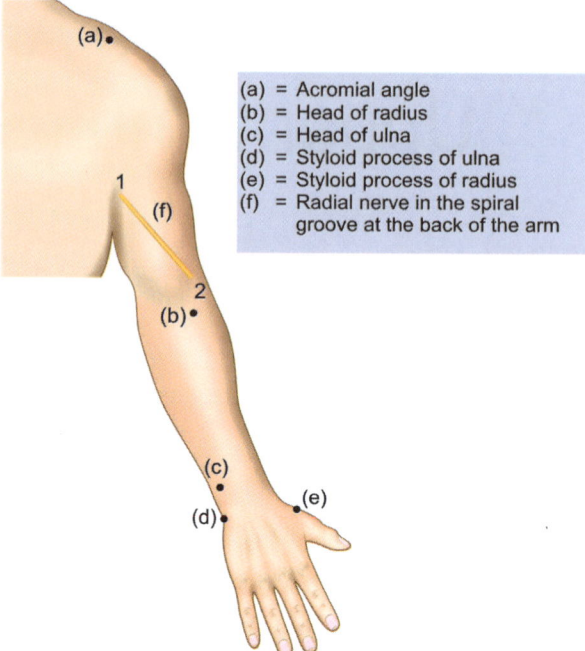

Fig. 7.1: Surface markings on the back of upper limb

2. Does it articulate with other bones/Is it articular?

Ans. Yes. It is articular. The upper concave surface articulates with the capitulum of the lower end of the humerus and forms the humeroradial part of elbow joint.
The peripheral margin is broader on the medial side and this broad medial margin articulates medially with the radial notch of the ulna and forms superior radioulnar joint. The rest of the peripheral margin is encircled by the annular ligament.

3. What is cubital joint?

Ans. The superior radioulnar joint together with the elbow joint is called cubital joint because of their common synovial membrane.

4. What is the type of humeroradial part of elbow joint?

Ans. Synovial. . (Elbow joint is a hinge variety of synovial joint).

5. What is the type and subtype of superior (proximal) radioulnar joint?

Ans.   i. Type—synovial.
     ii. Subtype—uniaxial pivot joint.

6. What is the time of appearance of ossification center for the head of radius?

Ans. The secondary center for the head appears in the 5th year and fuses with the shaft in 18th year.

2. HEAD OF ULNA (FIG. 7.1)

Flex the wrist and pronate the forearm. A rounded bony elevation is seen and palpated on the medial side of the back of the wrist. A point is given on the rounded bony elevation.

[Note that radial head is at its upper end and ulnar head is at its lower end]

PROBABLE QUESTIONS AND ANSWERS

1. What is the shape of the ulnar head?

Ans. Round.

2. Is it articular?

Ans. Yes. Its lateral surface articulates laterally with the ulnar notch of the radius and forms the inferior radioulnar joint.

3. What is the type of inferior radioulnar joint?

Ans. Synovial:
 i. Superior (proximal) radioulnar joint → Synovial
 ii. Middle radioulnar joint → Fibrous (syndesmosis)
 iii. Inferior (distal) radioulnar joint → Synovial.
All these three joints are concerned for supination and pronation of the forearm.

4. Does it form the wrist joint?

Ans. No. The articular disk of the inferior radioulnar joint separates the inferior surface of the head of the ulna from the lunate bone (one of the carpal bones), thus preventing the head to participate in the formation of wrist joint.

5. What is articular disk?

Ans. It is a fibrocartilage, triangular in shape. Its base is attached to the margin between the ulnar notch and the carpal articular surface of the lower end of the radius. The apex is attached to a depression at the junction between the styloid process and inferior articular surface of the head of the ulna.

6. What is the time of appearance of the ossification center?

Ans. The secondary ossification center for the lower end of the ulna appears is the 5th year and unites with the shaft in the 18th year.

3. STYLOID PROCESS OF RADIUS (FIG. 7.1)

Palpate the lateral border of the radius and follow it distally (i.e. towards its lower end). Otherwise, it can be felt following the lateral border of the flexed wrist with the forearm pronated. Radial styloid process is about ½" below the ulnar styloid process.

PROBABLE QUESTIONS AND ANSWERS

1. Which part of the lower end of the radius is continued as styloid process?

Ans. The lateral surface of the lower end of the radius projects downwards as the styloid process.

2. What are the attachments of this process?

Ans. i. Tip of the styloid process—lateral carpal ligament.
ii. Proximal part of styloid process—insertion of brachioradialis.

3. What are the tendons crossing overs this process?

Ans. The tendons of abductor pollicis longus and extensor pollicis brevis cross this process obliquely downwards and forwards.

4. STYLOID PROCESS OF ULNA (FIG. 7.1)

Styloid process of ulna is felt by following the medial border of the flexed wrist with the forearm pronated. This process projects downwards from the head of the ulna on the medial side of the wrist.

PROBABLE QUESTIONS AND ANSWERS

1. What are the structures attached to the ulnar styloid process?

Ans. i. Proximal attachment of the ulnar collateral ligament (medial ligament of the wrist) from its tip. The distal attachment of this ligament is the triquetral and pisiform bone (carpal bones).
ii. The apex of the articular disk is attached to a depression between the styloid process and the lower surface of the head of the ulna.

Surface Anatomy: Upper Limb (Superior Extremity)

?2. Where does the tendon of extensor carpi ulnaris (ECU) lodge?

Ans. The tendon of ECU lodges in a groove between the ulnar styloid process and posterior surface of the ulnar head.

?3. Which styloid process lies at a lower level?

Ans. The radial styloid process lies about 1.25 cm, below the ulnar styloid process.

?4. Name the bones which are having styloid process?

Ans.
i. Styloid process of radius (forearm bone)
ii. Styloid process of ulna (forearm bone)
iii. Styloid process of fibula (leg bone)
iv. Styloid process of temporal bone (cranial bone).

5. PISIFORM BONE (FIG. 7.2)

It is felt as a bony elevation at the medial side of the base of the hypothenar eminence on the palmar aspect of the hand.

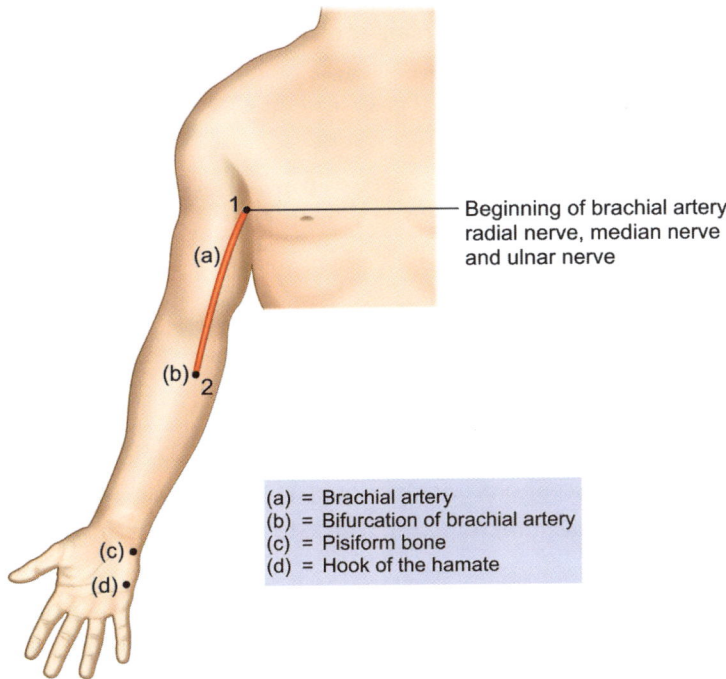

Fig. 7.2: Surface markings on the front of upper limb

PROBABLE QUESTIONS AND ANSWERS

1. What type of bone the pisiform is and what is the shape of the bone?

Ans. The pisiform is a **sesamoid** bone and it is pea-shaped.

2. What are the characteristics of a sesamoid bone?

Ans.
 i. Devoid of periosteum.
 ii. No haversian system.
 iii. Ossifies after birth (secondary center of ossification).
 iv. Develops in a tendon.
 v. No separate arterial supply. They are supplied by the arteries of the muscles in which they develop.

3. Under which tendon the pisiform develops?

Ans. It develops under the tendon of flexor carpi ulnaris.

4. Name some sesamoid bones in the body.

Ans.
 i. Patella (develops under quadriceps femoris).
 ii. Pisiform (develops under flexor carpi ulnaris).
 iii. Two sesamoid bons on the plantar aspect of 1st metatarsal bone (under flexor hallucis brevis).
 iv. Fabella (in the lateral head of gastrocnemius).
 v. Rider's bone (tendon of adductor longus).

5. Which is the largest sesamoid bone is the body?

Ans. Patella.

6. How many carpal bones are there?

Ans. There are eight carpal bones in each hand.

7. How they are arranged?

Ans.
 i. Proximal row: Scaphoid, lunate, triquetral and **pisiform** (from lateral to medial).
 ii. Distal row: Trapezium, trapezoid, capitate and hamate (from lateral to medial).

8. With which bone the pisiform articulates?

Ans. The oval facet on the posterior surface articulates with the triquetral bone.

9. What are the structures attached to the pisiform?

Ans.
i. Origin of abductor digiti minimi.
ii. Insertions of flexor carpi ulnaris.
iii. Flexor and extensor retinaculum.
iv. Pisohamate and pisometacarpal ligaments.

10. When do the pisiform ossify?

Ans. It ossifies after birth. The center appears at the age of 10 years (9–12 years). It appears last among the carpal bones.

11. Which carpal bone appears first?

Ans. Capitate (2nd month). (According to Mc Gregor—1st year).

12. Which is the largest carpal bone?

Ans. Capitate.

13. Which is the smallest carpal bone?

Ans. Pisiform.
Note that all the carpal bones ossify from the primary center of ossification after birth. This violates the law of ossification which says that the primary center of ossification appears before birth.

6. TIP OF CORACOID PROCESS (FIG. 7.3)

A point is given 2.5 cm below the junction of medail 3/4th and lateral 1/4th of the clavicle.

PROBABLE QUESTIONS AND ANSWERS

1. Which bone possesses coracoid process?

Ans. Scapula.

2. Name the muscle which covers the tip of this process?

Ans. Anterior fibers of deltoid.

3. What are the parts of the coracoid process?

Ans.
i. Vertical part.
ii. Horizontal part. **The end of the horizontal part is the tip of coracoid process.**

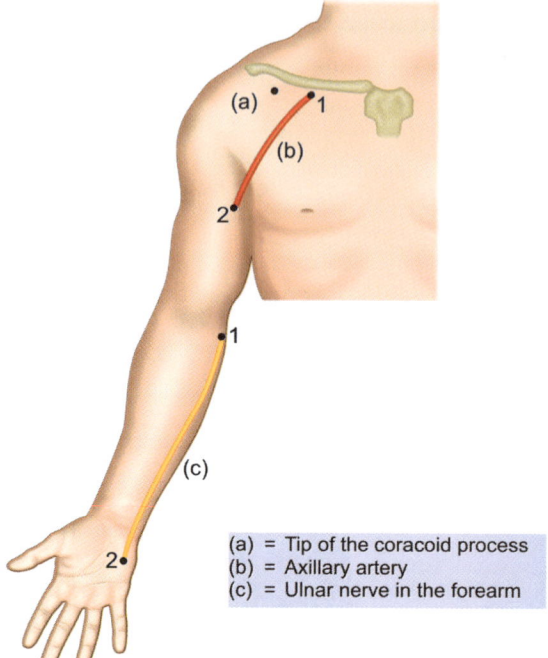

Fig. 7.3: Surface markings of: Tip of coracoid process, axillary artery and ulnar nerve in the forearm

4. What is conoid tubercle?

Ans. At the junction between the two parts, there is a tubercle which is known as conoid tubercle.

5. What are the muscles attached (origin) at the tip of the coracoid process?

Ans.
  i. Origin of **coraco brachialis—medially**.
  ii. Origin of **short head of biceps brachialis—laterally**.

6. What is the site of origin of long head of biceps?

Ans. Supraglenoid tubercle (intracapsular origin).

7. Name the structures that are attached to the coracoid process other than coracobrachialis and short head of biceps.

Ans.
  a. Muscles—insertion of pectoralis minor.
  b. Ligaments:
    i. Coracoclavicular ligament (conoid part and trapezoid part).
    ii. Coracoacromial ligament.
    iii. Suprascapular ligament (at the medial margin of root).

Surface Anatomy: Upper Limb (Superior Extremity)

8. What type of epiphysis the coracoid process is?

Ans. Atavistic epiphysis.

9. Name another example of atavistic epiphysis.

Ans. Posterior tubercle of talus.

10. When does the center of ossification for coracord process appear?

Ans. It appears in the 1st year.

7. ACROMIAL ANGLE (FIG. 7.1)

Palpate the spine of the scapula on the back of the chest. Feel the lower border of the crest of the spine and trace it laterally. The bend where this border meets the lateral border of acromion process is the acromial angle. Make a point over this bend.

PROBABLE QUESTIONS AND ANSWERS

1. What is the importance of this angle?

Ans. Acromial angle is an important landmark for the measurement of the length of the upper limb.

2. What are the presenting parts of acromion process?

Ans.
  i. 2 Borders—lateral border and medial border.
  ii. 2 Surfaces—upper or dorsal surface and lower surface.
  iii. Tip of acromion process.

3. What is the joint formed by acromion process?

Ans. Acromioclavicular joint (with the oval facet in the anterior part of the medial border of acromion process). This is a plane type of synovial joint.

4. Name the muscles that are attached to the acromion process?

Ans.
  i. Origin of deltoid (middle fibers) along its lateral border.
  ii. Insertion of trapezius (middle fibers) along its medial border.

5. What important ligament is attached to the acromion process?

Ans. The tip of the acromion process gives attachment to the coracoacromial ligament which together with the undersurface of the acromial process and horizontal part of coracoid process forms coracoacromial arch. This arch forms a secondary socket for the shoulder joint and provides protection to the joint.

8. HOOK OF HAMATE (FIG. 7.2)

Press over the hypothenar eminence and feel the pisiform bone. Give a point about 2.5 cm below the pisiform in a line along the ulnar border of the ring finger.

PROBABLE QUESTIONS AND ANSWERS

1. **What is the shape of the hamate bone?**

Ans. Wedge-shaped.

2. **What type of bone the hamate is?**

Ans. The hamate is a short bone.

3. **What is the relative position of the hamate among the carpal bones?**

Ans. It lies on the medial side of the distal row of the carpal bones.

4. **From which part of the hamate the hook projects?**

Ans. The hook of the hamate projects from its anterior surface.

5. **What are the attachments of the hook of hamate?**

Ans. i. Flexor retinaculum.
ii. Pisohamate ligament.

6. **What are the points of attachment of flexor retinacum?**

Ans. i. Pisiform and hook of the hamate on the medial side.
ii. Tubercle of scaphoid and crest of trapezium on the lateral side.

7. **What is the time of appearance of center of ossification?**

Ans. The center of ossification for the hamate appears at 3rd year.

9. BIFURCATION OF BRACHIAL ARTERY (FIG. 7.2)

A point is given 1 cm below the bend of the elbow just medial to the tendon of biceps brachii.

PROBABLE QUESTIONS AND ANSWERS

1. **What are the branches of brachial artery at its bifurcation?**

Ans. Radial artery and ulnar artery.

Surface Anatomy: Upper Limb (Superior Extremity)

2. What is the level of bifurcation in relation to radius?
Ans. The point of bifurcation corresponds the level of the neck of radius.

3. What is the relative position of the three important structures in the cubital fossa?
Ans. From lateral to medial: **Tendon of biceps brachii, brachial artery and median nerve (TAN).**

4. Name the branches of brachial artery other than its terminal branches?
Ans.
  i. Arteria profunda brachii.
  ii. Superior ulnar collateral artery.
  iii. Inferior ulnar collateral artery.
  iv. Nutrient artery for the humerus.
  v. Muscular branches for the muscles of the anterior compartment of the arm.

10. BEGINNING OF BRACHIAL ARTERY, RADIAL NERVE, MEDIAN NERVE AND ULNAR NERVE (FIG. 7.2)

For these structures, the beginning point is same.

The arm is abducted at right angle and rotated laterally so that palm faces upwards. Feel the distal ends of anterior and posterior axillary folds. Give a point at the junction of the anterior 2/3rd and posterior 1/3rd of the line joining the distal ends of the two axillary folds.

PROBABLE QUESTIONS AND ANSWERS

Brachial Artery

1. What is the level where the axillary artery continues as brachial artery?
Ans. At the distal border of teres major muscle.

2. What are the branches of brachial artery?
Ans. Discussed earlier (point number 9).

Radial Nerve

1. From which cord of brachial plexus the radial nerve arises?
Ans. Posterior cord.

2. What is the root value of radial nerve?

Ans. Dorsal branches of the ventral rami of C_5, C_6, C_7, C_8 and T_1 spinal nerves.

3. What are the branches from the posterior cord?

Ans.
  i. Upper subscapular nerve.
  ii. Lower subscapular nerve.
  iii. Thoracodorsal nerve.
  iv. Axillary nerve (C_5, C_6).
  v. Radial nerve.

Which is the largest branch of brachial plexus?

Ans. Radial nerve.

5. What are the muscles supplied by the radial nerve?

Ans. All extensor muscles of the arm and forearm.

6. What is the common clinical finding of radial nerve injury?

Ans. Wrist drop.

Median Nerve

1. How the median nerve is formed?

Ans. Median nerve is formed by two roots:
  i. Medial root from medial cord (C_8, T_1).
  ii. Lateral root from lateral cord (C_5, C_6, C_7).

2. What is the root value of median nerve?

Ans. Ventral branches of ventral rami of C_5-T_1 spinal nerves.

3. What are the muscles supplied by the nerve?

Ans.
  i. All the muscles of the flexor compartment of the forearm except flexor carpi ulnaris and medial half of flexor digitorum profundus.
  ii. Five intrinsic muscles of the hand—3 thenar muscles, 1st and 2nd lumbricals.

4. What are the digits supplied by the median nerve?

Ans. Lateral 3½ digits.

Surface Anatomy: Upper Limb (Superior Extremity)

5. Why the median nerve is called the laborer's nerve or workman's nerve?

Ans. This is because of its supply to the most of the large flexor muscles of the forearm.

6. What happens if the nerve is compressed in the carpal tunnel?

Ans. The manifestation is called carpal tunnel syndrome.

Ulnar Nerve

1. Which cord of brachial plexus forms ulnar nerve?

Ans. Continuation of medial cord of brachial plexus forms ulnar nerve.

2. What is the root value of this nerve?

Ans. (C_7), C_8, T_1 spinal nerves.

3. Is there any muscular branch in the arm?

Ans. No.

4. What are its muscular branches?

Ans. i. Flexor carpi ulnaris and medial half of flexor digitorum profundus in the forearm.
 ii. All intrinsic muscles of the hand (15 muscles) except thenar muscles and 1st, 2nd lumbricals.

5. Why the ulnar nerve is called musicians nerve?

Ans. This is because of its supply to the most of the intrinsic muscles of the hand which are used by a violinist.

6. What are the digits (fingers) supplied by this nerve?

Ans. It supplies the skin of the medial 1½ digits.

7. When does the cubital tunnel syndrome occur?

Ans. It occurs when the nerve is compressed behind the medial epicondyle of the humerus. Note that the deep branches of **radial nerve (posterior interosseous nerve** appears in the posterior compartment of forearm after passing between two strata of **supinator muscle. Median nerve** appears in the forearm after passing between superficial and deep heads of **pronator teres muscles. Ulnar nerve** appears in the forearm after passing between the two heads of **flexor carpi ulnaris.**

LESSON 2: LINES

1. RADIAL NERVE IN THE BACK OF THE ARM (FIG. 7.1)

i. 1st point is given on the medial side of the upper part of the arm at the junction of the posterior 1/3rd and anterior 2/3rd of the line joining the distal end of anterior and posterior axillary folds. This point is the proximal point for the radial nerve in the back of the arm.

ii. 2nd point (distal point) is given on the lateral border of the arm. To give this point, identify the site of insertion of deltoid at deltoid tuberosity and lateral epicondyle of humerus. Then take a point at the junction of upper 1/3rd and lower 2/3rd of the line joining the above two points.

Then join the 1st and 2nd point in a single line across the back of the arm which represents radial nerve in the spiral groove in the back of the arm.

> Note that the nerves are drawn in single line but arteries, veins, ducts, etc. are drawn in parallel lines (double lines).

PROBABLE QUESTIONS AND ANSWERS

1. What is the root value of radial nerve?

Ans. Dorsal branches of ventral rami of C_5, C_6, C_7, C_8, T_1 spinal nerves.

2. Which cord of brachial plexus gives rise to radial nerve?

Ans. The continuation of posterior cord of brachial plexus in the axilla is the radial nerve. It is the largest branch of brachial plexus.

3. Where does the radial nerve lodge in the back of the arm?

Ans. The nerve lodges in the spiral groove (radial groove) on the dorsal surface of the humerus between the lateral and medial head of triceps.

4. Which structure accompanies the radial nerve in spiral groove?

Ans. **Arteria profunda brachii,** branch of brachial artery.

5. What are the branches of radial nerve in spiral groove?

Ans.
  i. Muscular branches to medial and lateral heads of triceps. Nerve to medial head of triceps ends by supplying anconeus muscle.
  ii. Posterior cutaneous nerve of forearm.
  iii. Lower lateral cutaneous nerve of the arm.

6. What happens if the radial nerve is compressed against the spiral groove by placing the out stretched arm on a armchair under drunken condition?

Ans. This may lead to temporary palsy of the radial nerve which is commonly termed as **Saturday night palsy.**

Surface Anatomy: Upper Limb (Superior Extremity)

2. ULNAR NERVE IN FOREARM (FIG. 7.3)

i. A point on the posterior aspect of the medial epicondyle of humerus.
ii. 2nd point is given on the lateral side of the pisiform bone over the hypothenar eminence. Then join these two points and make single line which represents the ulnar nerve in the forearm.

PROBABLE QUESTIONS AND ANSWERS

1. How the ulnar nerve is formed?

Ans. The ulnar nerve is formed as a continuation of the medial cord of brachial plexus in the axilla.

2. What is its root value?

Ans. (C_7), C_8, T_1 spinal nerves.

3. What are the branches of ulnar nerve in the forearm?

Ans. i. **Muscular branches** to **flexor carpi ulnaris** and **medial half of flexor digitorum profundus.**
  ii. A dorsal cutaneous branch (arises about 5 cm above the wrist).
  iii. A palmar cutaneous branch (arises proximal to flexor retinaculum).

4. How does the ulnar nerve enter into the forearm from the arm?

Ans. It enters into the forearm between the two heads of flexor carpi ulnaris muscle.

5. How does the nerve enter into the palm of the hand?

Ans. It passes superficial to the flexor retinaculum of the hand and passes beneath the palmaris brevis and then divides into superficial and deep terminal branches.

6. What is the relationship between the ulnar nerve and ulnar artery in the lower part of the forearm?

Ans. The ulnar artery accompanies the ulnar nerve on its lateral (radial) side.

3. AXILLARY ARTERY (FIG. 7.3)

i. 1st point is given at the middle of the lower border of clavicle.
ii. 2nd point is given at the point of beginning of the brachial artery (vide point number 10). Join these two points in parallel lines (two lines) which represents axillary artery.

PROBABLE QUESTIONS AND ANSWERS

1. Where does the axillary artery begin?

Ans. The axillary artery begins at the outer border of the first rib as a continuation of 3rd part of subclavian artery.

2. Where does the axillary artery end?

Ans. The artery ends at the lower border of teres major muscle and continues as brachial artery.

3. Name the important structures which accompany the axillary artery?

Ans. The axillary artery is accompanied by the axillary vein (on its inferomedial aspect) and cords of brachial plexus with their branches.

4. How many parts of the axillary artery are there?

Ans. There are three parts of axillary artery.

5. Which structure divides the artery into three parts?

Ans. The **pectoralis minor muscle** crosses in front of the axillary artery and divides the artery into three parts—proximal to the muscle, behind the muscle and distal to the muscle.

6. What are these parts?

Ans. From proximal to distal:
   i. 1st part—from outer border of first rib to the proximal border of pectoralis minor muscle.
   ii. 2nd part—behind the pectoralis minor muscle.
   iii. 3rd part—from distal border of pectoralis minor to lower border of teres major muscle.

7. What are branches of the axillary artery from its different parts?

Ans. From the first part (1):
   a. Superior thoracic artery.
   From the 2nd part (2):
   a. Thoracoacromial artery.
   b. Lateral thoracic artery.
   From the 3rd part (3):
   a. Subscapular artery.
   b. Anterior humeral circumflex artery.
   c. Posterior humeral circumflex artery.

8. With which artery the axillary artery forms collateral anastomoses?

Ans. The branches of the third part of axillary artery makes a collateral anastomoses with the branches of first part of the subclavian artery around the scapula which is termed as scapular anastomoses.

4. BRACHIAL ARTERY (FIG. 7.2)

Two points are given; one at the beginning of the artery and another at the point of bifurcation. Beginning of the brachial artery (vide point number 10). Point of bifurcation of the artery (vide point number 9). Join these two points and make two parallel lines along the medial border of biceps brachii tendon.

PROBABLE QUESTIONS AND ANSWERS

1. Where does the brachial artery begin?

Ans. It begins at the distal border of teres major muscle as a continuation of axillary artery.

2. Where does it end?

Ans. It ends by dividing into radial artery and ulnar artery 1 cm below the bend of the elbow and corresponds to the neck of the radius.

3. What are the structures crossing in front of the artery?

Ans. Median nerve crosses from lateral to medial side at the level of midarm. Bicipital aponeurosis crosses over the artery at the cubital fossa and separates the artery from the median cubital vein which lies superficial to the aponeurosis.

4. What are the branches of the brachial artery?

Ans.
   i. Arteria profunda brachii.
   ii. Nutrient artery to humerus.
   iii. Superior ulnar collateral artery.
   iv. Inferior ulnar collateral artery.
   v. Muscular branches.
   vi. Terminal branches—radial artery and ulnar artery.

5. What is the clinical use of this artery?

Ans. This artery is ausculted in front of the elbow and medial to the biceps tendon as standard method of recording blood pressure.

5. RADIAL ARTERY IN THE FOREARM (FIG. 7.4)

i. 1st point is at the point of bifurcation of brachial artery.
ii. 2nd point is in front of the wrist between the anterior border of the lower part of radius and tendon of flexor carpi radialis.
iii. 3rd point is on the anatomical snuff box.
   Join these three points with two parallel lines which represents the radial artery in the forearm.

PROBABLE QUESTIONS AND ANSWERS

1. What is the point of beginning of radial artery?

Ans. The radial artery begins in the cubital fossa 1 cm below the bend of the elbow at the level of the neck of the radius as one of the terminal branches of the brachial artery.

2. Which terminal branch of the brachial artery is larger?

Ans. The ulnar artery is larger than the radial artery.

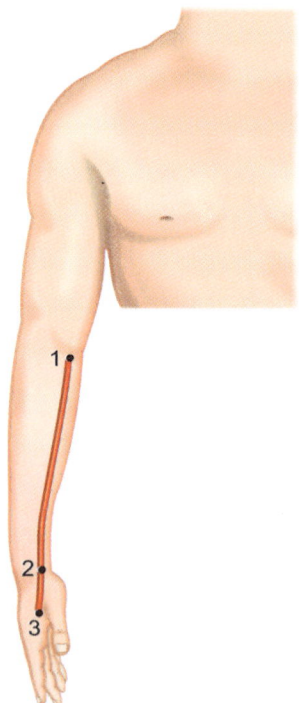

Fig. 7.4: Radial artery in the forearm

3. What may be the unusual origin of radial artery?

Ans. The radial artery may arise from axillary artery or it may arise from the brachial artery at a higher level, i.e. in the arm.

4. Which part of the artery runs superficial in the forearm?

Ans. The lower part of the artery runs superficially, i.e. deep to deep fascia in the forearm but the upper part of the artery is overlapped by the brachioradialis muscle.

5. How does it enter into the palm?

Ans. It enters into the palm between the two heads of first dorsal interosseous muscle.

6. What is the clinical importance of radial artery?

Ans. The radial artery is very important for the clinicians. The radial pulse is recorded between the flexor carpi radialis and base of the styloid process of the radius.

7. What is the position of the radial artery at the wrist?

Ans. At the wrist, the artery lies in the anatomical snuff box.

8. What is anatomical snuff box?

Ans. It is a space bounded laterally by the tendons of **abductor pollicis longus** and **extensor pollicis brevis**; medially by the tendon of **extensor pollicis longus**; floor is formed by the styloid process of radius, scaphoid, trapezium and base of 1st metacarpal bone.

9. What is the relationship of the radial artery with the radial nerve in the forearm?

Ans. In the middle 1/3rd of the forearm the radial nerve lies in close relation to the lateral side of the artery.

10. What are the branches of radial artery in forearm?

Ans.
i. Radial recurrent artery.
ii. Muscular branches.
iii. Superficial palmar branch.

6. ULNAR ARTERY IN FOREARM (FIG. 7.5)

i. 1st point is at the point of bifurcation of brachial artery (vide point number 9)
ii. 2nd point is given at the junction of upper 1/3rd and lower 2/3rd of a line joining the base of medial epicondyle of the humerus with a point on the lateral side of the pisiform on the hypothenar eminence or at the styloid process of the ulna.

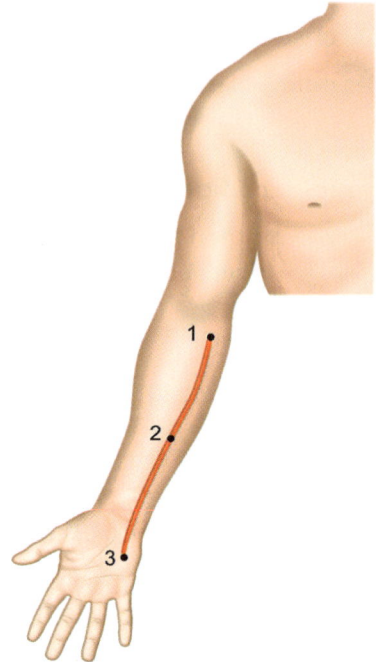

Fig. 7.5: Ulnar artery in forearm

iii. 3rd point is on the hypothenar eminence just lateral to the pisiform bone.

Now, join the 1st point with the 2nd point with the convexity medially and join the 2nd point with the 3rd point in a straight vertical line.

These points are joined in two parallel lines which represents the ulnar artery in forearm.

PROBABLE QUESTIONS AND ANSWERS

1. What is the point of beginning of ulnar artery?

Ans. The ulnar artery begins in the cubital fossa 1 cm below the bend of the elbow at the level of neck of the radius as one of the terminal branches of the brachial artery.

2. How does the artery enter into the front of the forearm from the cubital fossa?

Ans. The ulnar artery arises from the brachial artery in the cubital fossa and then passes downwards and medially deep to the ulnar head (deep head) of the pronator teres muscle to appear in the front of the forearm.

3. Which structure separates the ulnar artery from the median nerve while passing from cubital fossa to front of the forearm?

Ans. Ulnar head (deep head) of pronator teres muscle.

Q4. What is the relationship of the unlar artery with the ulnar nerve in the forearm?

Ans. In the lower 2/3rd of the forearm the ulnar nerve lies medial to the ulnar artery. Note that both the ulnar artery and radial artery are accompanied by a pair of venae comitantes along their entire extent.

Q5. How does the ulnar artery enter into the palm?

Ans. The ulnar artery and ulnar nerve enter the palm between the superficial part and main part of the flexor retinaculum and then the artery ends by dividing into superficial branch and deep branch.

Q6. What are the branches of ulnar artery in forearm?

Ans.
 i. Anterior ulnar recurrent artery.
 ii. Posterior ulnar recurrent artery.
 iii. Common interosseous artery which divides into anterior interosseous artery and posterior interosseous artery.
 iv. Muscular braches.

Q7. What is the largest branch of ulnar artery?

Ans. Common interosseous artery.

Q8. Interosseous recurrent artery belongs to which artery?

Ans. Interosseous recurrent artery is a branch of posterior recurrent artery.

7. SUPERFICIAL PALMAR ARCH (FIG. 7.6)

 i. 1st point is given on the lateral side of the pisiform bone
 ii. 2nd point is given on the hook of the hamate about 2 cm. below the 1st point.

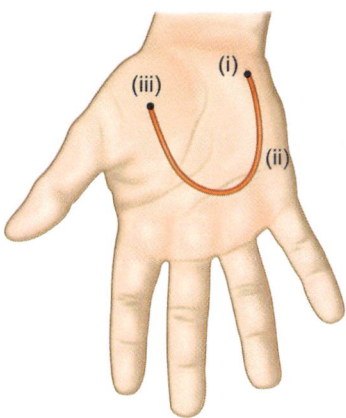

Fig. 7.6: Superficial palmar arch

iii. 3rd point is given on the thenar eminence at its middle in a line which corresponds the web between the index and middle finger.

Now join these three points with the convexity distally not extending below the level of the distal border of the extended thumb. The lines are drawn in two parallel lines which represents the superficial palmar arch.

PROBABLE QUESTIONS AND ANSWERS

1. How the superficial palmar arch is formed?

Ans. Superficial palmar arch is formed **mainly** by the superficial terminal branch of ulnar artery and **partly** by superficial palmar branch of radial artery.

2. Which terminal branch of ulnar artery is the continuation of ulnar artery?

Ans. Superficial terminal branch is the direct continuation of the ulnar artery.

3. Which artery provides main contribution to form the superficial palmar arch?

Ans. Superficial palmar branch of the ulnar artery provides main contribution to form the arch.

4. What is the direction of convexity of the arch?

Ans. The arch is convex distally at the level with the distal border of outstretched thumb.

5. What are the branches of superficial palmar arch?

Ans.
  i. Palmar digital artery for ulnar side of little finger.
  ii. Three common palmar digital arteries for the adjacent sides of index—middle, middle-ring and ring-little fingers.

6. How the deep palmar arch is formed?

Ans. Deep palmar arch is formed mainly by the radial artery and a little contribution from the deep branch of the ulnar artery.

Note that superficial palmar arch lies in between palmar aponeurosis and flexor tendons and deep palmar arch lies between flexor tendons and interossei.

8. FLEXOR RETINACULUM (FIG. 7.7)

i. 1st point is given at pisiform or (1).
ii. 2nd point is given at the tubercle of scaphoid (2).
iii. 3rd point is given on the crest of the trapezium (3).
iv. 4th point is given at the hook of the hamate (4).

Surface Anatomy: Upper Limb (Superior Extremity)

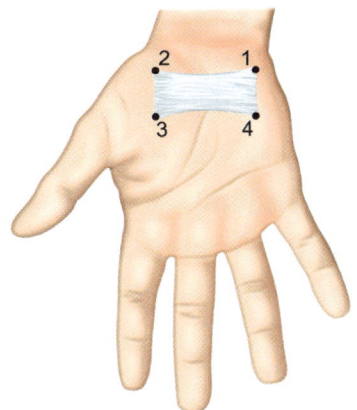

Fig. 7.7: Flexor retinaculum

Join these points so that the line joining the first two points forms a curve line with concavity upward and the line joining the last-two joints forms a curve line with concavity downwards.

PROBABLE QUESTIONS AND ANSWERS

1. What is flexor retinaculum?

Ans. It is the condensation of antebrachial fascia below the wrist.

2. What is its function?

Ans. It retains the flexor tendons in position while they act.

3. What is its shape?

Ans. It is a square-shaped structure, each side being about 1.25 cm.

4. What are its parts?

Ans. Two parts:
  i. Superficial part.
  ii. Deep (main) part.

5. What are the attachments of flexor retinaculum?

Ans. Medially: Pisiform bone and hook of the hamate.
Laterally: Tubercle of scaphoid and the crest of trapezium.

6. What are the structures passing superficial to the flexor retinaculum?

Ans.
i. Tendon of palmaris longus.
ii. Palmar cutaneous branch of median nerve.
iii. Palmar cutaneous branch of ulnar nerve.
iv. Ulnar vessels.
v. Ulnar nerve.

7. What are the structures passing deep to flexor reticulum (carpal tunnel)?

Ans.
i. The median nerve.
ii. Tendon of flexor digitorum superficialis et profundus.
iii. Tendon of flexor pollicis longus.
iv. Ulnar bursa.
v. Radial bursa.

8. What is carpal tunnel syndrome?

Ans. The compression of the median nerve in the carpal tunnel by long continued swelling of the synovial sheaths or bony pathology gives rise to the motor and sensory symptoms in the hand which constitute the carpel tunnel syndrome.

9. What are the motor and sensory symptoms of carpal tunnel syndrome?

Ans. **Motor:** Weakness and wasting of the thenar muscles with loss of power of opposition.
Sensory: Loss of cutaneous sensation of the palmar surface of lateral 3½ fingers.

10. Do you find any other retinaculum of hand?

Ans. Yes, it is extensor retinaculum on the back of the wrist.

CHAPTER 8
Surface Anatomy: Lower Limb (Inferior Extremity)

- ❏ Points
    1. Adductor Tubercle
    2. Tuberosity of Navicular
    3. Medial Malleolus
    4. Lateral Malleolus
- ❏ Lines
    1. Popliteal Artery
    2. Anterior Tibial Artery
    3. Posterior Tibial Artery
    4. Arteria Dorsalis Pedis
    5. Tibial Nerve in Popliteal Fossa
    6. Common Peroneal Nerve
    7. Deep Peroneal (Anterior Tibial) Nerve

LESSON 1: POINTS

1. ADDUCTOR TUBERCLE (FIG. 8.1)

i. The thigh is abducted and laterally rotated with the hip and knee slightly flexed.
ii. Now, follow the cord-like tendon of adductor magnus downwards with the tip of the middle finger.
iii. The finger comes in contact with the adductor tubercle a little above the medial condyle of the femur.

PROBABLE QUESTIONS AND ANSWERS

1. **What is adductor tubercle?**
   Ans. It is a small bony projection above the medial epicondyle of the femur.

2. **Which bony part of the femur is continuous with the tubercle?**
   Ans. The tubercle is continuous above with the medial supracondylar line and then with the medial lip of the linea aspera of the shaft of the femur.

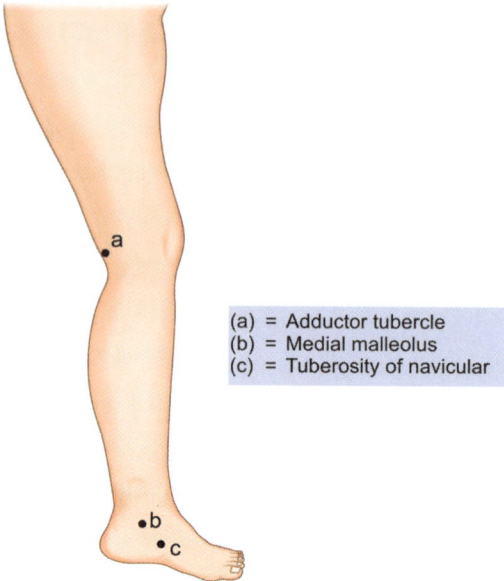

Fig. 8.1: Bony landmarks of lower limb

(a) = Adductor tubercle
(b) = Medial malleolus
(c) = Tuberosity of navicular

3. What is linea aspera?

Ans. It is the posterior border of the middle-third of the shaft of the femur.

4. What are the structures attached to the adductor tubercle?

Ans.
i. Insertion of the ischial part (hamstring part) of adductor magnus.
ii. Attachment of the tibial collateral ligament.

5. What are the importances of the adductor tubercle?

Ans.
i. The epiphyseal line of the lower end of the femur passes through the tubercle.
ii. It is an important bony landmark for surface anatomy.

6. What is epiphyseal line?

Ans. It is the line of union between the diaphysis (which forms the shaft) and epiphysis (which forms the ends) of the bone.

2. TUBEROSITY OF NAVICULAR (FIG. 8.1)

Take a point 2.5 cm below the tip of the medial malleolus.
Then another point is taken 2.5 cm in front of the previous point. This point is the point for tuberosity of navicular bone.

Surface Anatomy: Lower Limb (Inferior Extremity)

PROBABLE QUESTIONS AND ANSWERS

1. What is navicular bone?

Ans. It is one of the tarsal bones of the foot.

2. How many tarsal bones are there?

Ans. There are seven tarsal bones in each foot.

3. In which row of the tarsal bones the navicular bone lies?

Ans. It lies in the middle row of the tarsal bones.

4. What are the bones of the proximal and distal row?

Ans. Talus and calcaneum are in the proximal row. Cuboid and three cuneiforms are in the distal row.

5. On which surface of the navicular bone do you get its tuberosity?

Ans. Tuberosity of navicular is a projection on its medial surface.

6. Which muscle is attached to the tuberosity?

Ans. 2/3rd of the tendon of tibialis posterior are attached to the tuberosity of the navicular. Remaining 1/3rd is attached to the all tarsal and metatarsal bones except talus and 1st metatarsal bone.

7. When does the navicular bone ossify?

Ans. Navicular and three cuneiforms ossify within 2–3 years after birth (talus, calcaneus and cuboid ossify before birth).

3. MEDIAL MALLEOLUS (FIG. 8.1)

Mark a point on the lower end of the subcutaneous medial surface of the shaft of the tibia.

PROBABLE QUESTIONS AND ANSWERS

1. What is medial malleolus?

Ans. Medial malleolus as a downward and medially projected part from the lower end of the tibia.

2. What are the muscles related to medial malleolus?

Ans. Tibialis anterior passes anterior to it and tibialis posterior passes posterior to it.

13. Which important ligament is attached to medial malleolus?

Ans. The **deltoid ligament** is attached at its lower margin.

14. What is the relation of medial malleolus with great saphenous vein?

Ans. The vein passes upward about 2.5 cm, in front of the medial malleolus (short saphenous vein passes below and behind the lateral malleolus).

15. Through which part of the lower end of tibia the lower epiphyseal line passes?

Ans. The lower epiphyseal line passes through the base of the medial malleolus and is represented by a circular line.

4. LATERAL MALLEOLUS (FIG. 8.2)

Take a point on the projected part of the lateral side of the ankle.

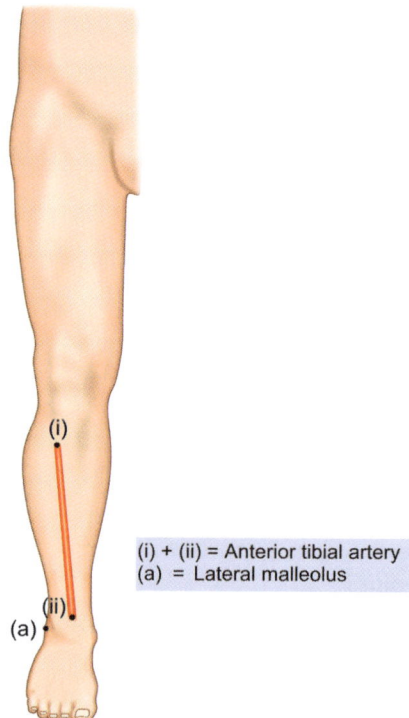

(i) + (ii) = Anterior tibial artery
(a) = Lateral malleolus

Fig. 8.2: Surface markings of lateral malleolus and anterior tibial artery

Surface Anatomy: Lower Limb (Inferior Extremity)

PROBABLE QUESTIONS AND ANSWERS

1. What is lateral malleolus?

Ans. It is the expanded lower end of the fibula.

2. What are the presenting parts of lateral malleolus?

Ans.
i. Three surfaces—lateral, medial and posterior.
ii. Border—anterior.
iii. Tip.

3. Is this malleolus articular?

Ans. **Yes**, there is a triangular articular facet on the its medial surface which articulates with the lateral surface of the body of the talus.

4. What is malleolar fossa?

Ans. It is a rough depression on the medial surface of the lateral malleolus posteroinferior to the articular facet.

5. What are the ligaments attached to the malleolus?

Ans.
i. Anterior and posterior talofibular ligaments.
ii. Posterior tibiofibular ligament (inferior transverse ligament)
iii. Lateral calcaneofibular ligament.
iv. Superior peroneal retinaculum.

6. What are the structures lodging in the groove on its posterior surface?

Ans. The groove lodges the tendons of peroneus longus and peroneus brevis.

7. What is the function of lateral malleolus in eversion of foot?

Ans. It acts as a pully for the tendons of peroneus longus and peroneus brevis.

LESSON 2: LINES

1. POPLITEAL ARTERY (FIG. 8.3)

i. Draw a horizontal line on the back of the thigh at the junction of the upper 2/3rd and lower 1/3rd.

↓

Take midpoint of that line

↓

Put a point 2.5 cm medial to the mid point (i)

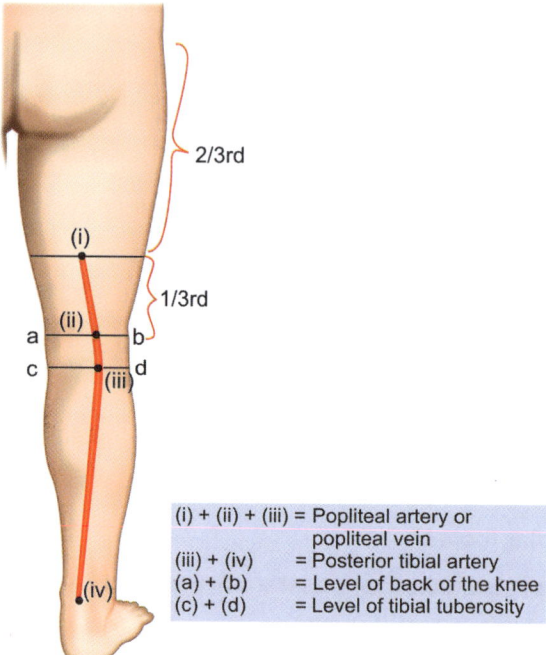

Fig. 8.3: Popliteal artery and posterior tibial artery

    ii. Second point is the midpoint of the transverse line drawn on the back of the knee joint (ii)

    iii. The third point is the mid point of the transverse line drawn on the back of the leg at the level of tibial tuberosity (iii).

Now, join the points (i), (ii) and (iii) by two parallel lines passing downwards and laterally.

PROBABLE QUESTIONS AND ANSWERS

1. What is the origin of popliteal artery?

Ans. Popliteal artery is the continuation of femoral artery.

2. What is the extent of the popliteal artery?

Ans. It extends from the 5th osseoaponeurotic opening of adductor magnus to the lower border of popliteus.

3. What is the length of this artery?

Ans. About 20 cm.

Surface Anatomy: Lower Limb (Inferior Extremity)

4. How does it end?

Ans. It ends by dividing into anterior and posterior tibial arteries.

5. What are the variations of its termination?

Ans.
i. It may terminate at the upper border of popliteus instead of its lower border.
ii. It may be terminated by dividing into anterior tibial and peroneal arteries.
iii. It may trifurcate into anterior tibial, posterior tibial and peroneal arteries.

6. What are the branches of popliteal artery?

Ans.
i. Cutaneous branches.
ii. Muscular branches.
iii. Genicular:
   a. Superior medial and superior lateral genicular arteries.
   b. Inferior medial and inferior lateral genicular arteries.
   c. Middle genicular artery (pierces the oblique popliteal ligament)

7. What is the relation of popliteal artery with the popliteal vein and tibial nerve in the popliteal fossa?

Ans. Tibial nerve is superficial, popliteal vein is in the middle and the popliteal artery is at the deep.

8. What is the relationship between the popliteal artery and popliteal vein?

Ans. In the upper part of its course, the vein lies **lateral** to the artery → then **behind** the artery → in the lower part, the vein lies **medial** to the artery. In other words, the vein crosses the artery from lateral to medial side passing superficial or behind the artery.

9. How the artery is palpated?

Ans. The popliteal artery is palpated by flexing the knee with deep pressure on the middle of the popliteal fossa.

2. ANTERIOR TIBIAL ARTERY (FIG. 8.2)

i. Palpate the head of the fibula → then take a point 2.5 cm to the medial side of the head (i)
ii. Take another point midway between two malleoli (ii)
Now, join these two points (i) and (ii) by two lines. This represents anterior tibial artery.

PROBABLE QUESTIONS AND ANSWERS

1. What is the origin of anterior tibial artery?

Ans. It is one of the terminal branches of popliteal artery.

2. What is the site of beginning of the anterior tibial artery?

Ans. At the lower border of popliteus muscle.

3. How does this artery appear in the anterior crural region?

Ans. After arising from the popliteal artery, this artery passes forward between the two heads of tibialis posterior and through an oval gap above the interosseous membrane. Thus, it appears in the anterior crural region.

4. What is the relation of this artery with the neck of the fibula?

Ans. This artery runs on the medial side of the neck of the fibula.
(Lateral side of the neck is related to common peroneal nerve)

5. How many anterior tibial veins are there?

Ans. There are two anterior tibial veins which run one on either side of the anterior tibial artery as **venae comitantes**.

6. Which tendon crosses this artery in front of the leg?

Ans. The tendon of extensor hallucis longus crosses this artery from lateral to medial side above the ankle.

7. Name the branches of the artery.

Ans.
  i. Anterior tibial recurrent artery.
  ii. Posterior tibial recurrent artery.
  iii. Muscular branches.
  iv. Medial malleolar artery.
  v. Lateral malleolar artery.

8. What is the fate of this artery or how does this artery end?

Ans. It continues **as arteria dorsalis pedis** on the dorsum of the foot.

3. POSTERIOR TIBIAL ARTERY (FIG. 8.3)

i. Take a midpoint of the transverse line drawn on the back of the leg at the level of tibial tuberosity (iii). It is the point of ending of popliteal artery.

Surface Anatomy: Lower Limb (Inferior Extremity)

ii. Put another point midway between the medial malleolus and insertion of tendo calcaneous (identified by the prominence of the heel) (iv). Now join (iii) and (iv) by two lines which represents posterior tibial artery.

PROBABLE QUESTIONS AND ANSWERS

1. What is the origin of posterior tibial artery?

Ans. It is one of the terminal branches of popliteal artery.

2. Which terminal branch of popliteal artery is larger?

Ans. Posterior tibial artery is larger than anterior tibial artery.

3. What is the level of beginning of posterior tibial artery?

Ans. It begins at the lower border of popliteus at the level of the neck of the fibula, in line with the tibial tuberosity.

4. Does this artery accompany venae comitantes?

Ans. Yes, it is accompanied by a pair of venae comitantes.

5. What is the relation of this artery with the tibial nerve?

Ans. In the upper part of the leg, the tibial nerve lies medial to the artery. Then the nerve crosses the artery superficially and finally in the lower part of the leg, it lies lateral to the artery. In other words, the tibial nerve crosses the posterior tibial artery from medial to lateral side passing superficial (behind) to the artery.

6. How does the artery end?

Ans. It ends by dividing into two terminal branches:
  i. Medial plantar artery
  ii. Lateral plantar artery.

7. What are the branches of posterior tibial artery?

Ans.
  i. Circumflex artery.
  ii. Nutrient artery to tibia.
  iii. Peroneal artery.
  iv. Muscular branches.
  v. Communicating branch.
  vi. Medial malleolar branch.
  vii. Calcaneal branch.
  viii. Two terminal branches:
    a. Medial plantar
    b. Lateral planter arteries.

8. What is the nutrient artery to fibula?

Ans. It is a branch from the peroneal artery.

4. ARTERIA DORSALIS PEDIS (FIG. 8.4)

i. Take a midpoint between the two malleoli (a).
   (**Remember** that this point must be lateral to the tendon of extensor hallucis longus)
ii. Put another point at the proximal end of the first metatarsal space (b).
   Now, join these two points (a) and (b) by two parallel lines with little breadth which represents the artery.

PROBABLE QUESTIONS AND ANSWERS

1. How does this artery begin?

Ans. It begins as a continuation of anterior tibial artery distal to the ankle.

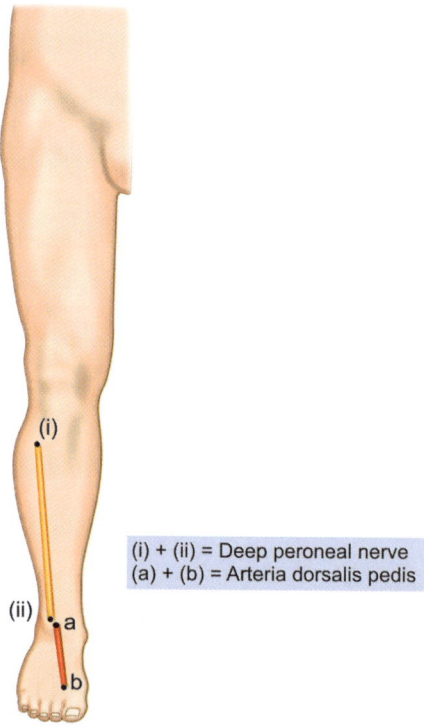

(i) + (ii) = Deep peroneal nerve
(a) + (b) = Arteria dorsalis pedis

Fig. 8.4: Deep peroneal nerve and arteria dorsalis pedis

2. Where do you palpate this artery?

Ans. It is palpated on the medial side of the dorsum of the foot between the tendon of extensor hallucus longus medially and tendon of extensor digitorum longus for the second toe laterally.

3. What are the branches of this artery?

Ans.
i. Lateral tarsal artery.
ii. Medial tarsal artery.
iii. First dorsal metatarsal artery.
iv. Arcuate artery.

4. Which artery gives off 2nd, 3rd, and 4th dorsal metatarsal arteries?

Ans. These are the branches of **arcuate artery**.

5. What is the fate of this artery?

Ans. It passes between the two heads of first dorsal interosseous muscle and enters the sole. Here it forms plantar arch by joining with the deep branch of lateral plantar artery.

5. TIBIAL NERVE IN POPLITEAL FOSSA (FIG. 8.5)

i. Take a midpoint of the transverse line drawn on the back of the thigh at the junction of its upper 2/3rd and lower 1/3rd (a)
ii. Take a midpoint of the transverse line drawn on the back of the leg at the level of tibial tuberosity (b)
Now, join points (a) and (b) by a single line which represents tibial nerve in popliteal fossa.

PROBABLE QUESTIONS AND ANSWERS

1. Whose branch is the tibial nerve?

Ans. It is the larger terminal branch of sciatic nerve.

2. What is the other terminal branch of sciatic nerve?

Ans. Common peroneal or lateral popliteal nerve.

3. What is the other name of tibial nerve?

Ans. Medial popliteal nerve.

4. What is the root value of tibial nerve?

Ans. Ventral branches of ventral rami of L_4, L_5, S_1, S_2, S_3 spinal nerves.

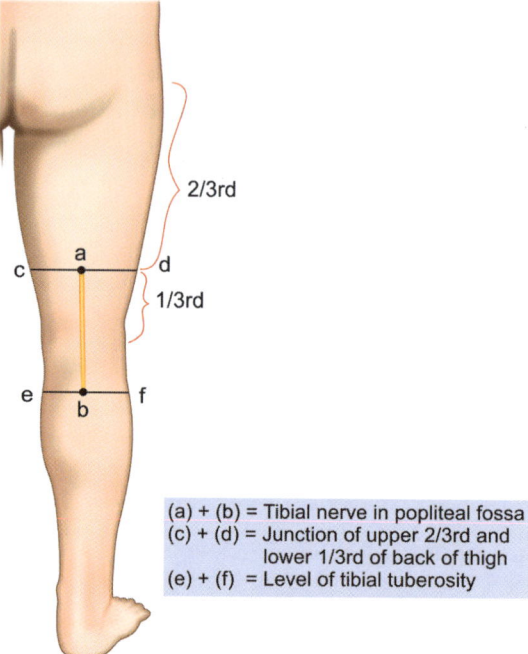

Fig. 8.5: Tibial nerve in popliteal fossa

5. What is the position of this nerve in popliteal fossa?

Ans. This nerve passes vertically downward along the midline of popliteal fossa. During the course, the nerve crosses behind the popliteal vessels in the middle of the fossa from lateral to medial side.

6. What are the branches of tibial nerve in popliteal fossa?

Ans.
a. Muscular branches to:
   i. Medial head of gastrocnemius.
   ii. Lateral head of gastrocnemius.
   iii. Soleus.
   iv. Popliteus.
   v. Plantaris.
b. **Cutaneous** branch: Sural nerve.
c. **Articular** (genicular):
   i. Superior medial genicular.
   ii. Inferior medial genicular.
   iii. Middle genicular.
d. **Vascular** branch.

Surface Anatomy: Lower Limb (Inferior Extremity)

7. What is the mode of supply of popliteus muscle?

Ans. The nerve to popliteus, a branch from tibial nerve in the popliteal fossa, winds round the distal border of the muscle to supply from its deep surface.

8. What is the mode of supply of middle genicular nerve?

Ans. This nerve pierces the oblique popliteal ligament and supplies the interior of knee joint.

9. What is the fate of sural nerve?

Ans. It pierces the deep fascia of the leg and joins with the sural communicating branch of common peroneal nerve.

10. What is the fate of tibial nerve?

Ans. The tibial nerve enters the posterior crural region under cover of the tendinous origin of soleus muscle.

6. COMMON PERONEAL NERVE (FIG. 8.6)

i. Take a midpoint of the transverse line on the back of the thigh at the junction of its upper 2/3rd and lower 1/3rd (a).

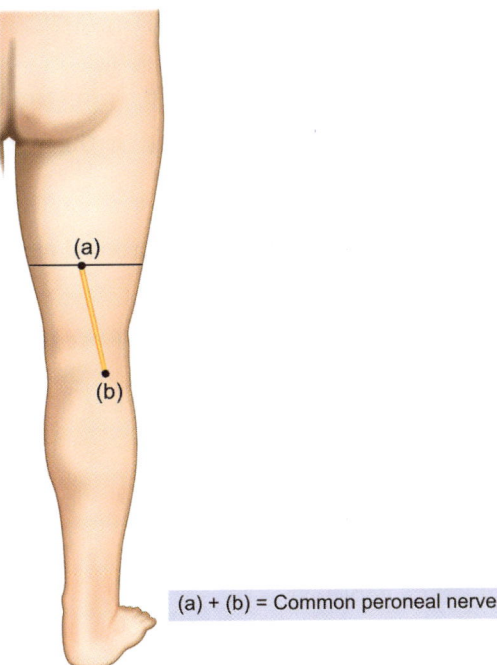

Fig. 8.6: Common peroneal nerve

ii. Put another point at the neck of the fibula (b).
Now, join these points (a) and (b) by a single line along the medial side of the tendon of biceps femoris which represents common peroneal nerve.

PROBABLE QUESTIONS AND ANSWERS

1. Whose branch is the common peroneal nerve?

Ans. It is the smaller terminal branch of sciatic nerve.

2. What is the other terminal branch of sciatic nerve?

Ans. Tibial nerve.

3. Where does the sciatic nerve divide into two terminal branches?

Ans. Sciatic nerve is divided on the back of the thigh at the junction of its upper 2/3rd and lower 1/3rd.

4. What is the other name of common peroneal nerve?

Ans. Lateral popliteal nerve.

5. What is the root valve of this nerve?

Ans. Dorsal branches of ventral rami of L_4, L_5, S_1 and S_2 spinal nerves.

6. Where does the common peroneal nerve divide into terminal branches?

Ans. The nerve curves forward on the **lateral side of the neck of the fibula** deep to **peroneus longus** and divides into two terminal branches.

7. What are the terminal branches of common peroneal nerve?

Ans. i. **Deep peroneal** (anterior tibial) nerve
ii. **Superficial peroneal** (musculocutaneous) nerve.

8. What are the branches of common peroneal nerve?

Ans. a. **Cutaneous** branches:
  i. Sural communicating nerve.
  ii. Lateral cutaneous branch of the calf (lateral sural nerve)
b. **Articular** (genicular) branches
  i. Superior lateral genicular.
  ii. Inferior lateral genicular.
  iii. Recurrent genicular.

9. What are the structures supplied by the recurrent genicular nerve?

Ans.
i. Tibialis anterior muscle.
ii. Anterolateral part of the capsule of the knee joint.
iii. Superior tibiofibular joint.

10. What are the manifestations observed in injury of the common peroneal nerve due to fracture of the neck of the fibula?

Ans.
i. Motor paralysis causing 'foot drop'
ii. Sensory loss on the dorsum of the foot and outer surface of the lower 1/3rd of the front of the leg.

11. What are the muscles paralyzed due to injury of common peroneal nerve?

Ans. All the muscles of extensor compartment of the leg including extensor digitorum brevis and the muscles of peroneal compartment.

12. What are the muscles of anterior and peroneal compartment of the leg?

Ans.
a. **Anterior (extensor) compartment:**
i. Tibialis anterior.
ii. Extensor hallucis longus.
iii. Extensor digitorum longus.
iv. Peroneus tertius.
b. **Peroneal compartment:**
i. Peroneus longus.
ii. Peroneus brevis.

7. DEEP PERONEAL (ANTERIOR TIBIAL) NERVE (FIG. 8.4)

i. A point is taken on the lateral side of the neck of fibula (i)
ii. Midpoint on a transverse line joining the two malleoli (ii)
Now, join these two points (i) and (ii) by a single line which represents deep peroneal nerve.

PROBABLE QUESTIONS AND ANSWERS

1. Whose branch is the deep peroneal nerve?

Ans. It is a branch of common peroneal nerve.

2. What is the bony landmark where the common peroneal nerve divides into deep and superficial peroneal nerve?

Ans. On the lateral side of the neck of the fibula.

3. What is the positional relation of the anterior tibial nerve with the anterior tibial vessels?

Ans. The nerve lies lateral to the vessels in the leg.

4. Why the deep peroneal nerve is known as nervi hesitans?

Ans. In the upper 1/3rd of the front of the leg, the deep peroneal nerve lies lateral to the anterior tibial vessels; in the middle third, the nerve lies in front of the vessels and then in the lower third, again it lies on the lateral side of the vessels. So, this nerve is known as **nervi hesitans** because of the course of the nerve; as if this nerve hesitates to cross the anterior tibial vessels from lateral to medial side.

5. What are the branches of deep peroneal nerve in the leg?

Ans.
    a. **Muscular branches** to:
        i. Tibialis anterior.
        ii. Extensor hallucis longus.
        iii. Extensor digitorum longus.
        iv. Peroneus tertius.
    b. **Articular branches** to ankle joint.

6. How does the deep peroneal nerve reach the dorsum of the foot?

Ans. The nerve reaches the dorsum of the foot beneath the extensor retinacula and lies on the lateral side of the arteria dorsalis pedis.

7. What are the terminal branches of deep peroneal nerve?

Ans.
    i. **Lateral terminal branch.**
    ii. **Medial terminal branch.**

8. What are the muscles supplied by the lateral terminal branch?

Ans.
    i. Extensor digitorum brevis.
    ii. Second dorsal interosseous muscle.

9. Which muscle is supplied by medial terminal branch?

Ans. First dorsal interosseous muscle.

10. Which terminal branch of deep peroneal nerve form pseudoganglion?

Ans. Lateral terminal branch.

11. Name other nerves which have pseudoganglion.

Ans.
    i. Posterior interosseous nerve in the forearm.
    ii. Posterior division of axillary nerve which supplies the teres minor muscle.

CHAPTER 9

Surface Anatomy: Abdomen

- Points
  1. Cardiac Orifice
  2. Pyloric Orifice
  3. Fundus of Gallbladder
  4. Appendicular Orifice
  5. McBurney's Point
  6. 4th Lumbar Spine
  7. Origin of Celiac Artery
  8. Origin of Superior Mesenteric Artery
  9. Duodenojejunal Flexure
- Lines
  1. Fundus of Stomach
  2. Lesser Curvature of Stomach
  3. Lower Border of Liver
  4. Root of the Mesentery
  5. Kidney from Back

LESSON 1: POINTS

1. CARDIAC ORIFICE (FIG. 9.1)

Put a point on the left 7th costal cartilage, 2.5 cm away from the midline.
↓
Draw two short parallel lines 1.25 cm apart sloping downwards and slightly to the left from this point.
↓
This represents the cardiac orifice.

PROBABLE QUESTIONS AND ANSWERS

1. What is the cardiac orifice?

Ans. It is an opening by which the esophagus communicates with the stomach.

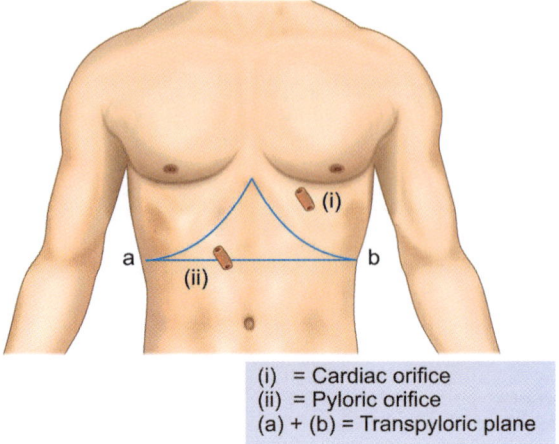

(i) = Cardiac orifice
(ii) = Pyloric orifice
(a) + (b) = Transpyloric plane

Fig. 9.1: Cardiac orifice and pyloric orifice

2. What are the orifices of stomach?

Ans. There are two orifices of stomach:
  i. Cardiac orifice
  ii. Pyloric orifice.

3. Which of these orifices is at the upper level?

Ans. Cardiac orifice is at the upper level than the pyloric orifice.

4. What is the vertebral level of the cardiac orifice?

Ans. T_{11} vertebra.

5. How much it is deep to the anterior abdominal wall?

Ans. It is about **10 cm** deep to the anterior abdominal wall.

6. How far it is away from the median plane?

Ans. It is about **2.5 cm** away from the median plane on the left side.

7. How far it is away from the incisor teeth?

Ans. It is about **40 cm** away from the incision teeth.

8. Which number of rib lies at the level of cardiac orifice?

Ans. The cardiac orifice lies behind the **left 7th costal cartilage**.

Surface Anatomy: Abdomen

9. How do you count the 7th costal cartilage?

Ans. First of all, locate the **sternal angle** by its prominence in the median plane, which is at the level of the 2nd costal cartilage. Then, count the costal cartilages from 2nd downwards and identify the 7th one.

10. What are the peritoneal relations at the level of cardiac orifice?

Ans. Anteriorly, it is covered with peritoneum, but posteriorly it is nonperitoneal over a triangular bare area.

11. What are the visceral relations at this level?

Ans. Anteriorly, it is overlapped by the **left lobe of the liver**. Posteriorly, it is connected with the **left crus of the diaphragm by the gastrophrenic ligament**.
(Remember that a part of left suprarenal gland may intervene between the cardiac orifice and the left crus).

12. Is there any sphincter around the cardiac orifice?

Ans. **No.** There is no anatomical sphincter around the cardiac orifice, but physiological sphincter is there.

2. PYLORIC ORIFICE (FIG. 9.1)

Draw the transpyloric plane
↓
Then, put a point on this plane about 1.2 cm to the right of the midline.
↓
Draw two short parallel lines from this point about 2 cm apart, directing upwards and to the right
↓
This represents the pyloric orifice.

PROBABLE QUESTIONS AND ANSWERS

1. What is pyloric orifice?

Ans. The opening by which the stomach communicates with the duodenum is the pyloric orifice.

2. How do you indicate the position of the pyloric orifice?

Ans. Its position is usually indicated by a circular groove on its surface. This circular groove is termed as **pyloric constriction** which indicates the position of the **pyloric sphincter**.

13. How do you identify its position in the living subject during operation?

Ans. It is identified by the **prepyloric vein of Mayo** which runs vertically across its anterior surface.

14. What is the vertebral level of the pyloric orifice?

Ans. It lies at the level of the lower border of L1 vertebra (transpyloric plane), provided the stomach is empty and the subject is in supine position.

15. How do you draw transpyloric plane?

The **transpyloric plane (line) of Addison** is drawn by one of the following ways:
  i. A horizontal line on the upper part of the anterior abdominal wall passing through the tips of both 9th costal cartilages.
  ii. A horizontal line passing through the midway between the xiphisternal joint and the umbilicus.
  iii. A horizontal line passing through one hand's breadth below the xiphisternal joint.

(Remember that the hand's breadth is taken of that person on whose body the line is drawn).

16. What are the relations of this pyloric area?

Ans.
  i. Anteriorly, it is covered with peritoneum of greater sac and related to the quadrate lobe of the liver.
  ii. Posteriorly, it is covered with the peritoneum of the lesser sac and related to the neck of the pancreas.

17. How is the pyloric sphincter formed?

Ans. The pyloric sphincter is formed by the deep fibers of the longitudinal muscle and the circular muscle of the stomach.

18. How does the pyloric end differ from the cardiac end of stomach?

Ans. The pyloric end is thicker, more movable and more superficial than the cardiac end. Moreover, the pyloric end is provided with an anatomical sphincter (pyloric sphincter), whereas there is no anatomical sphincter at the cardioesophageal junction though a physiological sphincter is observed. Sometimes, pyloric end is stained with bile after death.

19. Does the pyloric orifice remain open all the time?

Ans. No, it remains open only in empty stomach.

Surface Anatomy: Abdomen

10. What are the presenting parts of the stomach?

Ans. The stomach presents the following anatomical parts:
   a. Two orifices:
      i. Cardiac orifice.
      ii. Pyloric orifice.
   b. Two curvatures:
      i. Greater curvature.
      ii. Lesser curvature.
   c. Two surfaces:
      i. Anterosuperior surface.
      ii. Posteroinferior surface.
   d. Three subdivisions:
      i. Fundus (upper part)
      ii. Body (middle part)
      iii. Pyloric part (lower part).

11. What are the presenting parts of the pyloric part?

Ans. **Three parts**:
   i. Pyloric antrum (7.5 cm)
   ii. Pyloric canal (2.5 cm)
   iii. Pylorus.

12. How much is the capacity of the stomach?

Ans. Average capacity in adult is 1,000–1,500 mL.

13. How much is the capacity of the fundus of the stomach?

Ans. The fundus contains only **air** which is aspirated into the stomach with food. The capacity of the fundic air is about 50 mL.

14. How does the fundus look in a plain radiograph?

Ans. It looks black due to its content of air.

15. Where is the situation of the stomach?

Ans. The stomach is situated in the epigastrium, left hypochondrium and umbilical region.

16. What is the shape of stomach?

Ans. In cadaver, it is sickle-shaped and in the living, it is J-shaped.

3. FUNDUS OF GALLBLADDER (FIG. 9.2)

Put a point on the transpyloric plane just below the junction of the lateral border of rectus abdominis with the right costal margin.

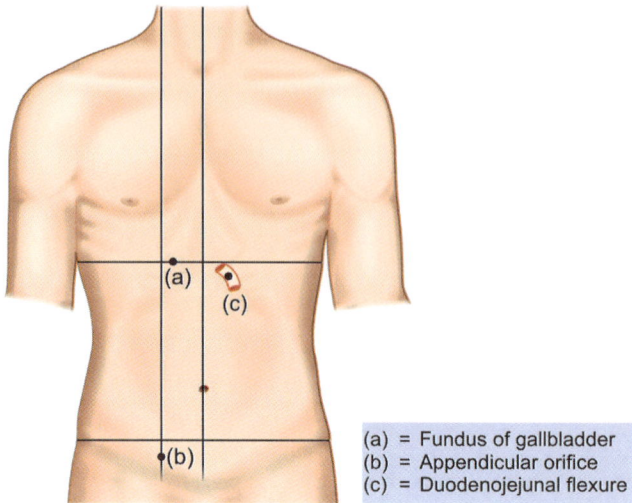

Fig. 9.2: Surface markings of fundus of gallbladder: Appendicular orifice and deodenojejunal flexure

PROBABLE QUESTIONS AND ANSWERS

1. What is gallbladder?

Ans. Gallbladder is a pyriform (pear) shaped hollow viscus situated in the fossa for the gallbladder on the undersurface of the right lobe of liver.

2. What is its function?

Ans. Its function is to store and concentrate hepatic bile by about 10 times.

3. What are its dimensions?

Ans. Its length is 7–10 cm and maximum breadth is about 3 cm.

4. What is the capacity of gallbladder?

Ans. 30–50 mL.

5. What are the parts of gallbladder?

Ans. It presents **three** parts:

    i. Fundus.   ii. Body.   iii. Neck.

6. What do you mean by fundus of gallbladder?

Ans. The fundus is the lower expanded part of gallbladder which projects beyond the inferior border of the liver.

Surface Anatomy: Abdomen

7. What is the direction of the fundus?

Ans. Fundus is directed downwards forwards and to the sight.

8. What is the vertebral level of the fundus?

Ans. Fundus lies at the level of lower border **of L_1 vertebra,** i.e. at the level of **transpyloric plane.**

9. How does the peritoneal relation of the fundus differ from the body and neck?

Ans. The fundus is surrounded by peritoneum on all sides, whereas the superior surface of the body and neck of the gallbladder is nonperitoneal.

10. What are the relations of the fundus?

Ans. i. Anteriorly: The diaphragm and the anterior abdominal wall.
ii. Posteriorly: Transverse colon (right colic flexure).

11. What is the arterial supply of the gallbladder?

Ans. The cystic artery, a branch of right branch of hepatic artery.

12. Name some other visceras of the abdomen where do you get fundus as a part of these visceras?

Ans. i. **Fundus of the stomach:** It is the part above the horizontal plane at the level of the cardiac notch.
ii. **Fundus of urinary bladder:** The base or posteroinferior surface of the urinary bladder is also called fundus.
iii. **Fundus of the uterus:** The part of the uterus lying above the imaginary horizontal line passing through the entrance of the two uterine tubes.

4. APPENDICULAR ORIFICE (FIG. 9.2)

Put a point 2 cm below the intersection between the transtubercular and right lateral planes.

PROBABLE QUESTIONS AND ANSWERS

1. How do you define appendicular orifice?

Ans. Appendicular orifice is a small circular opening through which the lumen of the appendix communicates with the interior of the cecum.

2. Is there any valve at this orifice?

Ans. **Yes,** this orifice is guarded by a semicircular mucous fold which is attached to the lower margin.

3. What is the other name of this valve?

Ans. This valve is known as **valve of Gerlach**.

4. How far it is away from the iliocecal orifice?

Ans. It is about 2 cm below and slightly behind the iliocecal orifice.

5. Which part of the appendix corresponds to the appendicular orifice?

Ans. Base of the appendix corresponds to the orifice.

6. What is the base of the appendix?

Ans. The base is the proximal part of the appendix which is attached to the posteromedial wall of the cecum about 2 cm below the ileocecal junction.

7. What is the anatomical and clinical importance of the base?

Ans. All teniae of the large gut converge to the base and thereby serve as a guide for the identification of appendix.

8. What are the parts of appendix?

Ans. It presents **three parts**: Base, body and the tip.

9. What is the length of the appendix?

Ans. The length varies between 2 cm and 20 cm, average being 9 cm.

10. What is the arterial supply of the appendix?

Ans. The appendix is supplied by **appendicular artery**, a branch of inferior division of ileocolic artery.
(Remember that appendicular artery is an end-artery)

5. MCBURNEY'S POINT (FIG. 9.3)

Locate the right anterior superior iliac spine.
↓
Then draw a line joining the right anterior superior iliac spine to the umbilicus (**right spino-umbilical line**).
↓
Put a point on this line at the junction of its lateral 1/3rd and medial 2/3rds
↓
This is McBurney's point

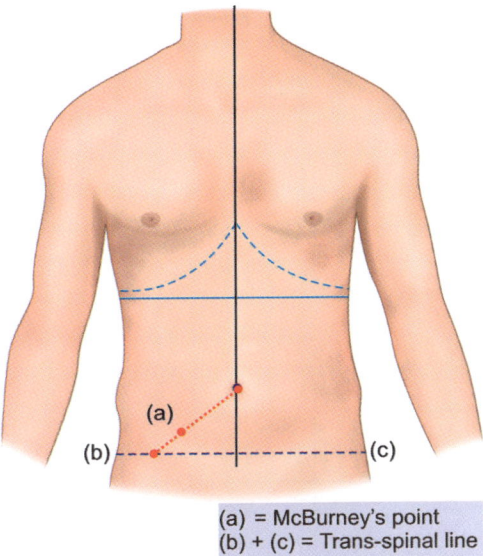

(a) = McBurney's point
(b) + (c) = Trans-spinal line

Fig. 9.3: McBurney's point

PROBABLE QUESTIONS AND ANSWERS

1. What is the clinical importance of McBurney's point?

Ans. Maximum tenderness is elicited on this point in a patient suffering from appendicitis.

2. Which part of the appendix is usually pressed to elicit maximum tenderness on McBurney's point in appendicitis?

Ans. It is the **base of the appendix**.

3. What are the anatomical factors responsible for the inflammation of the appendix?

Ans.
  i. It is a **blind tube**. So, a fecolith may obstruct its lumen precipitating the attack of appendicitis.
  ii. The appendix is supplied by an **end-artery** (appendicular artery).
  iii. Presence of numerous lymphatic follicles in the submucous coat of the appendix (hence the appendix is considered as the **abdominal tonsil**).
  iv. Muscular layer is deficient at certain places (**hiatus muscularis**) through which the infection from the lumen spreads outside.

6. 4TH LUMBAR SPINE (FIG. 9.4)

Locate the highest point of iliac crest of hip bones of both sides.
↓
A transcrestal line is drawn jouning the highest point of both iliac crests.
↓
Then, put a point on this line in the midline on the back, which represents the L_4 spine.

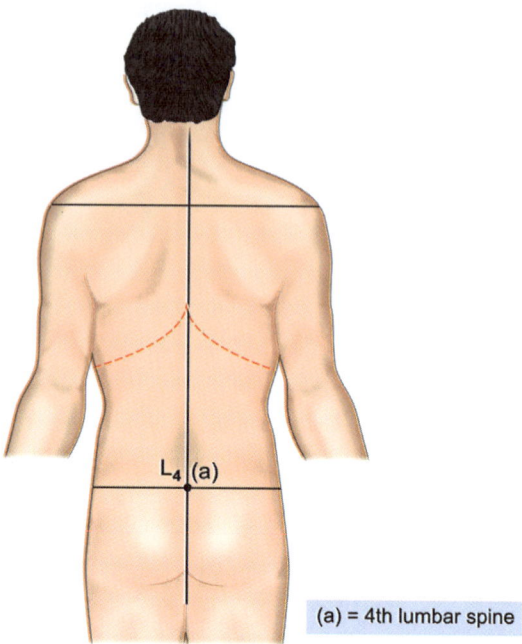

Fig. 9.4: 4th lumbar spine

PROBABLE QUESTIONS AND ANSWERS

1. What is the importance to locate the L_4 spine?

Ans. In lumbar puncture, the needle is introduced into the subarachnoid space between the spines of L_3 and L_4 vertebrae. So, it is useful in counting the lower vertebral spines.

2. What is transtubercular line?

Ans. It is a transverse line, drawn by joining the iliac tubercles of the outer lip of the iliac crests of both sides. This line passes through the upper border of L_5 spine.

3. What is trans-spinal line?

Ans. It is a transverse line, drawn by joining the anterosuperior iliac spines of both sides. Its vertebral level is at the junction of S_2 and S_3 vertebra.

4. What is supracrestal line/plane?

Ans. The **transcrestal** line is also called supracrestal line which passes between the highest point of iliac crests of both sides.

7. ORIGIN OF CELIAC ARTERY (FIG. 9.5)

Draw transpyloric line by joining the tips of both 9th costal cartilages.
↓
Then, draw a small circle about ½" to 1" above the transpyloric plane, slightly to the left of the midline.
↓
This represents the origin of **celiac artery.**

PROBABLE QUESTIONS AND ANSWERS

1. What is the origin of celiac artery?

Ans. Celiac artery is one of the ventral branches of abdominal aorta.

2. What are the other ventral branches of aorta?

Ans. Superior mesenteric artery and inferior mesenteric artery.

3. What is the vertebral level of origin of celiac artery?

Ans. The celiac artery arises opposite the lower border of T_{12} vertebra.

4. What is the length of this artery?

Ans. The celiac artery is about 1.25 cm long.

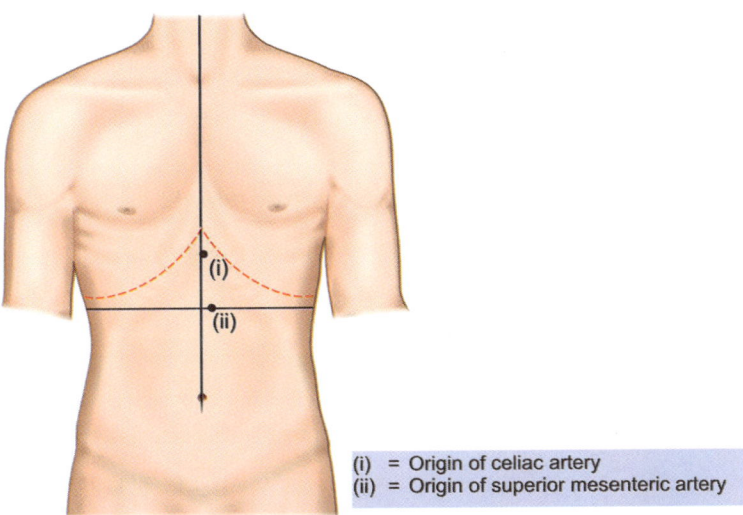

(i) = Origin of celiac artery
(ii) = Origin of superior mesenteric artery

Fig. 9.5: Origin of celiac artery and superior mesenteric artery

5. What is celiac plexus?

Ans. The celiac plexus is an autonomic nerve plexus which surrounds the celiac trunk (artery).

6. What are the branches of celiac artery?

Ans. Three branches:
    i. Left gastric artery.   ii. Common hepatic artery.   iii. Splenic artery.

7. Of these three branches which one is the smallest branch?

Ans. The left gastric artery is the **smallest branch** of celiac trunk.

8. Which one is the largest branch of the celiac trunk?

Ans. The splenic artery is the **largest branch** of celiac trunk.

9. What is the extent of supply of the celiac trunk?

Ans. The celiac artery (trunk) is the **artery of foregut**.

10. What may be the anomalous origin of celiac trunk?

Ans. Sometimes, the celiac artery and the superior mesenteric artery arise as a common trunk from the abdominal aorta.

8. ORIGIN OF SUPERIOR MESENTERIC ARTERY (FIG. 9.5)

Put a point on the transpyloric plane, slightly to the left of the midline.

PROBABLE QUESTIONS AND ANSWERS

1. What is the origin of superior mesenteric artery (SMA)?

Ans. The SMA is one of the ventral branches of abdominal aorta.

2. What is the vertebral level of origin of SMA?

Ans. It arises opposite the lower border of L_1 vertebra.

3. What is the distance between the origin of the celiac trunk and SMA?

Ans. It is about 1 cm. The SMA arises about 1 cm below the origin of the celiac trunk.

Surface Anatomy: Abdomen

4. What is the extent of supply of SMA?

Ans. The SMA is the **artery of midgut**.

5. What is the relation of the SMA close to its origin with the left renal vein?

Ans. The artery passes downwards in front of the left renal vein. It means that the left renal vein is sandwiched between the SMA and the abdominal aorta.

6. What is the relation of the SMA with the 3rd part duodenum?

Ans. The SMA passes in front of the uncinate process of the pancreas and the third part of duodenum, accompanied by the **superior mesenteric vein on its right side**.

7. What are the branches of the SMA?

Ans. **Branches from the convex (left) side of the artery**:
  i. Jejunal branches.
  ii. Ileal branches.

  Branches from the concave (right) side of the artery:
  i. Inferior pancreaticoduodenal artery.
  ii. Middle colic artery.
  iii. Right colic artery.
  iv. Ileocolic artery.

8. What is the first branches of the SMA?

Ans. Inferior pancreaticoduodenal artery is the first branch of SMA.

9. What is the terminal branch of SMA?

Ans. Ileocolic artery.

9. DUODENOJEJUNAL FLEXURE (FIGS 9.2 AND 9.6)

Draw the flexure 1.25 cm below the transpyloric plane and 2.5 cm to the left of the midline.

PROBABLE QUESTIONS AND ANSWERS

1. Which part of the duodenum meets the jejunum to form duodenojejunal flexure?

Ans. The 4th part of duodenum joins with the jejunum to form the flexure.

2. What is the vertebral level of this flexure?

Ans. This flexure lies at the level of L_2 vertebra.

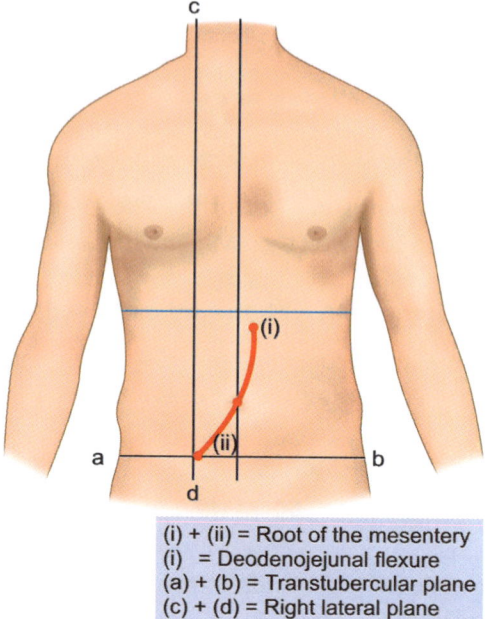

(i) + (ii) = Root of the mesentery
(i) = Deodenojejunal flexure
(a) + (b) = Transtubercular plane
(c) + (d) = Right lateral plane

Fig. 9.6: Duodenojejunal flexure and the root of the mesentery

13. How far the flexure is away from the midline?

Ans. The flexure is about 2.5 cm to the left of the midline.

14. Which factor keeps the flexure in position?

Ans. The suspensory muscle of **duodenum** (also known as **ligament of Treitz**) keeps the flexure in position.

15. What is suspensory muscle of duodenum?

Ans. It is a fibromuscular band developed from the **superior retention band** of the U-shaped loop of primitive midgut.

16. What is the attachment of this ligament?

Ans. It arises from the right crus of the diaphragm and attached to the posterosuperior surface of the duodenojejunal flexure.

17. Which artery may be encircled by this ligament?

Ans. The ligament sometimes encircles the celiac artery.

18. What is the structural feature of the suspensory muscle of duodenum?

Ans. The upper third consists of striated muscle, middle-third consists of elastic tissue and the lower-third consists of unstriped (smooth) muscle.

Surface Anatomy: Abdomen

LESSON 2: LINES

1. FUNDUS OF STOMACH (FIG. 9.7)

Draw the cardiac orifice (procedure already described in Lesson 1 (Point 1) of this chapter)—
(i)
↓
Put a point over the apex of the heart—(ii)
This is represented by a point in the left 5th intercostal space 9 cm away from the midline.
↓
Then, join the left margin of the cardiac orifice with the point of apex beat which arches upwards and to the left.
↓
This convex curve above the imaginary horizontal line extending from the cardiac orifice towards the left margin of the stomach (greater curvature) represents the fundus of the stomach.

PROBABLE QUESTIONS AND ANSWERS

 1. **Which part of the stomach is called its fundus?**

Ans. The part of the stomach above an imaginary horizontal line passing through the cardiac notch is known as fundus of stomach.

2. **What are the parts of stomach?**

Ans. There are **three parts** (subdivisions) of stomach:
    i. Fundus.   ii. Body.   iii. Pyloric part.

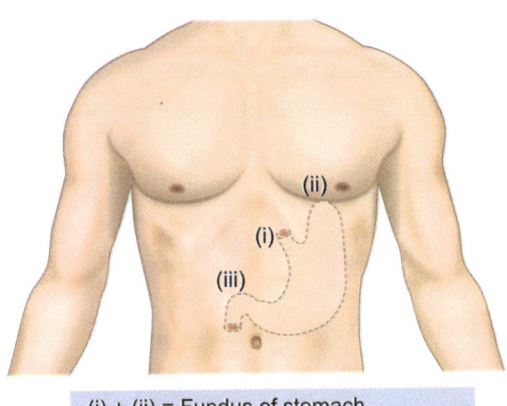

(i) + (ii) = Fundus of stomach
(i) + (iii) = Lesser curvature of stomach

Fig. 9.7: Fundus of stomach and lesser curvature of stomach

3. What does the fundus contain?

Ans. Fundus of stomach contains gas (air).

4. How much is the capacity of the fundus?

Ans. About 30 mL.

5. Where from the air appears in the fundus?

Ans. It is the swallowed atmospheric air which appears in the fundus of the stomach during eating.

6. Where is the fundus situated?

Ans. It is situated in the **Traube's space** beneath the left cupola of the diaphragm.

7. What is Traube's space?

Ans. It is a topographic area overlying the fundus of the stomach.

8. What are the boundaries of Traube's space?

Ans. **Above**: Lower border of the left lung.
Below: Left costal margin.
To the right: Inferior border of the liver.
To the left: Lateral end of spleen.

9. What are the conditions where this space may be obliterated?

Ans. This space may be obliterated in pleural effusion, hepatomegally and in splenomegaly.

10. Why a person feels cardiac distress in excessive accumulation of gas in the fundus?

Ans. If the fundus is hugely distended with gas, the apex of the heart may be pushed, as the apex of the heart lies in close relation to the fundus of the stomach.

11. How do you identify fundus of stomach in a straight radiograph of lower chest/upper abdomen taken in erect posture?

Ans. The fundic region looks black below the left cupola of the diaphragm.

12. Why it looks black in straight X-ray?

Ans. Fundus contains air and air is radiolucent. Radiolucent areas look black in a straight radiograph. So, it looks black.

13. Which artery mainly supplies the fundus?

Ans. Short gastric arteries (3–4), branches of splenic artery supply the fundus of the stomach.

14. What do you mean by fundic and body glands?

Ans. Fundic and body glands are the types of gastric glands which consist of short duct connected with two or more glands.

15. What are the types of the cells present in these glands?

Ans. **Three types of cells**:
  i. Zymogenic or chief cells.　　ii. Oxyntic cells.　　iii. Mucous neck cells.

2. LESSER CURVATURE OF STOMACH (FIG. 9.7)

Draw the transpyloric line
(This line passes through the tips of both 9th costal cartilages).
↓
Then, draw the cardiac and pyloric orifice
(Already described in Lesson 1 of this chapter).
↓
Then, join the inner borders of these two orifices in a J-shaped line which extends a little below the transpyloric plane.
↓
This line represents the lesser curvature of stomach.

PROBABLE QUESTIONS AND ANSWERS

1. How many curvatures of the stomach are there?

Ans. There are **two curvatures of stomach**:
  i. Lesser curvature.　　ii. Greater curvature.

2. How much longer is the greater curvature than the lesser curvature?

Ans. The greater curvature is about 4–5 times longer than the lesser curvature.

3. Which border of the stomach is formed by the lesser curvature?

Ans. The lesser curvature forms the right border of the stomach.

4. What is the shape of this curvature?

Ans. J-shaped.

5. What is the extent of lesser curvature?

Ans. It extends from the right margin of the esophagus up to the pylorus of the stomach.

6. Which ligament is attached to the lesser curvature?

Ans. The two layers of **lesser omentum**, also called **gastrohepatic ligament** are attached to the lesser curvature.

7. What structures lie between the two layers of lesser omentum?

Ans. The anastomosis between the right and left gastric vessels lie between the two layers of lesser omentum.

8. What is incisura angularis?

Ans. There is an angular notch at the **most dependent part of the lesser curvature,** which extends slightly below the transpyloric plane. This is known as **incisura angularis**.

9. Why the peptic ulceration is more common along the lesser curvature?

Ans.
  i. The submucosal arterial plexus is absent along the lesser curvature where long slender mucosal vessels arise straight from the intramuscular plexus and pierce the muscular layers of stomach obliquely. Occlusion of these blood vessels may produce ischemia of the gastric mucosa along the lesser curvature and forms ulceration.
  ii. Gastric canal extends along the lesser curvature. This canal allows to pass all kinds of fluid contents of the stomach including irritant fluids which may increase acid secretion.
  iii. The oxyntic cells are more abundant along the lesser curvature.

10. Do you find the both curvatures in the same plane?

Ans. No. The lesser curvature lies on a more posterior plane than the greater curvature. Moreover, the lesser curvature is more fixed.

3. LOWER BORDER OF LIVER (FIG. 9.8)

  i. A point is taken in the left 5th intercostal space, 9 cm to the left of the midline (apex of the heart).
  ii. The second point is put on the left 8th rib about 9 cm from the midline.
  iii. The third point is put on the transpyloric plane in the midline.
  iv. Put a point on the transpyloric plane just below the junction of lateral border of rectus abdominis with the right costal margin.
  v. The 5th point is put 1.25 cm below the tip of the right 10th costal cartilage.

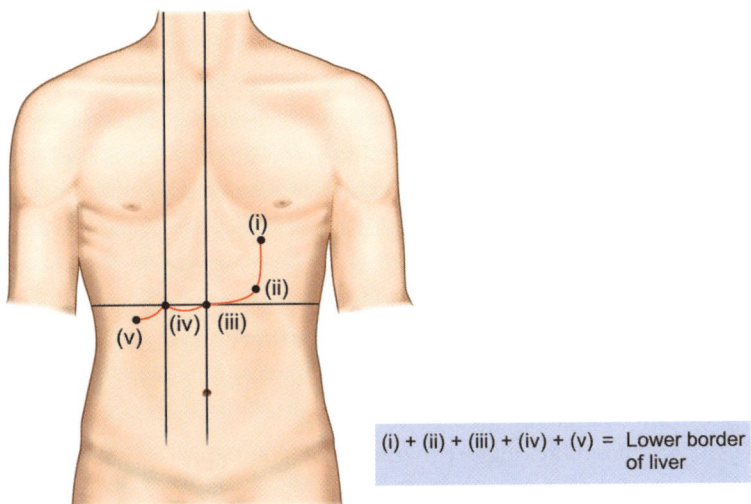

Fig. 9.8: Lower border of liver

vi. Now, join these five points with slight convexity downwards. This line represents the lower border of liver.

PROBABLE QUESTIONS AND ANSWERS

1. How many borders of liver are there?

Ans. There are **three borders** of liver. These are:
   i. Inferior border.
   ii. Posterosuperior border.
   iii. Posteroinferior border.

2. Which of these borders is sharp and prominent?

Ans. The inferior border is sharp and prominent.

3. What is the extent of this border?

Ans. It extends from the attachment of right triangular ligament to the free margin of left triangular ligament.

4. What is right triangular ligament?

Ans. The upper and lower layers of coronary ligament are continuous on the right side with right triangular ligament which connects the right lateral surface of the liver to the diaphragm.

5. What is left triangular ligament?

Ans. It is a short ligament which extends from the upper surface of left lobe of liver to the diaphragm.

6. What are the surfaces of the liver which are separated by the inferior border?

Ans. The inferior border separates the inferior surface from the anterior and right lateral surfaces of liver.

7. How many notches of the inferior border are there?

Ans. i. **Interlobar notch** which lodges the ligamentum teres hepatis lies a little to the right of the middle of the inferior border.
 ii. **Cystic notch** which lodges the fundus of gallbladder lies about 5 cm to the right of the interlobar notch.

8. What is ligamentum teres hepatis?

Ans. It is the obliterated left umbilical vein.

9. What is the fate of right umbilical vein?

Ans. It disappears in intrauterine life.

10. Do you know any other ligamentum teres?

Ans. In the hip joint we get the **ligamentum teres femoris**.

4. ROOT OF THE MESENTERY (FIG. 9.6)

 i. Put a point on the duodenojejunal flexure which is represented by a point 1.25 cm below the transpyloric plane and 2.5 cm to the left of the median plane.
 ii. Put another point on the ileocecal orifice which is represented by a point at the junction between the transtubercular plane and right lateral line.
 iii. Now, joint these two points by two lines with little convexity towards the left. This represents the root of the mesentery.

PROBABLE QUESTIONS AND ANSWERS

1. What is the mesentery?

Ans. The mesentery is a double fold of peritoneum which suspends the jejunum and ileum from the posterior abdominal wall.

2. What is its shape?

Ans. Fan-shaped.

Surface Anatomy: Abdomen

3. How many borders of the mesentery are there?

Ans. There are **two borders**:
   i. **Vertebral border or root.**
   ii. **Intestinal border or free border**.

4. What are the lengths of these borders?

Ans. Root (vertebral border) is about 15 cm long and the free border is about 20 feet long.

5. What is the breadth of the mesentery?

Ans. Maximum breadth is about 20 cm, average being 15 cm.

6. What is the extent of the root of the mesentery?

Ans. It extends from the left side of L_2 vertebra (site of duodenojejunal flexure) to the right sacroiliac joint (site of ileocecal junction).

7. What are the structures crossed by the root of the mesentery?

Ans. (From above downwards):
   i. Third part of duodenum.
   ii. Abdominal aorta.
   iii. Inferior vena cava.
   iv. Right gonadal vessels.
   v. Right ureter.
   vi. Right psoas major muscle.
   vii. Right genitofemoral nerve.
   viii. Right sacroiliac joint.

8. What are the contents of the mesentery?

Ans.
   i. Jejunum and ileum.
   ii. Jejunal and ileal arteries.
   iii. Mesenteric lymph nodes.
   vi. Plexus of nerves.
   v. Loose areolar tissue.
   vi. Fat.

9. What are the functions of the mesentery?

Ans.
   i. It suspends the jejunum and ileum from the posterior abdominal wall.
   ii. It conveys nutrition and innervation to the small gut.

10. Where from the layers of the mesentery are derived?

Ans. Both the layers of the mesentery are derived from the **greater sac**.

5. KIDNEY FROM BACK (FIG. 9.9)

   i. Draw a **horizontal line** through the T_{11} **spine**.
   ii. Draw another **horizontal line** through the L_3 **spine**.
   iii. Draw a vertical line 2.5 cm from the midline.
   iv. Draw another vertical line 10 cm from the midline.

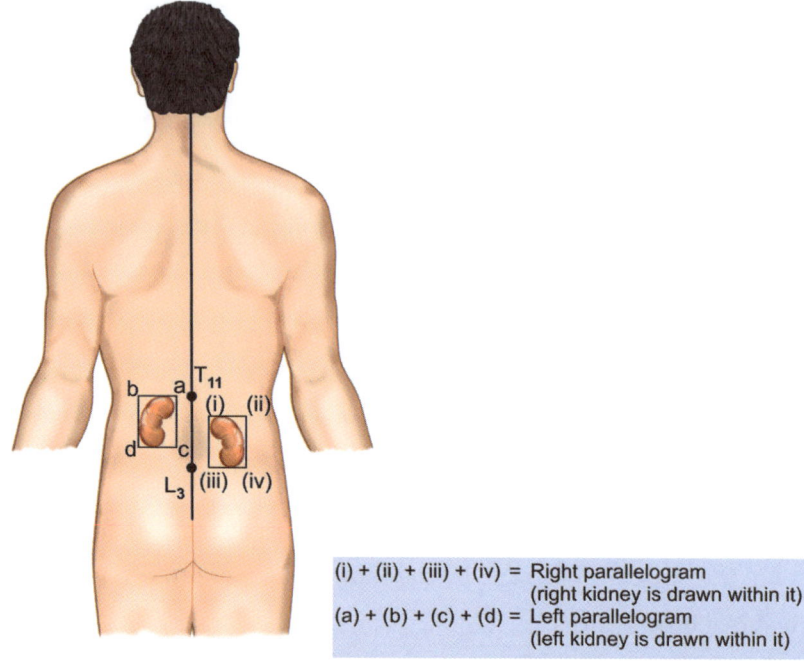

Fig. 9.9: Kidney from back

(i) + (ii) + (iii) + (iv) = Right parallelogram (right kidney is drawn within it)
(a) + (b) + (c) + (d) = Left parallelogram (left kidney is drawn within it)

The area between the horizontal lines and the vertical lines is known as **Morris parallelogram**. Now, the outline of the kidney is drawn within this parallelogram with the hilum facing medially at the level of L_1 spine, about 5 cm from the median plane. The upper pole is blunt and directed upwards and medially about 2.5 cm from median plane whereas the lower pole is pointed and directed downwards and laterally about 7.5 cm from the median plane.
(Remember that the parallelogram in the left side is 1.25 cm above the parallelogram of the right side)

PROBABLE QUESTIONS AND ANSWERS

1. What are the layers to come across to expose the kidney from back?

Ans. *From superficial to deep*:
  i. Skin.
  ii. Superficial fascia.
  iii. Posterior lamella of thoracolumbar fascia (lumbar fascia).
  iv. Erector spinae (sacrospinalis muscle).
  v. Middle layer of thoracolumbar fascia.
  vi. Quadratus lumborum muscle.
  vii. Anterior layer of thoracolumbar fascia.
  viii. Paranephric pad of fat.
  xi. Renal fascia (false capsule of kidney)
  x. Perinephric fat.

Surface Anatomy: Abdomen

2. Name the muscles between the layers of lumbar fascia?

Ans.
i. **Quadratus lumborum** between the anterior and middle layers of lumbar fascia.
ii. **Erector spinae** muscle between the middle and posterior layers of lumbar fascia.

3. What are the muscles attachmed to thoracolumbar fascia?

Ans.
i. From **anterior layer**—few fibers of the diaphragm.
ii. From the fusion of **middle layer** and **anterior layer**:
    a. Transversus abdominis
    b. Internal oblique.
iiii. From **posterior layer**:
    a. Latissimus dorsi
    b. Serratus posterior inferior.

4. Which of the ribs come in relation to the back of the kidney?

Ans.
i. On the back of the right kidney—12th rib.
ii. On the back of the left kidney—11th and 12th ribs.
(This is because of the fact that the right kidney lies at a lower level than the left kidney due to the large right lobe of the liver).

5. Name the important structures found on the back of the kidney.

Ans. *From above downwards*:
i. Subcostal vessels and nerve.
ii. Iliohypogastric nerve.
iii. Ilioinguinal nerve.
iv. 4th lumbar artery (only on right side).

6. Name the coverings of kidney.

Ans. *From within outwards*:
i. Fibrous capsule (true capsule).
ii. Perinephric fat.
iii. Renal fascia (false capsule).
iv. Paranephric fat.

CHAPTER 10

Surface Anatomy: Thorax

- **Points**
  1. Tip of 9th Costal Cartilage
  2. Sternal Angle
  3. Apex of Heart
  4. Tracheal Bifurcation

- **Lines**
  1. Anterior Border of Left Lung
  2. Right Border of Heart
  3. Left Border of Heart
  4. Arch of Aorta
  5. Superior Vena Cava

LESSON 1: POINTS

1. TIP OF 9TH COSTAL CARTILAGE (FIGS 10.1A AND B)

Put a point over the lower costal margin where the lateral border of rectus abdominis cuts it.

PROBABLE QUESTIONS AND ANSWERS

1. How do you identify the lateral border of rectus abdominis?

Ans. The lateral border is made prominent by contracting the anterior abdominal wall.

2. What is linea semilunaris?

Ans. It is a curved fibrous line forming the lateral margin of the rectus sheath.

3. What is the extent of the linea semilunaris?

Ans. It extends from the pubic tubercle to the tip of 9th costal cartilage.

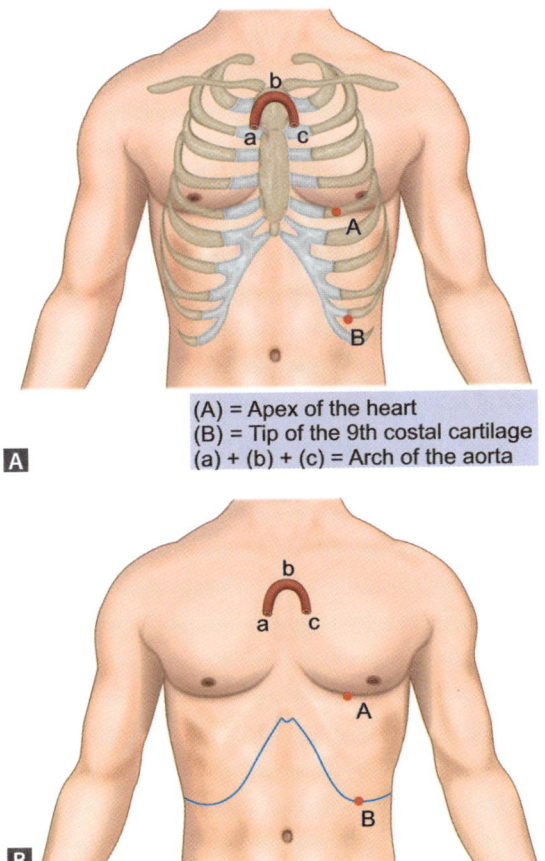

Figs 10.1A and B: Tip of the 9th costal cartilage, apex of the heart and arch of aorta

4. What is the importance of this anatomical landmark?
Ans. The transpyloric plane passes through the tips of both 9th costal cartilages.

5. Which important viscera lies below the tip of right 9th costal cartilage?
Ans. Fundus of the gallbladder.

6. Name some important structures which lie at the level of transpyloric plane.
Ans.
 i. Pyloric end of stomach.
 ii. Fundus of gallbladder.
 iii. Origin of superior mesenteric artery.
 iv. Lower end of spinal cord.

2. STERNAL ANGLE (FIGS 10.2A AND B)

Palpate the manubrium sterni and feel the transverse ridge at the junction of the manubrium and body of the sternum and then put a point over it.

PROBABLE QUESTIONS AND ANSWERS

1. Which rib articulates with the sides of the sternal angle?

Ans. Second costal cartilage articulates with the sternum at the level of sternal angle on its each side.

2. What is the vertebral level of the sternal angle?

Ans. Lower border of T_4 vertebra lies at the level of sternal angle.

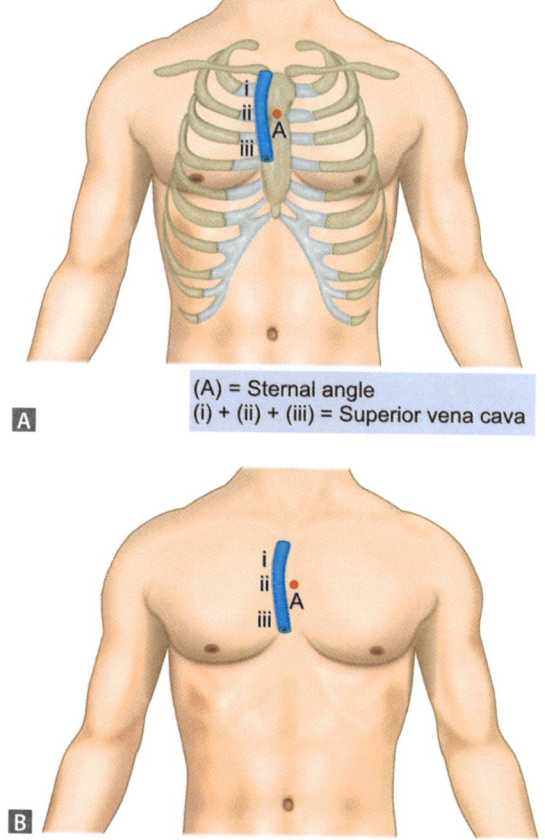

Figs 10.2A and B: Sternal angle and superior vena cava

3. What are the important features at the level of sternal angle?

Ans.
i. Second rib (second costal cartilage) articulates with the sternum at this level.
ii. It helps in counting of the ribs and intercostal spaces.
iii. It is the junction between superior and inferior mediastium.
iv. Trachea bifurcates at this level.
v. Arch of aorta begins and ends at this level.
vi. Thoracic duct deviates from right to left close to this level.
vii. It is at the level of junction between the extra- and intrapericardial part of superior vena cava (SVC).
viii. Arch of azygos vein terminates in SVC at this level.
ix. This level lies at a junction of two discontinuous dermatomes, C_4 segment above and T_2 segment below.
x. During respiration, the sternum moves forward and backward at this level.

3. APEX OF HEART (FIGS 10.1A AND B)

Put a point in the left 5th intercostal space, about 9 cm away from the midline, slightly below and medial to the left nipple.

PROBABLE QUESTIONS AND ANSWERS

1. Which chamber of the heart forms the apex?
Ans. Left ventricle forms the apex.

2. Which ventricle is thicker?
Ans. Musculature of left ventricle is three times thicker than that of right ventricle.

3. What is the thickness of musculature of left ventricle?
Ans. About 8–12 mm.

4. At which part of left ventricle the musculature is thin?
Ans. At the apex of the heart.

5. What is the shape of the apex of the heart?
Ans. Conical.

6. What is the direction of the apex?
Ans. Downwards, forwards and to the left.

7. Which part of the ventricle is most mobile?

Ans. Apex of the heart is the most mobile part of the ventricle.

8. What are the structures separating the apex from the anterior thoracic wall?

Ans. Anterior border of left lung and pleura.

9. What is apex beat?

Ans. It is a downward and outermost thrust by the heart in the left 5th intercostal space during ventricular systole.

10. What is relative position between the apex of the heart and apex beat?

Ans. Apex beat is 1.25 cm medial to the apex of heart.

4. TRACHEAL BIFURCATION (FIGS 10.3A AND B)

Palpate the sternal angle
↓
Draw vertically two parallel lines 2 cm apart with bifurcation at the level of sternal angle

PROBABLE QUESTIONS AND ANSWERS

1. What is length and breadth of trachea?

Ans.
a. **Length**: About 10 cm (cervical part 5 cm and thoracic part 5 cm).
b. **Breadth**:
  i. About 2 cm in adult male.
  ii. About 1.5 cm in adult female.
c. **Internal diameter**:
  i. 12 mm in adult
  ii. 3 mm newborn.

2. What is the extent of trachea?

Ans. From lower border of cricoid cartilage to the level of tracheal bifurcation.

3. What is the vertebral extent of trachea?

Ans. Opposite C_6 vertebra to lower border of T_4 vertebra (sternal angle).

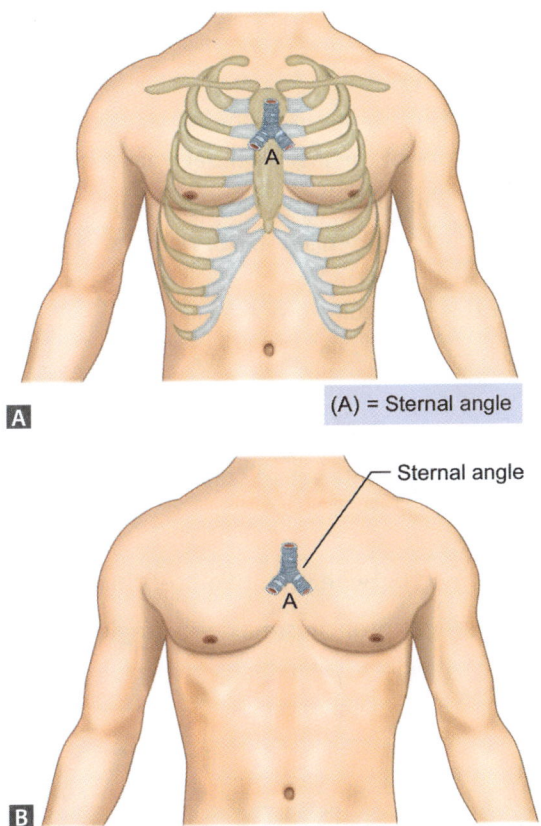

Figs 10.3A and B: Tracheal bifurcation

4. What is the vertebral level of tracheal bifurcation?

Ans.
i. **In cadaver and supine position:** Lower border of T_4 vertebra.
ii. **In living and standing position:** Lower border of T_6 vertebra.
iii. **In the newborn:** T_3 vertebra.

5. How many cartilaginous rings of trachea are there?

Ans. There are about 16–20 cartilaginous rings of trachea.

6. What is the type of the cartilage?

Ans. Hyaline cartilage.

7. Which of the rings is broadest?

Ans. First tracheal ring.

8. What is the special feature of the last tracheal ring at its bifurcation?

Ans. The last ring presents a triangular process which hooks upwards from the lower margin and surrounds the commencement of right and left bronchi.

9. What is the name of this triangular process?

Ans. Carina.

10. What is the clinical importance of carina?

Ans. Carina presents a ridge in the interior of tracheal bifurcation which acts as a guide during bronchoscopic examination.

11. How much angle is formed between the two bronchi at tracheal bifurcation?

Ans. About 70° (Fig. 10.3C). Right bronchus makes an angle of 25° and the left bronchus makes an angle of 45° with the median plane.

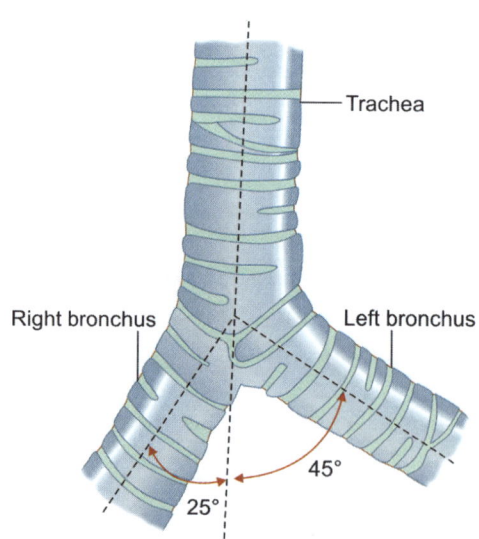

Fig. 10.3C: Tracheal bifurcation

Surface Anatomy: Thorax

12. Name the important structures related to the site of tracheal bifurcation.

Ans.
i. Deep cardiac plexus—in front of bifurcation.
ii. Three groups of tracheobronchial lymph nodes—one group below and the other two groups at the sides.

13. In which mediastinum the trachea lies?

Ans. Trachea lies in superior mediastinum.

14. Why the trachea deviates slightly towards to right close to its bifurcation?

Ans. Due to the arch of aorta.

15. What are the divisions of trachea at its bifurcation?

Ans.
i. Right bronchus: Shorter (2.5 cm), wider and more vertical than left bronchus.
ii. Left bronchus: Longer (5 cm), narrower and more oblique than right bronchus.

LESSON 2: LINES

1. ANTERIOR BORDER OF LEFT LUNG (FIGS 10.4A AND B)

Palpate the left clavicle
↓
Demarcate the junction of medial 1/3rd and lateral 2/3rd
↓
Take a mid-point of medial 1/3rd of the clavicle.
↓
Then put a point 2.5 cm above the mid-point of medial 1/3rd of clavicle (a_2)
↓
A second point is marked on left sternoclavicular joint (b_2)
↓
Third point is put on the **sternal angle** slightly to the left of the midline (c_2)
↓
Then, **4th point** is put over the body of the sternum, slightly to the left of the midline, at the level of the 4th costal cartilage (d_2)
↓
5th point is taken on the left 4th intercostal space 3.5 cm away of the sternal margin (f)
↓
Put a 6th point on the left 6th costal cartilage
4 cm away from the midline (g)

Now, join these points starting from (a_2) to (g) which represents the anterior border of left lung.

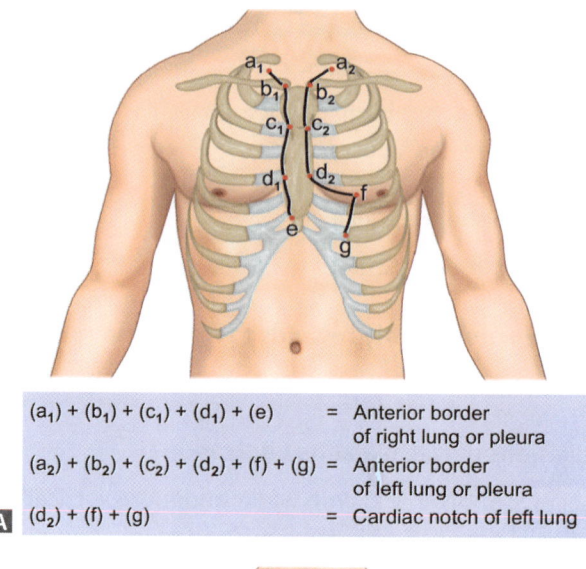

$(a_1) + (b_1) + (c_1) + (d_1) + (e)$ = Anterior border of right lung or pleura
$(a_2) + (b_2) + (c_2) + (d_2) + (f) + (g)$ = Anterior border of left lung or pleura
$(d_2) + (f) + (g)$ = Cardiac notch of left lung

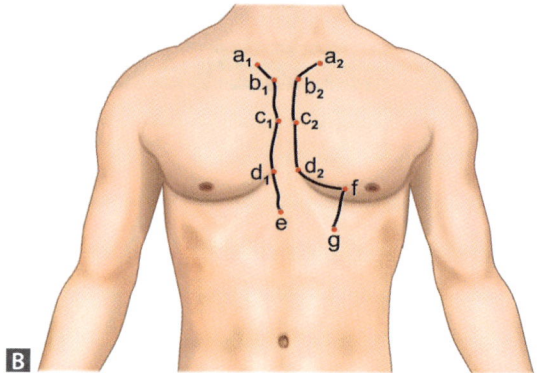

Figs 10.4A and B: (A) Anterior border of left lung; (B) Right lung and cardiac notch

PROBABLE QUESTIONS AND ANSWERS

1. What does the anterior border of lung demarcate?

Ans. The anterior border of lung demarcates the costal surface from the medial surface of the lung.

2. What is its extent?

Ans. It extends from the apex to the base of the lung.

Surface Anatomy: Thorax

?3. Why the anterior border of the lung is thin and sharp but its posterior border is thick and rounded?

Ans. i. The anterior border is thin and sharp, because it is sandwiched between the heart with pericardium inside and the anterior thoracic wall outside.
ii. The posterior border occupies the paravertebral gutter.

?4. What is the difference of line of pleural reflexion along the anterior border of the lung of two sides?

Ans. i. **On the right side**, the anterior border of the right lung corresponds to the costo-mediastinal line of pleural reflexion and it is more or less vertical.
ii. **On the left side**, the anterior border of the left lung follows the costomediastinal reflexion up to the 4th costal cartilage and then deviates laterally forming a notch called **cardiac notch**.

?5. What is the importance of this cardiac notch?

Ans. Due to the formation of the cardiac notch, the heart with pericardium in this region is uncovered by lung and called an area of **superficial cardiac dullness**.

?6. What is lingula?

Ans. Lingula of the lung is a small, tongue like projection from the lower part of the cardiac notch at the anterior border of the left lung.

Remember that cardiac notch and lingula are absent in the right lung.

2. RIGHT BORDER OF HEART (FIGS 10.5A AND B)

Put a point on the upper border of right third costal cartilage 1" away from the midline (b)
↓
2nd point is taken on the right 4th intercostal space 4 cm away from the midline (c)
↓
3rd point is put on the right 6th costal cartilage close to the lateral margin of the sternum (d)
↓

Now join these points (b), (c) and (d) with maximum convexity at the point (c) which represents the right border of heart.

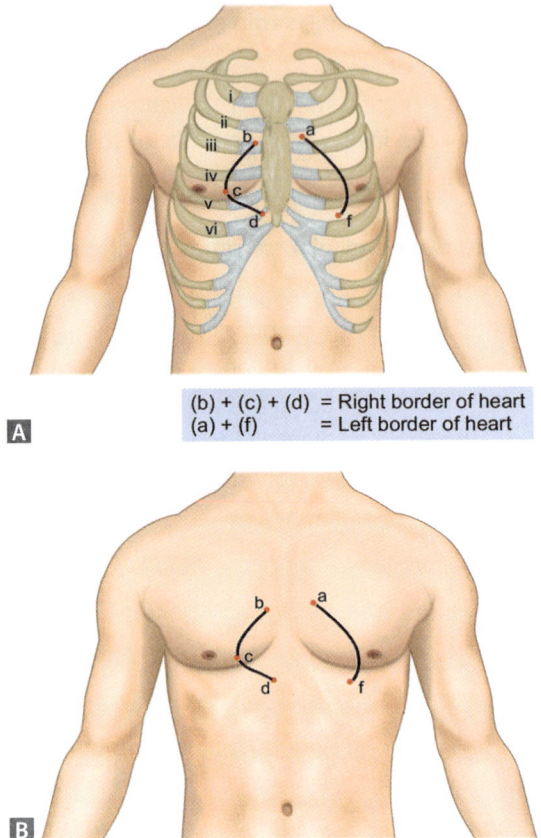

Figs 10.5A and B: (A) Right and left border of heart

PROBABLE QUESTIONS AND ANSWERS

1. What are the borders of the heart?

Ans. **Three borders:**
   i. Right border.
   ii. Inferior border (Margo acutus).
   iii. Left border (Margo obtusus).

2. How the right border is formed?

Ans. Right border is formed only by the **right atrium**.

Surface Anatomy: Thorax

3. What is its appearance?

Ans. It is rounded and convex.

4. What is the extent of right border?

Ans. It extends from the right side of the superior vena caval opening to the opening of inferior vena cava.

5. Name the surfaces of the heart which are separated by the right border?

Ans. The right border separates the sternocostal surface in front from the base of the heart behind.

6. What lies along the right border?

Ans. A shallow vertical groove known as **sulcus terminalis** runs along the right border.

7. What corresponds with the sulcus terminalis?

Ans. A smooth muscle ridge known as **crista terminalis** in the interior of the right atrium of heart corresponds with the sulcus terminalis.

8. What is the importance of crista terminalis?

Ans. It separates the interior of right atrium into two parts.

9. What are the two parts in the interior of right atrium?

Ans. Anterior rough part (**atrium proper**) and posterior smooth part (**sinus venarum**).

10. What important structure lies in the sulcus terminalis?

Ans. At the upper end of the sulcus, sinoatrial node is present.

3. LEFT BORDER OF HEART (FIGS 10.5A AND B)

Put a point on the lower border of the left second costal
cartilage 1" away from the midline (a)
↓
Then, take a point on the apex beat in the
left 5th intercostal space (f)
Now, join these two points with convexity upwards and to the
left which represents the left border of the heart.

PROBABLE QUESTIONS AND ANSWERS

1. How the left border is formed?

Ans. Left border is formed mainly by the left ventricle and partly by the left auricle.

2. What is the extent of the left border?

Ans. It extends from the left auricle to the apex of the heart with convexity upwards and to the left.

3. Which structure accompanies the left border?

A. The marginal branch of left coronary artery.

4. What are the surfaces separated by the left border?

Ans. Sternocostal surface from the left lateral surface.

5. How do the right and left borders form cardiophrenic angle?

Ans. The right and left border of heart form an angle with the diaphragm called right and left cardiodiaphragmatic angle.
(Note that right-sided angle is acute but the left angle is obtuse).

6. What is the inferior border of heart?

Ans. It is a sharp border extending from the opening of inferior vena cava to the apex of the heart.

7. What is incisura apicis cordis?

Ans. Incisura apicis cordis is a notch on the inferior border of heart close to its apex.

8. How the inferior border is formed?

Ans. It is formed mainly by the right ventricle and partly by the left ventricle near the apex.

9. Which structure runs along the inferior border?

Ans. Right marginal artery, a branch of right coronary artery and right marginal vein.

10. What are the surfaces that are separated by the inferior border?

Ans. Sternocostal surface from the diaphragmatic surface.

4. ARCH OF AORTA (FIGS 10.1A AND B)

Palpate the sternal angle.
Then put a point at the right end of the sternal angle (a)
↓
Then, **second point** is in the middle of manubrium sterni (b)
↓
Third point is put at the left end of the sternal angle (c)
Then, join these points (a), (b) and (c) which form a convex line.
Now, draw two convex lines one above the other
keeping a breadth of about 2.5 cm between these two lines.
These two convex lines will represent arch of aorta.

PROBABLE QUESTIONS AND ANSWERS

1. What is the extent of the arch of aorta?

Ans.
  i. It begins as a continuation of ascending aorta behind the right 2nd chondrosternal joint at the level of sternal angle.
  ii. It ends by continuing as descending aorta at the level of lower border of T4 vertebra.

2. What is the maximum limit of its convexity?

Ans. Center of manubrium sterni.

3. How many curvatures of the arch of aorta are there?

Ans. **Two**: i. Convexity upwards.   ii. Convexity forwards and to the left.

4. What are the branches of arch of aorta?

Ans. From right to left:
  i. Brachiocephalic trunk.   ii. Left common carotid artery.
  iii. Left subclavian artery.

5. What are the abnormal branches of arch of aorta?

Ans.
  i. Arteria thyroidea ima.   ii. Left vertebral artery.
  iii. Left bronchial artery.   iv. Inferior thyroid artery.
  v. Internal thoracic artery.

6. What is aortic knuckle?

Ans. It is a convex bulging on the left side of the sternal angle in a plain X-ray of chest, formed by the distal part of the arch of aorta.

7. What is the situation of arch of aorta?

Ans. Superior mediastium.

5. SUPERIOR VENA CAVA (FIGS 10.2A AND B)

Draw two parallel lines 2 cm apart starting from the lower border of right 1st costal cartilage to upper border of right 3rd costal cartilage.

PROBABLE QUESTIONS AND ANSWERS

1. How the superior vena cava is formed?

Ans. It is formed by the union of right and left brachiocephalic veins.

2. What is the site of formation of superior vena cava (SVC)?

Ans. It is formed behind the lower border of right 1st costal cartilage.

3. What is its length and width?

Ans. **Length**: About 7 cm, **width**: About 2 cm.

4. What are the parts of SVC?

Ans. **Two parts**: i. Extrapericardial part   ii. Intrapericardial part.

5. What is the situation of SVC?

Ans. i. Extrapericardial part in superior mediastium.
ii. Intrapericardial part in middle mediastium.

6. Where does the SVC terminate?

Ans. It terminates in the smooth part of right atrium (sinus venarum) behind the right 3rd costal cartilage.

7. What are its tributaries?

Ans.   i. Azygos vein.   ii. Pericardial veins.   iii. Mediastinal veins.

Surface Anatomy: Thorax

8. At which level the SVC pierces the fibrous pericardium?

Ans. At the level of sternal angle.

9. Is there any valve of SVC?

Ans. No, it is devoid of valves.

10. What is double superior vena cava?

Ans. This developmental anomaly may occur due to failure of development of oblique communication between two anterior cardinal veins. In such condition, right SVC directly opens in right atrium and the left SVC reaches right atrium via the coronary sinus.

11. How does the SVC develop?

Ans.
i. **Extrapericardial part**: From right anterior cardinal vein lying between the oblique anastomosis between the two anterior cardinal veins and the **right duct of Cuvier**.
ii. **Intrapericardial part**: From the right duct of Cuvier (formed by the union of anterior and posterior cardinal veins).

CHAPTER 11

Surface Anatomy: Head and Neck

- ❏ Points
    1. Supraorbital Notch
    2. Bifurcation of Common Carotid Artery
    3. Arch of Cricoid Cartilage
    4. Spine of 7th Cervical Vertebra
    5. Nasion
    6. Infraorbital Foramen
    7. Thyroid Prominence (Laryngeal Prominence)
    8. Tips of Greater Cornu of Hyoid
- ❏ Lines
    1. Isthmus of Thyroid Gland
    2. Lateral Lobe of Thyroid Gland
    3. Frontal Air Sinus
    4. Parotid Duct
    5. Right Common Carotid Artery
    6. Internal Carotid Artery
    7. Internal Jugular Vein
    8. External Jugular Vein
    9. Facial Artery in the Face
    10. Spinal Accessory Nerve
    11. Palatine Tonsil

LESSON 1: POINTS

1. SUPRAORBITAL NOTCH (FIG. 11.1)

Mark a point at the junction of medial 1/3rd and lateral 2/3rd of the supraorbital margin.

PROBABLE QUESTIONS AND ANSWERS

1. How far this point is away from the midline?

Ans. 2.5 cm away from the midline.

2. What is the relation between supraorbital notch and frontal notch?

Ans. Frontal notch is medial to the supraorbital notch.

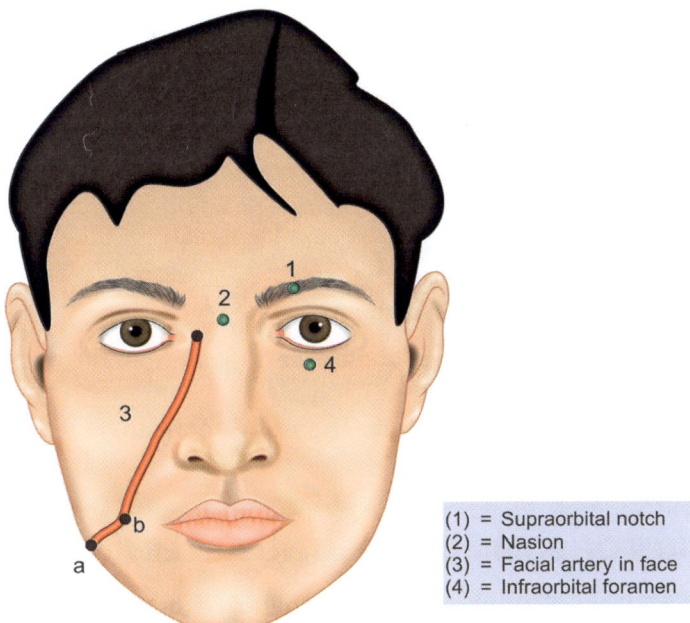

Fig. 11.1: Surface markings on face

(1) = Supraorbital notch
(2) = Nasion
(3) = Facial artery in face
(4) = Infraorbital foramen

3. What are the structures related to the supraorbital notch?

Ans. Supraorbital vessels and supraorbital nerves.

4. What is the origin of the supraorbital artery?

Ans. Supraorbital artery is the branch of ophthalmic artery, branch of internal carotid artery.
(Remember that the infraorbital artery is the branch of 3rd part of maxillary artery branch of external carotid artery)

5. Which nerve gives off the supraorbital nerve?

Ans. Supraorbital nerve is one of the branches of frontal nerve. Other branch of frontal nerve (branch of ophthalmic division of trigeminal nerve) is supratrochlear nerve.
(Remember that the infraorbital nerve is the branch of maxillary division of trigeminal nerve).

6. What is supraorbital foramen?

Ans. Sometimes, the supraorbital notch is converted into a foramen called supraorbital foramen.

7. **What are the structures passing medial to the supraorbital notch?**

Ans. Supratrochlear vessels and supratrochlear nerves pass medial to the supraorbital notch.
(Note that both the supraorbital and supratrochlear nerves are the branches of frontal nerve; the supraorbital and supratrochlear arteries are the branches of ophthalmic artery).

8. **Where do you find infraorbital foramen?**

Ans. Just below the infraorbital margin of maxillae 1" away from the midline.

9. **What are structures passing through infraorbital foramen?**

Ans. Infraorbital vessels and nerve.

2. BIFURCATION OF COMMON CAROTID ARTERY (FIG. 11.2)

Palpate the anterior border of sternocleidomastoid muscle and a point is given on this anterior border at the level of upper border of lamina of thyroid cartilage.

PROBABLE QUESTIONS AND ANSWERS

1. **What is the origin of common carotid artery?**

Ans. Right common carotid artery is the branch of brachio cephalic trunk (branch of arch of aorta) and left common carotid artery is the branch of arch of aorta.

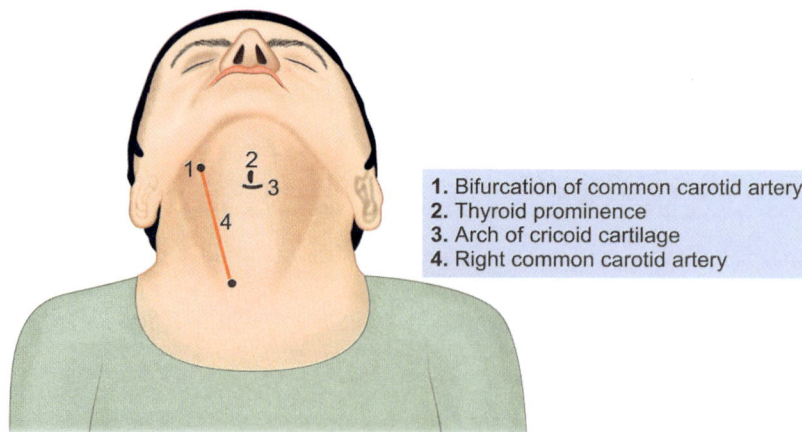

Fig. 11.2: Surface markings on the front of neck

Surface Anatomy: Head and Neck

2. What is the vertebral level of this bifurcation?

Ans. Opposite the disk between C_3 and C_4 vertebrae.

3. In which triangle of the neck this bifurcation takes place?

Ans. The bifurcation takes place in the carotid triangle of the neck.

4. What is carotid sinus?

Ans. It is a fusiform dilatation at the bifurcation of common carotid artery.

5. What is the importane of carotid sinus?

Ans. It acts as a **baroreceptor** for controlling intracranial blood pressure.

6. What is carotid body?

Ans. It is a small, oval neurovascular structure situated close to the posterior wall of the carotid sinus.

7. What is the function of carotid body?

Ans. It acts as a **chemoreceptor** monitoring oxygen tension within the artery.

8. What is potato tumor?

Ans. It is an enlargement of the carotid body.

9. What are the branches of common carotid artery at its bifurcation?

Ans.
  i. External carotid artery.
  ii. Internal carotid artery.

10. What is the inter-relation of these two large arteries in the neck?

Ans. The external carotid artery is anteromedial to the internal carotid artery.

11. What are the structures passing between internal and external carotid artery?

Ans.
  i. Styloid process.
  ii. Styloglossus and stylopharyngeus muscles.
  iii. Glossopharyngeal and pharyngeal branch of vagus nerves.
  iv. Sometimes, a part of parotid gland.
Remember that the internal carotid artery provides no branch in the neck

3. ARCH OF CRICOID CARTILAGE (FIG. 11.2)

Extend the neck and feel the thyroid angle of the thyroid cartilage in the midline of the neck. The prominence below the thyroid cartilage is the cricoid cartilage. It is marked as a broad transverse line.

PROBABLE QUESTIONS AND ANSWERS

1. What is cricoid cartilage?

Ans. Cricoid cartilage is one of the laryngeal cartilages and hyaline in structure.

2. How many laryngeal cartilages are there?

Ans. There are nine cartilages:
   a. Three unpaired cartilages
      i. Epiglottis.
      ii. Thyroid cartilage.
      iii. Cricoid cartilage.
   b. Three paired cartilages
      i. Arytenoid cartilages.
      ii. Corniculate cartilages.
      iii. Cuneiform cartilages.

3. What is the nature of these cartilage?

Ans. All are hyaline cartilage except epiglottis, corniculate, cuneiform, vocal process and apex of the arytenoid cartilages which are made up of elastic fibrocartilage.

4. What is the shape of cricoid cartilage?

Ans. Signet ring shaped.

5. What are the parts of cricoid cartilage?

Ans. Mainly two parts:
   i. Narrow anterior arch.
   ii. Broad posterior lamina.

6. What is the vertebral level of cricoid cartilage?

Ans. Lower border of C_6 vertebra.

7. What are the joints formed by cricoid cartilage?

Ans. i. Cricothyroid joint (synovial joint).   ii. Cricoarytenoid joint (synovial).

8. What are the muscles attached to the cricoid cartilage?

Ans. i. Cricoarytenoideus posterior.
   ii. Cricothyroid.
   iii. Cricopharyngeus part of inferior constrictor.
   iv. Cricoarytenoideus lateralis.

Surface Anatomy: Head and Neck

9. Which joint is situated at the junction of two parts of cricoid cartilage?

Ans. Cricothyroid joint.
Remember that the cricoid cartilage is the foundation stone of the larynx.

4. SPINE OF 7TH CERVICAL VERTEBRA

Flex the head and feel the most prominent spine on the back of the neck at the lower end of the nuchal furrow.

PROBABLE QUESTIONS AND ANSWERS

1. What is the features of the 7th cervical spine?

Ans. i. It is not bifid unlike other cervical vertebral spine.
ii. It is long and horizontal.

2. Why the 7th cervical spine is called vertebral prominence?

Ans. Because of the long and horizontal spine of the C_7 vertebra it causes a prominence on the back of the neck in the midline at the lower end of the nuchal furrow.

3. What are the ligaments attached to the spine?

Ans. i. Ligamentum nuchae. ii. Supraspinous ligament.

4. What is the extent of ligamentum nuchae?

Ans. i. **Upper limit** of the ligament is at the external occipital protuberance and occipital crest of occipital bone.
ii. **Lower limit** is at the tip of C_7 spine.

5. What is the extent of supraspinous ligament?

Ans. It connects the tips of two adjacent spinous process:
a. Upper limit of this ligament is at the tip of C_7 spine.
b. Lower limit is at the tip of S_1 spine.
Note that:
i. At the tip of spine of C_7 vertebra, there is ending of ligamentum nuchae and beginning of the supraspinous ligament.
ii. Ligamentum nuchae is present only in cervical region and supraspinous ligament is present in thoracolumbar region.

6. What are the muscles attached to the C₇ spine?

Ans.
i. Trapezius.
ii. Rhomboideus minor.
iii. Serratus posterior superior.
iv. Splenius capitis.
v. Spinalis cervicis.
vi. Simispinalis thoracis.
vii. Interspinales multifidus.

7. What is its importance as an anatomical landmark?

Ans.
i. In the anatomical position of the body, the inferior angle of the scapula is at the level of C₇ spine.
ii. Vertebral spines can be counted from the prominent C₇ spine (spine above the C₇ spine is C₆ spine and below it T₁ spine).

5. NASION (FIG. 11.1)

It is a point in the depression at the root of the nose. This point overlies the frontonasal and internasal sutures.

PROBABLE QUESTIONS AND ANSWERS

1. What are the bones underlying the nasion?

Ans.
i. Nasal notch of the frontal bone.
ii. Nasal bones of both sides.

2. What are the articulations deep to the nasion?

Ans. Frontonasal and internasal sutures.

3. What is the importance of nasion?

Ans. It is an important bony landmark to measure the length of the skull.

4. How the length of the skull is measured?

Ans. Length of the skull is measured from the nasion to inion.

5. What is inion?

Ans. Inion is the highest point of the external occipital protuberance of the occipital bone.

6. How the maximum breadth of the skull is measured?

Ans. It is measured from one parietal eminence of the parietal bone to the other.

7. What is cephalic index (CI)?

Ans. The maximum breath and length of the skull is measured. Then it is calculated by:

$$\frac{\text{Maximum breadth}}{\text{Maximum length}} \times 100$$

8. What is the importance of cephalic index?

Ans. Cephalic index has racial variation.
Indian skulls are mesaticephalic (cephalic index between 75 and 80).

9. What are the types of skull according to CI?

Ans.
i. Brachycephalic (CI >80)
ii. Mesaticephalic (CI → 75 to 80)
iii. Dolichocephalic (CI <75).

10. What is root of the nose?

Ans. Root of the nose is the junction of the external nose with the forehead.

6. INFRAORBITAL FORAMEN (FIG. 11.1)

It is a point just below the infraorbital margin 1" away from the midline.

PROBABLE QUESTIONS AND ANSWERS

1. Infraorbital foramen belongs to which bone?

Ans. Maxillae.

2. Where do you find supraorbital notch or foramen?

Ans. It is at the junction of medial 1/3rd and lateral 2/3rd of the supraorbital margin of the frontal bone.

3. What are the structures passing through the infraorbital foramen?

Ans. Infraorbital vessels and nerve.

4. How does the infraorbital nerve appear in the face?

Ans. Infraorbital nerve is the continuation of the maxillary nerve and appears in the face through the infraorbital foramen between the origins of levator labii superioris and levator anguli oris muscle.

5. What are the branches of infraorbital nerve in the face?

Ans. These sets of branches: (i) Palpebral, (ii) Labial, (iii) Nasal.

6. What is the fate of infraorbital nerve in the face?

Ans. The branch of infraorbital nerve joins with those of facial nerve and form **infraorbital plexus**.
Note that infraorbital nerve is a sensory nerve and facial nerve is a motor nerve. So, the infraorbital plexus is formed by both the branches of motor and sensory nerves.

7. Where from the infratrochlear nerve arises?

Ans. Infratrochlear nerve is a branch of nasociliary nerve which is a branch of ophthalmic division of trigeminal nerve.

8. What is the origin of infraorbital artery?

Ans. Infraorbital artery is the branch of 3rd part of maxillary artery.

9. What are the structures passing through supraorbital foramen?

Ans. Supraorbital vessels and nerves.
Note that supraorbital artery is the branch of ophthalmic artery whereas the **infraorbital artery** is the branch of maxillary artery and **supraorbital nerve** is the branch of frontal nerve which is the branch of ophthalmic division of trigeminal nerve whereas the **infraorbital nerve** is the continuation of maxillary nerve which is the branch of trigeminal nerve.

7. THYROID PROMINENCE (LARYNGEAL PROMINENCE) (FIG. 11.2)

Extend the neck. The most prominent structure in the midline of the neck is the thyroid prominence.

PROBABLE QUESTIONS AND ANSWERS

1. What is thyroid prominence?

Ans. It is the most prominent part of the thyroid cartilage seen in front of the neck in the midline.

2. How the prominence is formed?

Ans. The anterior border of each quadrilateral lamina of thyroid cartilage meet with each other in front and in the midline.

Surface Anatomy: Head and Neck

3. What is the vertebral level of the prominence?

Ans. It lies opposite the C_4 and C_5 vertebrae.

4. What is the other name of laryngeal prominence?

Ans. Adam's apple.

5. What is thyroid angle?

Ans. The fusion of anterior border of two lamina of thyroid cartilage form an angle called thyroid angle.

6. How does the thyroid angle differ in males from females?

Ans. The thyroid angle is about 90° in males and 120° in females. So, the laryngeal or thyroid prominence is more prominent in males than in females.

7. What are the structures attached to the thyroid angle?

Ans. The inner surface of the angle receives the attachments of the following structures from above downwards:
   i. Thyroepiglottic ligament (unpaired).
   ii. Vestibular ligaments (paired).
   iii. Vocal ligaments (paired).
   iv. Vocalis, thyroarytenoideus and thyroepiglotticus muscles.

8. Where do you find the cricoid cartilage?

Ans. Cricoid cartilage is just below the thyroid prominence in the midline of the neck.

9. What is the joint formed between thyroid and cricoid cartilage?

Ans. The inferior horn of the posterior border of the thyroid cartilage articulates with the cricoid cartilage forming a synovial cricothyroid joint.

10. What important structures is related to this joint?

Ans. Recurrent laryngeal nerve.

11. What are the presenting parts of thyroid cartilage?

Ans. It consists of two quadrilateral lamina. Each lamina consists of 4 borders → (i) Anterior, (ii) Posterior, (iii) Upper, (iv) Lower, and 2 surfaces → (i) Outer, (ii) Inner. Thyroid angle is the fusion of two anterior borders.

8. TIPS OF GREATER CORNU OF HYOID (FIG. 11.3)

Slightly extend the neck and feel the body of the hyoid bone by running a finger downwards from the symphysis menti up to the junction of the floor of the mouth and front of the neck. Then trace laterally (either right or left) till it ends at the tip of greater cornu.

PROBABLE QUESTIONS AND ANSWERS

1. Where do you find the hyoid bone?

Ans. It is below the mandible and lies in front of the neck.

2. What are the presenting parts of hyoid?

Ans. Body, two greater cornu and two lesser cornu.

3. What is the shape of hyoid bone?

Ans. U-shaped.

4. What is the prominence below the hyoid?

Ans. Thyroid prominence (thyroid angle).

5. What is the vertebral level of hyoid?

Ans. C_3 vertebra.

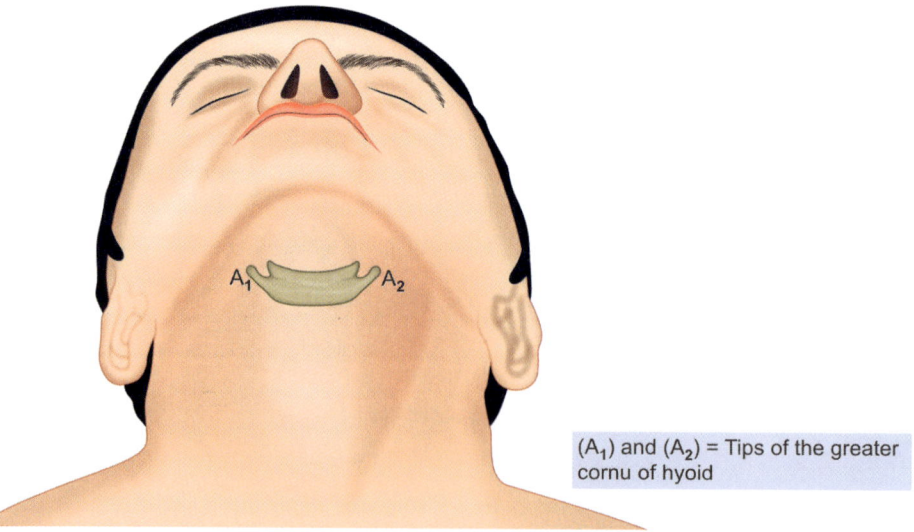

(A_1) and (A_2) = Tips of the greater cornu of hyoid

Fig. 11.3: Tips of the greater cornu of hyoid

Surface Anatomy: Head and Neck

6. How the hyoid is suspended from the base of the skull?

Ans. By stylohyoid ligament.

7. What are the presenting parts of greater cornu?

Ans. Two surfaces (superior and inferior) and a tubercle.

8. What are the structures attached to the greater cornu?

Ans. Muscles: i. Middle constrictor ii. Hyoglossus.
Ligaments: i. Stylohyoid ii. Fibrous loop from the tendon of digastric.

9. What is the ossification of hyoid bone?

Ans. Cartilaginous in ossification. It is ossified from the cartilages of 2nd and 3rd branchial arches.

10. Where from the greater cornu is ossified?

Ans. From the cartilage of 3rd branchial arch.

11. What are the suprahyoid and infrahyoid muscles?

Ans. Suprahyoid muscles (4 pairs):
 i. Digastric. ii. Stylohyoid.
 iii. Mylohyoid. iv. Geniohyoid.
Infrahyoid muscle (4 pairs):
 i. Sternohyoid. ii. Omohyoid.
 iii. Sternothyroid. iv. Thyrohyoid (upward extension of sternothyroid).

LESSON 2: LINES

1. ISTHMUS OF THYROID GLAND (FIG. 11.4)

Locate the cricoid cartilage. It is located by extending the neck and the prominence below the thyroid cartilage is the cricoid cartilage. Now, take a point about 1.2 cm below the cricoid cartilage. Then draw a 1.25 cm long horizontal line through this point across the trachea. It represents the upper border of the isthmus. Take another point 1.25 cm below the previous point and then draw a 1.25 cm long horizontal line through this point. It represents the lower border of the isthmus. Now join these two lines forming a square-shaped area (1.25 cm x 1.25 cm) which represents the isthmus.

2. LATERAL LOBE OF THYROID GLAND (FIG. 11.4)

i. 1st point at the anterior border of sternocleidomastoid muscle at the level of laryngeal prominence (upper pole) point (i).
ii. 2nd point—1 cm below the lateral end of the lower border of the isthmus point (ii).
iii. 3rd point—2.5 cm lateral to the 2nd point (iii).

Now join the point (i) and lateral end of the upper border of the isthmus with concavity forwards represents the **anterior border of the lateral lobe**.

Join the point (i) with the (iii) with convexity outwards represents the **lateral border** of the lateral lobe.

A line extending from the lateral end of the lower border of the isthmus and then joining the points (b) and (c) with convexity below represents the **lower pole** of the lateral lobe.

PROBABLE QUESTIONS AND ANSWERS

1. What are the presenting parts of thyroid gland?

Ans. Two lateral lobes and an isthmus connecting the lateral lobes.
Each lateral lobe presents:
i. An apex.
ii. A base.
iii. Three surfaces—anterolateral, posterolateral and medial (deep)
iv. Three borders—anterior, posterior and lateral.

2. What is the dimension of the lateral lobe and isthmus?

Ans. i. Lateral lobe → 5 cm × 3 cm × 2 cm
ii. Isthmus → 1.25 cm × 1.25 cm.

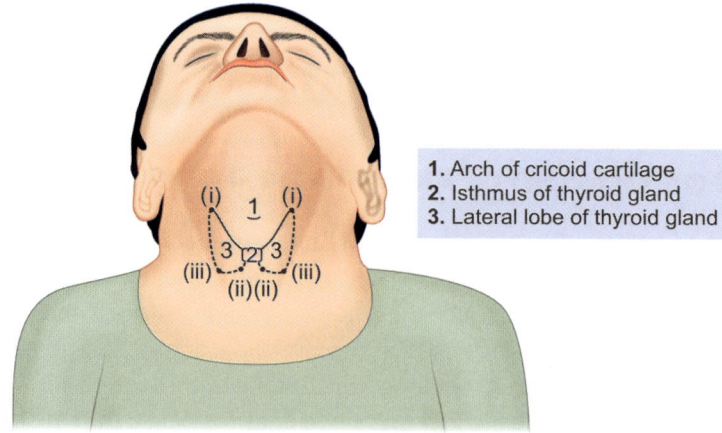

1. Arch of cricoid cartilage
2. Isthmus of thyroid gland
3. Lateral lobe of thyroid gland

Fig. 11.4: Lateral lobe and isthmus of thyroid gland

Surface Anatomy: Head and Neck

3. What are the covering of the thyroid gland?

Ans. Inner true capsule formed by fibrous stroma of the gland and outer false capsule formed by the splitting of pretracheal fascia.

4. How much is the weight of the gland?

Ans. About 25 g.

5. What are the muscles related to the lateral lobes?

Ans.
i. Sternothyroid.
ii. Sternohyoid.
iii. Omohyoid.
iv. Sternocleidomastoid.

6. What important structures related to the posterolateral surface?

Ans. Carotid sheath containing common carotid artery, internal jugular vein and vagus nerve.

7. What are the structures related to the medial surface of the gland?

Ans.
a. Two tubs:
  i. Lower part of larynx and upper part of trachea.
  ii. Lower part of pharynx and upper part of esophagus.
b. Two muscles:
  i. Inferior constrictor.
  ii. Cricothyroid.
c. Two nerves:
  i. External laryngeal nerve.
  ii. Recurrent laryngeal nerve.

8. What are the arteries supplying the thyroid gland?

Ans.
i. Superior thyroid branch of external carotid artery.
ii. Inferior thyroid branch of thyrocervical trunk.
iii. Accessory thyroid arteries from esophageal and tracheal branches.
iv. Arteria thyroidea ima (if present) branch of arch of aorta.

9. What are the veins draining the gland?

Ans.
i. Superior and middle thyroid veins drain into the internal jugular vein.
ii. Inferior thyroid vein drains into brachiocephalic vein.
iii. 4th thyroid vein (**Kocher's vein**) drain into internal jugular vein.

10. Which vein does not have corresponding artery?

Ans. Middle thyroid vein.

11. Where do you find the thyroid venous plexus?

Ans. It lies deep to the inner true capsule.
(Prostatic venous plexus lies between its true and false capsule).

12. Where do you locate the parathyroid glands?

Ans. Parathyroid glands (two superior and two inferior) are situated along the posterior border of the lateral lobes of the thyroid gland.

13. What is goiter?

Ans. Enlargement of thyroid gland is called goiter.

14. What is physiological goiter?

Ans. Enlargement of thyroid gland due to the physiological needs of the body is called physiological goiter.
Examples: i. Puberty goiter.   ii. Pregnancy goiter.   iii. Lactational goiter.

15. Where from does the thyroid gland develop?

Ans. i. Median thyroid diverticulum (major part of the gland)
ii. Lateral thyroid rudiment.

16. What is the levator glandulae thyroideae?

Ans. It is the remnant of the median thyroid diverticulun.

17. What is pyramidal lobe of the gland?

Ans. It is said to be the 3rd lobe of the gland and lies above the isthums.

18. What is lingual thyroid?

Ans. Presence of thyroid tissue in the foramen cecum of the tongue is called lingual thyroid. It occurs due to the failure of caudal migration of the median thyroid diverticulum (thyroglossal duct).

19. What is retrosternal goiter?

Ans. Enlargement of retrosternal thyroid (presence of thyroid tissue is the thorax) is called (retrosternal goiter which may press on the superior vena cava causing superior vena caval syndrome.

Surface Anatomy: Head and Neck

20. What are the structures to be saved in partial thyroidectomy?

Ans.   i. Parathyroid glands.
      ii. External laryngeal nerves.
     iii. Recurrent laryngeal nerves.

3. FRONTAL AIR SINUS (FIG. 11.5)

a. Take a point at the nasion (depression at the root of the nose).
b. A point 1.25 cm lateral to the nasion at the supraorbital margin.
c. The third point 2.5 cm above the nasion in the midline.

Now, join these three points, so that a triangle is formed which represents the frontal air sinus.

PROBABLE QUESTIONS AND ANSWERS

1. What do you mean by frontal air sinus?

Ans. Frontal air sinus is a mucous diverticula of the nasal cavity invading between the two tables of the squamous part of the frontal bone at the expense of the diploic tissue.

2. How many frontal sinuses are there?

Ans. Two (right and left).

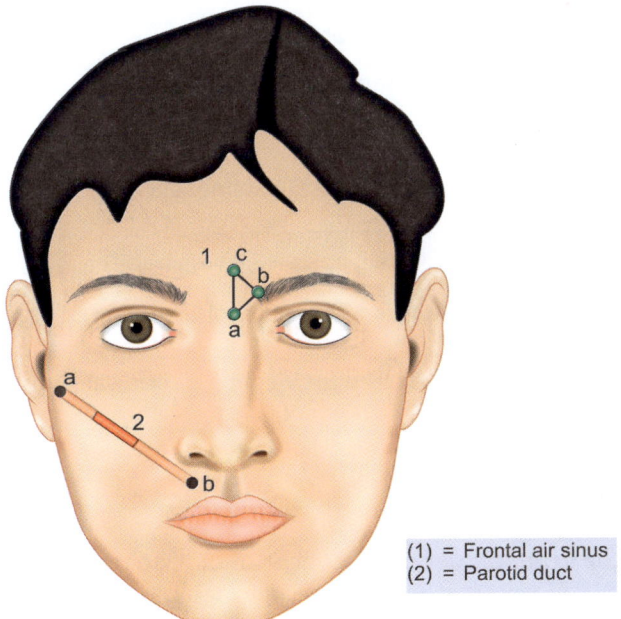

(1) = Frontal air sinus
(2) = Parotid duct

Fig. 11.5: Frontal air sinus and parotid duct

3. What are the measurements of each frontal sinus?

Ans. 3 cm (vertical) × 2.5 cm (transverse) × 1.8 cm (anteroposterior).

4. Where do the frontal sinus open?

Ans. It opens in the infundibulum of the middle meatus of the nose via the frontonasal duct.

5. When do the frontal air sinuses develop?

Ans. Frontal air sinuses start development at 2–3 years after birth.

6. What are paranasal air sinuses?

Ans. The paranasal air sinuses are the air containing bony spaces around the nasal cavity and lined by mucous membrane of ciliated columnar epithelium.

7. Name the paranasal air sinuses?

Ans.
  i. Frontal (2).
  ii. Ethmoidal—variable in number (3–18).
  iii. Maxillary (2).
  iv. Sphenoidal (2).

8. Which one is the largest paranasal air sinus?

Ans. The maxillary air sinus (present within the body of each maxilla).

9. Do you find all the paranasal sinus at birth?

Ans. No. All the sinuses are rudimentary at birth except the frontal air sinuses which start development at 2–3 years after birth.

10. What are the functions of the paranasal air sinuses?

Ans.
  i. They make the cranial bones lighter.
  ii. Act as resonance box.
  iii. Regulate the temperature and humidity of the inspired air.

11. What radiological view is preferred for viewing paranasal sinuses?

Ans. Occipitomental view (OM view).

4. PAROTID DUCT (FIG. 11.5)

a. Mark a point at the lower border of the concha of the ear.
b. Mark another point midway between the ala of the nose and upper margin of the upper lip.

Surface Anatomy: Head and Neck

Draw two parallel lines joining these two points (a) and (b). Middle 1/3rd of this line represents the parotid duct.

PROBABLE QUESTIONS AND ANSWERS

1. What is the other name of the parotid duct?

Ans. Stenson's duct.

2. How and where the parotid duct is formed?

Ans. It is formed by the union of two vertical ducts within the parotid gland.

3. How does the duct emerge from the parotid gland?

Ans. It emerges through the anterior border of the gland.

4. What is the length and breadth of the duct?

Ans. Length = 5 cm; width = 3 mm.

5. Where does the parotid duct open?

Ans. It opens in the vestibule of the mouth on a papilla opposite the crown of upper second molar tooth.

6. How many bents does the duct take from its beginning to ending?

Ans. Three bents, each being a right angle bent, a total of 270° (3 × 90°).

7. Where does the duct bend?

Ans. It bends at three areas:
  i. Anterior border of masseter.
  ii. Submucous coat.
  iii. Mucous membrane of the cheek.

8. Is there any valve of the duct?

Ans. The oblique course of the submucosal part of the duct acts as a valve. The suctorial pad of fat and buccinator muscle through which the duct passes also act as valve of the duct.

9. What are the structures related above and below the duct?

Ans. Above, the duct is related to accessory parotid gland and upper buccal branch of facial nerve. Below, it is related to the lower buccal branch of facial nerve.

10. What are the structural layers of the parotid duct?

Ans.   i. Outer fibroelastic coat containing smooth muscle.
        ii. Inner mucous membrane lined by simple cubical epithelium.

11. What is the source of development of parotid gland and its duct.

Ans. Ectoderm.

5. RIGHT COMMON CAROTID ARTERY (FIG. 11.2)

a. A point at right sternoclavicular joint.
b. A second point at the anterior border of sternocleidomastoid at the level of upper border of thyroid cartilage.

Now, join these two points (a) and (b) by two lines making a breadth of about 6 mm.

PROBABLE QUESTIONS AND ANSWERS

1. What is the origin of right common carotid artery?

Ans. Brachiocephalic trunk which divides into right common carotid and right subclavian artery.

2. What is the origin of left common carotid and left subclavian artery?

Ans. Both are the branches from **arch of aorta**.

3. What are the main branches of arch of aorta?

Ans. From right to left:
   i. Brachiocephalic trunk.
   ii. Left common carotid.
   iii. Left subclavian artery.

4. What is the level of bifurcation of common carotid artery?

Ans. Opposite the disk between C_3 and C_4 vertebrae at the level of upper border of thyroid cartilage.

5. What are the terminal branches of common carotid artery?

Ans. External carotid artery and internal carotid artery.

6. What is carotid sinus?

Ans. Carotid sinus is a fusiform dilatation at the bifurcation of the common carotid artery.

7. What is the function of carotid sinus?

Ans. It acts as a baroreceptor.

8. What are the collateral branches of common carotid artery?

Ans. It does not provide any collateral branches. Only two terminal branches—external and internal carotid artery.

9. How does the right common carotid artery differ form the left one?

Ans.
  i. Right artery starts behind the right sternoclavicular joint and left artery starts behind the manubrium sterni.
  ii. Right artery is the branch of brachiocephalic trunk whereas the left common carotid arises from the arch of aorta.
  iii. Right artery is present only in the neck but the left artery has thoracic and cervical parts.
  iv. Right artery is shorter in length than the left common carotid artery.

10. What is the developmental source of common carotid artery?

Ans. Each common carotid artery is developed from the ventral part of the third aortic arch.

6. INTERNAL CAROTID ARTERY (FIG. 11.6A)

a. A point at the point of bifurcation of common carotid artery, i.e. at the anterior border of sternomastoid muscle at the level of upper border of the lamina of thyroid cartilage.
b. Another point at the posterior border of the condyle of the mandible (condyle of the mandible can be felt immediately in front of the lower part of the tragus when the person is asked to open his mouth).

Now, join these two points (a) and (b) by double lines which represent the internal carotid artery.

PROBABLE QUESTIONS AND ANSWERS

1. What is the origin of internal carotid artery?

Ans. It is one of the two terminal branches of common carotid artery.

2. What is the other branch of common carotid artery?

Ans. External carotid artery.

3. What are the structures passing between internal and external carotid artery?

Ans.
  i. A part of parotid gland.
  ii. Styloid process of temporal bone.
  iii. Styloglossus and stylopharyngeus muscles.
  iv. Glossopharyngeal nerve.
  v. Pharyngeal branch of vagus nerve.

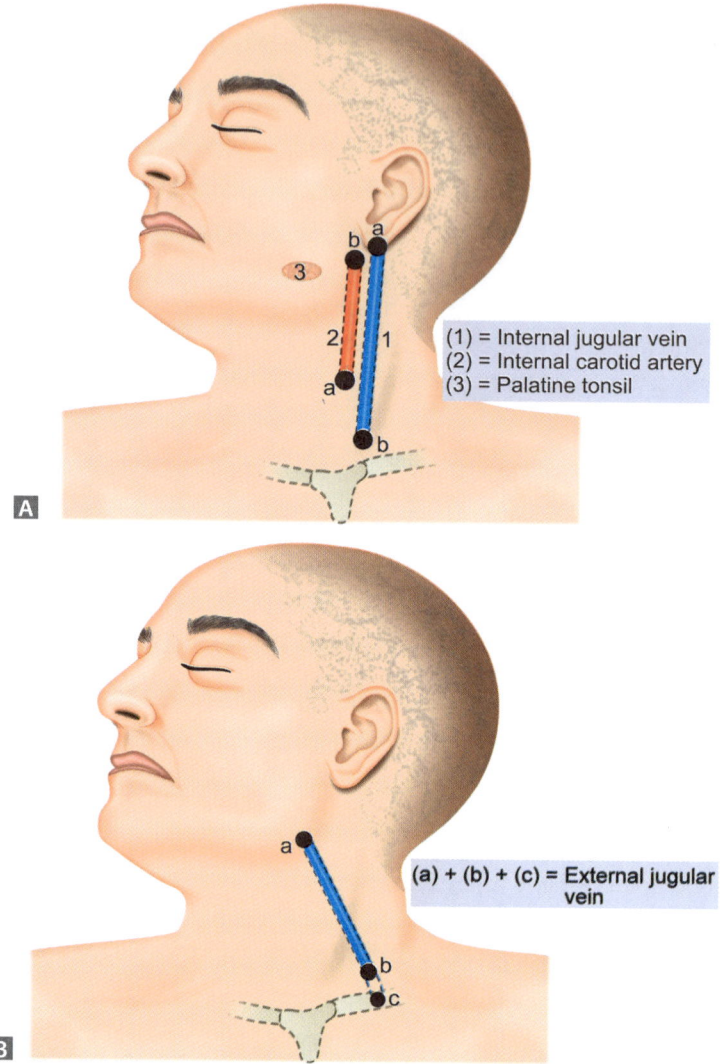

Figs 11.6A and B: (A) Internal carotid artery, internal jugular vein and palatine tonsil; (B) External jugular vein

4. What is the relation of the internal carotid artery with the external carotid artery in the neck?

Ans. The internal carotid artery lies external and dorsal to the external carotid artery. This is explained by the fact that the external carotid artery develops mainly from the ventral aorta and the internal carotid artery arises mainly from the dorsal aorta.

Surface Anatomy: Head and Neck

5. How does it pass in the neck?

Ans. It passes straight upwards through the neck within the carotid sheath along with internal jugular vein and vagus nerve.

6. How does it enter the skull?

Ans. It enters the skull through the lower opening of carotid canal in the petrous part of temporal bone.

7. How does it come out the petrous part of temporal bone?

Ans. It comes out through the anterior opening of carotid canal in the petrous part of temporal bone.

8. How does the artery end?

Ans. The internal carotid artery ends in the middle cranial fossa by dividing into anterior and middle cerebral arteries.

9. What is the origin of posterior cerebral artery?

Ans. Basilar artery divides into two posterior cerebral arteries.

10. What are the branches of internal carotid artery in the neck?

Ans. It does not provide any branch in the neck.

11. Where does the internal carotid artery supply?

Ans. This artery supplies the pituitary gland, orbit and major part of the brain.

12. Where does the external carotid artery supply?

Ans. The external carotid artery supplies the structures in the neck and face by giving off eight branches.

13. What are the branches of external carotid artery?

Ans.
  i. Superior thyroid.
  ii. Ascending pharyngeal.
  iii. Facial.
  iv. Lingual.
  v. Posterior auricular.
  vi. Occipital.
  vii. Maxillary.
  viii. Superficial temporal.

14. What are the different parts of internal carotid artery according to its position?

Ans. Four parts:
  i. Cervical.
  ii. Petrous branches are:
      a. Caroticotympanic.
      b. Branches to pterygoid canal.
  iii. Cavernous branches are:
      a. Meningeal.
      b. Inferior hypophyseal.
  iv. Cerebral branches are:
      a. Superior hypophyseal.
      b. Ophthalmic.
      c. Posterior communicating.
      d. Anterior choroids.
      e. Anterior cerebral.
      f. Middle cerebral.

15. Give an example of collateral circulation between internal and external carotid artery.

Ans. Anastomosis between the dorsal nasal branch of ophthalmic artery (branch of internal carotid artery) and the terminal branch of facial artery (branch of external carotid artery).

16. What is carotid siphon?

Ans. The U-shaped loop of the internal carotid artery while passing through and above the cavernous sinus is known as carotid siphon.

7. INTERNAL JUGULAR VEIN (FIG. 11.6A)

a. Take a point on the lobule of the ear.
b. Take another point on the sternal end of the clavicle.

Join these two points (a) and (b) by two lines making a breadth of about 1.25 cm. Draw two dilatations, one at its upper end (point a) and another at its lower end (point b) between the sternal and clavicular heads of sternocleidomastoid muscle. These two dilatations represent the superior and inferior bulbs of internal jugular vein.

PROBABLE QUESTIONS AND ANSWERS

1. How does this vein begin?

Ans. Internal jugular vein begins as a continuation of sigmoid sinus in the posterior comportment of jugular foramen.

2. How many bulbs (dilations) does it possess?

Ans. It possesses one superior bulb at its commencement and one inferior bulb close to its termination.

3. Do these bulbs have valves?

Ans. Only inferior bulb is guarded above by a pair of valves.

4. What are the drainage areas of internal jugular vein?

Ans. It collects the blood form the brain, superficial part of the face and neck.

5. How does it terminate?

Ans. It terminates by joining with the subclavian vein and forms brachiocephalic vein.

6. Where is the site of formation of brachiocephalic vein?

Ans. Brachiocephalic vein is formed behind the sternal end of clavicle.

7. What is the first tributary and last tributary of internal jugular vein?

Ans.
  i. First tributary is inferior petrosal sinus.
  ii. Last tributary is thoracic duct (left side) or right lymphatic duct (right side).

8. What are the other tributaries of this vein?

Ans. Pharyngeal vein, common facial vein, lingual vein, superior and middle thyroid vein, occipital vein, etc.

9. Does the internal jugular vein communicate with other important veins?

Ans. Yes.
  i. It communicates with the **external jugular** vein in the upper part of the neck by means of oblique jugular vein.
  ii. It communicates with the **cavernous sinus** via inferior petrosal sinus.

10. What is the difference of drainage area of right and left internal jugular vein?

Ans. Right internal jugular vein receives the blood from superficial cerebral vein, whereas the left internal jugular vein receives the blood from deep cerebral veins.

8. EXTERNAL JUGULAR VEIN (FIG. 11.6B)

a. A point is taken slightly below and behind the angle of the mandible (point of beginning).
b. A point 1" above the middle of the clavicle just lateral to the posterior border of sternomastoid muscle (point of piercing the investing layer of deep cervical fascia).

c. A third point is taken at the middle of the clavicle (point of termination into the subclavian vein).

Join the points (a) and (b) by double lines which extend obliquely downwards and backward crossing the sternocleidomastoid represent the external jugular vein.

PROBABLE QUESTIONS AND ANSWERS

1. How is the external jugular vein formed?

Ans. It is formed by the union of the posterior division of retromandibular vein and posterior auricular vein.

2. Where does it terminate?

Ans. It terminates in the subclavian vein by piercing the deep cervical fascia about 2.5 cm above the middle 1/3rd of the clavicle.

3. How many valves does the vein possess?

Ans. The vein possesses two pairs of valves. One pair at its termination and the other pair above 4 cm above the clavicle.

4. Do these valves prevent regurgitation of blood?

Ans. No.

5. What is the sinus of external jugular vein?

Ans. Between the two sets of valves, the vein is often dilated to form a sinus.

6. What are the tributaries of external jugular vein?

Ans. Posterior auricular, posterior division of retromandibular vein, transverse cervical, suprascapular, anterior jugular veins.

7. How this vein can be made prominent?

Ans. This vein can be made prominent by an attempt by blowing with mouth and nose closed.

8. What is the special feature of this vein to be noted?

Ans. The wall of this vein is adherent to the margins of the opening of the deep cervical fascia through which the vein passes. So, the vein can not collapse at this point.

9. FACIAL ARTERY IN THE FACE (FIG. 11.1)

a. A point at the base of the mandible at the anteroinferior angle of the masseter (masseter becomes prominent by clenching the teeth).
b. A point about 1.25 cm lateral to the angle of the mouth.
c. A point at the medial angle of the eye of that side.

Now, join these 3 points (a), (b) and (c) by wavy (tortuous) double lines.

PROBABLE QUESTIONS AND ANSWERS

1. What is the origin of facial artery?

Ans. Facial artery is the branch of external carotid artery in the neck.

2. Where does the external carotid artery give off the facial artery?

Ans. The facial artery arises in the carotid triangle of the neck just above the tip of greater cornu of hyoid bone.

3. What are the parts of the facial artery?

Ans. i. Cervical part (part in the neck).   ii. Facial part (part in the face).

4. What are the branches of cervical part of facial artery?

Ans. i. Ascending palatine artery.
ii. Tonsillar artery (principal artery of the tonsil).
iii. Glandular artery (supply the submandibular gland).
iv. Submental artery.

5. What are branches from facial part of facial artery?

Ans. i. Inferior labial artery.   ii. Superior labial artery.   iii. Lateral nasal artery.

6. How does the facial artery appear in the face?

Ans. The facial artery pierces the investing layer of deep cervical fascia and appears in the face after turning around the lower border of the body of the mandible at the antero-inferior angle of the masseter.

7. How does the facial artery end in the face?

Ans. It ends at the medial angle of the eye as the angular artery which anstomoses with the dorsal nasal branch of ophthalmic artery.

8. How far the facial artery lies from the angle of the mouth?

Ans. About 1.25 cm.

9. What is the relation of the facial artery with the facial vein in the face?

Ans. The facial vein lies above and behind the facial artery.

10. How does the facial artery differ from the facial vein in its course?

Ans. The facial artery is remarkably tortuous but the facial vein undergoes a straight course and is more superficial than the artery.

Why the facial artery is tortuous?

Ans. The facial artery is tortuous because:
  i. The tortuosity of this artery allows expansion of the pharynx (the facial artery lies on superior and middle constrictor of the pharynx).
  ii. The tortuosity allows movements of the mandible, cheeks and lips (the artery rests on the mandible, buccinator and levator anguli oris in the face)

12. Name some other tortuous arteries in the body?

Ans.
  i. Splenic artery.
  ii. Uterine artery (in females).
  iii. Nutrient arteries of long bones.
  iv. Posterior inferior cerebellar artery.

13. How is the facial vein formed?

Ans. **The facial vein is formed by the** union of supraorbital and supratrochlear veins as angular vein.

14. Where does the facial vein terminate?

Ans. The facial vein drains into the internal jugular vein.

10. SPINAL ACCESSORY NERVE (FIG. 11.7)

a. A point is taken just below and in front of tragus.
b. 2nd point is taken on the tip of the transverse process of atlas (C_1 vertebra) which is the mid point between the tip of the mastoid process and the angle of the mandible.
c. A point on the anterior border of sternocleidomastoid muscles at the junction of its upper 1/3rd and lower 2/3rd.
d. A point at the middle of the posterior border of sternomastoid muscle.
e. A point on the anterior border of the trapezius about 5 cm above the clavicle.

Now, join these points (a) to (e) which represent the spinal part of accessory nerve.

Surface Anatomy: Head and Neck

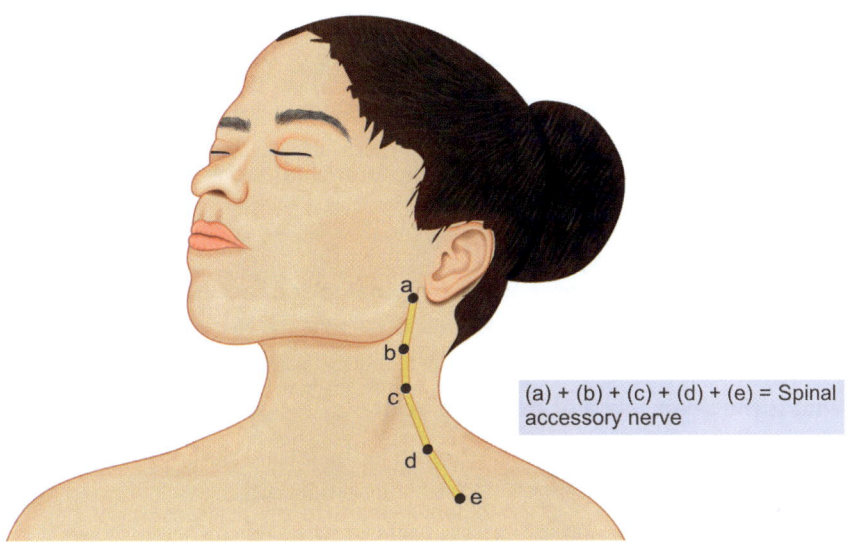

(a) + (b) + (c) + (d) + (e) = Spinal accessory nerve

Fig. 11.7: Spinal accessory nerve

PROBABLE QUESTIONS AND ANSWERS

1. What is the accessory nerve?

Ans. The 11th cranial nerve is the accessory nerve.

2. Why it is so called?

Ans. It is so called because it is accessory to vagus nerve. The cranial root of the accessory nerve represents the detached rootlets of the vagus.

3. What are the two roots of the accessory nerve?

Ans. (i) Cranial root, (ii) Spinal root. The combinations of these two roots are called spinal accessory nerve.

4. What is the deep origin of the cranial root?

Ans. Nucleus ambiguus in the medulla oblongata.

5. Name the cranial nerve nuclei which form nucleus ambiguus?

Ans. 9th (glossopharyngeal), 10th (vagus) and 11th cranial nerve nuclei form the nucleus ambiguus in the medulla.

6. Whether the accessory nerve is a motor nerve or sensory nerve?
Ans. It is an entirely motor nerve.

7. The cranial root belongs to which of the branchial arches?
Ans. It is the nerve of the 6th branchial arch.

8. What is the functional component of spinal root of accessory nerve?
Ans. Somatic efferent.

8. What is the deep origin of the spinal root?
Ans. It is an elongated motor nucleus situated in the lateral part of the anterior horn of the upper five cervical segments (C_1-C_5) of the spinal cord.

10. How do the fibers from these segments appear in the vertebral canal?
Ans. The fibers appear in the vertebral canal by the side of the spinal cord between the ventral root and dorsal root of upper five cervical nerves.

11. What is the relation of the spinal root of accessory nerve with the ligamentum denticulatum?
Ans. It ascends in the vertebral canal behind the ligamentum denticulatum.

12. How does the spinal accessory nerve enter into the cranium?
Ans. It enters through the foramen magnum.

13. How does it come out of the cranium?
Ans. Though the jugular foramen along with 9th and 10th cranial nerves.

14. What is the relation between the cranial and spinal roots of the accessory nerve in the jugular foramen?
Ans.
 i. At the jugular foramen, the spinal root joins with the cranial root of the accessory nerve and then passes in a common sheath of dura matter accompanied by the vagus.
 ii. At the exit from the jugular foramen, the spinal root is separated from the cranial root and passes downwards, backwards and laterally.

15. How does the spinal root appear in the posterior triangle of the neck?
Ans. The nerve pierces the deep surface of the sternomastoid muscle at the junction of its upper 1/4th and lower 3/4th, then it enters the posterior triangle of the neck at the middle of the posterior border of the sternomastoid.

Surface Anatomy: Head and Neck

16. What is the clue for identifying the spinal accessory nerve during dissection of the posterior triangle of the neck?

Ans. The nerve enters the posterior triangle at the middle of the posterior border of sternomastoid muscle and it is hooked superficially upward by the lesser occipital nerve.

17. What is the fate of the spinal accessory nerve?

Ans. It passes deep to the anterior border of trapezius about 5–6 cm above the clavicle and joins with C_3 and C_4 cervical nerves to form **subtrapezoid plexus**.

18. What are the muscles supplied by the spinal accessory nerve?

Ans. Sternocleidomastoid and trapezius.

19. What are the cervical nerves which supply the above two muscles?

Ans. Sternmastoid by C_2 and C_3 nerves and Trapezius by C_3 and C_4 nerves.

20. What is torticollis?

Ans. Irritation of spinal accessory nerve causes clonic spasm of sternomastoid and trapezius leading spasmodic torticollis.

11. PALATINE TONSIL (FIG. 11.6A)

Draw an oval area over the masseter muscle about 1.25 cm above and in front of the angle of the mandible.

PROBABLE QUESTIONS AND ANSWERS

1. What are the palatine tonsils?

Ans. The palatine tonsils are masses of lymphoid tissue situated in the lateral wall of the oropharynx.

2. Where does the palatine tonsils lodge?

Ans. It lodges in a triangular tonsillar sinus in the lateral wall of he oropharynx.

3. What is boundary of the tonsillar fossa (tonsillar sinus)?

Ans.
  i. Anteriorly: Palatoglossal arch containing palatoglossus muscle.
  ii. Posteriorly: Palatopharyngeal arch containing palatopharyngeus muscle.
  iii. Base: Dorsal surface of the posterior 1/3rd of the tongue.
  iv. Apex: Meeting of the palatoglossal and palatopharyngeal arch in the soft palate.

4. What is tonsillar bed?

Ans. The lateral wall of the tonsil on which it lies is called tonsillar bed.

5. What are the structures forming the tonsillar bed?

Ans. The structures from within outwards are as follows:
  i. Pharyngobasilar fascia.
  ii. Palatopharyngeus muscle.
  iii. Superior constrictor muscle.
  iv. Styloglossus muscle.
  v. Glossopharyngeal nerve.

6. What are the presenting parts of palatine tonsil?

Ans.
  i. Two surfaces: Medial and lateral.
  ii. Two borders: Anterior and posterior.
  iii. Two ends: Upper and lower.

7. What are the features on the medial surface?

Ans.
  i. Tonsillar pits (12–15).
  ii. Intratonsillar cleft (supratonsillar fossa).
  iii. Plica triangularis and plica semilunaris (both are embryonic folds).

8. What is the lining epithelium of the medial surface?

Ans. Nonkeratinized stratified squamous epithelium.

9. Is there any mucous lining on the lateral surface?

Ans. No, the lateral surface is covered by a fibrous capsule.

10. Where do you find the paratonsillar vein?

Ans. Paratonsillar vein is related to the lateral surface of tonsil.

11. Name some important structures related to its lateral surface?

Ans.
  i. Paratonsillar vein.
  ii. Pharyngobasilar fascia.
  iii. Superior constrictor fascia.
  iv. Boccopharyngeal fascia.
  v. Styloglossus, stylopharyngeus, posterior belly of digastric and stylohyoid and medial pterygoid muscles.
  vi. Glossopharyngeal nerve.

Surface Anatomy: Head and Neck

12. What is the suspensory ligament of palatine tonsil?

Ans. The lower end of the tonsil is connected to the side of the tongue by a band of fibrous tissue called the suspensory ligament of the tonsil.

13. What are the structures keeping the tonsil in position?

Ans.
i. The suspensory ligament of the tonsil.
ii. Perivascular stalks.
iii. Attachment of the fibrous capsule of the tonsil with the palatopharyngeus and palatoglossus muscle.

14. What are the arteries supplying the tonsil?

Ans.
i. Anterior tonsillar, branch of dorsal lingual artery (branch of lingual artery)
ii. Posterior tonsillar, branch of ascending palatine and ascending pharyngeal arteries.
iii. Superior tonsillar, branch of greater palatine artery.
iv. Inferior tonsillar, branch of facial artery.

15. Which one is the principal artery of the tonsil?

Ans. The inferior tonsillar branch of facial artery.

16. What is the developmental origin of the tonsil?

Ans. The ventral part of 2nd pharyngeal pouch.

17. The damage of which blood vessel may be the cause of excessive bleeding during tonsillectomy operation?

Ans. Paratonsillar vein.

18. Why pain from the infected tonsil is referred to middle ear?

Ans. Because, both are supplied by the glossopharyngeal nerve.

19. Name the other tonsil (mass of lymphoid tissue) found around the commencement of the digestive and respirators tubes.

Ans.
a. Anteriorly: Lingual tonsils (on the posterior 1/3rd of the tongue.
b. Laterally:
  i. Tubal tonsils (present bilaterally on the lateral wall of nasopharynx).
  ii. Palatine tonsils (present bilaterally in the lateral wall of oropharynx).
c. Posteriorly: Pharyngeal tonsil (on the posterior wall of nasopharynx).

20. Imaginary lines joining these tonsils from a ring. What is that ring called?

Ans. Waldeyer's ring.

SECTION 3

Histology

Section Outline

- Histology: Introduction
- Histology: Musculoskeletal System
- Histology: Blood Vascular System
- Histology: Gastrointestinal System
- Histology: Liver and Pancreas (Both Exocrine and Endocrine Types of Gland)
- Histology: Salivary Glands (Exocrine Glands)
- Histology: Endocrine Glands
- Histology: Lymphatic System
- Histology: Respiratory System
- Histology: Urinary System
- Histology: Male Reproductive System
- Histology: Female Reproductive System
- Histology: Nervous System
- Histology: Integumentary System
- Histology: Identification of Histological Slides at a Glance

CHAPTER 12

Histology: Introduction

- Microscope
- Preparation of Tissue for Histological Study
- Procedure of Hematoxylin and Eosin Staining
- Epithelial Tissue (Epithelium)

LESSON 1: MICROSCOPE

The microscope is an instrument for examining histological and cytological specimens (Flowchart 12.1).

1. What is the principle of light microscope?

Ans. The day light or artificial light is used to illuminate the object, while the lens magnifies it.

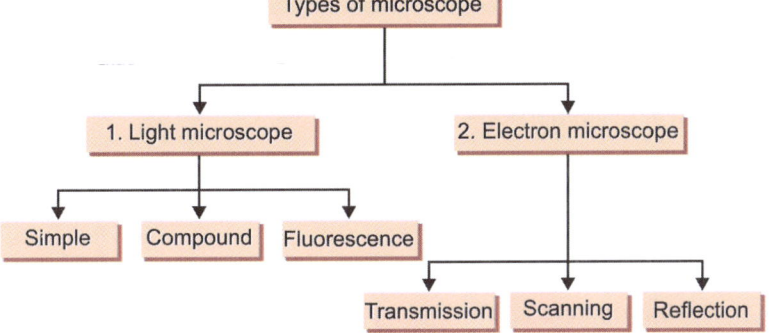

Flowchart 12.1: Types of microscope

2. **What are the basic differences between a simple microscope and a compound microscope?**

Ans. Differences between simple and compound microscopes are shown in Table 12.1.

Table 12.1: Differences between simple and compound microscopes

	Simple microscope	Compound microscope
Lens system	Single lens system	Double lens system (objective lens and eyepiece lens)
Magnification	Less than 200 times	100 to 1500 times

3. **Which type of microscope is commonly used by the students?**

Ans. Compound light microscope.

4. **How many objective lenses are there?**

Ans. There are three objective lenses with different magnifications.
   i. 10X—low power objective
   ii. 40X—high power objective
   iii. 100X—oil immersion lens

5. **What is the magnification of eyepiece?**

Ans. Usually 10X. It may be 5X, 8X or 15X.

6. **What is the principle of electron microscope?**

Ans. High speed electron beams are used in place of light.

7. **What is the magnification of electron microscope?**

Ans. More than 2,00,000 times.

8. **How is the magnification calculated in a light compound microscope?**

Ans. Magnification in low power = 10X (objective lens) × 10X (eyepiece lens) = 100X.
Magnification in high power = 40X (objective lens) × 10X (eyepiece lens) = 400X.
Magnification in oil immersion = 100X (objective lens) × 10X (eyepiece lens) = 1000X.

9. **What are the types of light compound microscope?**

Ans. i. Uniocular (single eyepiece),   ii. Binocular (double eyepiece).

10. **What are the parts of a light compound microscope?**

Ans. a. Mechanical parts (Fig. 12.1):
   i. Base
   ii. Limb or arm
   iii. Body tube
   iv. Nose piece carrying three objectives (10X, 40X, 100X)

Histology: Introduction

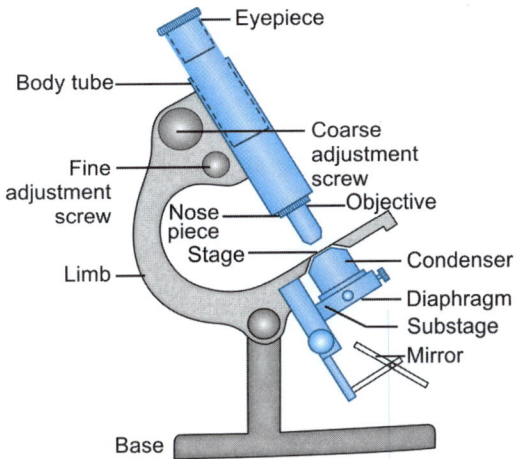

Fig. 12.1: Light compound microscope

    v. Focusing adjustment screw:
       a. Coarse adjustment screw
       b. Fine adjustment screw
   vi. Stage (platform for slides)
  vii. Substage (to carry condenser).
b. Optical parts (Fig. 12.1):
    i. Condenser
   ii. Objectives (Low power, high power, oil immersion)
  iii. Eyepiece
  iv. Iris diaphragm
   v. Mirror (one surface is plane and the other surface is concave).

LESSON 2: PREPARATION OF TISSUE FOR HISTOLOGICAL STUDY

STEPS

1. Collection of Tissue
    i. Tissue is collected from healthy bodies, preferably within 6–8 hours after death.
   ii. Tissue size must not exceed 1 cm in any dimension.
  iii. Tissue must not be allowed to dry.

2. Fixation
    i. Tissue is fixed with **10% formal saline** (most commonly used fixative) for 12–24 hours.
   ii. Volume of fixative must be 10–15 times more than the volume of the tissue.
  iii. Fixation hardens the tissue and prevents autolysis, bacterial putrefaction and decomposition.

iv. Composition of 10% formal saline = 10% formalin (10 mL) + NaCl (0.85 mg) + water (90 mL).

3. **Dehydration**
   i. Gradual dehydration by ascending grades of alcohol as shown in Flowchart 12.2.
   ii. Dehydration is necessary to impregnate the tissue with xylene in which paraffin dissolves. *Remember that xylene mixes well both with alcohol and paraffin.*
   iii. Gradual dehydration prevents distortion of cells of the tissue.
   iv. **If extended periods of dehydration or a stopping point is needed, most histologists choose 70% alcohol. The tissue no longer shrinks as much by 70% alcohol.**

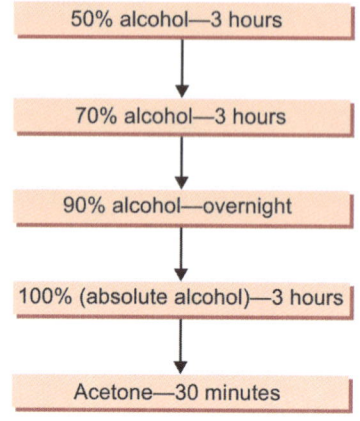

Flowchart 12.2: Gradual dehydration by ascending grades of alcohol

4. **Clearing**
   i. Commonly used clearing agent is **xylene.**
   ii. Two changes of xylene are done:
      a. Xylene 1–30 minutes
      b. Xylene 2–30 minutes
   iii. Xylene replaces the alcohol in the tissue and makes the tissue clear (transparent).
   iv. Some other clearing agents are: Toluene, benzene, chloroform, cedar wood oil, etc.

5. **Wax impregnation or wax infiltration**
   i. Xylene treated tissue is passed through three successive changes of pure paraffin for a period of ½ hour in each (embedding temperature—58 to 60°C).
   ii. Xylene is removed and replaced by paraffin.

6. **Paraffin embedding**
   i. Paraffin infiltrated tissue is placed in moulds containing melted paraffin (melting point—58°C).
   ii. The mould is allowed to cool.
   iii. Paraffin block containing tissue hardens and becomes ready for section cutting.

7. **Section cutting**
   i. Commonly used microtome is **rotary microtome.**
   ii. Some other microtomes are: Rocking, Sledge, Sliding, Freezing, etc.
   iii. **Thickness of sections**: 5–7 microns.

8. **Mounting**
   i. The thin sections of tissue is allowed to float in warm water in **hot water bath (45–50°C).**
   ii. Sections are affixed to albuminized slides.

Histology: Introduction

iii. **Egg albumin** is prepared by equal parts of glycerin and white of egg, mixed with few crystals of thymol as preservative.
iv. Slides containing sections are allowed to dry.
v. Now, the tissue on the slide is ready for staining.

LESSON 3: PROCEDURE OF HEMATOXYLIN AND EOSIN STAINING

HEMATOXYLIN AND EOSIN (H&E) STAINING PROTOCOL

1. Choose slides to be stained from boxes in the cold room or the 4° fridge (usually 3–4 slides from the same block—choose slide numbers ~ evenly spaced throughout).
2. Place slides to be stained in slide rack and allow to warm to RT for ~ 5–10 minutes.
3. **Deparaffinization**:
    i. Xylene or xylol (30 seconds)
    ii. Xylene or xylol (30 seconds)
    iii. Xylene or xylol (30 seconds)—if necessary.
4. **Hydration: To remove xylol from the tissue:**
    i. 100% alcohol (30 seconds).
    ii. **Descending grades of alcohol:**
        90% (30 seconds), 70% (5–6 dips), 50% (5–6 dips).
5. **Washing** slides as follows:
    Tap water for 15 seconds.
6. **Stain** slides as follows: **Hematoxylin** for 2–5 minutes.
7. **Washing:** Rinse slides in running tap water for 2–3 minutes.
8. **Differentiation** by 1% acid alcohol (3–4 dips).
    Rinse slides in running tap water for 2–3 minutes.
9. **1% eosin:** 1–2 minutes.
10. **Washing** (wash off surplus stain).
11. **Dehydration** (by ascending grades of alcohol): 50% (5–6 dips) → 70% (5–6 dips) → 90% (1 minute) → absolute alcohol (100%) (2–3 minutes).
12. **Clearing in** xylene for 2–3 minutes (2 changes).
13. **Mounting:** By D.P.X.
14. **Coverslipping:** Carefully add a coverslip by holding the coverslip at about a 45° angle and then touching the bottom edge of the coverslip to the edge of the slide, allowing the mounting media to form a line between the slide and coverslip. Lower the coverslip onto the slide, making sure to eliminate any air bubbles.

LESSON 4: EPITHELIAL TISSUE (EPITHELIUM)

1. What do you mean by a tissue?

Ans. A tissue is a collection of cells together with their intercellular substance performing a similar function.

2. What are the basic types of tissues?

Ans. Four basic types of tissues are:
  i. Epithelial tissue or epithelium
  ii. Connective tissue
  iii. Muscular tissue
  iv. Nervous tissue

3. What do you mean by epithelium?

Ans. **Epithelium** is a type of tissue which is formed by continuous sheets of epithelial cells resting on a basement membrane.

4. What are the characteristics of epithelium?

Ans.
  i. The epithelial cells are closely packed together by cell junctions.
  ii. Avascular and devoid of lymphatics.
  iii. Epithelial cells rest on a basement membrane.
  iv. All epithelial cells have one basal surface and one free surface, i.e. polarized.
  v. Cell turn over occurs in epithelial cells.
  vi. Epithelial cells may be arranged in a single row or in multiple rows.
  vii. Epithelium covers the body surface and external surface of solid organs.
  viii. Epithelium lines the luminal surface of hollow tubular structures and forms glands.
  ix. Minimum intercellular substances.
  x. Developed from all three types of germinal layers according to their location.
     a. **Endoderm**: Lining epithelium of gastrointestinal and respiratory tracts.
     b. **Mesoderm**: Endothelium of blood vessels and lymphatics.
     c. **Ectoderm**: Epidermis of skin.

5. How do you classify epithelium?

Ans. Flowchart 12.3 shows the classification of epithelium.

Histology: Introduction

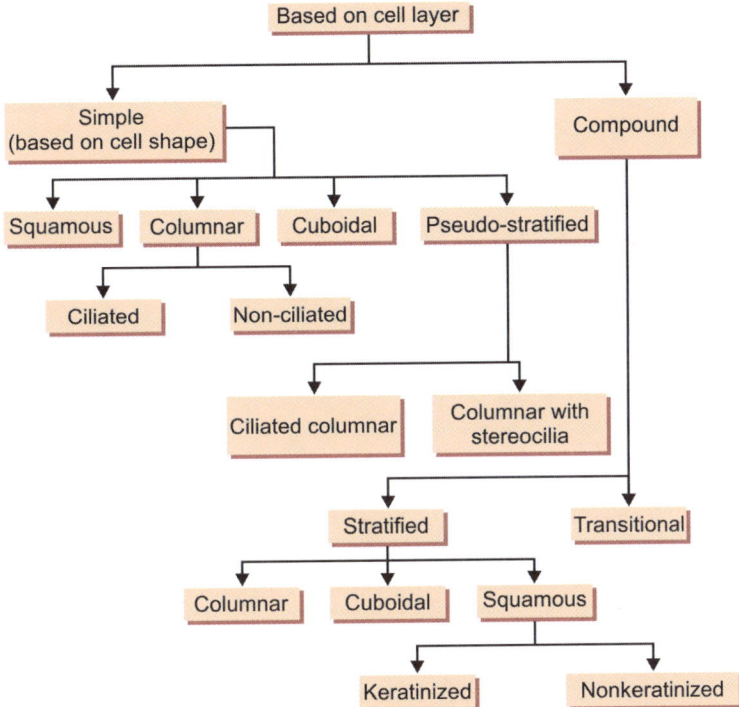

Flowchart 12.3: Classification of epithelium

6. What are the differences between simple and stratified epithelium?

Ans. Table 12.2 shows the differences between simple and stratified epithelium.

Table 12.2: Differences between simple and stratified epithelium

	Simple epithelium	*Stratified epithelium*
Layers of cells	Single	Multiple
Size and shape of cells	Uniform	Variable
Basement membrane	All the cells rest on basement membrane	Only the basal cells rest on the basement membrane

7. What are the points for identification of different types of epithelium and their distribution at a glance?

Ans. Table 12.3 shows the different types of epithelium and their distribution.

Table 12.3: Different types of epithelium and their distribution

Types of epithelium	Points for identification	Distribution
Simple epithelium (Fig. 12.2)	Flat cells with centrally located oval nucleus	i. Lung alveoli ii. Bowman's capsule of kidney iii. Lining of blood vessels and serous cavities
Simple cuboidal (Fig. 12.3)	Cuboid-shaped cells with centrally located round nucleus	i. Thyroid follicle ii. Tubules of kidney iii. Germinal epithelium of ovary iv. Ducts of exocrine glands
Simple columnar	Column-like tall cells, i.e. height is more than width with elongated nucleus toward the base	
(a) Nonciliated (Fig. 12.4)		i. Stomach ii. Intestine iii. Gallbladder iv. Uterus
(b) Ciliated (Fig. 12.5)		i. Respiratory tract ii. Uterine tubes iii. Olfactory epithelium iv. Ventricles of brain
Pseudostratified columnar (Fig. 12.6)	i. Cells are of different height but all the cells touch the basement membrane ii. Nuclei lie at different levels due to variable shape and height of the cells	i. Nasal cavity ii. Trachea iii. Bronchi iv. Vas deferens v. Membranous and spongy parts of male urethra vi. Epididymis
Stratified squamous (nonkeratinized) (Fig. 12.7)	i. Multilayered cells ii. Surface cells are squamous and living iii. No keratin layer	i. Mouth cavity ii. Tongue iii. Pharynx iv. Esophagus v. Lower part of anal canal vi. Vagina vii. Cornea
Stratified squamous (keratinized) (Fig. 12.8)	i. Multilayered cells. • Basal cells are columnar • Intermediate cells are cuboidal/polyhedral • Superficial cells are squamous ii. Superficial cells are dead without nuclei, containing **keratin**	Epidermis of skin
Stratified columnar (Fig. 12.9)	i. Multilayered (2–3) cells ii. Surface cells are columnar iii. Basal cells are columnar or cuboidal	i. Palpebral conjunctiva ii. Urethra
Stratified cuboidal (Fig. 12.10)	i. Multilayered (2–3) cells ii. Surface cells are cuboidal iii. Basal cells are columnar or cuboidal	Ducts of sweat glands

Contd...

Histology: Introduction

Contd...

Types of epithelium	Points for identification	Distribution
Transitional epithelium (Fig. 12.11) umbrella cells	i. Multilayered (3–6) cells ii. Deepest cells are columnar or cuboidal iii. Intermediate cells are pear-shaped iv. Superficial cells are large polyhedral (umbrella cells)	i. Renal calyces ii. Pelvis of ureter iii. Ureter iv. Urinary bladder v. Proximal part of urethra

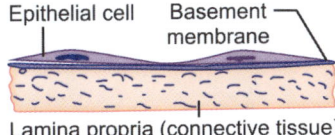

Fig. 12.2: Simple squamous epithelium

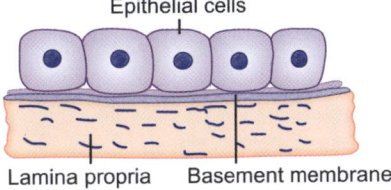

Fig. 12.3: Simple cuboidal epithelium

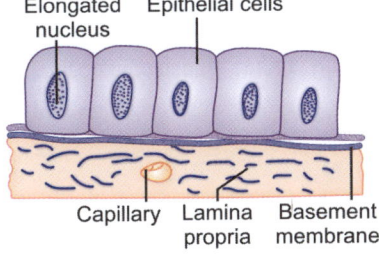

Fig. 12.4: Nonciliated simple columnar epithelium

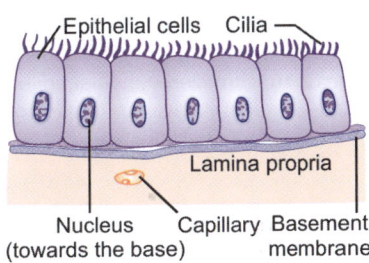

Fig. 12.5: Ciliated columnar epithelium

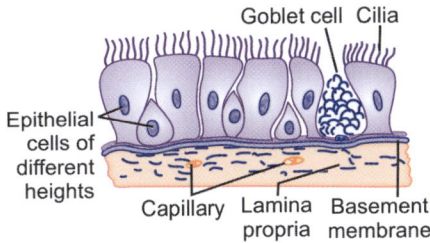

Fig. 12.6: Pseudostratified ciliated columnar epithelium

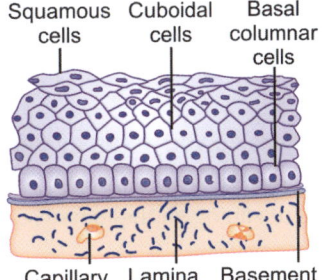

Fig. 12.7: Nonkeratinized stratified squamous epithelium

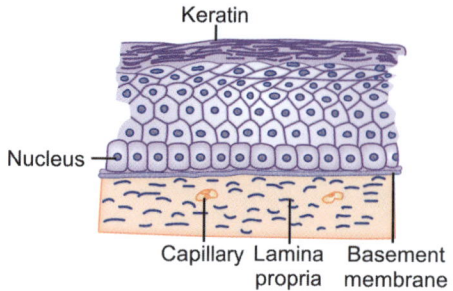

Fig. 12.8: Keratinized stratified squamous epithelium

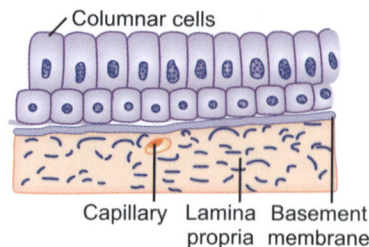

Fig. 12.9: Stratified columnar epithelium

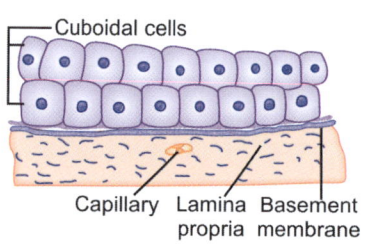

Fig. 12.10: Stratified cuboidal epithelium

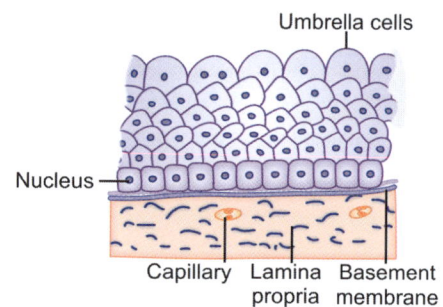

Fig. 12.11: Transitional epithelium

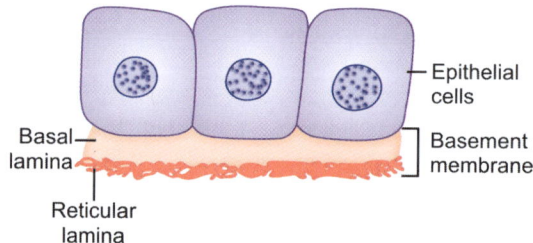

Fig. 12.12: Components of a basement membrane

8. What do you mean by endothelium and mesothelium?

Ans. **Endothelium** is the lining epithelium of heart, blood vessels and lymphatics, **mesothelium** is the lining epithelium of serous membranes of body cavities (peritoneum, pleura).

9. What is the composition of basement membrane?

Ans. Basement membrane consists of (Fig. 12.12):
    i. Basal lamina, consisting of:
        a. Lamina densa   b. Lamina lucida
    ii. Fibroreticular lamina

Histology: Introduction

10. What is the ideal staining for basement membrane?

Ans. **PAS stain** (periodic acid-Schiff stain).

11. What is the composition of mucous membrane?

Ans. Mucous membrane consists of:
   i. Epithelium (epithelial cells + basement membrane).
   ii. Lamina propria (layer of connective tissue).
   iii. Muscularis mucosae (layer of smooth muscle).

CHAPTER 13

Histology: Musculoskeletal System

- Compact Bone
- Skeletal Muscle
- Cardiac Muscle

LESSON 1: COMPACT BONE

Key points for identification (Figs 13.1 and 13.2)

1. Presence of Haversian system.
2. Presence of Haversian canal at the center of Haversian system.
3. Haversian canal containing blood vessels surrounded by concentric lamella.
4. Presence of lacunae filled with osteocytes.
5. Presence of radiating canaliculi containing processes of bone cells.

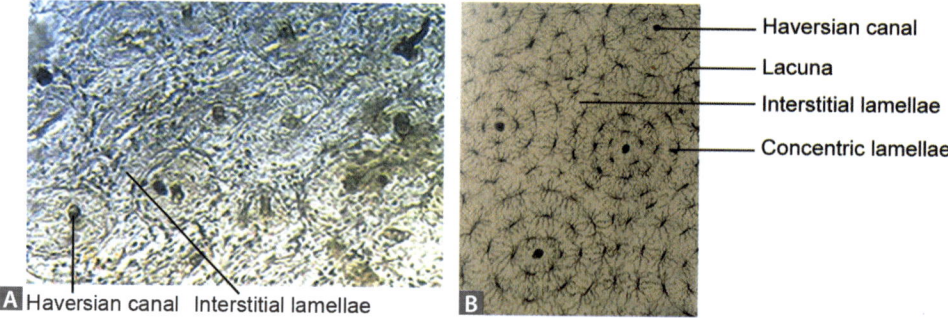

Figs 13.1A and B: Compact bone

Histology: Musculoskeletal System

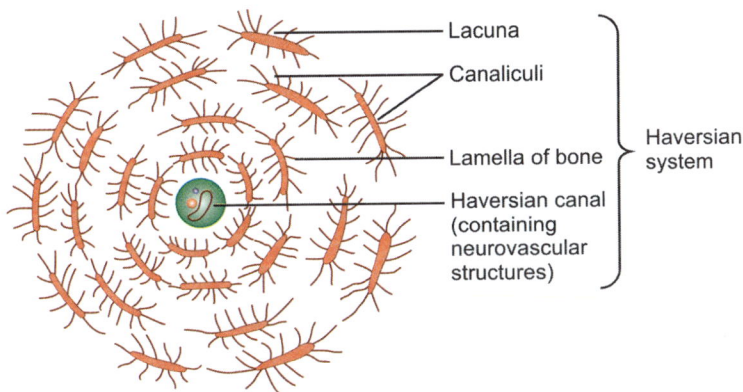

Fig. 13.2: Haversian system

PROBABLE QUESTIONS AND ANSWERS

1. What are the types of bone on the basis of its structural components?

Ans. **Two types**:
   i. Mature bone or lamellar bone.
   ii. Immature bone or Woven bone or non-lamellar bone.

2. What are the varieties of mature bone?

Ans. **Two types**:
   i. Compact bone
   ii. Spongy or cancellous bone.

3. What are the components of a bone?

Ans. **Bone consists of two components**:
   i. Bone cells
   ii. Bone matrix or intercellular substance.

4. What are the different types of bone cells?

Ans. **Four types**:
   i. **Osteoprogenitor cells**: Spindle-shaped stem cells which are derived from mesenchyme, distributed in periosteum and endosteum of bone.

ii. **Osteoblasts**: Cuboidal cells with eccentric round nucleus and numerous cytoplasmic processes, derived from osteoprogenitor cells and act as precursor cells of osteocytes.

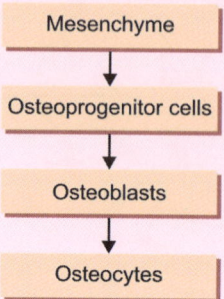

iii. **Osteocytes**: Flattened cells with central nuclei and numerous cytoplasmic processes, derived from osteoblasts and found in lacunae in the bone matrix.

iv. **Osteoclasts**: Large cells (20–100 μm) with multiple nuclei formed by fusion of monocytes, lying in the depressions of bone surface called **Howship's lacunae** but devoid of any processes.

15. What are the types of osteoprogenitor cells?

Ans. Two types:
  i. **Committed osteoprogenitor cells** which are committed to differentiate into osteoblasts.
  ii. **Inducible osteoprogenitor cells** which give rise to heterotopic bone formation.

16. What are the staining properties of different bone cells?

Ans.
  i. **Osteoprogenitor cells:** Light staining.
  ii. **Osteoblasts:** Strongly basophilic.
  iii. **Osteocyte:** Faintly basophilic.
  iv. **Osteoclasts:** Acidophilic.

17. What do you mean by lacunae?

Ans. Lacunae are the spaces in the bony matrix in which lie the osteocytes.

18. How many osteocytes are occupied in one lacuna?

Ans. Each lacuna occupies one osteocyte.

19. What are canaliculi?

Ans. Canaliculi are fine canals that spread out from each lacuna and communicate with the canals from other lacunae.

Histology: Musculoskeletal System

10. What do these lacunae and canaliculi contain?

Ans. Each lacuna contains one osteocyte and the canaliculi contain thin and delicate cytoplasmic processes of the osteocytes, called **filopodia.**

11. What is the composition of bony matrix?

Ans. Bony matrix or intercellular substance consists of two components:
   i. **Organic matrix:**
      a. Ground substance, consisting of:
         - Glycosaminoglycans
         - Proteoglycans
         - Glycoproteins (osteonectin, osteocalcin, osteopontin)
         - Water
      b. Type 1 collagen fibers
   ii. **Inorganic matrix:**
      a. Hydroxyapatite crystals of calcium phosphate.
      b. Potassium, magnesium, citrate, chloride, fluoride, sodium, strontium, plutonium, lead, etc.

12. How much dry weight of bone is contributed by organic substance of bony matrix?

Ans. About 35%.

13. How much dry weight of bone is contributed by inorganic salts of bony matrix?

Ans. About 65%.

14. Which type of inorganic salt is maximally found?

Ans. Calcium phosphate (about 85%), followed by calcium carbonate (about 10%).

15. Where do you find maximum calcium in the body?

Ans. Bone contains about 97% of total calcium of the body.

16. What is the source of collagen fibers in the organic matrix of bone?

Ans. Osteoblasts synthesize collagen fibers.

17. What do you mean by osteoid?

Ans. Osteoid is the organic matrix of the bone before it is mineralized.

18. What do you mean by lamellar bone?

Ans. Lamellar bone is a mature bone which is made up of lamellae or layers.

19. What is a lamellus?

Ans. A lamellus (singular form of lamellae) is a thin plate of bone consisting of mineral salts and collagen fibers that are embedded in a gelatinous ground substance.

20. What is responsible for the lamellar appearance of bone?

Ans. The arrangement of collagen fibers which run parallel to each other in one lamellus contribute to the lamellar appearance of bone.

21. What is the important characteristic feature of a compact bone?

Ans. The presence of **Haversian system** is the most important feature.

22. What are the components of a Haversian system?

Ans.
  i. Haversian canal
  ii. Lacunae
  iii. Canaliculi
  iv. Volkmann's canal
  v. Interstitial lamellae
  vi. Circumferential lamellae

23. What are the features of a Haversian system?

Ans.
  i. **Shape:** Cylindrical.
  ii. **Diameter:** About 100–400 µm, average being about 150 µm.
  iii. **Disposition:** Longitudinal (along the long axis of the bone).
  iv. **Number:** About 21 million in adult human skeleton.
  v. Also called **secondary osteons.**
(Circumferential lamellae are known as **primary osteons**).

24. What is Haversian canal?

Ans. Haversian canal is the central canal of the Haversian system containing blood vessels, nerves and lymphocytes, diameter being about 20–50 µm.

25. What are Volkmann's canals?

Ans. **Volkmann's canals** are oblique or horizontal channels which communicate the Haversian canal with the medullary cavity and with the external surface of the bone. These canals also connect adjacent Haversian canals.

26. What do these canals contain?

Ans. Volkmann's canals contain blood vessels and nerves. These blood vessels permeate the bones from periosteum to endosteum to provide nutrition.

27. What are interstitial lamellae?

Ans. There are irregularly arranged lamellae occupying the angular intervals in between the Haversian system.

28. What do you mean by circumferential lamellae?

Ans. These lamellae are arranged circumferentially around the outer and inner surfaces of the bone.

29. What is cement line?

Ans. It is the non-collagenous strongly basophilic line which demarcates one Haversian system from the neighboring system.

30. What are the types of lamellae?

Ans. Two types:
  i. **Cementing lamellae** consisting of high mineral salts and less collagen fibers.
  ii. **Fibrillary lamellae** consisting of less mineral salts and high collagen fibers.

31. What do you mean by spongy or cancellous bone?

Ans. These bones are also mature bones (like compact bones), formed by trabeculae.

32. What are trabeculae?

Ans. Trabeculae are made up of a number of superimposed lamellae and enclose wide spaces which are filled in by bone marrow.

33. Do these bones possess Haversian system?

Ans. No, spongy bones usually do not form Haversian system but their trabeculae are lamellated.

34. What are the structural similarities and differences between compact bone and spongy bone?

Ans. Both the types of bone are lamellar, i.e. made up of lamellae.
Differences are: Compact bone is composed of abundant bone tissue with intervening small spaces. But, the spongy or cancellous bone is composed of less bone tissue with intervening large spaces.

35. What do you mean by woven or immature bone?

Ans. These are newly formed bone without having any lamellar structure.

36. Can you give some examples of this type of bone?

Ans.
i. Embryonic bone.
ii. Young fetal bone.
iii. Remodeling bone following fracture.

37. What is the structural feature of woven bone?

Ans. Woven bone is composed of:
i. Randomly arranged collagen bundles.
ii. Less minerals and cement substance.
iii. Absence of lamellae (but later, it is replaced by lamellar bone).

38. Why such bone is termed as woven bone?

Ans. The bundles of collagen fibers in this bone run randomly in different directions interlacing with each other, **hence called woven bone.**

39. What is periosteum?

Ans. Periosteum is a thickened fibrocellular/fibrovascular covering the outer surface of bone except sesamoid bones which are devoid of periosteum.

40. What are the layers of periosteum?

Ans. Two layers:
i. **Outer fibrous layer**, made up of collagen fibers.
ii. **Inner vascular or cellular layer** containing osteoblasts, hence called osteogenic layer.

41. What is endosteum?

Ans. **Endosteum** is a fibrovascular membrane which lines the medullary cavity of bone.

42. What do you mean by perforating fibers of Sharpey?

Ans. These are fibers of tendons which continue into the outer layers of bone to get attached firmly.

43. What are the types of bone marrow?

Ans. Two types:
i. **Red marrow,** composed of blood vessels and blood forming cells.
ii. **Yellow marrow,** made up mainly of fat cells.

Histology: Musculoskeletal System

44. What is the location of marrow in a bone?

Ans.   i. **Red marrow** is located at the ends of bone (epiphysis), ribs, sternum, vertebrae, flat bones, etc.

   ii. **Yellow marrow** is located within the medullary cavity of adult long bones.

Remember that the entire bone marrow is red in the bones of fetus or of a young child. The yellow marrow appears gradually in the shaft of bones by replacing the red marrow with the advancement of age.

LESSON 2: SKELETAL MUSCLE

Key points for identification (Figs 13.3 to 13.7)

1. Long cylindrical muscle fibers.
2. Multinucleated muscle cells (muscle fibers) and nuclei are placed peripherally.
3. Prominent transverse striations.
4. Muscle fibers do not branch.

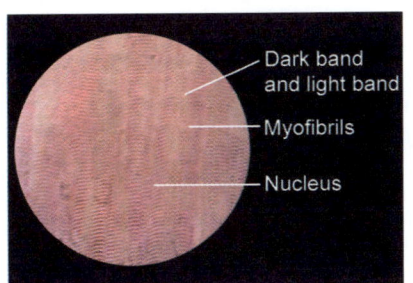

Fig. 13.3: Microstructure of skeletal muscle

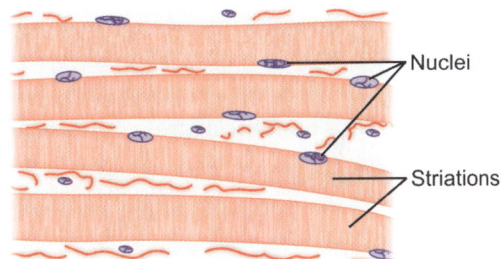

Fig. 13.4: Skeletal muscle fibers

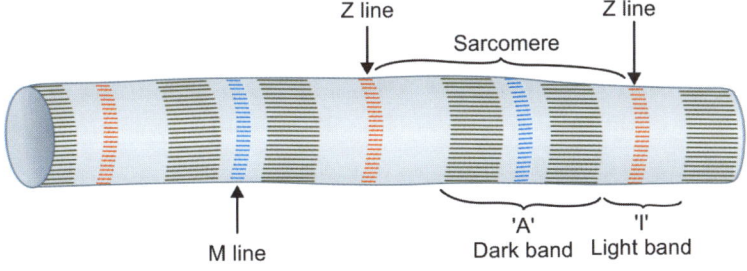

Fig. 13.5: Myofibril

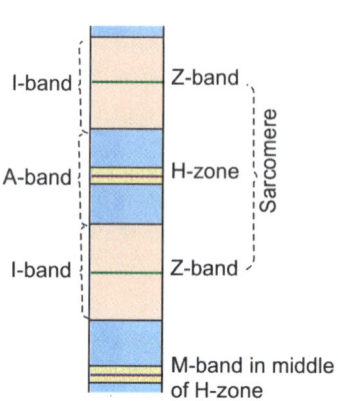

Fig. 13.6: Sarcomere

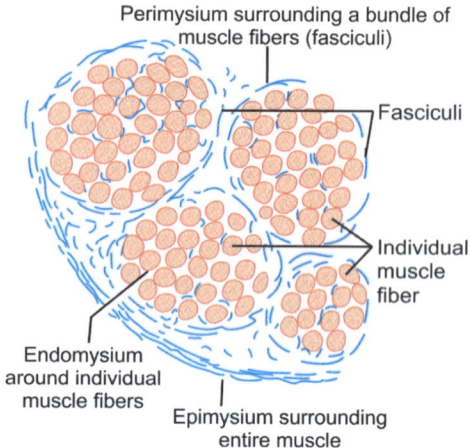

Fig. 13.7: Endomysium, perimysium and epimysium

PROBABLE QUESTIONS AND ANSWERS

1. **What are the types of muscles?**

Ans. Three types:
  i. Skeletal/striped/striated/voluntary muscle.
  ii. Smooth/unstriped/nonstriated/involuntary muscle.
  iii. Cardiac muscle.

2. **Are the nomenclatures of these three types of muscles completely satisfactory?**

Ans. No, because:
  i. Though the esophageal muscles and the muscles of the anal canal are skeletal muscles, they have no relation to the skeleton. Moreover, the esophageal muscle is voluntary in upper third, mixed in middle third and involuntary in the lower third.
  ii. Though the muscles of the diaphragm and the pharynx are striped, they are not strictly voluntary in action.
  iii. Though the skeletal muscles are striated in appearance under a microscope, the cardiac muscle also shows striations.
  iv. Though the smooth muscle is involuntary in action, the contraction of the muscle of the urinary bladder (smooth muscle) may be produced by voluntary effort during passing urine.

3. **What do you mean by the word "muscle"?**

Ans. Musculus (Latin) means a little mouse. The muscle resembles to mouse and the tendon of the muscle represents the tail of the mouse.

Histology: Musculoskeletal System

4. What are the components of muscle tissue?

Ans.
i. Muscle cells or myocytes.
ii. Intercellular matrix of connective tissue.

5. What are the developmental sources of muscles?

Ans. All the muscles of the body are **mesodermal** in origin except the following muscles which are ectodermal in origin.

Muscles of ectodermal origin are:
i. Arrectores pilorum.
ii. Muscles of iris.
iii. Myoepithelial cells of sweat glands, lacrimal glands and salivary glands.

6. What do you mean by a muscle fiber?

Ans. A muscle fiber is nothing but an elongated muscle cell (myocyte).

7. What are the dimensions of a skeletal muscle fiber?

Ans.
i. **Length:** Few mm to few cm depending on location, maximum length being up to 35 cm (found in sartorius muscle).
ii. **Diameter:** 10–100 μm.

8. What do you mean by fasciculi of muscle?

Ans. Fasciculi are groups of muscle fibers surrounded by a connective tissue sheath.

9. What is endomysium?

Ans. Endomysium is the delicate connective tissue sheath around individual muscle fiber, made up of type III and type IV collagen fibers.

10. What is perimysium?

Ans. Perimysium is a tough connective tissue covering of muscle fasciculi, made up of type I and type II collagen fibers.

11. What do you mean by epimysium?

Ans. Epimysium is the connective tissue covering of entire muscle, made up of type I collagen fibers.

Note that individual muscle fiber is covered by **endomysium**; fasciculi which are formed by groups of muscle fibers are covered by **perimysium** and the whole muscle which is formed by groups of fasciculi is covered by **epimysium** (Fig. 13.7).

12. How is a muscle fiber of skeletal muscle formed?

Ans. A skeletal muscle fiber is formed by the fusion of numerous myoblasts.

13. What are the components of a skeletal muscle cell?

Ans.
  i. Sarcolemma (cell membrane).
  ii. Sarcoplasm (cytoplasm).
  iii. Nuclei.
  iv. Constituents of sarcoplasm.

14. What do you mean by sarcolemma?

Ans. The sarcolemma is the cell membrane of a muscle cell.

15. What are the layers and thickness of sarcolemma?

Ans. It is about (75Å) thick and consists of three layers:
  i. Outer protein layer.
  ii. Intermediate lipid layer.
  iii. Inner protein layer.

16. What do you mean by sarcoplasm?

Ans. The sarcoplasm is the cytoplasm of a muscle cell.

17. What are the important constituents of sarcoplasm?

Ans.
  i. Sarcoplasmic reticulum.
  ii. Mitochondria, also called **sarcosomes**.
  iii. Golgi bodies.
  iv. Ribosomes.
  v. Lysosomes.
  vi. Glycogen.
  vii. Lipid vacuoles.

18. How many nuclei are present in a single muscle cell?

Ans. A skeletal muscle cell is **multinucleated**. A single muscle cell may possess hundreds of nuclei which are located peripherally just beneath the sarcolemma.

19. What is the most distinctive feature of skeletal muscle?

Ans. Unbranched myofibrils with cross-striations and peripherally located multiple nuclei.

20. What are myofibrils?

Ans. Myofibrils are longitudinally oriented unbranched contractile elements of muscle fiber.

21. How are the myofibrils composed of?

Ans. Myofibrils are composed of protein filaments, called **myofilaments** which are the ultimate contractile elements of a skeletal muscle.

22. What are the types of these protein filaments (myofilaments)?

Ans. Mainly two types:
    i. Thin filaments (about 8 nm in diameter).
    ii. Thick filaments (about 12 nm in diameter).

23. What are the proteins of thin filaments?

Ans.
    i. Actin
    ii. Troponin
    iii. Tropomyosin

24. What is the protein of thick filament?

Ans. Myosin. Each myosin filament (thick filament) is made up of about 274 myosin molecules.

25. How much of total proteins of a striated muscle is contributed by actin and myosin?

Ans. About 55% of the total proteins are actin and myosin.

26. What are the other proteins of muscle?

Ans.
    i. Actinin
    ii. Myomesin
    iii. Titin
    iv. Desmin
    v. Nebulin
    vi. Merosin

27. What are the two forms of actin filament?

Ans.
    i. Fibrillar actin or F-actin (polymerized form).
    ii. Globular actin or G-actin (unpolymerized form).

28. What is the function of actin and myosin?

Ans. Motile forces (movements) are generated by the interaction between actin and myosin proteins.

29. What is Conheim's area?

Ans. Though the myofibrils are distributed uniformly throughout the fiber, they may be arranged in groups, called the **Conheim's area or the field of Conheim**.

30. What is the reason of cross-striated appearance of a muscle fiber?

Ans. With hematoxylin stain, **alternate dark and white bands** along the length of each myofibril under polarized microscope gives the muscle fiber a cross-striated appearance.

31. What is the other term used for dark bands?

Ans. Dark bands are termed as **anisotropic or A bands**.

32. What do you mean by the term anisotropic?

Ans. The materials which do not refract light equally in different planes are anisotropic.

33. What is the other term used for light bands?

Ans. Light bands are termed as **isotropic bands or I-bands**.

34. What do you mean by the term isotropic?

Ans. The materials which refract light equally in all directions are said to be isotropic.

35. What is Z-band/Z-line/Z-disk/Krause's membrane?

Ans. It is a thin dark transverse line in the middle of I-band.

36. Why this line is termed as Z-line?

Ans. "Z" means **Zwischenschiebe** (German word)
- Zwischen means between
- Schiebe means disk

37. What is H-band/H-zone?

Ans. It is a clear/light area in the middle of A-band.

38. What do you mean by the letter "H"?

Ans. **H** stands for **Henson** who first described this zone. H-bands are obliterated in a state of contraction of myofibril.

39. What is M-line?

Ans. M-line is a thin dark line in the middle of H-band.

Histology: Musculoskeletal System

40. What does the letter M stand for?"

Ans. M stands for **Mittleschiebe** (German word) which means: **Mittle** = Middle and **Schiebe** = disk.

41. What is the importance of M-line?

Ans. The M-line is the site of interconnections between the adjacent thick myosin filaments. The major protein of this M-line is creatine kinase.

42. What is sarcomere?

Ans. The sarcomere is the contractile apparatus of myofibril, which is situated between two consecutive Z-bands of myofibrils.

43. What is the length of sarcomere?

Ans. About 2.5 µm in the resting state. It shortens during contraction of muscle.

44. What are the stages of development of a muscle fiber?

Ans. Flowchart 13.1 shows the stages of development of a muscle fiber.

45. How is the single-nucleated myoblast converted into multinucleated myotube?

Ans. This is due to the fusion of the myoblasts forming myotube which is later converted to muscle fiber.
So, each muscle fiber is a **syncytium** with hundreds of nuclei.

Flowchart 13.1: Stages of development of a muscle fiber

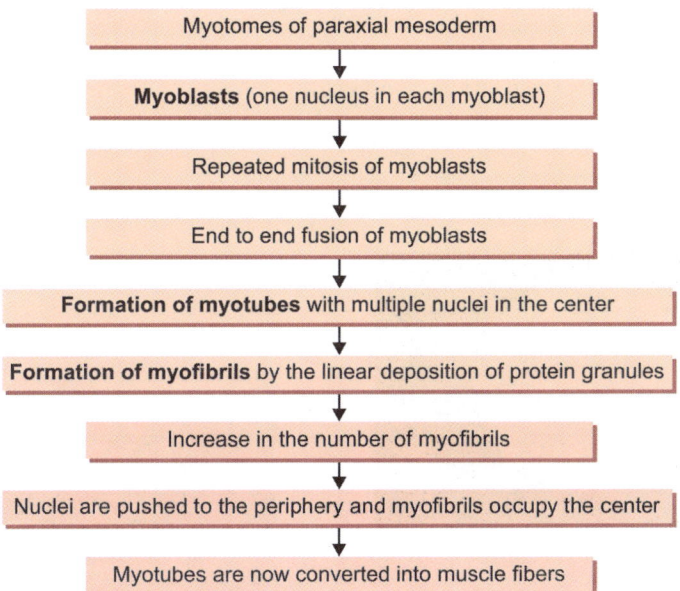

46. What do you mean by syncytium?

Ans. Syncytium means a multinucleate mass of protoplasm produced by the merging of cells.

47. Why do the nuclei of a muscle fiber occupy its peripheral region?

Ans. During the process of development of muscle fiber, the nuclei occupy its center. Later on, the myofibrils in the cytoplasm are increased in number and push the nuclei to the periphery. If the nuclei remain in the center, they may interrupt the continuity of the contractile mechanism of the muscle fiber which is contributed by the contractile elements (myofibrils).

48. What is the external lamina of a muscle fiber?

Ans. The external lamina or basement membrane is the covering of a muscle fiber outside the sarcolemma (plasma membrane).

49. What are satellite bodies?

Ans. These are mononucleated cells which are found between the basement membrane and the plasma membrane of muscle fiber.

50. What is the function of these satellite cells?

Ans. The satellite cells can undergo mitosis and form new muscle fibers after damage.

LESSON 3: CARDIAC MUSCLE

Key points for identification (Figs 13.8 to 13.10)

1. Cylindrical fibers, interconnected by side branches.
2. Centrally placed nucleus.
3. Faintly striated.
4. Presence of intercalated disk.

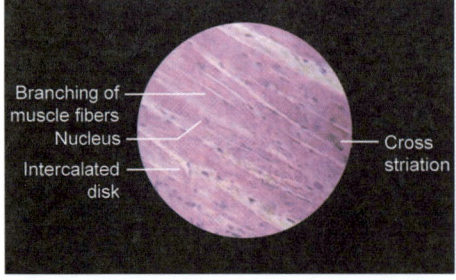

Fig. 13.8: Microstructure of cardiac muscle

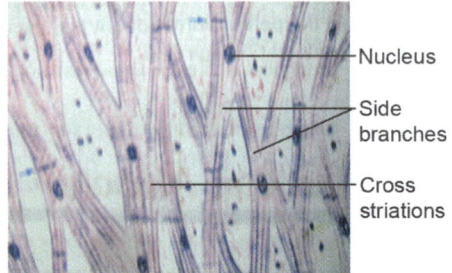

Fig. 13.9: Cardiac muscle fibers

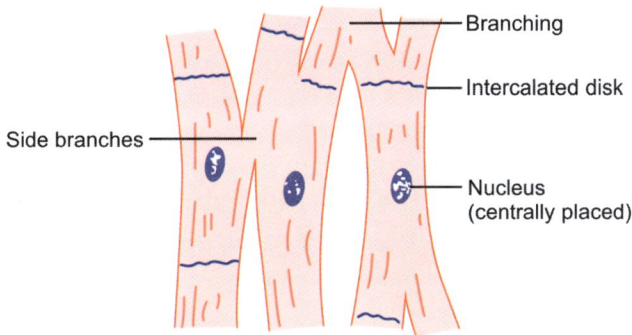

Fig. 13.10: Cardiac muscle fibers with branching

PROBABLE QUESTIONS AND ANSWERS

1. What is the shape of cardiac muscle fiber?

Ans. Short and cylindrical.

2. What is the length and diameter of cardiac myocytes?

Ans.
  i. **Length:** About 120 μm.
  ii. **Diameter:** About 20–30 μm.

3. What is the position of nucleus?

Ans. Nucleus is centrally located unlike skeletal muscle.

4. Do they form true syncytium?

Ans. No, they do not form true syncytium, but they form physiological syncytium.

5. What do you mean by physiological syncytium?

Ans. The plasma membrane of individual cardiac myocyte is separated from adjacent cardiac myocyte by gap junctions. The gap junctions allow spread of contractile stimulus from one cell to another and behave like a syncytium. Hence, functionally cardiac myocytes form syncytium, called physiological syncytium.

6. What are intercalated discs?

Ans. Intercalated discs are dark staining zigzag transverse lines at the junctions between adjoining cardiac myocytes.

7. Where are these disks found?

Ans. The **intercalated disks are** found opposite the I-bands of cardiac muscle fiber and also in the side branches of cardiac myocytes.

8. What is the interval between the consecutive disks?

Ans. The disk appears at an regular interval of 70–80 μm.

9. How is a disk formed?

Ans. A disk is formed by the apposition of the cell membranes of the adjoining myocytes which are connected by numerous desmosomes, gap junctions and tight junctions.

10. What is the interval between the cell membranes of adjacent myocytes?

Ans. About 20 nm.

11. What is the most distinctive feature of cardiac muscle?

Ans. **Side branches** from the cardiac myocytes and their anastomosis with the adjacent myocytes.
The ends of the myocytes split longitudinally to form branches.

12. Can cardiac myocytes regenerate following injury?

Ans. **No**, cardiac myocytes cannot regenerate following injury because of absence of satellite cells.

13. What are the different types of cardiac myocytes?

Ans. **Four types**:
  i. Nodal myocytes
  ii. Transitional myocytes
  iii. Purkinje myocytes
  iv. Ordinary or working myocytes

14. What are the similarities between skeletal and cardiac muscles?

Ans. i. Both are elongated fibers containing myofibrils.
  ii. Both show transverse striations.
  iii. Structure of myofibrils is same in both these muscle fibers.
  iv. Both are having similar connective tissue framework.

Histology: Musculoskeletal System

15. What are the similarities between smooth and cardiac muscles?

Ans.
i. Both are involuntary in function.
ii. Both are supplied by autonomic nerves.
iii. Nuclei in both these types of myocytes lie in the center of the cell.

16. What are the areas of distribution of smooth muscles?

Ans. Smooth muscles are distributed in the following areas:
i. Walls of gastrointestinal tract.
ii. Respiratory tract.
iii. Walls of ureter, urinary bladder and urethra.
iv. Walls of arteries, veins, ducts of exocrine glands.
v. Walls of uterus, uterine tubes, vas deferens.
vi. Intrinsic muscles of eye, etc.

17. Comparative chart between three types of muscle fibers.

Ans.

Features	Skeletal muscle	Cardiac muscle	Smooth muscle (Fig. 13.11)
Shape	Long, cylindrical	Short, cylindrical	Long, spindle-shaped
Length	Few mm to few cm	About 80–120 μm	15–500 μm
Nucleus	Multiple, peripheral	Single, central	Single, central
Striation	Prominent transverse striations	Prominent transverse striations	Longitudinal striations
Intercalated disks	Absent	Present	Absent
Satellite cells	Present	Absent	Absent
Branches	Absent	Present	Absent
Nerve supply	Somatic	Autonomic	Autonomic
Control	Voluntary	Involuntary	Involuntary

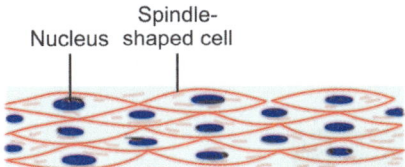

Fig. 13.11: Smooth muscle fiber

CHAPTER 14

Histology: Blood Vascular System

- Arteries
- Veins

LESSON 1: ARTERIES

Key points for identification (Figs 14.1 to 14.3)

1. Lumen is small, round and empty.
2. Wall is thick and consists of three layers (From outside inwards: Tunica adventitia, tunica media and tunica intima).
3. One-third thickness of the wall is formed by tunica adventitia.
4. Two-thirds thickness of the wall is formed by tunica media.
5. Well-defined endothelial lining.
6. Distinct internal and external elastic lamina.

PROBABLE QUESTIONS AND ANSWERS

1. What are the layers of an arterial wall?

Ans. From inside outwards:
   i. Tunica intima.
   ii. Tunica media.
   iii. Tunica externa (adventitia).

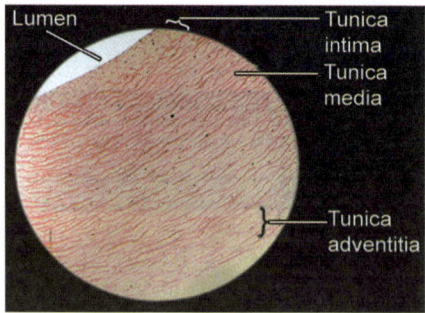

Fig. 14.1: Elastic artery

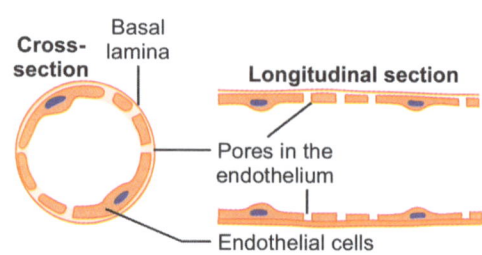

Fig. 14.2: Fenestrated capillary

Histology: Blood Vascular System

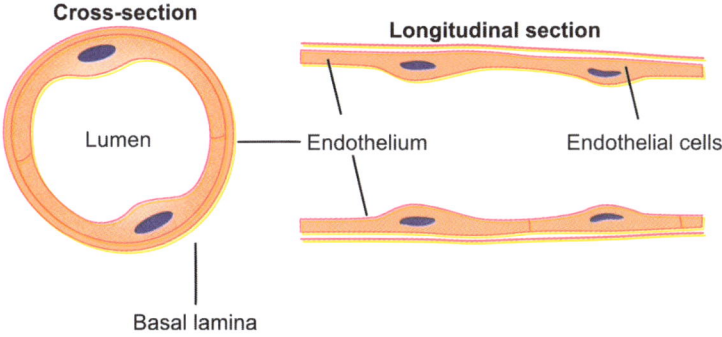

Fig. 14.3: Continuous capillary

2. What are the components of tunica intima?

Ans. Tunica intima consists of **(from inside outwards)**:
  i. An endothelial lining, lined by flattened endothelial cells (endotheliocytes).
  ii. Basal lamina (a thin layer of glycoprotein).
  iii. Subendothelial connective tissue.
  iv. Internal elastic lamina, formed by elastic fibers.

3. What is the type of endothelial cells?

Ans. Simple squamous epithelium.

4. What is the composition of tunica media?

Ans. Tunica media is composed of:
  i. Smooth muscle.
  ii. Elastic fibers.
  iii. External elastic lamina, formed by elastic fibers.

5. What is the composition of tunica adventitia?

Ans. Tunica adventitia consists of:
  i. Collagen fibers.
  ii. Elastic fibers.

6. Which structures intervene between the three layers of arterial wall?

Ans.
  i. **Internal elastic lamina** intervenes between tunica intima and tunica media.
  ii. **External elastic lamina** intervenes between tunica media and tunica adventitia.

7. What is the important anatomical feature of this elastic lamina?

Ans. Both the internal and external elastic laminae are formed by elastic fibers and present fenestrations.

8. What is the utility of these fenestrations?

Ans. Fenestrations help in diffusion of nutritive material.

9. Which of these three layers of the arterial wall is the thickest?

Ans. Tunica media is the thickest layer.

10. Which of these three layers of the arterial wall is the strongest?

Ans. Tunica adventitia is the strongest layer.

11. What is the nutritional source of the arterial wall?

Ans. The tunica adventitia and the outer part of the tunica media receive nutrition from the capillaries of the vasa vasorum, while the inner part of the tunica media and tunica intima get nutrition from the blood in the arterial lumen by diffusion.

12. What are the types of arteries?

Ans. Two types:
  i. Elastic artery.
  ii. Muscular artery.

13. What is the basis of this classification?

Ans. This classification is based on the kind of tissue which predominates in the tunica media. It means that the tunica media consists of more elastic fibers and less muscle fibers in elastic arteries and vice versa in muscular arteries.

14. What are the differences between elastic and muscular arteries?

Ans. Table 14.1 shows the differences between elastic and muscular arteries.

Table 14.1: Differences between elastic and muscular arteries

	Elastic artery	Muscular artery
Tunica media	More elastic tissue and less muscle fibers	More muscle fibers and less elastic tissue
Tunica intima	Subendothelial connective tissue contains more elastic fibers	Subendothelial connective tissue contains less elastic fibers
Internal elastic lamina	Not well-defined	Well-defined
Size of the arteries	Large	Medium and small
Known as	Conducting vessels	Distributing vessels
Examples	Aorta, pulmonary trunk, brachiocephalic trunk, common carotid arteries, subclavian arteries, etc.	Femoral artery, popliteal artery, brachial artery, radial artery, ulnar artery, etc.

Histology: Blood Vascular System

15. What is an arteriole?

Ans. An arteriole is the smallest division of muscular artery.

16. What is the average diameter of an arteriole?

Ans. About 50–100 µm.

17. What are the varieties of arterioles?

Ans.
i. **Muscular arterioles**, having larger diameter (50–100 µm).
ii. **Terminal arterioles,** having smaller diameter (12–50 µm).

18. What do you mean by terminal arterioles?

Ans. The arterioles which have a diameter of less than 50 µm and devoid of internal elastic lamina are called terminal arterioles.

19. What do you mean by meta-arterioles?

Ans. The meta-arterioles are thin lateral branches of terminal arterioles.

20. What are capillaries?

Ans. Capillaries are narrow exchange vessels where the arterioles empty, forming a network.

21. What is the average diameter of a capillary?

Ans. About 8 µm.

22. What is the length of a capillary?

Ans. About 0.5–1 mm long.

23. What is the structure of a capillary wall?

Ans. **From inside outwards**:
i. Capillary endothelium, lined by simple squamous epithelium.
ii. Basal lamina composed of glycoprotein.
iii. Pericytes.

24. What are the types of capillaries?

Ans. **Two types**:
i. **Continuous capillaries**, in which both the endothelial cells and basal lamina form a continuous wall.
ii. **Fenestrated capillaries**, in which the walls of the capillaries have fenestrae (pores).

25. What is the size of these pores?

Ans. About 20–100 microns.

26. What do you mean by muscular capillaries?

Ans. Continuous capillaries are also called muscular capillaries as they are mostly found in muscles.

27. Do the muscular capillaries contain muscle in their walls?

Ans. No, the muscular capillaries do not contain muscle in their walls unlike the muscular arteries and muscular arterioles.

28. Are the pores of the fenestrated capillaries always open?

Ans. No, they are closed by electrodense basal lamina.

29. What are the sites where capillaries are absent?

Ans. Capillaries are absent in:
  i. Epidermis of skin.
  ii. Hairs.
  iii. Nails.
  iv. Cornea.
  v. Articular hyaline cartilage.
  vi. Epithelial cells.

30. What are the sites of continuous capillaries?

Ans.
  i. Skin
  ii. Muscles
  iii. Lungs
  iv. Brain
  v. Connective tissue

31. What are the sites of fenestrated capillaries?

Ans.
  i. Intestinal villi
  ii. Endocrine glands
  iii. Renal glomeruli
  iv. Pancreas

32. What is the blood pressure at the arterial end of a capillary?

Ans. About 30 mm of Hg.

33. What is the blood pressure at the venous end of a capillary?

Ans. About 12 mm of Hg.

Histology: Blood Vascular System

34. What is the filtration presser at the arterial end of a capillary?

Ans. Outward driving force – Inward pulling force = Filtration pressure, i.e. (30 – 25) = 5 mg of Hg.

35. What is the resultant inward force of blood at the venous end?

Ans. Inward pulling force – Outward driving force = Resultant inward force, i.e. (25 – 13) = 12 mm of Hg.

36. What is the blood pressure of an artery?

Ans. About 120 mm of Hg.

37. What is the blood presser of an arteriole?

Ans. About 60 mm of Hg.

38. What is the function of capillaries?

Ans. Exchange of oxygen, carbon dioxide, various molecules and fluid take place between blood and tissue through the walls of the capillaries, hence called **exchange vessels**.

LESSON 2: VEINS

Key points for identification (Figs 14.4 and 14.5)

1. Wall is thin.
2. Tunica media forms one-third of the venous wall.
3. Two-thirds thickness of the wall is formed by tunica adventitia.
4. Both internal and external elastic laminae are ill-defined.
5. Larger quantities of collagen fibers and smaller quantities of elastic or muscle tissue in tunica media.

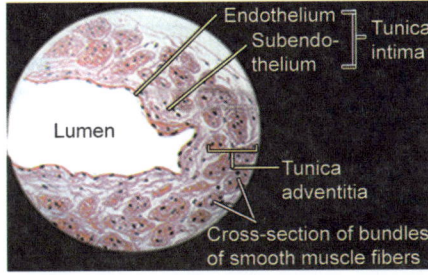

Fig. 14.4: Large vein

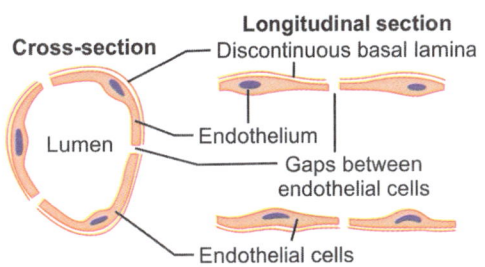

Fig. 14.5: Sinusoid

PROBABLE QUESTIONS AND ANSWERS

1. How many types of venous system are there in our body?

Ans. **Four types:**
   i. Caval or systemic venous system.
   ii. Portal venous system.
   iii. Azygos venous system.
   iv. Paravertebral veins of Batson.

2. How are the veins of the caval system arranged?

Ans. They are arranged in three sets:
   i. **Superficial veins** which lie in the superficial fascia.
   ii. **Deep veins** which lie deep to the deep fascia.
   iii. **Connector or perforating veins** which connect the superficial veins with the deepveins by perforating the deep fascia.

3. How many layers are there in a venous wall?

Ans. Three layers like that of an arterial wall:
   i. Tunica intima.
   ii. Tunica media.
   iii. Tunica adventitia.

4. What are the structural differences between an artery and a vein?

Ans. Table 14.2 shows the differences between artery and vein.

Table 14.2: Differences between artery and vein		
	Artery	**Vein**
Wall	Thicker	Thinner
T. media	Two-thirds of the wall thickness	One-third of the wall thickness
T. adventitia	One-third of the wall thickness	Two-thirds of the wall thickness
Collagen fiber in T. media	Large in quantity	Less collagen fiber
Lumen	Small and patent	Large, irregular or even collapsed
Internal and external elastic lamina	Well-defined	Ill-defined

5. Why the venous wall is easily collapsed?

Ans. This is because of:
   i. Thinner venous wall.
   ii. Large quantity of collagen fiber and less quantity of elastic and muscle tissue in tunica media.

Histology: Blood Vascular System

16. Why does the metastasis of malignant tumors frequently invade the venous wall?

Ans. The lymphatic capillaries ramify in the venous wall more closely than that on the arterial wall due to low venous blood pressure and this explains the lymphatic spread of malignant tumors into the venous wall.

17. What is the blood pressure of large veins?

Ans. About 5 mm of Hg.

18. Do all the veins contain valves?

Ans. Most of the veins are provided with valves, particularly the veins of the extremities.

19. How is the venous valve formed?

Ans. Each venous valve is made up of two semilunar cusps which are formed by reduplication of tunica intima.

20. What are the veins that are devoid of valves?

Ans. Valves are absent in the following veins:
  i. Very small veins.
  ii. Emissary veins.
  iii. Veins within the vertebral canal.
  iv. Superior vena cava (SVC) and inferior vena cava (IVC).

21. What are sinusoids?

Ans. Sinusoids are exchange vessels which possess **wider and irregular lumen** than that of capillaries.

22. What are the differences between capillaries and sinusoids?

Ans. Table 14.3 shows the differences between capillaries and sinusoids.

Table 14.3: Differences between capillaries and sinusoids

Features	Capillaries	Sinusoids
Lumen	Narrow and regular	Wider and irregular
Diameter	About 8 µm	About 20–30 µm
Endothelium	Continuous	Fenestrated
Lining cells	Only endothelial cells	Endothelial cells and phagocytic reticuloendothelial cells such as Kupffer cells in liver
Basal lamina	Continuous	Discontinuous

13. What are the different forms of deficiencies in the walls of sinusoids?

Ans. Two forms of deficiencies:
- i. In the form of fenestrations, called **fenestrated sinusoids**.
- ii. In the form of long slits, called **discontinuous sinusoids**.

14. What are the organs which contain sinusoids?

Ans.
- i. Spleen (discontinuous sinusoids).
- ii. Bone marrow.
- iii. Liver.
- iv. Hypophysis cerebri.
- v. Suprarenal glands.
- vi. Parathyroid glands.
- vii. Carotid bodies.

15. Why the blood flow through the sinusoids is sluggish?

Ans. The lumen of the sinusoids may be irregular which causes sluggish blood flow.

Chapter 15

Histology: Gastrointestinal System

- Tongue
- Esophagus
- Stomach
- Duodenum
- Jejunum and Ileum
- Appendix
- Rectum (Large Intestine)

LESSON 1: TONGUE

Key points for identification (Figs 15.1 to 15.3)

1. Mucosa lined by nonkeratinized stratified squamous epithelium with papillae.
2. Various types of lingual papillae with taste buds.
3. Presence of seromucous glands in between skeletal muscles.

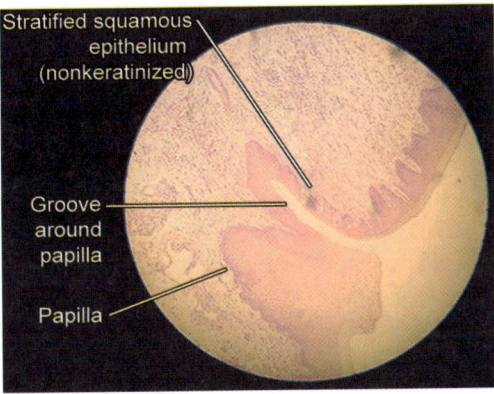

Fig. 15.1: Microstructure of tongue

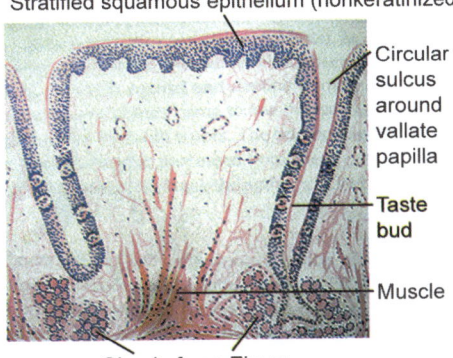

Fig. 15.2: Vallate papilla

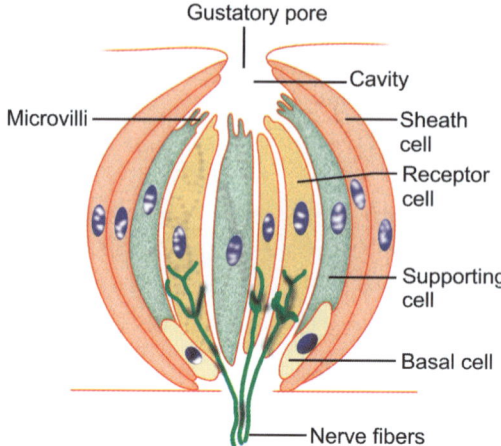

Fig. 15.3: Taste bud

PROBABLE QUESTIONS AND ANSWERS

1. **What is the lining epithelium of tongue?**

Ans. **Nonkeratinized** stratified squamous epithelium.

2. **Do you get keratinization in any parts of the tongue?**

Ans. Yes, keratinization is found at the tips of the **filiform papillae.**

3. **Why keratinization of the epithelium is found on filiform papilla?**

Ans. This keratinization provides a rough surface which helps in the manipulation and processing of foods.

4. **What are lingual papillae?**

Ans. Lingual papillae are the projections of epithelium with a core of subepithelial connective tissue (lamina propria).

5. **What are the different types of papillae?**

Ans. There are **four types** of papillae **on the basis of their appearance.**
   i. Filiform papillae.
   ii. Fungiform papillae.
   iii. Circumvallate papillae.
   iv. Foliate papillae.

Histology: Gastrointestinal System

6. Which of these papillae are smallest and largest?

Ans. **Filiform papillae** are the **smallest** and most numerous papillae. **Circumvallate papillae** are the **largest** papillae (1–2 mm in diameter).

7. What is the location of fungiform papillae?

Ans. Fungiform papillae are more prevalent at the apex and along the lateral margins of the tongue.

8. What is the shape of fungiform papillae?

Ans. Mushroom-like shape (rounded top and narrow base).

9. How do you differentiate them from filiform papillae?

Ans. Fungiform papillae are larger, broader, taller and less numerous than filiform papillae. Unlike filiform papillae, the epithelium of fungiform papillae is **nonkeratinized** and provided with **taste buds**.

10. How do you identify circumvallate (vallate) papillae?

Ans. These papillae are characterized by a circular top, surrounded by deep furrows/sulcus.

11. How many vallate papillae are there in a tongue?

Ans. These are **8–12** in number.

12. What is the location of these papillae in the tongue?

Ans. These are located in the ventral 2/3rd of the tongue on its dorsal surface, in a single row in a V-shaped manner just anterior and parallel to the sulcus terminalis.

13. Why these papillae are called vallate papillae?

Ans. These papillae are surrounded by a circular sulcus. The outer wall of this sulcus is known as vallum (**vallum means wall**), hence the name vallate or circumvallate papillae.

14. Where do you find taste buds in these papillae?

Ans. Taste buds are found on both the inner and outer wall of the sulcus.

15. What are glands of Von Ebner?

Ans. These are underlying serous glands located in the connective tissue.

16. Where do their ducts open?

Ans. The excretory ducts of these glands open into the base of the circular sulcus surrounding the vallate papillae.

17. What is the function of the secretion of these glands?

Ans. The secretion makes the food soluble to stimulate the taste buds. **Secondly**, the continuous flow of secreted fluid over the taste buds washes the food particles and prepares them to receive new taste stimulus.

18. Do you get taste buds in all types of papillae?

Ans. No. Taste buds are **absent** in filiform papillae.

19. What is a taste bud?

Ans. Taste bud is barrel-shaped microscopic structures, made up of modified epithelial cells occupying the full thickness of the epithelium (Fig. 15.3).

20. Name the types of cells present in the taste bud?

Ans. Three types of cells:
   i. Gustatory cells or receptor cells or taste cells.
   ii. Supporting or sustentacular cells.
   iii. Basal cells.

21. What is the appearance of gustatory cells?

Ans. Gustatory calls are slender fusiform cells, occupying the central part of the taste bud, showing light cytoplasm with centrally located oval and lighter nucleus. The apices of these cells possess numerous microvilli which protrude through the gustatory pore.

22. What is the function of gustatory cells?

Ans. These cells are **sensory cells**. Afferent nerve endings end at the basal side of these cells.

23. What is the appearance of supporting cells?

Ans. They surround the gustatory cells, showing darker cytoplasm with dark nucleus and exhibit numerous microvilli (similar to gustatory cells) which project through the gustatory pores.

24. What is the function of supporting cells?

Ans. They have supporting function without any sensory role.

Histology: Gastrointestinal System

25. What are basal cells?

Ans. These are present at the base of the taste buds. They are believed to act as **stem cells** and multiply to produce specialized cells of the taste buds.

26. What is gustatory pore or taste pore?

Ans. Gustatory pore is an opening of each taste bud, which opens apically on the surface.

27. Which structures protrude through the gustatory pore?

Ans. Microvilli of gustatory and supporting cells of taste bud.

28. Where do you get taste buds other than tongue?

Ans.
  i. Lingual aspect of soft palate.
  ii. Epiglottis.
  iii. Palatoglossal arch.
  iv. Posterior wall of oropharynx.

29. Do the different types of taste buds respond to different types of taste?

Ans. Though a particular taste bud can respond to different types of taste, the sweet and salt are best appreciated at the tip of the tongue (filiform and fungiform papillae), sour at the sides and bitter taste by the circumvallate papilla and in the pharyngeal part of the tongue.

30. What are foliate papillae?

Ans. These are 3–4 mucous folds, arranged vertically in front the sulcus terminalis of the tongue.

31. Do they contain taste buds?

Ans. Yes.

32. Are they always present?

Ans. No. They are rudimentary or poorly developed in humans. But they are well-developed in some animals, particularly in rabbits.

33. Does the whole of the tongue present same histological features?

Ans. No, the differences of histological features on the dorsal surface of the tongue are:

	Ventral 2/3rd	Dorsal 1/3rd
Submucous coat	Absent	Present
Lingual papillae	Present	Absent
Lymphatic follicles	Absent	Present in the submucous coat (lingual tonsils)

34. Name the muscles of the tongue.

Ans. Table 15.1 shows the extrinsic and intrinsic muscles of tongue.

Table 15.1: Extrinsic and intrinsic muscles of tongue

Extrinsic muscles (5 pairs)	Intrinsic muscles (4 pairs)
Genioglossus	Superior longitudinal
Hyoglossus	Inferior longitudinal
Chondroglossus	Transversus linguae
Styloglossus	Verticalis linguae
Palatoglossus	
(All are skeletal muscles)	

35. What are papillae simplex?

Ans. Papillae simplex are projections of subjacent connective tissue into the epithelium without showing any surface projections. These papillae are always microscopic unlike other 4 varieties of papillae (Table 15.2).

Table 15.2: Features of different types of papillae at a glance

Features	Filiform	Fungiform	Circumvallate	Foliate
Number	Most numerous	More than vallate but less than filiform	8–12	3–4 (on each side)
Shape	Conical	Mushroom shaped	Truncated conical	Leaf-like
Size	Very small (**smallest papillae**)	Larger than filiform but smaller than vallate	**Largest** (1–2 mm in diameter)	
Location	Entire dorsal surface of the ventral 2/3rd of the tongue	Sides and tip of the tongue	In front and parallel to the sulcus terminals	Margin of the tongue, in front of sulcus terminals
Taste buds	Absent	Present	Present	Present
Overlying mucosa	Tips are keratinized	Non-keratinized	Non-keratinized	Non-keratinized
Functions	Facilitates movements of food particles	Perception of taste	Perception of taste	Perception of taste

Histology: Gastrointestinal System

LESSON 2: ESOPHAGUS

Key points for identification (Figs 15.4A and B)
1. Mucosa lined by nonkeratinized stratified squamous epithelium.
2. Longitudinal mucosal folds.
3. Submucosal glands.

PROBABLE QUESTIONS AND ANSWERS

1. What are the layers of esophageal wall?

Ans. Four layers (**from outside inwards**):
   i. Serosa/adventitia.
   ii. Muscularis externa:
      a. Outer longitudinal layer.
      b. Inner circular layer.
   iii. Submucosa
   iv. Mucosa.

2. What are the areas of the gut where you get stratified squamous epithelium?

Ans. The lining epithelium of the esophagus and lower part of the anal canal are stratified squamous, whereas the mucosa in other parts of the gut is lined by simple columnar epithelium.

3. What is the function of stratified squamous epithelium?

Ans. This epithelium has a **protective role**. But the columnar epithelial cells are either absorptive or secretory.

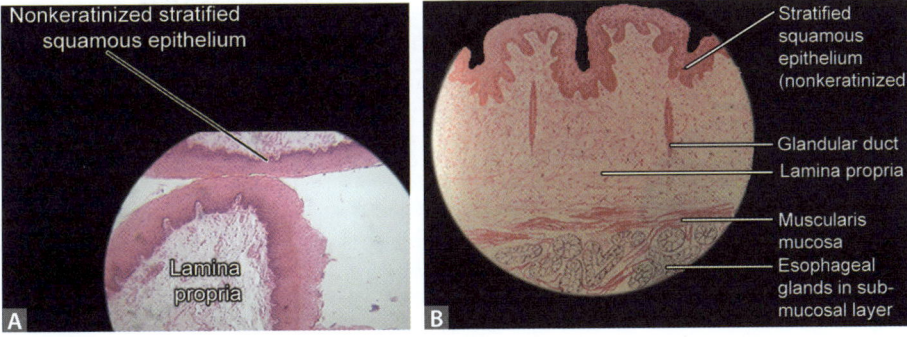

Figs 15.4A and B: Esophagus

4. Why the mucosa is thrown into longitudinal folds?

Ans. The folds increase the surface area of the lumen of the esophagus and allow distension during the passage of food.

5. Is the muscularis mucosa uniform throughout the esophagus?

Ans. No, it is distinct in the lower part of the esophagus but absent or poorly developed in its upper part.

6. What is the special feature of the submucosal layer?

Ans. Presence of submucosal glands.

7. What types of glands they are?

Ans. The glands are **compound tubuloalveolar mucous** glands.

8. What do these glands secrete?

Ans. These glands secrete mucous which lubricates the lumen of the esophagus and aids the passage of food.

9. What is the disposition of muscularis externa?

Ans. It has got two layers:
  i. Outer longitudinal.
  ii. Inner circular.

10. What are the types of these muscles?

Ans.
  i. In the upper 1/3rd of esophagus—striated variety.
  ii. In the middle 1/3rd—both striated and nonstriated (smooth) variety.
  iii. In the lower 1/3rd—nonstriated (smooth) variety.

11. What is the outermost coat of the esophagus?

Ans. Fibrous coat (adventitia).

12. Do you get the fibrous coat as the outermost layer throughout the length of the esophagus?

Ans. No, in the lowermost 1.25 cm of the esophagus (intra-abdominal part of esophagus) the outermost coat is lined by serosa (peritoneum).

Histology: Gastrointestinal System

LESSON 3: STOMACH

Key points for identification (Figs 15.5 and 15.6)

1. Mucosa lined by simple columnar epithelium.
2. Gastric pits lead to gastric glands.
3. Presence of tubular glands in lamina propria.
4. Three layers of smooth muscles.

PROBABLE QUESTIONS AND ANSWERS

 1. **What are the layers of the wall of stomach?**

Ans. Four layers (**from outside inwards**):
  i. Serous (peritoneal) coat.
  ii. Muscular coat:
     a. Outer longitudinal.
     b. Middle circular
     c. Inner oblique.
  iii. Submucous coat.
  iv. Mucous coat.

2. **Do you get the serous layer in every part of the stomach wall?**

Ans. **No**, this layer is absent along the greater and lesser curvatures and also at the bare area of the stomach.

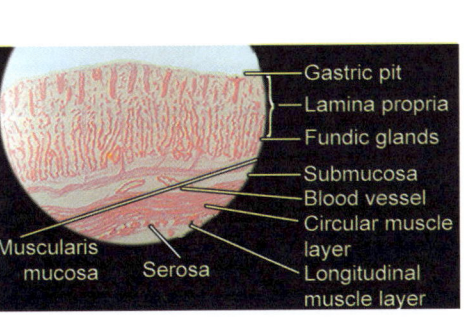

Fig. 15.5: Fundus of stomach

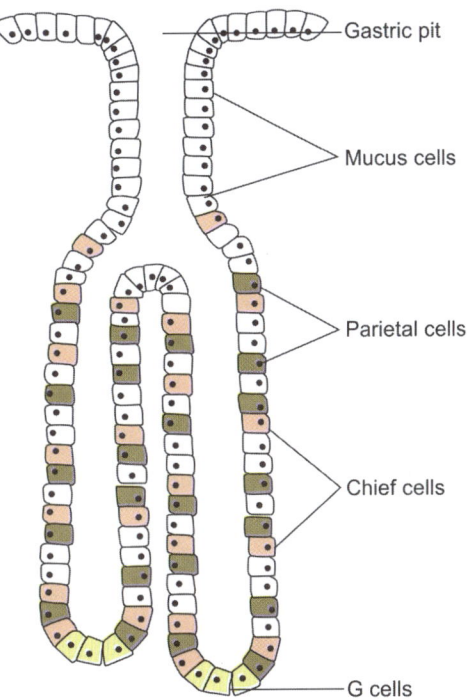

Fig. 15.6: Gastric gland

3. What are the differences of disposition between the longitudinal and circular muscles?

Ans.

Longitudinal fibers	Circular fibers
i. More thick along the curvatures	i. More thick at the pylorus
ii. These fibers are continuous with the corresponding muscle fibers of the esophagus at the cardiac end but at the pyloric end only the superficial fibers of longitudinal muscle are continuous with the corresponding muscle of duodenum	ii. These fibers are continuous with the corresponding muscle fibers of the esophagus at the cardiac end but at the pyloric end the fibers are not continuous with the corresponding muscle of duodenum due to presence of a connective tissue septum

Note: *The deep fibers of the longitudinal muscle join with the circular muscle at the pylorus to form pyloric sphincter.*

4. Which muscle layer plays a role to form gastric canal?

Ans. Contraction of the right free margin of the **oblique muscle** plays a role to form gastric canal.

5. What does the submucous coat contain?

Ans. The submucous coat contains blood vessels, nerves, lymphatics and fat.

6. How does the muscular coat of the stomach wall differ from the muscular coat of the other parts of the gut wall?

Ans. The stomach consists of three layers of muscle fibers (longitudinal, circular and oblique), whereas the rest of the gut wall consists of only two layers of muscle fibers (outer longitudinal and inner circular).

7. What is the lining epithelium of stomach?

Ans. Simple columnar epithelium.

8. What do you mean by gastric rugosities?

Ans. These are prominent but temporary mucosal longitudinal folds which permit distension of stomach.

9. What are the layers of the mucous membrane?

Ans. It consists of **three layers** (from within outwards):
 i. Surface epithelium (Epithelial cells + Basement membrane)
 ii. Lamina propria.
 iii. Muscularis mucosa.

Histology: Gastrointestinal System

10. How the gastric mucosa is protected against acid and enzyme produced by the gastric mucosa?

Ans. The mucous, secreted by the lining epithelial cells protects the gastric mucosa.

11. What are gastric pits?

Ans. Gastric pits are funnel-shaped depressions, formed by the dipping down of the lining epithelium into the lamina propria.

12. What does the gastric pit receive?

Ans. Each gastric pit receives the ducts gastric glands.

13. How many glands open in each gastric pit?

Ans. 1–7 glands open in each pit.
(3–7 fundic/principal glands open into each gastric pit, whereas 2 to 3 pyloric glands open into each pit).

14. How much is the depth of gastric pit?

Ans.
  i. Gastric pits of fundic and body region extend up to 1/4th thickness of the gastric mucosa (so, the glands in this region are long → 3/4th depth).
  ii. Gastric pits in pyloric region extend up to 2/3rd thickness of the mucosa (so, the glands in this region are short → 1/3rd depth).

15. How the gastric glands are formed?

Ans. Gastric glands are formed by the invagination of the epithelial cells deep inside the mucosal gastric pits.

16. What are the types of gastric glands?

Ans. **Three types**:
  i. Cardiac glands—close to cardiac orifice.
  ii. Fundic and body glands—confined to fundus and body.
  iii. Pyloric glands—in pyloric region.

17. What are the types of cells in cardiac glands?

Ans.
  i. Mainly mucous secreting mucous cells.
  ii. Few oxyntic cells and zymogenic cells.

18. What type of the fundic and body glands are?

Ans. These glands are simple or branched tubular type.

19. What are the types of cells in fundic and body glands?

Ans. Fundic and body glands are also known as **main gastric glands** or **principal glands**. The cell types of these glands are (Fig. 15.6):
  i. Chief cell or peptic cells or zymogen cells.
  ii. Parietal cells or oxyntic cells.
  iii. Mucous neck cells.
  iv. Argentaffin cells.
  v. Stem cells (undifferentiated cells).

20. What are the parts of a fundic gland tube?

Ans. Three parts:
  i. Isthmus   ii. Neck   iii. Base

21. What are the features of chief (zymogen) cells?

Ans.
  i. Most numerous cells.
  ii. Mostly found in the **basal parts of the glands.**
  iii. The cells are **cuboidal or low columnar** in type.
  iv. Their cytoplasm is basophilic (purple stained).
  v. They secrete pepsinogen which is converted into active pepsin in presence of acidic medium (HCl) of stomach.

22. What are the features of oxyntic (parietal) cells?

Ans.
  i. Most numerous in the **isthmus** of gastric glands.
  ii. The cells are large polyhedral.
  iii. Total number of cells are about **one billion**.
  iv. Their cytoplasm is **acidophilic** (pink in H&E stain).
  v. Secrete:
     a. HCl which helps in the conversion of inactive pepsinogen to active pepsin.
     b. **Intrinsic factor** of castle which helps in absorption of vitamin B_{12} (called **extrinsic factor**) in the terminal part of ileum, thereby preventing **pernicious anemia.**

23. Why the oxyntic cells are so named?

Ans. These cells stain strongly with **eosin** in H&E staining. Hence, the name oxyntic.

24. Why the oxyntic cells are also known as parietal cells?

Ans. These cells lie against the basement membrane, hence called parietal cells.

25. Where do the mucous neck cells lie?

Ans. They lie in the neck of the gland tube.

Histology: Gastrointestinal System

26. What do they secrete?

Ans. These cells secrete **mucous** but this mucous is less **alkaline** and **thinner** than the mucous secreted by the mucous cells lining the surface of gastric mucosa.

27. What are argentaffin cells?

Ans. These are **neuroendocrine cells** which lie between chief cells and the basement membrane near the basal parts of the gastric glands. They never do not reach the lumen.

28. Why these cells are called argentaffin cells?

Ans. These endocrine cells contain membrane bound neurosecretory granules which stain with **silver salts**. So, they are called argentaffin cells.

29. Name the secretions of argentaffin cells?

Ans.
   i. Gastrin.
   ii. Serotonin.
   iii. Somatostatin.

30. What is the function of stem cells?

Ans. These cells multiply by mitosis to replace other cells of the gastric glands. It means that they have regenerative function.

Remember that the zymogenic cells are found at the base, oxyntic cells in the isthmus and mucous neck cells in the neck region of the gastric glands.

31. What is the appearance of pyloric glands?

Ans. They are simple or branched tubular glands and occupy the deeper 1/3rd of the mucosa.

32. What are the cell types of pyloric glands?

Ans.
   i. Mucous-secreting cells (mucous cells).
   ii. Few oxyntic cells and argentaffin cells.
   iii. **G-cells** (gastrin-secreting cells).

33. How do the G-cells act?

Ans. G-cells → Secrete gastrin → Released in the bloodstream → Acting locally on gastric glands → Secretion of HCl and pepsin.
Hence, G-cells belong to **APUD system** and **paracrine in action**.

34. What is the composition of gastric juice?

Ans. It is mainly composed of mucous, HCl, pepsin and other substances.

35. How the gastric epithelium is protected from autodigestion?

Ans. The mucous cells lining the luminal surface of the stomach (gastric mucosa) secrete mucous and bicarbonate ions which protect the gastric epithelium from autodigestion.

LESSON 4: DUODENUM

Key points for identification (Figs 15.7A and B)

1. Mucosa—presence of villi, lined by simple columnar epithelium.
2. Crypts of Lieberkuhn.
3. Presence of submucosal Brunner's gland.

PROBABLE QUESTIONS AND ANSWERS

1. What are the layers of duodenal wall?

Ans. **Four layers** (from outside inwards):
  i. Serous/peritoneal coat.
  ii. Muscular coat:
     a. Outer longitudinal layer.
     b. Inner circular layer.
  iii. Submucous coat.
  iv. Mucous coat.

2. What is the most important histological feature of duodenum?

Ans. Submucosal Brunner's gland.

3. What is the type of the Brunner's gland?

Ans. These are compound tubuloalveolar glands (compound racemose).

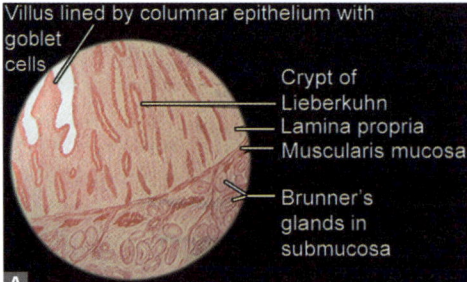 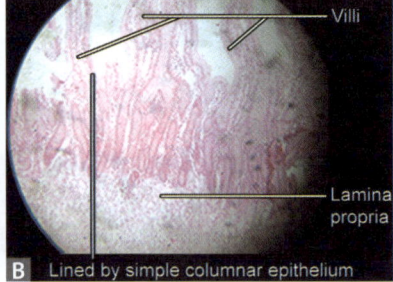

Figs 15.7A and B: Duodenum

Histology: Gastrointestinal System

4. Where do their duct open?

Ans. Their ducts open into the bases of intestinal crypts of Lieberkuhn.

5. What are crypts of Lieberkuhn?

Ans. Crypts of Lieberkuhn are simple tubular intestinal glands which are formed by the invaginations of epithelium into the lamina propria.

6. Are the glands uniform in number throughout the duodenum?

Ans. No, they are most numerous in the proximal part but few or even missing in the distal part.

7. What are the secretions of Brunner's glands?

Ans. They secrete:
  i. **Mucous (rich in bicarbonate)** that neutralizes gastric acid.
  ii. **Trypsinogen activating factor** (an enzyme) which activates trypsinogen (produced by the pancreas) to trypsin.

8. What are valves of kerkring?

Ans. These are permanent circular mucosal folds containing all layers of mucosa.

9. What are the layers of mucosa?

Ans. Layers of mucosa are:
  i. Lining epithelium.
  ii. Lamina propria.
  iii. Muscularis mucosae.

10. What are the functions of the circular folds?

Ans.
  i. Increase surface area for absorption.
  ii. Slow down the passage of intestinal contents, thus facilitating absorption.

11. Are these folds present throughout the duodenum?

Ans. No, they are absent in the proximal one to two inches of the duodenum.

12. What are villi?

Ans. Villi are finger-like processes, consisting of a core of connective tissue covered by a surface epithelium, projecting into the lumen.

13. What is the type of surface epithelium?

Ans. It is simple columnar epithelium.

14. What are the contents of the core of a villus?

Ans. The core of a villus is an extension of lamina propria containing:
  i. Lacteal.
  ii. Blood capillaries.
  iii. Smooth muscles derived from muscularis mucosae.
  iv. Fibroblasts, lymphocytes, plasma cells, eosinophils and macrophages.

15. What is lacteal?

Ans. It is a dilated blind lymph capillary forming the central axis of villi. The blind end projects toward the tip of the villus and the other end communicates with the submucous coat. It is concerned with fat absorption.

16. What is length of a villus?

Ans. About 0.5–1.0 mm.

17. What is the density of villi?

Ans. About 10–40/mm^2

18. Do you find goblet cells in duodenal mucosa?

Ans. Yes, they are few in number.

19. What is a goblet cell?

Ans. Goblet cell is an unicellular, inverted flask-shaped mucous-secreting cell.

20. What are paneth cells?

Ans. Paneth cells or zymogen cells are exocrine cells which are found only in the deeper parts of intestinal crypts of Lieberkuhn. They contain acidophilic (eosinophilic) zymogen granules. They are also rich in zinc content. The paneth cells secrete lysozyme which destroys bacteria.
Remember that paneth cells at the bottom of the intestinal crypts are characteristics of small intestine.

Histology: Gastrointestinal System

21. What is urogastrone?

Ans. It is a polypeptide hormone, produced by the duodenal glands of Brunner and this hormone increases epithelial proliferation of small intestine by inhibiting HCl secretion by the gastric parietal cells.

LESSON 5: JEJUNUM AND ILEUM

Key points for identification (Figs 15.8 and 15.9)

1. Mucosa lined by simple columnar epithelium.
2. Presence of villi and crypts of Lieberkuhn.
3. Absence of submucosal Brunner's glands.
4. Presence of solitary lymphatic follicles and aggregated lymphatic follicles in the lamina propria and submucous coat of jejunum and ileum respectively.

HISTOLOGICAL COMPARISON BETWEEN DUODENUM, JEJUNUM AND ILEUM

Features	Duodenum	Jejunum	Ileum
Mucosal villi	Tongue-like (broad)	Leaf-like (tall)	Finger-like (short)
Lamina propria	–	Solitary lymphatic follicles	Aggregated lymphatic follicles (Peyer's patches)
Submucosa	Brunner's glands	Solitary lymphatic follicles	Peyer's patches

PROBABLE QUESTIONS AND ANSWERS

1. Do you find villi in the entire length of jejunum and ileum?

Ans. No, villi are absent over the solitary lymphatic follicles and Peyer's patches.

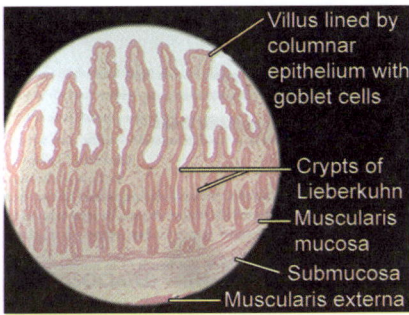

Fig. 15.8: Jejunum

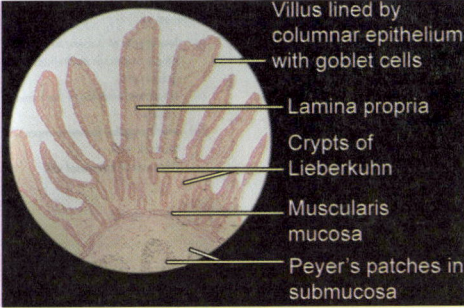

Fig. 15.9: Ileum

2. Where are the Peyer's patches maximally found?

Ans. They are most numerous in the terminal part of ileum.

3. How are the Peyer's patches formed? What is the function of these patches?

Ans. They are formed by the aggregation of about 260 solitary lymphatic follicles. They analyze and respond to the pathogenic bacteria and prevent their growth in the ileum.

4. How many patches are there?

Ans. They are about 20–30 in number, may be even more.

5. What is the length of a Peyer's patch?

Ans. About 2–10 cm long.

6. What are paneth cells?

Ans. Paneth cells are exocrine cells that lie at the bottom of the intestinal glands (crypts of Lieberkuhn).

7. What are M-cells?

Ans. M-cells (microfold cells) are follicle-associated epithelial cells, found in the epithelium overlying the lymphatic follicles.

8. What is the function of M-cells?

Ans. These cells take up antigens from the intestinal lumen and transferred them to the underlying lymphoid tissue to produce antibodies against the antigens.

9. Is there any difference in size and number of villi between jejunum and ileum?

Ans. In jejunum, the villi are large and numerous but in ileum, these are smaller and fewer.

10. What are the layers of muscular coat of jejunum and ileum?

Ans. The muscular coat of the entire intestine consists of outer longitudinal and inner circular layers.

LESSON 6: APPENDIX

Key points for identification (Fig. 15.10)

1. Poorly developed mucous membrane which is lined by simple columnar epithelium.
2. Small lumen.
3. Profuse numbers of lymphoid tissue in lamina propria and submucosa.
4. Presence of hiatus muscularis in the muscular coat.

Histology: Gastrointestinal System

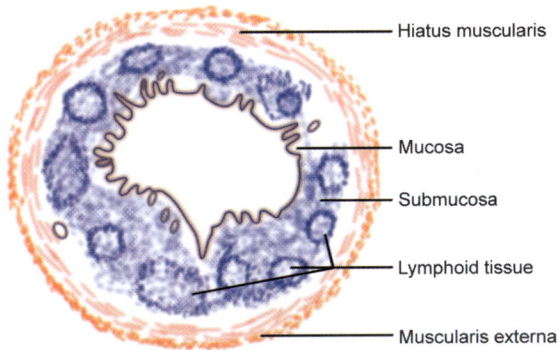

Fig. 15.10: Appendix

PROBABLE QUESTIONS AND ANSWERS

1. What are the layers of the appendix?

Ans. Appendix presents **four layers** from outside inwards:
  i. Serous or peritoneal coat.
  ii. Muscular coat.
  iii. Submucous coat.
  iv. Mucous coat.

2. What are the muscle layers in muscular coat?

Ans.
  i. Outer thin layer of longitudinal fibers.
  ii. Inner thick layer of circular fibers.

3. What is hiatus muscularis?

Ans. These are gaps is the muscle coat.

4. What is the importance of hiatus muscularis?

Ans. Infection from the appendicular lumen may extend to the peritoneal surface through these gaps.

5. Why appendix is known as abdominal tonsil?

Ans. The submucous coat of appendix contains numerous lymphatic follicles, hence called abdominal tonsil.

6. Does the lamina propria contain lymphoid tissue?

Ans. The lamina propria of mucous membrane is also rich in lymphoid tissue which extends into submucosa.

7. Do the lymphoid follicles contain germinal center?

Ans. Yes, germinal centers are present. These lymphoid tissues in the submucosa and lamina propria are considered as **GALT** (gut-associated lymphoid tissue).

8. What is the special finding in muscularis mucosae?

Ans. Muscularis mucosae are interrupted at places through which the lymphoid tissue of lamina propria extends into submucosa.

LESSON 7: RECTUM (LARGE INTESTINE)

Key points for identification (Fig. 15.11)
1. Mucosae are thrown into folds and lined by simple columnar epithelium.
2. Absence of villi.
3. Presence of numerous goblet cells.
4. Presence of crypts of Lieberkuhn.
5. Solitary lymphatic follicles in lamina propria.

PROBABLE QUESTIONS AND ANSWERS

1. What are the coats of rectal wall?

Ans. Rectal wall consists of **4 coats (from outside inwards):**
  i. Serous (peritoneal)/adventitia.
  ii. Muscular
     a. Outer longitudinal fibers.
     b. Inner circular fibers.
  iii. Submucous.
  iv. Mucous.

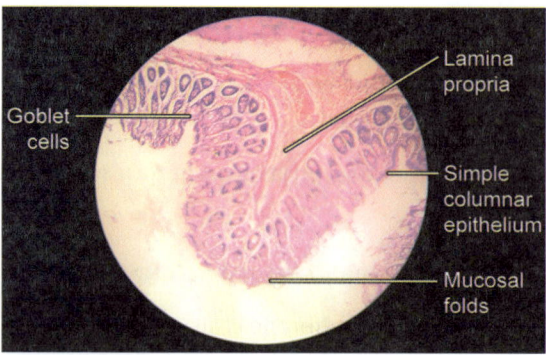

Fig. 15.11: Rectum

Histology: Gastrointestinal System

2. Do you get the peritoneal covering throughout the length of the rectum?

Ans. No. The visceral peritoneum covers only the **sides and front** of upper one-third and only the **front** of the middle third of the rectum. The lower third of the rectum is devoid of peritoneal covering. So, the parts of the rectum which have no serous covering are covered with **adventitia**.

3. What is the difference of disposition of muscles between rectum and colon?

Ans. **In colon**, the longitudinal fibers of muscular coat are gathered to from **three thick bands** called **taenia coli**. But **in rectum**, there is no such taenia.

4. What are the names of these taenia coli?

Ans.
  i. Taenia libera.
  ii. Taenia mesocolica.
  iii. Taenia omentalis.

5. What are the areas of large gut where taenia coli are absent?

Ans. Taenia coli are **absent** in:
  i. Rectum.
  ii. Anal canal.
  iii. Appendix.

6. What are the cardinal features of large gut?

Ans. Three cardinal features of large gut are:
  i. Taenia coli.
  ii. Haustrations or sacculations.
  iii. Appendices epiploicae.

7. Why are the colon sacculated?

Ans. The length of taenia is shorter (about 4 feet) than the length of the other layers of the wall of the colon (about 6 feet). This results in the formation of haustrations or sacculations.

8. What are the appendices epiploicae?

Ans. These are peritoneal pouches containing fat.

9. Are they present in all parts of the large gut?

Ans. **No**, the appendices epiploicae are absent in cecum, appendix and rectum.

10. What is a goblet cell?

Ans. A goblet cell is an **unicellular exocrine gland.**

11. What does it secrete?

Ans. It secretes **mucous.**

12. What is the meaning of a goblet?

Ans. A goblet means a drinking glass which is broad above and narrow below.

13. What type of cell it is?

Ans. A goblet cell is a modified simple columnar epithelial cell, the height of which is four times more than that of its width.

14. What is the method of secretion of a goblet cell?

Ans. It secretes mainly in **merocrine method** of secretion.

15. What is merocrine method of secretion?

Ans. In this method, the secretion is discharged by **exocytosis** without any loss of cellular components.

16. Do they secrete in any other method of secretion?

Ans. Yes, they may secrete in **apocrine method** under stress.

17. What is the apocrine method of secretion?

Ans. In this method, the cell disintegrates and dies to liberate its secretion.

18. What are the histological features of a goblet cell?

Ans.
    i. Expanded upper part which is distended with **mucous granules.**
    ii. The flattened **nucleus** lies near the basal part of the cell.
    iii. The luminal surface of the cell, i.e. the apical plasma membrane projects irregular microvilli to increase the surface area for secretion.

19. Why the apical or upper part of the cell is expanded?

Ans. The upper part of the goblet cell is expanded due to the accumulation of mucous granules near the apical surface of the cell. This causes the displacement of nucleus towards the base of the cell. The Golgi apparatus lies between the mucous granules and the nucleus.

Histology: Gastrointestinal System

20. How does a goblet cell look in H&E staining?

Ans. The **basal part** of the cell gives a basophilic staining (purple) because of nucleic acids within the nucleus and presence of rough endoplasmic reticulum.
But the **apical part** of the cell containing large mucous granules stains **pale** or appears to be unstained.

21. Why the apical part of the cell remains pale or unstained in H&E staining?

Ans. This is primarily due to the fact that the mucous granules which are carbohydrate-rich proteins are washed out in the preparation of slides for microscopic examination.

22. Which stain is preferred to stain mucous granules?

Ans. **PAS stain** which colors them magenta.

23. What is the full form of PAS?

Ans. **Periodic acid-Schiff**.

24. What are the organs where the goblet cells are found?

Ans.
  i. Mucosa of stomach, small and large intestines.
  ii. Mucosa of trachea, bronchus and larger bronchioles in respiratory tract.
  iii. Conjunctiva in the upper lid (chief source of mucous in tears).

25. What is the function of goblet cell?

Ans. Their main function is to secrete mucous in order to protect the mucous membranes where they are found. It is interesting to know that the secreted mucous forms about a 200 μm thick layer on the inner surface of human intestine which lubricates and protects the mucous membrane.

CHAPTER 16

Histology: Liver and Pancreas (Both Exocrine and Endocrine Types of Gland)

- Liver
- Pancreas

LESSON 1: LIVER

Key points for identification (Figs 16.1 to 16.4)

1. Presence of hepatic lobules.
2. Central vein with radiating sinusoids.
3. Lamina of hepatocytes in radiating manner.
4. Portal triad, consisting of interlobular branch of portal vein, hepatic artery and bile ductule.

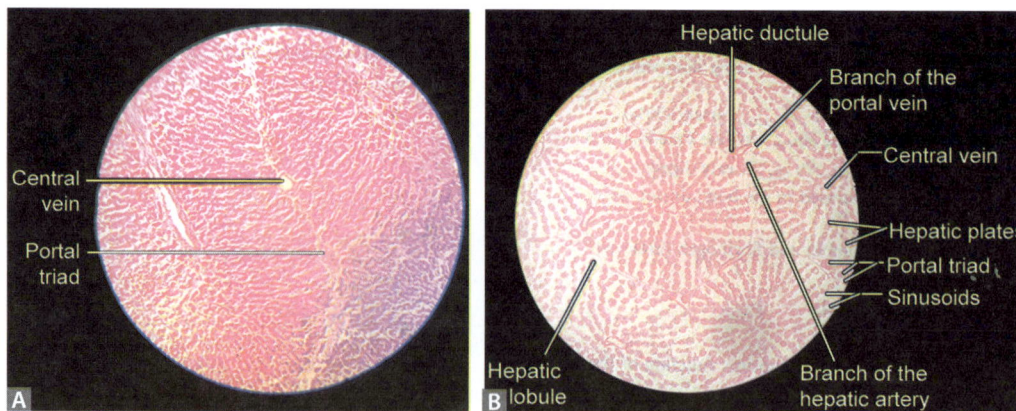

Figs 16.1A and B: Liver

Histology: Liver and Pancreas (Both Exocrine and Endocrine Types of Gland)

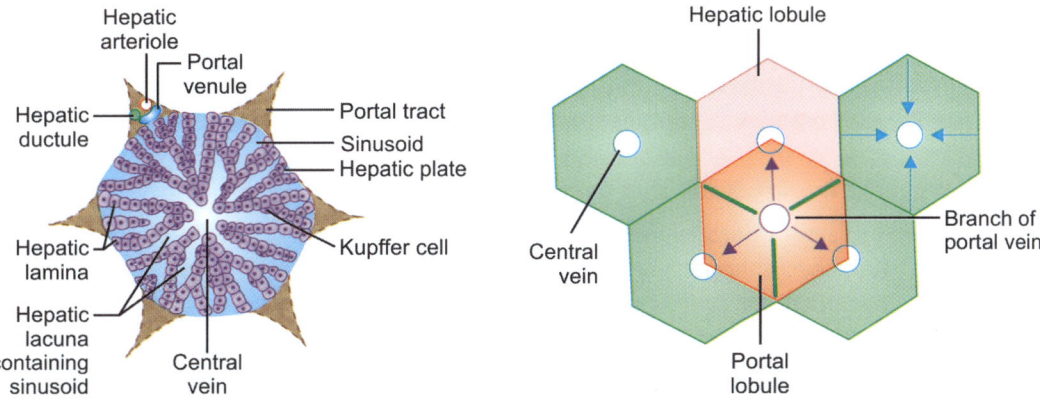

Fig. 16.2: Classical liver lobule

Fig. 16.3: Hepatic and portal lobule

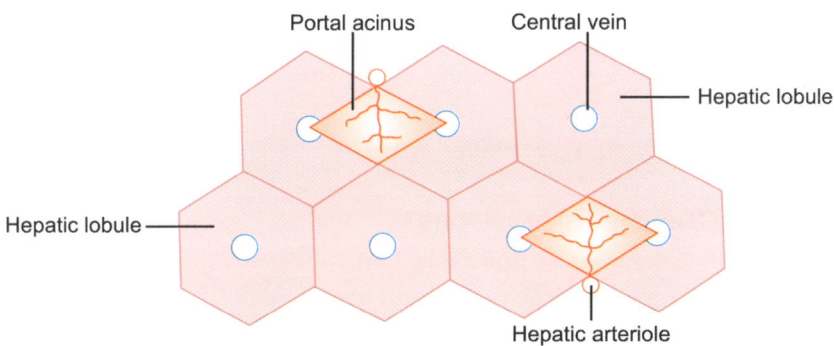

Fig. 16.4: Portal acinus

PROBABLE QUESTIONS AND ANSWERS

1. What type of gland is the liver?

Ans. The liver is **both an endocrine and exocrine gland**.

2. What is the size of the liver?

Ans. The liver is about 1.5 kg in weight. It is the second largest viscera of the body (next to skin) and it is the largest gland of the body.

3. What are the coverings of the liver?

Ans.
i. Outer covering is the visceral peritoneum.
ii. Inner covering is the dense fibroelastic connective tissue known as Glisson's capsule which is 50–100 mm thick.

4. Does this Glisson's capsule extend into the liver?

Ans. Yes, this capsule extends into the interior of the liver as connective tissue septae and undergoes branching, forming numerous hepatic lobules.

5. What are the structures that pass through the septae?

Ans.
i. Radicles of portal vein.
ii. Branches of hepatic artery.
iii. Bile ductules.

6. What is the composition of liver?

Ans. The liver is composed of:
i. Hepatocytes (liver cells).
ii. Connective tissue stroma.
iii. Numerous portal triads.
iv. Bile canaliculi.
v. Sinusoids.
vi. Hepatic veins and their tributaries.

7. What are the features of a hepatocyte?

Ans.
i. Shape: Polyhedral (6 or more surfaces)
ii. Nucleus: Large spherical, central and euchromatic.
iii. Cytoplasm: Eosinophilic.
iv. Some cells are binucleate.

8. What is hepatic lobule?

Ans. A hepatic lobule is a small hexagonal vascular unit of liver consisting of hexagonal masses of liver cells and the central axis being the central vein (Fig. 16.3).

9. What is the size of a hepatic lobule?

Ans. About 1 mm in width.

10. Do you know any other name for hepatic lobule?

Ans. Hepatic **lobule of Kiernan**.

11. What do you mean by portal lobule?

Ans. A portal lobule is a polygonal mass of liver tissue comprising of portions of three adjacent hepatic lobules with a branch of portal vein at its center. This portal lobule is also known as portal **lobule of Mall** and referred to as true functional unit of liver (Fig. 16.3).

12. What are hepatic laminae?

Ans. Hepatic laminae are plates of liver cells radiating from the central vein toward the periphery of the hepatic lobules (Fig. 16.2).

13. How many hepatic cells form a hepatic lamina?

Ans. Each hepatic lamina is formed by about 20 hepatic cells which radiate from the center to the periphery.

14. What is the arrangement of the cells of hepatic lamina?

Ans. The cells of one hepatic lamina are arranged in multiple rows which are placed one above the other (one—cell thick) and are often branched and irregular.

15. What do you mean by the limiting plate?

Ans. On reaching the periphery of hepatic lobule the adjacent laminae join with one another forming a plate of hepatic cells, called limiting plate.

16. Does any structure pass through this limiting plate?

Ans. **Yes, the inlet venules,** branches from the transverse terminal branches of portal vein enter into the hepatic sinusoids through the perforations of the limiting plate.

17. Why are the hepatic lobules not considered as a functional or structural unit of liver?

Ans. The liver cells, from the limiting plate of one hepatic lobule, sprout outwards around the portal tract to become continuous with the adjacent hepatic lobules. This means that the liver cells are not confined to any particular hepatic lobule, rather they form a continuous wall-work. Therefore, the hepatic lobules are not considered as a functional or structural unit of the liver rather they are considered as an independent vascular unit.

18. What is hepatic lacunae?

Ans. The space between the hepatic laminae are called hepatic lacunae (Fig. 16.2).

19. What do these lacunae contain?

Ans. The lacunae contain hepatic sinusoids.

20. What are hepatic sinusoids?

Ans. Hepatic sinusoids are vascular channels in the hepatic lacunae, lined by fenestrated endothelium.

21. What is the source of blood of these sinusoids?

Ans. The sinusoids receive blood from the branches of portal vein as well as from the branches of hepatic artery. So sinusoidal blood is a mixture of venous and arterial blood.

22. How do the vein and artery open into the sinusoids?

Ans. Inlet venules enter into the sinusoids through the perforations of the limiting plate of the hexagonal hepatic lobules. But the radicals of the hepatic artery enter into the sinusoids directly through these perforations, or indirectly after joining with the inlet venules.

23. What do you mean by inlet venules?

Ans. Inlet venules are those small veins which arise from the transverse terminal branches of portal veins and enter into the hepatic sinusoids.

24. What is the pressure of these inlet venules (portal venules)?

Ans. About 8–10 mm of Hg.

25. What is the pressure of the radicles of hepatic artery?

Ans. About 90 mm of Hg.

26. What do you mean by venules?

Ans. Venules are smallest veins which receive blood from capillaries and about 20–30 µm in diameter.

27. What is the structure of a sinusoid?

Ans. A sinusoidal wall is formed only by endothelial cells without having any basal lamina. At some places of the sinusoidal wall the endothelial cells are replaced by phagocytic cells.

28. What is the size of the lumen of sinusoid?

Ans. About 20 µm.

29. What are the basic differences between a capillary and a sinusoid?

Ans. A capillary is lined by endothelial cells resting on a basal lamina but the sinusoid is lined only by endothelial cells without any basal lamina. Secondly, a capillary is narrower (about 8 µm in diameter) than a sinusoid (about 20 µm).

Histology: Liver and Pancreas (Both Exocrine and Endocrine Types of Gland)

30. Name some other organs where sinusoids are also present.

Ans.
i. Adrenal gland
ii. Pituitary gland
iii. Parathyroid glands
iv. Spleen
v. Bone marrow
vi. Carotid body

31. How is the intersinusoidal pressure maintained?

Ans. The lumen of a sinusoid communicates with its adjacent sinusoids through the holes in the hepatic lamina. By these communications the sinusoids act as a huge vascular sponge and maintain inter-sinusoidal pressure.

32. Why the blood flow to the sinusoids is sluggish?

Ans. The lumen of the sinusoid is wider (about 20 µm) and irregular. Because of these factors the blood flows slowly across the sinusoids. The average transit time of blood from the inlet venules to the central vein of liver is about 8.4 seconds.

33. What are the lining cells of hepatic sinusoids?

Ans.
i. Flattened endothelial cells which are fenestrated and do not have basement.
ii. Kupffer cells which belong to mononuclear phagocyte system (MPS).
iii. Pit cells or natural killer cells which have large granular lymphocytes.

34. What are Kupffer's cells?

Ans. Kupffer's cells are large stellate or pyramidal shaped phagocytic cells with several processes lining the walls of hepatic sinusoids, derived from bone marrow.

35. What is the role of these cells?

Ans. Being a major part of MPS these cells remove cellular and microbial debris from the circulation (phagocytic function).

36. What do you mean by the spaces of Disse?

Ans. The spaces between the cells of hepatic lamina and the wall of the hepatic sinusoid are known as spaces of Disse. They are also termed as perisinusoidal spaces.

37. What is the width of this space?

Ans. 0.2–0.5 µm wide.

38. What are the contents of this space?

Ans.
  i. Blood plasma
  ii. Chylomicrons
  iii. Ito cells

39. What are Ito cells?

Ans. Ito cells are:
  i. Stellate shaped in outline.
  ii. Perisinusoidal in location.
  iii. Mesenchymal in origin.
  iv. Attached to the cells of hepatic laminae.
  v. They are lipocytes containing small lipid vesicles.
  vi. Store vitamin A and varieties of microfilaments.
  vii. Secrete intralobular collagenous matrix.

40. What is the role of Ito cells in liver injury (cirrhosis)?

Ans. Ito cells produce increased amount of collagen which replace the damaged liver cells, causing fibrosis.

41. How does the perisinusoidal space remain patent?

Ans. The interdigitations between the projecting microvilli from the cells of hepatic laminae towards the perisinusoidal space and the processes of Kupffer's cells keep the space patent.

42. What do you mean by the spaces of Mall?

Ans. The space of Mall is a potential space in the interlobular area between the Glisson's capsule of portal canal and the hepatic plates of cells.

43. What is the importance of this space?

Ans. This space is thought to be the site of origin of hepatic lymphatics.

44. Are the spaces of Disse and the spaces of Mall completely separated from each other?

Ans. No, the spaces are continuous with each other at the periphery of the hepatic lobules.

45. Why the hepatic lymph is rich in protein content?

Ans. Due to the continuity of the spaces of Disse with the spaces of Mall the excess blood plasma with rich protein content in the Disse's space is reabsorbed by the lymphatics of Mall's space.

Remember: That the spaces of Disse are intralobular and perisinusoidal but the spaces of Mall are interlobular and peri-portal. At the periphery of the hepatic lobules, both the spaces are continuous with each other.

46. How are the bile canaliculi formed?

Ans. The bile canaliculi are nothing but the spaces lined by the plasma membrane of adjacent liver cells of hepatic laminae excluding the surfaces of the liver cells facing towards the hepatic sinusoids.

47. How many surfaces of hepatic cells of hepatic laminae are involved to form bile canaliculi?

Ans. The plasma membranes of the adjacent four surfaces (excluding two surfaces facing toward sinusoids) of hepatic cells are separated to form bile canaliculi.

48. What is the special feature of the walls of the adjacent laminar hepatocytes?

Ans. The central part of the wall which forms the bile canaliculi possesses numerous microvilli projecting toward the lumen of canaliculi and the rest of the wall is sealed with the corresponding wall of the adjacent cell by tight junctions.

49. What is the source of liver bile?

Ans. Hepatic bile is secreted from the hepatic cells of hepatic laminae through their microvilli into the bile canaliculi.

50. What is the difference between the bilirubin in the hepatic sinusoids and the bilirubin in bile canaliculi?

Ans. Sinusoidal bilirubin is lipid soluble but the canalicular bilirubin is water soluble (bilirubin glucuronide).

51. How does the lipid soluble bilirubin become water soluble?

Ans. Lipid soluble bilirubin in sinusoids → taken up by the intralaminar hepatic cell through the perisinusoidal space → conjugation of bilirubin by glucuronyl transferase within the hepatic cells → delivered into the bile canaliculi through the microvilli of hepatic cells (water soluble bilirubin glucuronide).

Remember: That sinusoids are inter-laminar and centripetal in direction but the bile canaliculi are intralaminar and centrifugal in direction.

52. What do you mean by canal of Herings?

Ans. The bile canaliculi run in centrifugal direction and at the periphery of hepatic lobules they join to form intralobular canal of Herings.

53. What is the fate of the canal of Herings?

Ans. The canal of Herings drain into the bile ductile of portal triad which are interlobular in location.

Remember: That bile canaliculi are intralobular and intralaminar, canals of Herings are intralobular and the bile ductules are interlobular in location.

54. What do you mean by acinus of Rappaport?

Ans. Acinus of Rappaport is a diamond-shaped area of hepatic parenchyma supplied by one hepatic arteriole running along the line of junction of two adjacent hepatic lobules. Central veins of these two adjacent hepatic lobules lie at the ends of the acinus (Fig. 16.4).

55. What are the zones of these acini?

Ans. There are three zones according to the gradient of blood supply;
  i. Zone 1 or peri-portal zone: Most oxygenated
  ii. Zone 2 or intermediate or transitional zone: Moderately oxygenated.
  iii. Zone 3 or peri-central zone: Least oxygenated

56. What is the clinical importance of these zones?

Ans. Zone 1 is more susceptible to toxic injury and zone 3 is more susceptible to ischemic changes and fat accumulation.

57. How many acini are there in the human liver?

Ans. About 100,000.

Remember that:
  i. Hepatic lobules are vascular units of liver.
  ii. Portal lobules are functional and nutritional units of liver.
  iii. Acini of Rappaport are structural and metabolic units of liver.

Histology: Liver and Pancreas (Both Exocrine and Endocrine Types of Gland)

LESSON 2: PANCREAS

Key points for identification (Figs 16.5 and 16.6)

1. Exocrine part containing dark staining serous acini arranged in small lobule.
2. Endocrine part consists of highly stained Islets of Langerhans.
3. Centroacinar cells within the lumen of acini.
4. Absence of striated ducts.

PROBABLE QUESTIONS AND ANSWERS

1. **What are the functional components of pancreas?**

   Ans.
   i. Exocrine part produces pancreatic juice which reaches duodenum via the pancreatic duct.
   ii. Endocrine part produces hormones which is directly delivered into the blood stream.

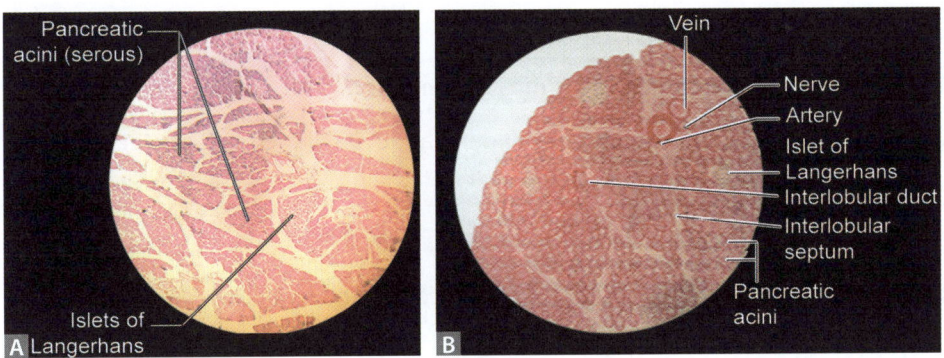

Figs 16.5A and B: Pancreas

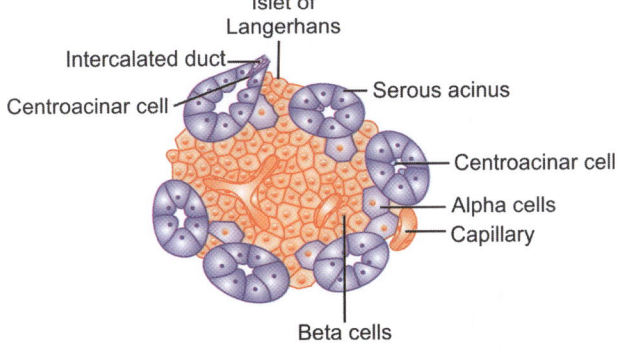

Fig. 16.6: Islets of Langerhans

2. What type of gland is the exocrine part?

Ans. The exocrine part of the pancreas is a compound tubuloalveolar (tubuloacinar) type of gland.

3. Does this part possess capsule?

Ans. No definite capsule exists but a layer of thin capsule, made up of thin collagen fibers is present.

4. What are lobules of exocrine part of the pancreas?

Ans. The fibrous capsule sends numerous interlobular septae inside the gland to form numerous lobules.

5. What are the contents of a lobule?

Ans. A lobule contains:
  i. Intralobular ducts
  ii. Intercalary ducts
  iii. Alveoli/acini
  iv. Islets of Langerhans

6. Which type of pancreatic cell is most abundant?

Ans. Cells of pancreatic acini are more abundant (about 99%) then cells of Islets of Langerhans (about 1%).

7. What is an acini?

Ans. An acini is a round cluster of exocrine cells connected to a duct.

8. What do the pancreatic acini secrete?

Ans. The cells of pancreatic acini secrete pancreatic juice which reaches duodenum.

9. How do the juice reach the duodenum from the acini?

Ans. Juice secreted from the cells of pancreatic acini → lumen of the acini (alveoli) → intercalary ducts → intralobular ducts → interlobular ducts → pancreatic duct (duct of Wirsung) → 2nd part of duodenum.

10. What type of acini are found in pancreas?

Ans. Pancreatic acini are serous in type.

Histology: Liver and Pancreas (Both Exocrine and Endocrine Types of Gland)

11. What are the features of the cells of serous acini?

Ans. The features are:
  i. Roughly truncated pyramidal in shape.
  ii. Presents a basal zone and an apical zone.
  iii. Basal zone is basophilic, striated, contains mitochondria and rough ER.
  iv. Nucleus of the cell is basal in position and round/spherical in shape.
  v. Apical zone is eosinophilic (acidophilic) and contains zymogen granules.
  vi. Lumen is small.

12. What is the method of secretion of the pancreatic acinar cells?

Ans. Merocrine method of secretion.

13. What do you mean by merocrine method?

Ans. It is a method of secretion of their substances by exocytosis.

14. What are the other methods of exocrine secretion?

Ans.
  i. Apocrine method in which the apical (luminal) part of the cell disintegrates to liberate the secretion (e.g. mammary gland).
  ii. Holocrine method in which the entire cell disintegrates to liberate secretion (e.g. sebaceous gland).

15. Give some examples of serous cells other than pancreatic serous cells.

Ans.
  i. Gastric cheif cells.
  ii. Paneth cells.
  iii. Acinar cells of parotid gland (predominantly serous).

16. Give some examples of mucous cells.

Ans.
  i. Brunner's glands in duodenum.
  ii. Esophageal glands.
  iii. Pyloric gland.
  iv. Goblet cells.
  v. Sublingual gland (predominantly mucous).

17. What are the component parts of the exocrine system of the pancreas?

Ans.
  i. Alveoli/acini (alveoli are flask-shaped and acini are rounded).
  ii. Intercalary ducts.
  iii. Intralobular ducts.
  iv. Interlobular ducts.
  v. Pancreatic duct (main and accessory).

18. What are the lining cells of these parts?

Ans.
  i. Alveolar or acinar cells → serous cells (tall columnar).
  ii. Intercalary (intercalated) ducts → low cuboidal or flattened epithelial cells.
  iii. Intralobular/interlobular → stratified cuboidal cells.
  iv. Pancreatic duct → stratified columnar cells.

19. What is centroacinar cells?

Ans. The cells of the intercalated duct penetrate partially into the acini forming pale staining centroacinar cells (Fig. 16.6).

20. Why are they called centroacinar?

Ans. This is because of their location near the center of the acini.

21. What type of cell is the centroacinar cell?

Ans. Low cuboidal cell (similar to the lining cells of intercalated duct).

22. Do these centroacinar cells cause any alteration in the size of the lumen of acini?

Ans. Yes, the lumen becomes smaller.

23. Why is the arrangement of duct system of pancreas called herringbone pattern?

Ans. The main pancreatic duct receives smaller ducts at regular angles in its course resembling herringbone pattern.

24. Do you find any striated duct in pancreas?

Ans. No, striated ducts are absent in pancreas unlike parotid gland.

25. What are the types of secretion in respect of composition?

Ans. Mainly two types of secretion are secreted by the secretory cells of pancreas:
  i. Watery secretion—rich in bicarbonate ions.
  ii. Digestive enzymes—trypsinogen, chymotrypsinogen, amylase, lipase.

Histology: Liver and Pancreas (Both Exocrine and Endocrine Types of Gland)

26. What is the stimulation and functions of these secretions?

Ans. Flowchart 16.1 and 16.2 show the biocarbonate-rich watery secretion and enzyme secretion.

27. Which element of pancreas forms its endocrine part?

Ans. Islets of Langerhans.

28. What are islets?

Ans. Islets are highly vascular rounded clusters of cells which are permeated by fenestrated capillaries.

29. How many islets are there in human pancreas?

Ans. About one to two million.

30. In which part of pancreas it is more numerous?

Ans. In the tail of pancreas.

31. How many types of cells are there in each islet?

Ans. There are three main types of cells in each islet.

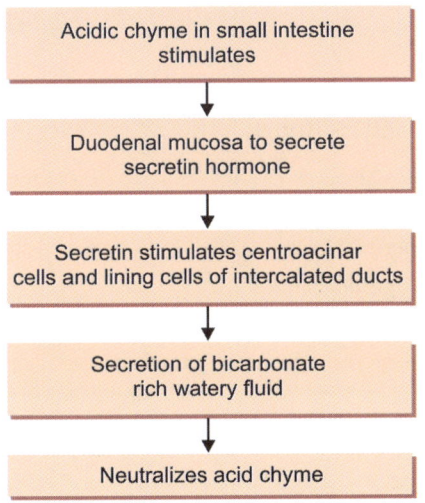

Flowchart 16.1: Bicarbonate-rich watery secretion

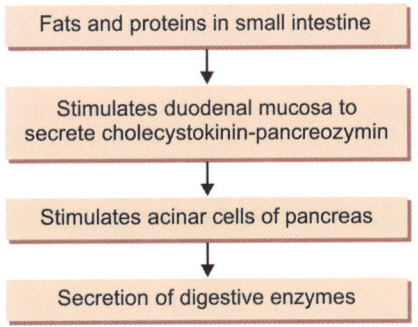

Flowchart 16.2: Secretion of enzymes

32. What are these cell types?

Ans.
i. Alpha (α) or A-cells (20% of total cells of Islets).
ii. Beta (β) or B-cells (70% of total cells of Islets).
iii. Delta (δ) or D-cells (5% of total cells of Islets).

33. What is the basis of the classification of these cells?

Ans. The cells are classified according to the staining property of their cytoplasmic granules.

34. What are the staining properties of these cells?

Ans.
i. A-cells are stained with acid fuchsin.
ii. B-cells are stained with aldehyde fuchsin.
iii. D-cells are stained with acid fuchsin.

35. Why are the D-cells sometimes called A_1 cells?

Ans. This is because of the reason that both D-cells and A-cells are stained with acid fuchsin.

36. What are A_2 cells?

Ans. A_2 cells are glucagon producing cells and simply they are also called A cells.

37. How do you differentiate A_1 and A_2 cells?

Ans.
i. A_1 cells are argyrophilic, i.e. they are capable of binding silver salts and stain black.
ii. A_2 cells are non-argyrophilic, i.e. incapable of binding silver salts.

38. What are the zones of an islet?

Ans. Outer heterocellular zone or cortex, containing both A(α) and D(δ) cells. Inner homocellular zone or medulla, containing predominantly B(β) cells.

39. What do these cells secrete?

Ans.
i. Alpha (α) cells secrete glucagon (diabetogenic)
ii. Beta (β) cells secrete insulin (anti-diabetogenic)
iii. Delta (D) cells secrete:
   a. Gastrin which increases gastric motility and stimulate gastric parietal cells to secrete HCl.
   b. Somatostatin which inhibits both α and β cells.

Histology: Liver and Pancreas (Both Exocrine and Endocrine Types of Gland)

40. Do you find any other types of cells in islets?

Ans.
i. P-P cells, secreting pancreatic polypeptide hormones
ii. VIP cells, secreting vasoactive intestinal peptide.

41. What is neuroinsular complex?

Ans. It is a complex mass, consisting of all types of Islet cells in the center and postganglionic parasympathetic neurons at the periphery. These neurons supply the islet cells.

42. What is the structural similarity of the pancreas with the parotid gland?

Ans. Both the pancreas and parotid gland are composed of large number of closely packed serous acini.

43. What are the differentiating features of the pancreas from the parotid gland?

Ans. The following histological features are found only in pancreas but not in parotid gland.
i. Islets of Langerhans.
ii. Centroacinar cells.
iii. Less connective tissue and only lobules without lobes.

But striated ducts are present in parotid gland, not in pancreas.

44. How many amino acids are present in insulin and glucagon?

Ans. Both insulin and glucagon are polypeptide hormones and having 51 and 29 amino acids, respectively.

CHAPTER 17

Histology: Salivary Glands (Exocrine Glands)

- ❑ Parotid Gland
- ❑ Submandibular Gland
- ❑ Sublingual Gland

LESSON 1: PAROTID GLAND

Key points for identification (Figs 17.1 and 17.2)

1. Presence of basophilic serous acini.
2. No serous demilunes.
3. Presence of intercalated and striated ducts.
4. Infiltration of adipocytes.

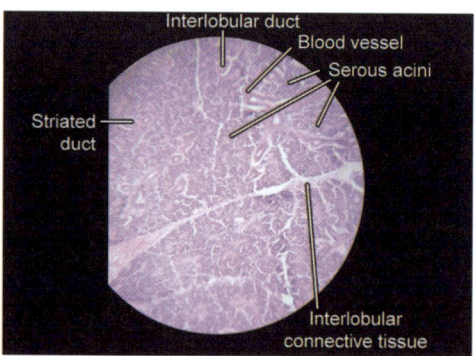

Fig. 17.1: Microstructure of parotid gland

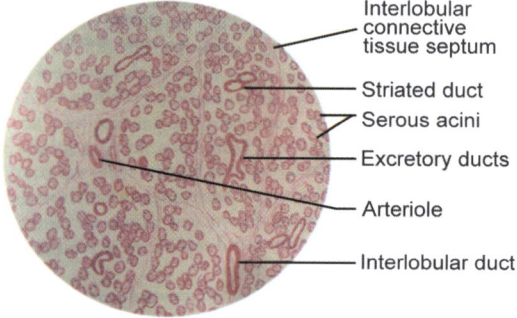

Fig. 17.2: Parotid gland (microstructure)

PROBABLE QUESTIONS AND ANSWERS

1. What is a gland?

Ans. A gland is a structure which is derived from epithelial tissue and produces secretion necessary for normal body functioning.

2. What are the types of glands on the basis of ducts?

Ans. Two types:
    i. Ductless glands or endocrine glands.   ii. Glands with ducts or exocrine glands.

3. What are major salivary glands?

Ans. Three paired major salivary glands are:
    i. Parotid glands
    ii. Submandibular glands
    iii. Sublingual glands

4. What type of glands are they?

Ans. They are exocrine glands.

5. What do they secrete?

Ans. The salivary glands secrete saliva.

6. What are the important functions of saliva?

Ans.
    i. It moistens the oral mucous membrane.
    ii. It lubricates food for easy swallowing.
    iii. Its lysozyme content controls bacterial flora.
    iv. It contains the digestive enzyme amylase
    v. It contains IgA which has a protective role.

7. What are the coverings of parotid gland?

Ans.
    i. **Inner true capsule**, formed by condensation of fibrous stroma of the gland.
    ii. **Outer false capsule**, formed by splitting of the investing layer of deep cervical fascia.

8. What is parotid fascia?

Ans. Outer false capsule is also known as parotid fascia.

9. What type of gland is the parotid gland on the basis of shape?

Ans. Compound tubuloalveolar or racemose gland.

10. What type of acini is found in parotid gland?

Ans. Mostly serous acini.

11. What do mean by acini?

Ans. Acini are groups of secretory cells.

12. What do you mean by serous acini?

Ans. Serous acini are made up of serous cells.

13. What does a serous cell secrete?

Ans. A serous cell secretes a clear watery fluid, rich is protein.

14. Can you name any other organ which is composed of serous acini?

Ans. Pancreas, it is identified by the presence of islets of Langerhans in addition to serous acini.

15. What are the other types of acini?

Ans.
   i. Mucous acini which secretes mucin are found in sublingual gland.
   ii. Mixed acini, found in submandibular gland.

16. What is mucous?

Ans. Mucin mixed with water becomes mucous.

17. What are the histological differences between serous and mucous acini?

Ans. Histological differences between serous and mucous acini are shown in Table 17.1, and Figures 17.3 and 17.4.

Table 17.1: Differences between serous and mucous acini

Features	Serous cell/serous acini	Mucous cell/mucous acini
Size	Smaller	Larger
Shape	Roughly pyramidal	Columnar
Nuclei	Rounded or slightly ovoid, placed basally	Flattened and pushed toward the base
Lumen	Very small	Larger and visible
Microvilli	Present at the apex	Absent
Staining (H&E)	Basal cytoplasm is basophilic and stains darkly due to the presence of zymogen granules and appears blue or bluish purple. Apical cytoplasm is eosinophilic and appears pinkish or reddish	Most of the cells appear empty. This is because of the fact that the secretory product of the cells is dissolved during the staining process or remains unstained

Histology: Salivary Glands (Exocrine Glands)

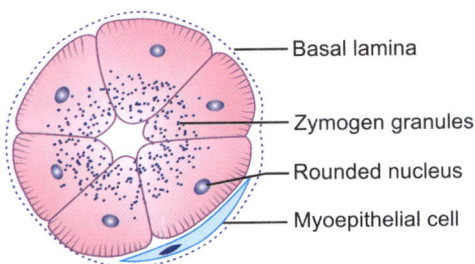

Fig. 17.3: Serous acini

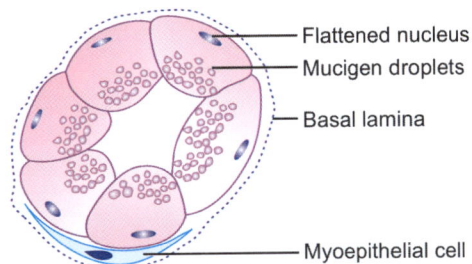

Fig. 17.4: Mucous acini

18. What are the ducts in salivary glands?

Ans. Two types of ducts:
   i. Intralobular ducts.
   ii. Interlobular ducts.

19. What are the types of intralobular ducts?

Ans. Two types:
   i. Smallest intralobular ducts are called intercalated ducts which are lined by cuboidal or simple squamous epithelium.
   ii. Striated ducts receive the intercalated ducts and are lined by simple columnar epithelium. These ducts are so named because of the vertical striations at the basal part of cells.

Note that striated ducts are absent in purely mucous glands.

20. What are interlobular ducts?

Ans. All excretory ducts are interlobular ducts.

21. What is the epithelium of the interlobular ducts?

Ans. Epithelium is variable. It may be simple columnar, simple cuboidal, stratified columnar, stratified cuboidal, pseudostratified columnar or stratified squamous.

22. What is type of the epithelium in the terminal duct, i.e. near the opening into the oral cavity?

Ans. Stratified squamous epithelium.

23. What is the sequence of the ducts of the salivary glands starting from its smallest branches to the terminal duct?

Ans. Sequence of the duct of salivary glands are shown in Flowchart 17.1.

Flowchart 17.1: Sequence of the duct of salivary glands

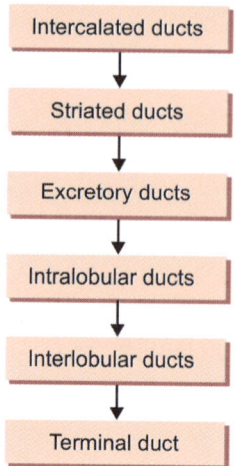

24. What are myoepithelial cell of salivary glands?

Ans. Myoepithelial cell are contractile cells which are present in relation to alveoli (acini) and intercalated ducts of salivary glands.

25. What is the location of these cells?

Ans. They are located between the epithelial cells and their basement membrane.

26. What is the shape of these cells?

Ans. Myoepithelial cells of the alveoli are stellate shaped and the cells of the intercalated ducts are fusiform.

27. Which is responsible for their contractile property?

Ans. Cytoplasmic myofilamets of smooth muscle cells provide contractile function.

28. What is the function of these myoepithelial cells?

Ans. Contraction of these cells helps to squeeze out secretion from the acini and ducts.

LESSON 2: SUBMANDIBULAR GLAND

Key points for identification (Fig. 17.5)

1. Presence of both serous and mucous acini.
2. Mucous acini are capped with serous demilunes.
3. Intra and interlobular ducts are seen.

Histology: Salivary Glands (Exocrine Glands)

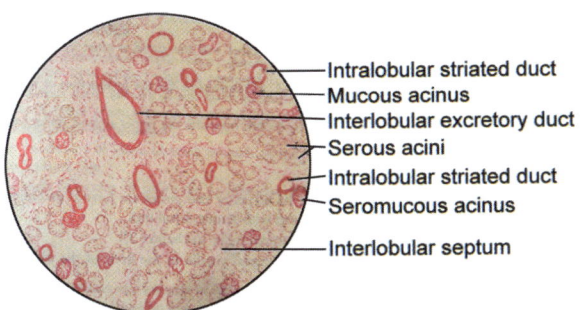

Fig. 17.5: Submandibular salivary gland

PROBABLE QUESTIONS AND ANSWERS

1. What is the location of submandibular gland?

Ans. Submandibular region.

2. What are the parts of submandibular gland?

Ans. Two parts:
   i. Large superficial part.
   ii. Small deep part.
Both parts are continuous around the posterior border of mylohyoid muscle.

3. Where does the submandibular duct open?

Ans. It opens in the floor of the mouth on a sublingual papilla.

4. What is the other name of the duct?

Ans. Wharton's duct.

5. What is its length?

Ans. About 5 cm.

6. Where does the parotid duct open?

Ans. Parotid duct opens in the vestibule of mouth opposite the crown of upper second molar tooth.

7. What is its length?

Ans. Parotid duct is about 5 cm in length.

8. What do mean by lobules of salivary glands?

Ans. The connective tissue divides the glands into numerous small compartments, called lobules.

9. What is the type of submandibular acini?

Ans. Submandibular acini are round alveolar in shape and made up of both serous and mucous cells.

10. What is serous demilunes?

Ans. Mucous acini are covered by groups of serous cells in the form of crescents called serous demilunes.

Remember that serous demilunes (demilunes of Giannuzzi) are the characteristic feature of submandibular gland.

11. What is the duct system of submandibular gland?

Ans. The duct system consists of:
  i. Intralobular ducts. ii. Interlobular ducts.

12. Does the submandibular gland possess myoepithelial cells?

Ans. Yes, the gland possesses myoepithelial cells.

13. Why is the submandibular gland called a mixed gland?

Ans. This gland consists of both serous and mucous acini.

14. What is homocrine gland?

Ans. A gland or its secretory unit which consists of only type of cell (either serous or mucous) is called a homocrine gland.

15. What is heterocrine gland?

Ans. If a gland or its secretory unit consists of more than one type of cells, the gland is said to be a heterocrine gland.

16. What is paracrine gland?

Ans. It is a variety of endocrine gland but its secretions are thrown into blood close to its target organs.
Example: D-cells of pancreas.

LESSON 3: SUBLINGUAL GLAND

Key points for identification (Fig. 17.6)

1. Made up of mucous acini.
2. May possess serous demilunes.
3. Intralobular and interlobular ducts.

PROBABLE QUESTIONS AND ANSWERS

1. Where does the sublingual gland lodge?

Ans. In the sublingual fossa of the mandible and it is located in the floor of the mouth.

2. What is its shaped

Ans. Almond in shape.

3. What is its weight?

Ans. About 3–4 g.

4. Which of the three major salivary glands is the largest?

Ans. Parotid gland is the largest of all major salivary glands.

5. Which of the three major salivary glands is the smallest?

Ans. Sublingual gland is the smallest.

6. How many ducts does the gland possess?

Ans. The gland possesses about 8 to 20 ducts.

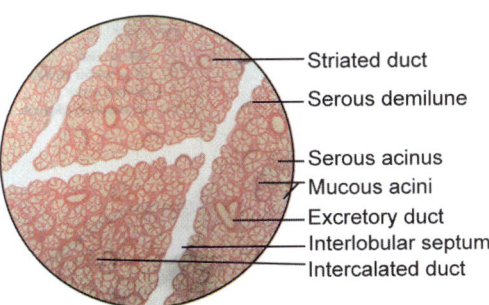

Fig. 17.6: Sublingual salivary gland

7. **Where do the ducts of sublingual gland open?**

Ans. Most of the ducts open in the floor of the mouth as **ducts of Rivinus**. Few ducts unite to from sublingual duct or duct of Bartholin to open into the submandibular duct.

8. **What are the histological differences of three major salivary glands?**

Ans. Histological differences between parotid, submandibular and sublingual glands are shown in Table 17.2.

Table 17.2: Histological differences between parotid, submandibular and sublingual glands

	Parotid gland	Submandibular gland	Sublingual gland
Acini	Mostly serous	Mixed (both serous and mucous)	Mostly mucous
Serous demilunes	Absent	Present	Very few
Striated ducts	Large in number	Many	Few
Staining (Color)	Darkly stained (Basophilic)	Mixer of dark and light stain	Lightly stained (Eosinophilic)

CHAPTER 18

Histology: Endocrine Glands

- Thyroid Gland
- Adrenal Gland

LESSON 1: THYROID GLAND

Key points for identification (Figs 18.1 to 18.3)

1. Made up of numerous follicles lined by cuboidal epithelium containing pink staining colloid.
2. Presence of parafollicular cells.

PROBABLE QUESTIONS AND ANSWERS

1. What are the microscopic constituents of thyroid gland?

Ans. Thyroid gland is composed of **parenchyma and stroma.**

2. What is parenchyma?

Ans. Parenchyma is the essential or functional elements of an organ.

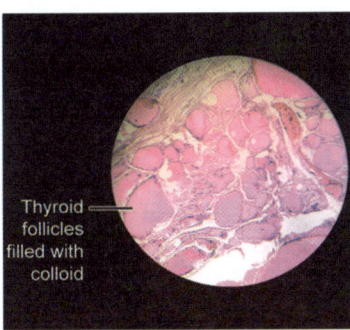

Fig. 18.1: Thyroid (low power)

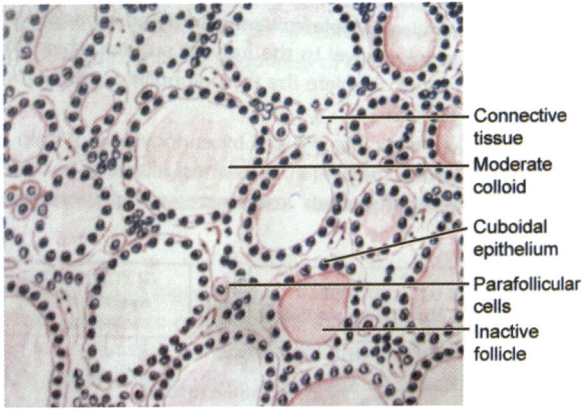

Fig. 18.2: Thyroid (high power)

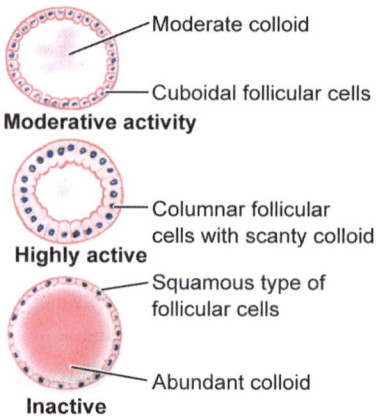

Fig. 18.3: Thyroid gland at different levels of activity

13. What is stroma?

Ans. The supporting tissue of an organ is its stroma.

14. What are the thyroid lobules?

Ans. The interior of the thyroid gland is divided into a number of compartments by numerous connective tissue trabeculae (septae) projecting from the true fibroelastic capsule of thyroid gland, called thyroid lobules.

15. What do the connective tissue trabeculae contain?

Ans. The trabeculae contain numerous blood vessels, arteries, venules and capillaries.

16. What does a lobule contain?

Ans. Each lobule contains about 40–60 thyroid follicles and stroma in which blood vessels lymphatics and nerve fibers exist.

17. What are these thyroid follicles?

Ans. Thyroid follicles are irregular and spherical, structural and functional units of thyroid gland containing **colloid**.

18. What is the size of a follicle?

Ans. The size of the follicles varies from 0.02 to 0.9 mm in diameter.

19. How many follicles are there in human thyroid?

Ans. There are about more than 3 million follicles in human thyroid.

Histology: Endocrine Glands

10. What is the structure of a follicle?

Ans. Each follicle is lined by a single layer of follicular cells, also called principal cells resting on a basement membrane and the lumen of the follicle contains colloid (Fig. 18.4).

11. What are the special features of follicular cells?

Ans.
 i. Apical or luminal surfaces of follicular cells are provided with microvilli which engulf droplets of colloid into the cell.
 ii. **Secondly**, adjacent follicular cells are connected by tight junctional complexes which act as an effective barrier between the follicular lumen and the surrounding capillaries.

12. What is the type of follicular epithelium?

Ans. The follicular epithelium is **cuboidal** in type and the colloid in the follicular lumen is moderate in amount at an average state of activity of thyroid gland.

13. How does the type of epithelium vary in different level of activity of the gland?

Ans.
 i. **In resting or inactive state** of activity the follicular cells are squamous with abundant colloid.
 ii. **In hyperactive state**, the cells are columnar with reduced amount of colloid.

	Resting or inactive follicles	*Hyperactive follicles*
Size of follicles	Large	Small
Type of epithelium	Squamous	Columnar
Amount of colloid	Abundant	Reduced
Level of TSH	Low	High

14. Do all the follicles show equal level of activity at a particular state of the gland?

Ans. No. Different follicles may show different levels of activity.

15. What is colloid?

Ans. Colloid is a homogeneous and gelatinous material occupying the lumen of thyroid follicles.

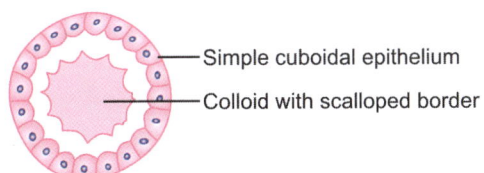

Fig. 18.4: Structure of a thyroid follicle

16. How does it appear?

Ans. It appears pink in Haematoxylin and Eosin (H&E) stain, i.e. colloid is eosinophilic (acidophilic).

17. What is the composition of colloid?

Ans. Colloid is an iodinated glycoprotein, called **thyroglobulin**.
Iodine (absorbed from blood) + glycoprotein (synthesized by follicular cells) = Iodinated glycoprotein (thyroglobulin).

18. What is the function of thyroglobulin?

Ans. Thyroglobulin is the inactive storage form of thyroid hormones.

19. What are the active forms of thyroid hormones?

Ans. T_3 (triiodothyronine) and T_4 (tetraiodothyronine) are active forms of thyroid hormones.

20. How T_3 and T_4 are produced?

Ans.

Thyroglobulin is taken back into follicular cells from the follicular lumen by **endocytosis**

↓

Hydrolysis of iodinated thyroglobulin by lysosomes within the follicular cells

↓

Release of principal thyroid hormones (T_3 and T_3) into the surrounding capillaries

21. Which of these two hormones is more active?

Ans. T_3 is more active than T_4 (thyroxin).

22. Which of these two hormones are more produced?

Ans. T_4 is more produced than T_3 (T_3 is produced less than 10%).

23. Apart from follicular cells, do you find any other type of cell in thyroid lobule?

Ans. Yes. These are **parafollicular cells** or C-cells or clear cells or light cells.

24. Where do you find these cells?

Ans. These cells lie between the follicular cells and their basement membrane. They may lie in the interfollicular spaces.

Histology: Endocrine Glands

25. Are they part of thyroid follicles?

Ans. No, they are neither a part of thyroid follicle nor reach the follicular lumen, i.e. not in contact with colloid.

26. How do they appear microscopically?

Ans. They appear as **pale-staining** polyhedral cells with eccentric nuclei. These cells are larger than follicular cells.

27. What do these cells secrete?

Ans. They secrete the hormone **thyrocalcitonin** which lowers the raised serum calcium level by suppressing bone resorption by osteoclasts.

28. Are the thyrocalcitonin and parathyroid hormone having same function?

Ans. No. They have opposite action on calcium metabolism.

29. Why the thyroid gland is unique as an endocrine gland?

Ans. Most endocrine organs store their secretory products within their cytoplasm but the follicular cells of thyroid gland store their products outside their cytoplasm.

30. What is the developmental difference between the follicular cells and parafollicular cells?

Ans. **The follicular cells** are developed from the **endoderm** of the thyroglossal duct but the **parafollicular cells** are developed from the **neural crest cells of neuroectoderm**.

LESSON 2: ADRENAL GLAND

Key points for identification (Figs 18.5A and B)

1. Outer cortex, covered by capsule.
2. Cortex consists of three layers—zona glomerulosa, zona fasciculata and zona reticularis.
3. Inner medulla, containing chromaffin cells separated by sinusoids.

PROBABLE QUESTIONS AND ANSWERS

1. What are the main regions of adrenal gland?

Ans. i. Outer cortex constitutes about 90% of the gland.
 ii. Inner medulla constitutes about 10% of the gland.

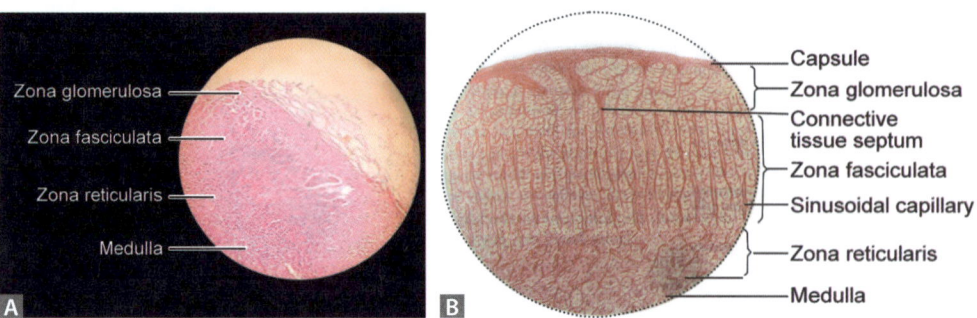

Figs 18.5A and B: Microstructure of suprarenal gland

2. What are the different zones of cortex?

Ans. Three zones (from outside inwards):
  i. Zona glomerulosa (about 15% of the cortex).
  ii. Zona fasciculata (about 75% of the cortex).
  iii. Zona reticularis (about 10% of the cortex).

3. How are the cells of zona glomerulosa arranged?

Ans. The cells of this zone are columnar or small polyhedral and are arranged in curved columns or in irregular clusters.

4. What is the staining property of the cells of zona glomerulosa?

Ans. The cytoplasm of these cells is scanty and basophilic, but their nuclei stain dark (deep staining).

5. What is the secretion of these cells?

Ans. The cells secrete mineralocorticoids (aldosterone) by the influence of renin-angiotensin mechanism (not by ACTH).

6. What is the role of mineralocorticoids?

Ans. It regulates acid-base balance of the body.

7. Which one is the thickest layer of the cortex?

Ans. Zona fasciculata.

8. How are the cells of zona fasciculata arranged?

Ans. The cells are arranged in long straight cords, 1–2 cell thick, separated by capillaries and sinusoids.

Histology: Endocrine Glands

9. What do the cells secrete?

Ans. They secrete glucocorticoids.

10. What is the function of glucocorticoids?

Ans. They regulate carbohydrate and protein metabolism of the body.

11. What is the active principle of glucocorticoids?

Ans. Cortisol (hydrocortisone) is the most active principle.

12. What is the staining property of the zona fasciculata?

Ans. The nucleus stains light and the cytoplasm is acidophilic. The cytoplasm often appears foamy due to its lipid content.

13. What are the important contents of the cellular cytoplasm of zona fasciculata?

Ans.
 i. Lipid droplets
 ii. Cholesterol
 iii. Vitamin C

14. What are spongiocytes?

Ans. The cells of this layer are also known as spongiocytes.

15. How are the cells of zona reticularis arranged?

Ans. The cells of this zone are arranged as irregular network of branching cords separated by capillaries and sinusoids.

16. What do the cells of this zone secrete?

Ans. The cells secrete gonadocorticoids/sex hormones.
These are:
 i. Estrogen
 ii. Androgen
 iii. Dehydroepiandrosterone
 iv. Progesterone

17. What is the role of the sex hormones?

Ans. They regulate the secondary sex characters in conjunction with the gonads (Testis/ovaries).

18. What is the staining property of the cells of this zone?

Ans. The cytoplasm is acidophilic (eosinophilic), i.e. stains pink with eosin and the nucleus stains deep.

19. What is the difference between the fetal cortex and definitive cortex of adrenal gland?

Ans.
i. Fetal cortex is larger than the definitive cortex. So the fetal adrenal gland is comparatively larger than the glands of the adult.
ii. Fetal cortical cells secrete dehydroepiandrosterone (a prehormone) which is converted into androgen or estrogen after it reaches placenta. These are placental androgen or estrogen.

20. What is the composition of adrenal medulla?

Ans. Adrenal medulla is composed of:
i. Chromaffin cells (pheochromocytes).
ii. Few sympathetic ganglion cells.
iii. Collagen fibers.
iv. Capillaries and sinusoids.

21. What is the nerve supply of the adrenal medulla?

Ans. Sympathetic nerves (T_8-L_1).

22. What is the nerve supply of the chromaffin cells?

Ans. Preganglionic sympathetic fibers innervate chromaffin cells which act as post-ganglionic neurons.

23. What are the features of chromaffin cells?

Ans.
i. They are tall columnar cells.
ii. Their cytoplasm is basophilic.
iii. Cytoplasm contains two types of secretory granules. Accordingly two types of cells are described:
   a. Adrenaline (epinephrine) producing cells (60%)
   b. Noradrenaline (norepinephrine) producing cells (20%)
   Remember that during emergency, chromaffin cells secrete more epinephrine than norepinephrine.
iv. Cells show chromaffin reaction.

24. What do you mean by chromaffin reaction?

Ans. The membrane bound secretory granules of chromaffin cells are stained brown/yellow when they are treated with a solution containing chromium salts (potassium bichromate) due to oxidation of epinephrine and norepinephrine.

25. What are the developmental origins of adrenal cortex and medulla?

Ans.
i. Cortex is mesodermal in origin.
ii. Medulla is developed from neural crest cells.

Histology: Endocrine Glands

26. What do you mean by suprarenal portal system?

Ans. Portal system means the connection between two sets of capillaries at their two ends. The vessels which connect the cortical sinusoids with the medullary sinusoids are known as suprarenal portal system.

27. What is the importance of this portal system?

Ans. This suprarenal portal system conveys some chemical substances from cortex to medulla to help in conversion of norepinephrine into epinephrine by the methylation of primary amines.

28. What are the other sites of portal system in the body?

Ans.
  i. Hepatic portal system (in liver).
  ii. Renal portal system (in kidney).
  iii. Hypophyseal portal system (in pituitary gland).

29. Can you find suprarenal cortical tissue in other sites of the body other than the suprarenal gland?

Ans. **Accessory suprarenal cortical tissues** may be found in the following sites:
  i. Within the spermatic cord or testis in males.
  ii. Within the broad ligament in females.
  iii. Within the mesenteries of gut.
  iv. Around the main suprarenal glands.

30. What are the salient distinguishing features of the different zones of adrenal cortex?

Ans. Salient features of different zones of adrenal cortex are shown in Table 18.1.

Table 18.1: Salient features of different zones of adrenal cortex

Features	Z. glomerulosa	Z. fasciculata	Z. reticularis
Location	Outer zone	Middle zone	Inner zone
Thickness	15% of the cortex	75% of the cortex	10% of the cortex
Arrangement of cells	Arched columns or rounded groups	Straight cords supported by sinusoids	Irregular network of branching cords
Staining (H&E) a. **Nucleus** b. **Cytoplasm**	Deep staining Basophilic	Light staining Eosinophilic	Deep staining Eosinophilic
Secretion	Mineralocorticoids	Glucocorticoids	Sex-hormones
Function	Regulates acid-base balance	Regulates carbohydrate and protein metabolism	Regulate secondary sex characters

CHAPTER 19

Histology: Lymphatic System

- Lymph Node
- Thymus
- Spleen
- Palatine Tonsil

LESSON 1: LYMPH NODE

Key points for identification (Figs 19.1 and 19.2)

1. Outer cortex covered by capsule.
2. Capsule sends processes into the substance of lymph node, known as trabeculae.
3. Presence of subcapsular sinus.
4. Presence of lymphatic nodules in cortex.
5. Inner medulla shows medullary cords and blood vessels.

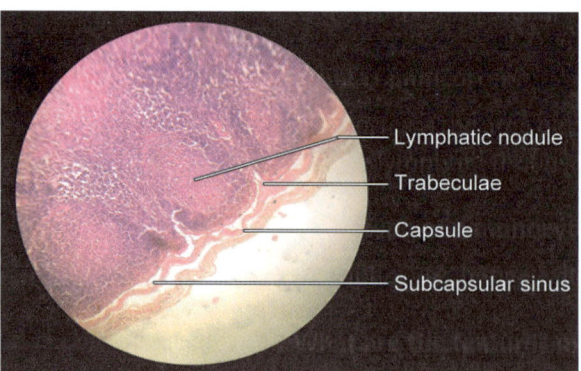

Fig. 19.1: Lymph node

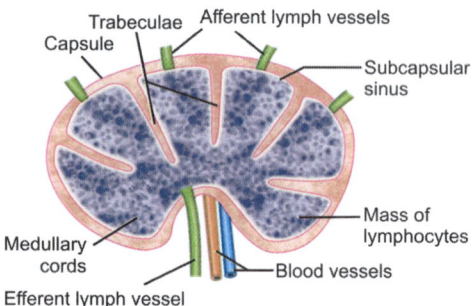

Fig. 19.2: Section of a lymph node

Histology: Lymphatic System

PROBABLE QUESTIONS AND ANSWERS

1. What is a lymph node?

Ans. A lymph node is a collection of lymphatic tissue, organized to form a small, encapsulated, oval or bean-shaped body in the course of lymph vessels. **Length of a lymph node** varies (0.1–2.5 cm).

2. What are the basic components of a lymph node?

Ans.   i. Outer thin fibrous capsule.   ii. Inner gland substance.

3. What are the constituent parts of gland substance?

Ans. Two parts:
  i. Outer cortex.   ii. Inner medulla.

4. How is the capsule of lymph node formed?

Ans. The capsule is formed by dense collagen fibers and few elastic fibers.

5. What are trabeculae?

Ans. Trabeculae are processes which extend inwards from the capsule into the substance of the node.

6. What are lobules of lymph node?

Ans. The trabeculae form incomplete septa and divide the interior of the lymph node into numerous small compartments, called lobules of lymph node.

7. What is subcapsular sinus or space?

Ans. It is the space between the capsule and the gland substance.

8. Do you find any structures in this space?

Ans. The subcapsular space is traversed by coarse reticular fibers and numerous afferent lymphatic vessels open in this space.

9. How many efferent lymphatic vessels are there in a single lymph node?

Ans. A single lymph node possesses a single efferent lymphatic vessel which exits through the hilum.

10. What are the structures that pass through the trabeculae?

Ans. The trabeculae convey blood vessels of the lymph node.

11. What are the paratrabecular space?

Ans. Paratrabecular spaces are the spaces that are found on either side of the trabeculae.

12. What structures lie in this space?

Ans. Paratrabecular space is also traversed by coarse reticular fibers which are continuous with the reticular fibers of subcapsular space.
(Note that the subcapsular space and paratrabecular space are continuous with each other).

13. What is intervened between the paratrabecular spaces of a lobule?

Ans. The area between the paratrabecular spaces of a lobule is occupied by a reticular mesh which is formed by fine reticular fibers.

14. What is present on this reticular mesh of cortex?

Ans. The reticular mesh of the cortex contains rounded or pear-shaped nodules called lymphatic nodules/lymphatic follicles.

15. What are the types of lymphoid follicles?

Ans. Two types:
  i. Primary follicles.
  ii. Secondary follicles.

16. What do you mean by a lymphatic follicle (nodule)?

Ans. A lymphatic nodule consists of a collection of lymphocytes supported by reticular fibers.

17. Are the number of lymphatic nodules are constant in a particular lymph node?

Ans. No, their number varies with the prevailing antigenic stimuli.

18. How do you differentiate a primary lymphatic nodule from a secondary nodule?

Ans.   i. A primary nodule consists mainly of small lymphocytes without any germinal center.
  ii. **A germinal center exists only in a secondary nodule.**

19. What is a germinal center?

Ans. Germinal center is a lightly stained reaction center in response to an antigen containing large lymphocytes (lymphoblasts) and plasma cells.

20. What are the types of lymphocytes?

Ans.
a. According to function:
   i. T-lymphocytes (thymus dependent).
   ii. B-lymphocytes (bursa equivalent).
b. According to size:
   i. Small (6–10 μm in diameter).
   ii. Medium (10–12 μm in diameter).
   iii. Larger (12–16 μm in diameter).

21. Where do the precursors of lymphocytes reside?

Ans. Lymphocyte precursors (lymphoblasts) are located in the bone marrow where they produce lymphocytes.

22. What are the cells of origin of these lymphocyte precursors (stem cells)?

Ans. Mesenchymal cells in the liver, in spleen and in the wall of the yolk sac are the cells of origin of stem cells.

23. Where are the lymphocytes produced?

Ans. Both T-lymphocytes and B-lymphocytes are formed in the bone marrow.

24. How do the T-lymphocytes circulate?

Ans. They circulate in the following path:

T-cells in bone marrow
↓
Bloodstream
↓
Thymus
↓
Re-enter bloodstream
↓
Reach different lymphoid tissue
(Lymph nodes, spleen, tonsils, intestines, etc.)
↓
Via lymph vessels
↓
Back into circulation

25. What is the basic difference of circulation of B-lymphocytes from T-lymphocytes?

Ans. Lymphocytes (B-cells) directly go to lymphatic tissues of lymph nodes, spleen, tonsils, intestines, etc. via bloodstream without passing through thymus (unlike T-cells).

26. Why are T-cells called thymus dependent cells?

Ans. T-cells reach the thymus where they are multiplied and processed to re-enter the circulation.

27. Why are the B-cells called bursa dependent cells?

Ans. B-cells are derived from 'Bursa of Fabricus', a diverticulum of the cloaca in birds.

28. What type of immunity is produced by T-cells and B-cells?

Ans. T-cells are concerned with cell-mediated immunity (cellular immunity) and delayed hypersensitivity reaction. B-cells are concerned with humoral immunity.

29. What is the mode of action of T-lymphocytes to show immune response?

Ans. T-lymphocytes can recognize foreign antigens (viruses, bacteria, fungi, parasites, tumor cells, cells of another individual). When the body is exposed to such foreign antigens, T-lymphocytes multiply and attack these foreign cells by direct contact and destroy them. Otherwise they may produce cytotoxic substances called cytokines or lymphokines and destroy abnormal cells. These types of T-cells are known as cytotoxic T-cells.

30. What is the mode of action of B-lymphocytes?

Ans. In response to specific antigen, B-cells proliferate by mitosis and get converted to plasma cells. These plasma cells secrete antibodies or immunoglobulins which are antigen specific. These antibodies enter the circulation and act against these antigens.

31. How much lymphocytes (Both T and B) are found in circulating blood?

Ans. 20–30% of total leukocytes are lymphocytes.

32. Which type of lymphocytes are more in blood circulation?

Ans. T-lymphocytes outnumber B-lymphocytes. About 80–85% of lymphocytes are T-lymphocytes in circulating blood.

Histology: Lymphatic System

33. What are the different types of T-lymphocytes (T-cells)?

Ans. i. Cytotoxic cells (T_c).
ii. Helper cells (T_H).
iii. Supressor cells (T_s).
iv. Natural killer cells (NL cells).
v. Delayed type hypersensitivity related T-cells.
vi. T memory cells.

Remember that all true T-lymphocytes are CD_3 positive (CD = Clusters of differentiation).

34. What do you mean by null cells?

Ans. These cells are believed to be the circulating stem cells, without possessing T or B lymphocyte surface antigens and constitute about 5% lymphocytes in blood.

35. What type of lymphocytes are found in the lymphatic nodules of cortex of a lymph node?

Ans. The germinal centers of lymphatic nodules are formed by actively dividing B-lymphocytes, surrounded by dense aggregation of B-lymphocytes.

36. How is a lymphatic follicle divided into different zones?

Ans. A lymphatic follicle is divided into three zones.
From within outwards:
i. Zone 3 represents the germinal center.
ii. Zone 2 is represented by dark rims of follicles, produced by dense aggregation of B cells.
iii. Zone 1 represents the region immediately around the follicle which is continuous with the medullary cords.

37. What is paracortex?

Ans. Paracortex is the area between the lymphoid follicles of cortex and medulla, containing diffuse lymphoid tissues.

38. What type of lymphocyte is found in paracortex?

Ans. T-lymphocyte, hence called thymus dependent cortex.

39. What are the features of medulla?

Ans. The medulla of a lymph node consists of:
i. Medullary cords.
ii. Poorly demarcated medullary sinuses.

40. What do you mean by medullary cords?

Ans. Medullary cords are the extensions of lymphatic tissue of the cortex and paracortex containing lymphocyte, plasma cells and macrophages in the form of branching and anastomosing pattern.

41. What are the types of cells that are found in medulla?

Ans.
i. Lymphocytes (Both and B-lymphocytes).
ii. Plasma cells (derived from B-lymphocytes).
iii. Macrophages.

Remember the different lymphocytes in different regions of a lymph node:
i. Cortex (lymphatic follicles) → B-lymphocytes.
ii. Paracortex → T-lymphocytes.
iii. Medulla → both B and T-lymphocytes.

42. What are medullary sinuses?

Ans. These are poorly demarcated cavities, representing minute lymphatic channels which join together at the hilum to deliver lymph to efferent lymphatic vessel.

43. Do you find cortex in the entire peripheral part of the lymph node?

Ans. No, the cortex does not extend into the hilum of the lymph node which is occupied by a mass of fibrous tissue.

44. What are the staining properties of different parts of a lymph node?

Ans.
i. The outer zone of lymph node (cortex and paracortex) containing densely packed lymphocytes stains dark (purple).
ii. The inner zone (medulla) containing loosely packed lymphocytes (fewer lymphocytes) stains lighter.

LESSON 2: THYMUS

Key points for identification (Fig. 19.3)

1. Surrounded by thin capsule.
2. Trabecular septa from the capsule divides the thymus into a large number of lobules.
3. Each lobule consists of outer cortex and inner medulla.
4. Aggregation of lymphocytes in darker cortex.
5. Presence of Hassall's corpuscles in inner medulla.

Histology: Lymphatic System

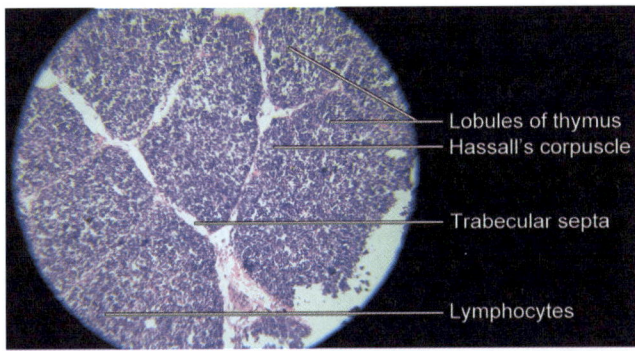

Fig. 19.3: Thymus

PROBABLE QUESTIONS AND ANSWERS

1. What type of organ the thymus is?

Ans. The thymus is primarily a lymphoreticular organ, situated in the superior and anterior mediastinum of thorax.

2. Is this organ consistant throughout life?

Ans. The thymus is present at birth and weighs about 10–15 g. After puberty it involutes and is converted into a fibrofatty mass and weighs about 10 g in mid adult life. Note that 90% weight of the thymus is contributed by the thymic lymphocytes.

3. How many lobes does it possess?

Ans. The thymus possesses two lobes connected by fibro areolar tissue.

4. What is the developmental source of thymus?

Ans. It develops from the endoderm of 3rd pharyngeal pouch.

5. Is there any covering of thymus?

Ans. Yes, the thymus is covered by a thin connective tissue capsule.

6. What are thymic lobules?

Ans. The trabeculae arising from the capsule enter inwards and divide the organ into a large number of lobules.

7. What is the size of a lobule?

Ans. Each lobule is about 1–2 mm in diameter.

8. To what extent the trabecular septa enters inside the thymus?

Ans. The septae extend only upto corticomedullary junction of thymic lobules and divide the thymus into numerous incomplete lobules.
(Note the feature of incomplete lobulation of thymus).

9. What are the regions of a thymic lobule?

Ans. Each lobule consists of a outer cortex and inner medulla. The medulla of adjoining lobules is continuous because of incomplete lobulation.

10. Does the thymus possess a hilum?

Ans. No, the thymus is devoid of hilum.

11. How do the blood vessels and nerves enter the thymus?

Ans. The blood vessels and nerves enter the thymus through the trabecular septae.

12. What are the special features of thymus?

Ans.
  i. Incomplete lobulation.
  ii. Absence of hilum.
  iii. Devoid of lymph capillaries.
  iv. No afferent lymph vessels but only efferent vessels.
  v. Replaced by a fibrofatty mass in adult life.

13. What does the cortex of thymus contain?

Ans.
  i. Closely packed numerous lymphocytes.
  ii. Epithelioreticular cells (epitheliocytes).
  iii. Macrophage cells.

14. What are the parts of the cortex?

Ans.
  i. Superficial part of cortex (subcapsular cortex).
  ii. Deep part of cortex (main cortex).

15. What is the source of lymphocytes of thymus?

Ans. Lymphocytes of thymus are derived from stem cells of bone marrow. Stem cells produce lymphocytes in bone marrow and then these lymphocytes reach thymus via bloodstream. Hence, these lymphocytes are called T-lymphocytes (T stands for thymus).

Histology: Lymphatic System

16. How are these lymphocytes arranged in the cortex?

Ans. The lymphocytes in the outer cortex are large and densely packed. These lymphocytes divide by mitosis into small lymphocytes which are pushed into deeper cortex and medulla. Medullary lymphocytes are small and densely packed.

17. Do all the thymic lymphocytes survive throughout life?

Ans. No, about 90% thymic lymphocytes have short life span (3–5 days). They react against self antigens of the body and undergo disintegration. These dead lymphocytes are phagocytosed by pale-stained macrophages.

18. How much lymphocytes leave the thymus to enter into blood circulation?

Ans. Only about 5% of the surviving lymphocytes leave the thymus via postcapillary venules and appear in the circulation through the hemothymic barrier as immunologically competent T-lymphocytes.

19. How is the hemothymic barrier formed?

Ans. The hemothymic barrier consists of (from outside inwards):
   i. A layer of continuous capillary endothelial cells.
   ii. A layer of thick basement membrane.
   iii. A tissue space between basement membrane of capillary endothelium and reticular epithelial cells.
   iv. A continuous layer of reticular epithelial cells.

20. Which is impermeable through this barrier?

Ans. The barrier is impermeable to antigens present in blood.

21. What are the substances and cells that can cross the hemothymic barrier?

Ans.
   i. Nutritive substances.
   ii. Stem cells from the bone marrow, which proliferate and differentiate to form lymphocytes.
   iii. Surviving thymic lymphocytes which appear in the circulation.

22. What is the important function of this barrier?

Ans. The barrier prevents circulating antigens from reaching the lymphocytes in the thymus. Thus the lymphocytes can proliferate in an antigen free environment.

23. What are the factors that influence lymphopoiesis?

Ans. Lymphopoiesis is the process of production and proliferation of lymphocytes. Thymic lymphopoiesis is autonomous and controlled by the hormones produced by the epitheliocytes of the thymus.

24. Does this lymphopoiesis occur only in thymus?

Ans. No, lymphopoiesis is influenced by the thymic hormones in other peripheral lymphoid tissue, i.e. in lymph nodes.

25. What are the hormones of thymus that influence lymphopoiesis?

Ans. These hormones are produced by the epitheliocytes of thymus. These are:
  i. Thymulin.
  ii. Lymphopoietin (thymopoietin).
  iii. Thymosin alfa 1.
  iv. Thymosin beta 4.
  v. Thymic humoral factor.

26. How are the epitheliocytes distributed in the thymus?

Ans. The epitheliocytes line the inner surface of fibrous capsule as a continuous sheet of cells. They also line the trabecular septae that arise from the capsule. Blood capillaries of cortex and medulla are also surrounded by the reticular epithelial cells.

27. How are the reticular epithelial cells (epitheliocytes) arranged?

Ans. These cells possess processes that join with similar processes of the outer cells forming a reticular or supporting framework.

28. What type of junctional complex lies between the epithelial cells?

Ans. The cells are connected with one another by desmosomes.

29. What is the difference between the reticulum of thymus and reticulum of lymph node and spleen?

Ans. The reticulum of thymus is formed by the irregular branching cords of reticular epithelial cells, i.e. the reticulum is cellular. But the reticulum of lymph nodes and spleen formed by the reticular fibers (not reticular cells).

30. What lie in the interstices of this reticulum?

Ans. Numerous lymphocytes and macrophages lie in this interstices of reticulum.

Histology: Lymphatic System

31. What are the different types of epitheliocytes?

Ans. Epitheliocytes are of six types (Type 1 to type 6) on the basis of their structural differences.

32. What is the distribution of the different types of epitheliocytes?

Ans.
i. Type 1 to type 4 epitheliocytes are found in the cortex called cortical epitheliocytes. Type 5 and 6 are medullary epitheliocytes.
ii. Type 1 is found beneath the capsule.
iii. Type 2 is present in the outer part of the cortex.
iv. Type 3 is found in the inner part of the cortex.
v. Type 4 lies in the deeper part of the cortex.
vi. Type 5 is present around Hassall's corpuscles.
vii. Type 6 is found in Hassall's corpuscles.

33. What are Hassall's corpuscles?

Ans. These are small rounded whorl like aggregates of type 6 epitheliocytes arranged concentrically with pink staining degenerating epithelial cells in the center.
Remember that Hassall's corpuscles are the characteristic feature of thymus.

34. What is the size of each Hassall's corpuscle?

Ans. 30–100 µm in diameter.

35. What are thymic nurse cells?

Ans. Cortical epitheliocytes (Type 1 to type 4) are called thymic nurse cells.

36. What are the functions of thymus?

Ans.
i. It is regarded as a primary lymphoid organ (central organ of lymphatic system).
ii. It regulates the growth of peripheral lymphatic tissue.
iii. It is an organ of lymphopoiesis.
iv. It produces some hormones.

37. What is the role of thymus in myasthenia gravis?

Ans. Myasthenia gravis is an autoimmune disease caused by disturbance of the immune system. In this condition there is great weakness of voluntary muscles. Thymus is enlarged. The weakness of the muscles is probably due to the liberation of curare-like inhibitor substance from the thymus which blocks neuromuscular transmission. Thymectomy (removal of thymus) may result in improvement of the symptoms of these cases.

LESSON 3: SPLEEN

Key points for identification (Figs 19.4 and 19.5)

1. Presence of thick capsule and trabeculae.
2. White pulp consists of lymphatic follicles with central arteriole in eccentric position.
3. Red pulp containing venous sinusoids and splenic cords of Billroth.

PROBABLE QUESTIONS AND ANSWERS

1. What type of organ is the spleen?

Ans. The spleen is the largest lymphoid organ of the body.
(Remember that the thymus is the primary lymphoid organ).

2. Is the spleen composed of only lymphoid tissue?

Ans. No, the spleen may be considered as a hemolymph organ.

3. Why is it considered as a hemolymph organ?

Ans. This is because of the facts that:
  i. Spleen manufactures blood cells (erythrocytes) in fetal life.
  ii. It filters blood.
  iii. It manufactures lymphocytes after birth.
  iv. It contains plenty of lymphatic follicles (white pulp).
  v. It contains red pulp which contains venous sinusoids.

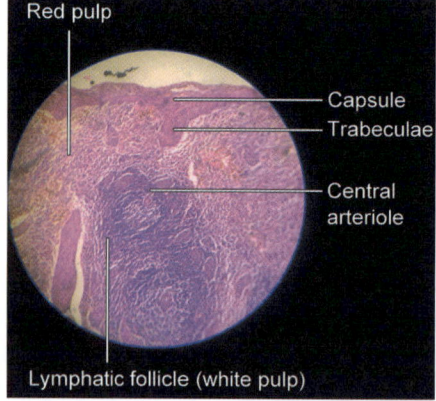

Fig. 19.4: Spleen

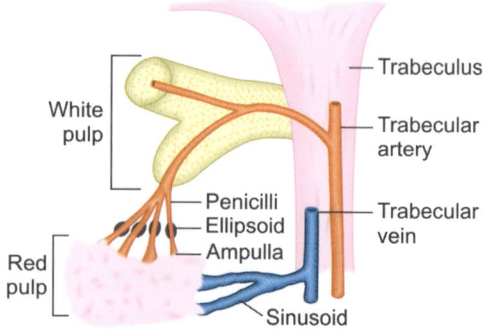

Fig. 19.5: Splenic circulation

Histology: Lymphatic System

4. What are the coverings of spleen?

Ans.
i. Outer covering of peritoneum (except at the hilum).
ii. Inner covering of fibroelastic thick capsule which covers the entire spleen.

5. What are trabeculae of spleen?

Ans. These are extensions of the capsule into the substance of the spleen.

6. What do you mean by lobules of spleen?

Ans. The trabecular septae divide and subdivide inside the spleen and these smaller branches of the septae anastomose with one another to form a trabecular framework. The area of splenic parenchyma bounded by trabecular septa is known as lobule of spleen.

7. What does the trabecular septa contain?

Ans. It contains blood vessels.

8. What are the contents of the interstices of the trabecular framework?

Ans.
i. Lymphocytes.
ii. Blood vessels and blood cells.
iii. Macrophages.
iv. Fine collagen and reticular fibers.

9. Does the spleen possess cortex and medulla?

Ans. No, spleen is not differentiated into cortex and medulla.

10. What are the types of parenchymal tissue in the spleen?

Ans. Two types:
i. White pulp.
ii. Red pulp.

11. What is white pulp?

Ans. An arteriole surrounded by lymphocytes is termed as a white pulp.

12. Why is the pulp called white?

Ans. It is not a histological finding. Splenic sections show numerous semiopaque white dots measuring about 0.25–1.0 mm in diameter on naked eye examination, hence called white pulp.

13. What is the other name for white pulp?

Ans. Malpighian body.

14. How are the cells arranged in a white pulp?

Ans. The cells of white pulp are arranged from within outwards as follows:
   i. T-lymphocytes, surrounding the central arteriole as periarteriolar lymphatic sheath (PALS).
   ii. B-lymphocytes in the mid-zone.
   iii. Antigen presenting cells (APCs) at the periphery.

15. What is the germinal center of white pulp?

Ans. B-lymphocytes proliferate on antigenic stimulation and form germinal center. These lymphocytes also differentiate into plasma cells.

16. Why the central arteriole is eccentric in position?

Ans. Primarily the arteriole is central in position in a white pulp. Due to proliferation and asymmetrical collection of B-lymphocytes the arteriole is pushed in eccentric position.

17. What is the function of APC?

Ans. These cells are concerned with immune response.

18. When do the white pulps become prominent?

Ans. White pulps become prominent during infection.

19. Do these pulps persist throughout life?

Ans. No, these may disappear in advanced age.

20. Which type of lymphocyte is more abundant in white pulp?

Ans. Most of the lymphocytes of white pulp are T-lymphocytes. Lymphocytes in the germinal center are dividing B-lymphocytes.

21. What is red pulp of splenic parenchyma?

Ans. Red pulp is the highly vascular sponge-like structure of splenic parenchyma, located between the white pulps and trabeculae.

Histology: Lymphatic System

22. What does red pulp contain?

Ans. It contains:
   i. Venous sinusoids.
   ii. Diffuse lymphoid tissue in the form of cords (Cords of Billroth).
   iii. Reticular cells.
   iv. Lymphocytes (both T and B).
   v. Macrophages.
   vi. Blood cells.
   vii. Penicillar arteries, etc.

23. What is the structure of a venous sinusoid?

Ans. A venous sinusoid is lined internally by fenestrated endothelial cells **(Stave cells)** resting on a perforated discontinuous basal lamina and externally wrapped by circularly oriented reticular fibers.

24. How are stave cells oriented?

Ans. The elongated stave cells are oriented in the long-axis of the venous sinusoids and present slit-like gaps (intercellular slits).

25. What is the function of the stave cells?

Ans. The cytoplasmic content of actin and myosin filaments of Stave cells modulate the shape and size of the intercellular slits in order to allow only healthy red blood cells to enter.

26. What is the size of the venous sinusoids?

Ans. About 50 μm in diameter.

27. Give a short description of splenic circulation.

Ans. **Splenic artery** → Segment arteries (about 5 or more) → Smaller branches (within the trabeculae) → Arterioles (in the splenic lobules, i.e. intertrabecular but intralobular) → Penicillar branches (3–5 straight branches) → Venous sinusoids in the red pulp → Trabecular veins → **Splenic vein.**

28. What type of circulation is the above-mentioned circulation?

Ans. Open circulation.

29. What is a closed circulation?

Ans. In this type of circulation, the blood is directly discharged into the venous sinusoids without passing through the red pulp. So the circulation in the red pulp is an open circulation.

30. Which type of circulation is faster?

Ans. Closed type of circulation is faster and takes only few seconds.

31. In which type of circulation maximum amount of blood flows?

Ans. About 90% of blood flows through closed circulation and rest of blood follows open circulation.

32. What is an ellipsoid?

Ans. An ellipsoid is an aggregation of fibroblasts and macrophages around a penicillar branch of arteriole.

33. What is ampulla?

Ans. An ampulla is a dilated part of the penicillar branch distal to the ellipsoid.

34. What is the marginal zone?

Ans. It is the interval zone between the red pulp and white pulp containing rich network of sinusoids, numerous APCs, dense reticular fibers and loosely arranged lymphocytes.

LESSON 4: PALATINE TONSIL

Key points for identification (Fig. 19.6)

1. Mucosa lined by nonkeratinized stratified squamous epithelium.
2. Subepithelial numerous lymphoid nodules.
3. Presence of tonsillar crypts.
4. Mucous and serous acini in the deeper part of the gland.

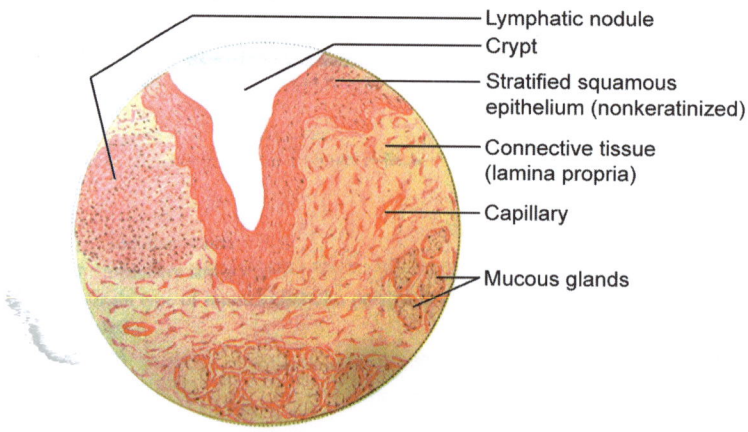

Fig. 19.6: Palatine tonsil

Histology: Lymphatic System

PROBABLE QUESTIONS AND ANSWERS

1. What is the lining epithelium of palatine tonsil?

Ans. The oral surface (medial surface) is lined by nonkeratinized stratified squamous epithelium but the rest of the surface is covered by a connective tissue capsule (hemicapsule) which separates the gland from the underlying muscles.

2. What are tonsillar crypts?

Ans. Tonsillar crypts are narrow tubular invagination the surface epithelium which extends into the substance of the tonsil.

3. What is the lining epithelium of these crypts?

Ans. The lining epithelium of thee crypts is similar to the surface epithelium of the tonsil, i.e. nonkeratinized stratified squamous epithelium, because the crypts are nothing but the invagination of the surface epithelium.

4. How many tonsillar crypts are there?

Ans. There are about 10–20 crypts.

5. What are tonsillar pits?

Ans. Tonsillar pits are the small openings of the tonsillar crypts on the medial surface of the tonsil.

6. What structures open into the tonsillar crypts?

Ans. Numerous mucous glands and serous glands open into the crypts.

7. Where do you find lymphatic follicles of tonsil?

Ans. Numerous lymphatic follicles are found in the lamina propria surrounding the tonsillar crypts.

8. What are the types of the follicles?

Ans. The lymphoid follicles may be primary or secondary.

9. What is secondary lymphatic follicle?

Ans. A lymphatic follicle which possesses germinal center is termed as secondary follicle.

10. When does the germinal center appear?

Ans. It appears following the exposure of the follicle to an antigen.

11. What does the germinal center contain?

Ans. It contains large lymphocytes and plasma cells.

12. What is the staining property of the germinal center?

Ans. It shows lighter staining in H&E method.

13. What types of lymphocytes are present in lymphatic follicles?

Ans. B-lymphocytes.

14. Do you find T-lymphocytes in tonsil?

Ans. Yes, T-lymphocytes are present in the interfollicular region.

15. What do you mean by salivary corpuscles?

Ans. Desquamated epithelial cells with lymphocytes and few bacteria appear in the lumen of tonsillar crypts and are washed out in the saliva as salivary corpuscles.

16. How do you histologically differentiate the following lymphoid organs?

Ans. Table 19.1 shows the histological differentiation of lymphoid organs.

Table 19.1: Histological differentiation of lymphoid organs

	Lymph node	Thymus	Spleen	Palatine tonsil
Capsule	Thin capsule	Thin capsule	Thick capsule	Hemicapsule
Subcapsular sinus	Present	Absent	Absent	Absent
Lining epithelium	Absent	Absent	Absent	Present on oral surface and lined by nonkeratinized stratified squamous epithelium
Cortex and medulla	Present	Present	Absent	Absent
Germinal center	Present	Present	Present	Absent
Hassall's corpuscles	Absent	Present	Absent	Absent

Chapter 20

Histology: Respiratory System

- Trachea
- Lungs

LESSON 1: TRACHEA

Key points for identification (Figs 20.1 and 20.2)

1. Mucous membrane lined by pseudostratified ciliated columnar epithelium.
2. Epithelium contains numerous goblet cells.
3. Presence of hyaline cartilage and tracheal glands outside the mucous membrane.

PROBABLE QUESTIONS AND ANSWERS

1. **What are the layers of trachea?**

Ans. Form within outwards:
   i. Mucous coat.
   ii. Submucous coat.
   iii. Musculocartilaginous coat.
   iv. Fibrous coat.

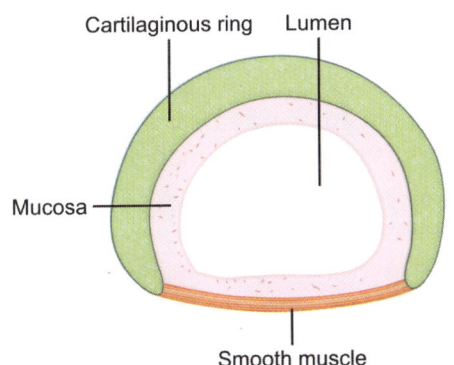

Fig. 20.1: Cross-section of trachea

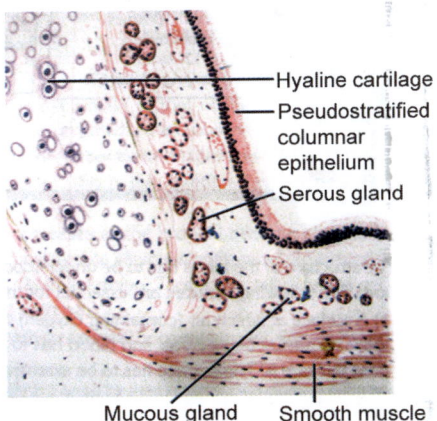

Fig. 20.2: Microstructure of trachea

2. **What are the layers of mucous membrane?**

Ans. Mucous membrane consists of:
   i. An **epithelium,** lined by **pseudostratified ciliated columnar cells** resting on a basement membrane.
   ii. Lamina propria.

Note that there is no muscularis mucosa.

 3. **Why is the epithelium called pseudostratified?**

Ans. Stratified means multilayered cells. But in this epithelium, the cells are single-layered with different cell heights and different nuclear positions and all the cells rest on a single (common) basement membrane. So, apparently these cells show stratification under light microscope. Hence, the epithelium is called pseudostratified.

 4. **What are the types of cells in the epithelium?**

Ans.
   i. Tall columnar ciliated cells.
   ii. Goblet cells.
   iii. Basal cells (stem or reserve cells).
   iv. Numerous lymphocytes (in the deeper part of epithelium).

5. **How many cilia are present in a single tall columnar cell?**

Ans. Each cell supports about 270 cilia with a basal body.

6. **What do the goblet cells secrete?**

Ans. Goblet cells secrete mucous.

7. **How does the mucous membrane of trachea help in the defensive mechanism of respiratory passage?**

Ans. Goblet cells secrete mucous → Secreted mucous entraps bacteria, dust or other particulate matters → Entrapped particles are swayed by the ciliary movements of the tall ciliated columnar cells → These particles are now ejected from the trachea by cough reflex. So, the mucous membrane acts as a **mucociliary barrier.**

 8. **What is the characteristic of tracheal epithelium?**

Ans. Thick basement membrane, because of increased collagen fiber bundles.

Histology: Respiratory System

9. What are the contents of lamina propria?

Ans.
i. Lymphoid tissue.
ii. Serous glands.
iii. Mucous glands.
iv. Elastic fibers.

10. What is bronchus associated lymphatic tissue (BALT)?

Ans. Aggregates of lymphoid tissue in lamina propria infiltrates the mucosa and submucosa, which is functionally equivalent to BALT.

11. What is the role of the basal cells?

Ans. These cells are small columnar cells resting on the basement membrane. These cells replace the damaged mucosa.

12. How are the tracheal cartilages arranged?

Ans. Tracheal cartilages are arranged in a number of incomplete C-shaped cartilaginous rings.

13. How many cartilaginous rings are there?

Ans. **16–20 rings** are there.

14. What type of cartilage is this?

Ans. The cartilage is **hyaline** in nature.

15. How do you identify hyaline cartilage?

Ans. **Hylos** means **glass**. So, it appears like a translucent glass. The cartilage cells **(chondrocytes)** lie singly or in groups, called **cell nests**. These cartilage cells secrete matrix which presents a **ground-glass appearance**.

16. How many cells are there in each cell nest?

Ans. Each cell nest may contain 2 to 4 or even more cartilage cells.

17. What is lacunae?

Ans. The cartilage cells secrete matrix which surrounds the cells. This surrounding matrix around the groups of cartilage cells (cell nests) is called **territorial matrix**. The space between this territorial matrix and cells is called lacuna.

18. Why are the wall of the adjoining cells flattened?

Ans. The chondroblasts (cartilage cells) secrete matrix. Each chondroblast is surrounded by a capsule. The matrix continues to be secreted and the cell may divide. Now the cells and matrix are enclosed in the lacuna. As the space in the lacuna is limited, there is mutual compression among the cells. So, the cellular wall facing the territorial matrix becomes rounded and the walls adjoining the neighboring cells become flattened.

19. What is the color of the territorial matrix in H&E staining?

Ans. It stains **dark purple**. The interterritorial matrix is clear and basophilic in staining because of its rich acid content **(chondroitin sulfate)**.

20. What is the staining property of chondroblasts?

Ans. They present highly basophilic cytoplasm.

21. Give some examples of hyaline cartilages of our body.

Ans.
  i. Articular cartilage.
  ii. Costal cartilage.
  iii. Tracheobronchial cartilage.
  iv. Thyroid and cricoid cartilages of larynx.

22. Are the cartilaginous rings are complete?

Ans. No, they are incomplete C-shaped cartilages which are deficient behind.

23. How are the gaps closed in the posterior aspect?

Ans. The gaps between the ends of the cartilaginous rings are bridged by fibroelastic membrane and involuntary trachealis muscle.

24. Why the cartilages are replaced by trachealis muscle on the posterior aspect of trachea?

Ans. Trachealis muscle is related to esophagus behind. The relaxation of trachealis allows esophagus to bulge into the lumen of trachea during swallowing.

25. What is the effect of contraction of trachealis muscle?

Ans. The trachealis muscle contracts during the cough reflex to narrow the tracheal lumen and thereby increases velocity of expelled air.

26. How are the gaps between the adjacent cartilaginous rings closed?

Ans. The gaps between the adjacent cartilages are closed by fibroelastic lamina which becomes continuous with the perichondrium covering the cartilages.

27. How much is the interval between the adjacent cartilaginous rings?

Ans. About 1 mm.

28. What is the average height of the cartilaginous rings?

Ans. About 4 mm.

29. Why is the trachea made up of cartilages?

Ans. The cartilaginous rings are elastic and rigid, which maintain the lumen patent.

30. What is the advantage of cartilages in trachea over bone?

Ans. Bones are rigid and non-elastic unlike cartilage which is rigid but elastic. Bones can prevent bending of neck but cartilages do not. Moreover, no separate blood supply is required for cartilages which get their nutrition by diffusion from the nearby capillary plexus.

LESSON 2: LUNGS

Key points for identification (Figs 20.3A and B)

1. Presence of alveoli in honey-comb appearance, lined by simple squamous epithelium.
2. Presence of intrapulmonary bronchi and bronchioles.

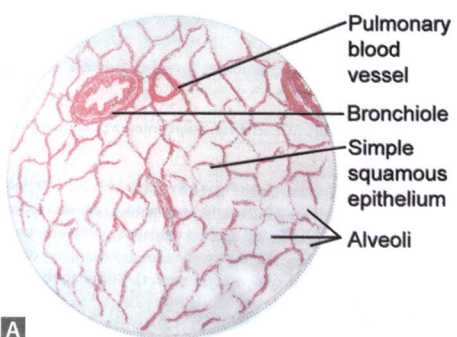
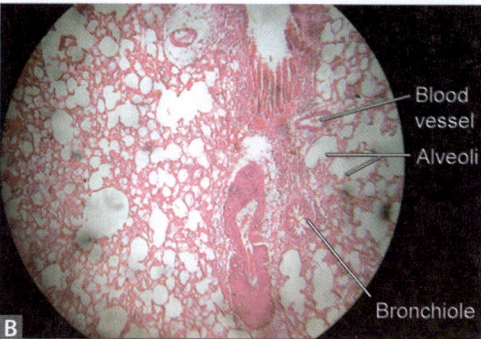

Figs 20.3A and B: Lung

PROBABLE QUESTIONS AND ANSWERS

1. What are the layers of lung?

Ans. From outside inwards:
   i. Serous coat.
   ii. Subserous coat.
   iii. Pulmonary substance (structures within lung lobules).

2. How is the serous coat formed?

Ans. The serous coat is formed by the **visceral pleura** and lined by a single layer of **flattened mesothelial cells.**

3. How is the subserous coat composed of?

Ans. It is composed of fibroelastic areolar tissue.

4. What are lung lobules?

Ans. The fibroelastic connective tissue of the subserous coat extends into the substance of the lung as numerous fibroelastic septae. The spaces between these septae are known as **lobules of lung.**

5. How many lobules are there in each lung?

Ans. About 20,000 lobules are there in each lung.

6. What is the shape of a lung lobule?

Ans. **Peripheral lobules** are pyriform in shape with the bases directed toward the surface and **central lobules** are somewhat irregular in shape.

7. What does the interlobular septa contain?

Ans. It contains:
   i. All branches of intrapulmonary **bronchi** (not bronchioles).
   ii. Corresponding branches of pulmonary vessels.
   iii. Corresponding branches of bronchial vessels.

8. Give on outline of branching pattern of tracheobronchial tree.

Ans. Flowchart 20.1 shows the branching pattern of tracheobronchial tree.

Histology: Respiratory System

Flowchart 20.1: Branching pattern of tracheobronchial tree

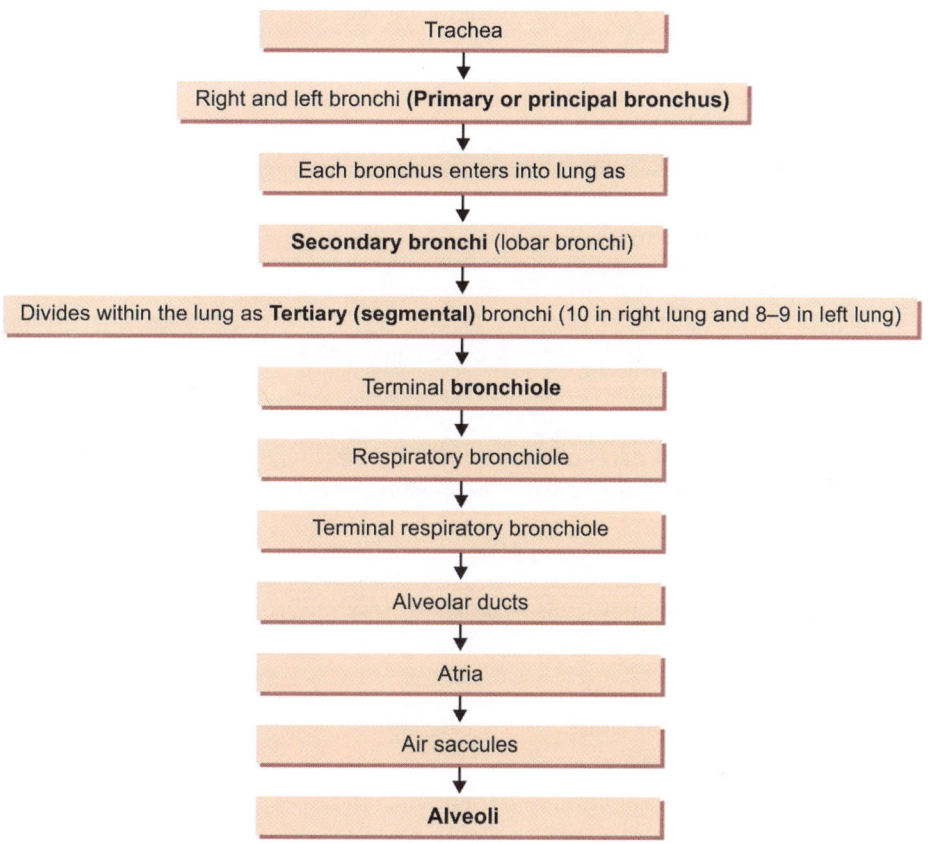

Remember that there are **23 orders of divisions** from principal bronchi to alveoli.

9. What are the layers of a bronchial wall?

Ans. From within outwards:
  i. **Mucosa, lined** by pseudostratified ciliated columnar epithelium.
  ii. **Muscular layer** (bronchial muscle).
  iii. **Cartilage layer**.
  iv. **Adventitia (fibroelastic coat)**.

10. What is the speciality of bronchial muscle?

Ans. Bronchial muscles are arranged in **helical pattern** in opposite direction and are increased in thickness toward the distal parts of the bronchi. They are **involuntary** (smooth muscle) in nature.

11. What is the difference of arrangement between bronchial muscle and tracheal muscle?

Ans. **Bronchial muscle** is arranged in a complete circular layer deep to the mucous membrane but the **tracheal muscle** (trachealis) is present only in the posterior aspect of trachea.

12. What is the type of bronchial cartilage and how are they arranged?

Ans. **Hyaline cartilage**.
Cartilages are arranged as irregular plates and in many pieces unlike trachea and extrapulmonary **bronchus** in which the cartilages are arranged in incomplete C-shaped manner.

Remember that in the distal parts of the bronchi the **cartilage plates** decrease in size and **glands** become fewer but the **musculature** increases in thickness and subepithelial lymphoid tissue increases in quantity.

13. What are the epithelial cells found in tracheobronchial passage in electron microscope studies?

Ans. **Ten types of cells**. These are:
  i. **Ciliated columnar cells:**
      a. Possess about 300 cilia/cell.
      b. Rate of ciliary beats: 12–16 beats/second.
      c. Cause mucociliary rejection current.
  ii. **Goblet cells:** Secretes mucous to trap foreign particles.
  iii. **Serous cells**: Secretes serous fluid to keep the epithelium moist.
  iv. **Basal cells:** Occupy the basal region of epithelium and replace damaged cells.
  v. **Brush cells:** Possess apical microvilli and absorptive in function.
  vi. **Intermediate cells:** Undifferentiated cells, may help in regeneration of ciliated and secretory cells.
  vii. **Clara cells:** Nonciliated cuboidal cells. **Their functions**:
      a. Production of components of surfactant.
      b. Act as reserve cells to replace the damaged cell types.
      c. Protect against noxious inhaled substances by detoxification due to its content of some enzymes.
      d. Act against proteases and protect the lung alveoli from destruction.
  viii. **Mast cells:** Occupy the basal region of epithelium and release histamine.
  ix. **Lymphocytes** and **other leukocytes:** Mainly T-lymphocytes and concerned with immunity.

x. **Argentaffin cells of Kulchitsky:**
   a. Belong to APUD cells (diffuse endocrine system).
   b. Secrete histamine or serotonin.
   c. Most numerous in fetal lungs.

14. What are the differences between extrapulmonary and intrapulmonary bronchus?

Ans. Table 20.1 shows the differences between extrapulmonary and intrapulmonary branches.

Table 20.1: Differences between extrapulmonary and intrapulmonary branches

Structures	Extrapulmonary bronchus	Intrapulmonary bronchus
Mucosal epithelial cells	Tall ciliated columnar cells	Less tall
Mucosal folds	Present only posteriorly where muscles are bridged between the ends of C-shaped bronchial cartilages	Present throughout the lumen, because a complete muscular layer is present between the mucous membrane and irregularly placed cartilages
Seromucous glands	Found between the mucous membrane and cartilage	Found between the cartilage and muscle layer
Muscle	Present only posteriorly	Complete layer of muscle throughout the circumference of the lumen
Cartilage	Incomplete C-shaped	Irregular and multiple plates of hyaline cartilage

15. What do you mean by bronchopulmonary segment?

Ans. It is a wedge-shaped independent respiratory district aerated by a **tertiary (segmental) bronchus**.

16. How many segments are there in a lung?

Ans. 10 segments in right lung and 8–9 segments in left lung.

17. What are the contents of a bronchopulmonary segment?

Ans.
   i. A tertiary bronchus.
   ii. A branch of pulmonary artery.
   iii. A branch of bronchial artery.

18. What is a bronchovascular unit (segment)?

Ans. The area of the lung drained by an intersegmental tributary of pulmonary vein is called a bronchovascular segment.

19. What is a pulmonary or lung unit?

Ans. The area of the lung supplied by one **terminal respiratory bronchiole** is called a lung unit.

20. What are the parts of bronchial tree that are found within a lung unit?

Ans. Terminal respiratory bronchiole, alveolar ducts, atria, air saccules and alveoli.

21. What is a bronchiole?

Ans. A bronchiole is a part bronchial tree, whose diameter is 1 mm or less and characterized by absence of submucosal glands and cartilages.

22. What are the differences between bronchus and bronchiole?

Ans. Differences between bronchus and bronchiole are shown in Table 20.2.

Table 20.2: Differences between bronchus and bronchiole

Differentiating parameters	Bronchus	Bronchiole
Diameter	More than bronchiole	1 mm or less
Lining epithelium	Pseudostratified ciliated columnar	Simple columnar or simple cuboidal
Goblet cells	Present	Absent (goblet cells are replaced by Clara cells)
Smooth muscle cells	Ill-defined	Well-defined
Seromucous glands	Present	Absent
Cartilage	Present as irregular plates	No cartilage
Location	Interlobular	Intralobular

23. What are sinuses of Lambert?

Ans. Lambert's sinuses are holes which connect terminal bronchioles with the alveoli. The sinus of Lambert is another term for alveolar sac.

24. What is the meaning of alveolus (plural: alveoli) ?

Ans. An alveolus (Latin) means **little cavity.**

25. What are pulmonary alveoli?

Ans. Pulmonary alveoli are very thin-walled blind sac, outcropping from either **alveolar sacs** or **alveolar ducts** or **respiratory bronchiole.**

26. Which part of the lung is called respiratory part?

Ans. The respiratory part of the lung is the area where gaseous exchange takes place. In other words, the gaseous exchange takes place through the alveoli of air sacs and alveolar outpockets of alveolar ducts and respiratory bronchioles.

27. How many alveoli are there is each lung?

Ans. About 150–200 million (including alveolar outpockets).

Histology: Respiratory System

28. How much surface area is covered by alveoli in each lung?

Ans. About 75 m².

29. How much capillary surface area is available for exchange of gas?

Ans. About 125 m².

30. What is diameter of an alveolus?

Ans. The average diameter is about 200 µm. In inflated alveoli, the diameter may be about 250 µm.

31. Which capillary plexus surrounds lung unit?

Ans. **Pulmonary capillary plexus** surrounds the structures of lung unit. But it is to be remembered that **lymphatic plexuses** are absent around the lung units.

32. What are the components of alveolar wall?

Ans. Three components:
   i. Alveolar epithelium.
   ii. Supporting connective tissue consisting of fine reticular, collagen and elastic fibers with occasional fibroblasts.
   iii. Pulmonary capillaries.

33. What is the lining epithelium of alveoli?

Ans. Alveoli is lined internally by **simple squamous epithelium** resting on a basement membrane.

34. What are the types of cells of alveolar epithelium?

Ans.
   i. Type I alveolar epithelial cells (type I pneumocytes).
   ii. Type II alveolar epithelial cells (type II pneumocytes).
   iii. Type III alveolar epithelial cells (Brush cells).
   iv. Macrophage cells.

35. What are the characteristic features of type I pneumocytes?

Ans.
   i. 90% of the alveolar surface is lined by type I pneumocytes.
   ii. Very thin (thickness varies between 0.05 and 0.2 µm).
   iii. Adjacent cells are connected by tight junction.
   iv. Gaseous exchange takes place through these cells by **diffusion**.

36. What are the features of type II pneumocytes?

Ans.
i. It forms about 10% of alveolar surface area.
ii. Cells are rounded with **microvilli** on their free surfaces.
iii. Contain secretory granules in their cytoplasm, hence also called **granular pneumocytes**.
iv. Secretory granules appear to be made up of several layers, hence called **multilamellar bodies**.
v. Secrete surfactant fluid which lowers surface tension.

37. What are type III alveolar cells?

Ans.
i. They are **columnar cells** without cilia.
ii. The cells have apical **microvilli**.
iii. Donot contain secretory granules.

38. What are macrophages or alveolar phagocytes?

Ans. These cells phagocytose bacteria, carbon particles, dust particles or other debris. They are also called **dust cells** as they phagocytose dust particles.

39. What is heart failure cells?

Ans. Alveolar phagocytes (large monocytes) are sometimes called heart failure cells as they engulf extravasated red blood cells (RBCs) in congestive cardiac failure and produce brick red sputum.

40. What does the surfactant fluid contain?

Ans. Surfactant contains:
i. Protein
ii. Phospholipid
iii. Glycosaminoglycan.

41. What is the principal constituent of phospholipid?

Ans. Dipalmitoyl phosphatidyl choline.

42. What is interalveolar septum?

Ans. Interalveolar septum is the partition between two adjacent alveoli.

43. What are the constituents of interalveolar septae?

Ans.
i. Epithelial lining of adjacent alveoli.
ii. Supporting connective tissue.
iii. Capillaries.

Histology: Respiratory System

44. What is alveolar pore of Kohn?

Ans. Adjacent alveoli communicate with each other through the holes in the interalveolar septum, called **pores of Kohn.**

45. Why these pores are named as pores of Kohn?

Ans. The pores are so named because these are first described by a German physician **Hans Kohn.**

46. When are these pores developed?

Ans. They develop at 3–4 years of age. They are absent in the lungs of newborns.

47. What are the functions of these pores?

Ans.
   i. They provide collateral ventilation and even distribution of air to the alveoli.
   ii. They equalize the pressure between adjacent alveoli.
   iii. They allow passage of fluid and bacteria in spread of infection.

48. What is air-blood barrier?

Ans. Air-blood barrier is the intervening structures between the air of lung alveoli and blood of lung capillaries.

49. What are the structures of air-blood barrier?

Ans.
   i. Flattened alveolar cells (type I pneumocytes).
   ii. Basement membrane of alveolar cells.
   iii. Basement membrane of capillary endothelial cells.
   iv. Flattened endothelial cells.

50. What is the thickness of air blood barrier?

Ans. About 0.2 µm.

51. What is alveolocapillary membrane?

Ans. The basement membrane of alveolar cells and the basement membrane of capillary endothelial cells are fused at some places. This fused basement membrane is known as alveolocapillary membrane.

52. What is the basis of elastic recoil of lung?

Ans. The elastic fibers of the alveolar wall are continuous with the elastic fibers in the rest of the bronchial tree. This provides the basis of elastic recoil of lung.

Chapter 21

Histology: Urinary System

- Kidneys
- Ureter
- Urinary Bladder

LESSON 1: KIDNEYS

Key points for identification (Figs 21.1 to 21.5)

1. Renal cortex contains glomeruli, Bowman's capsule and convoluted tubules.
2. Medulla contains loop of Henle, collecting tubules, ducts of Bellini.
3. Parenchyma contains blood vessels and scanty connective tissue.

PROBABLE QUESTIONS AND ANSWERS

1. What are the macroscopic parts of kidney?

Ans. Kidney consists of two parts:
  i. Outer renal parenchyma.
  ii. Inner renal sinus.

2. What are the parts of renal parenchyma?

Ans. Two parts:
  i. Outer cortex.
  ii. Inner medulla.

3. What are the parts of renal cortex?

Ans. Cortex consists of two parts:
  i. Outer cortical arches.
  ii. Inner renal columns of Bertin.

4. What is cortical arch?

Ans. Cortical arch is the outer part of the cortex which arches over the base of the renal pyramids.

Histology: Urinary System

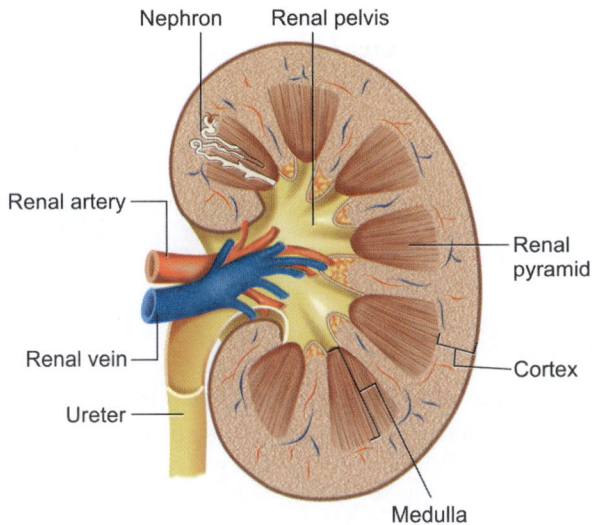

Fig. 21.1: Cut-section of kidney

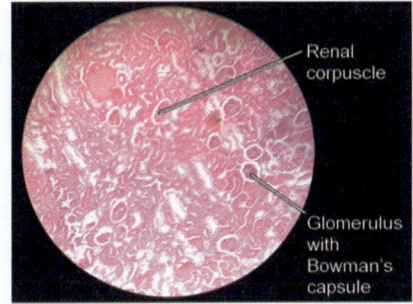

Fig. 21.2: Microstructure of kidney (cortex)

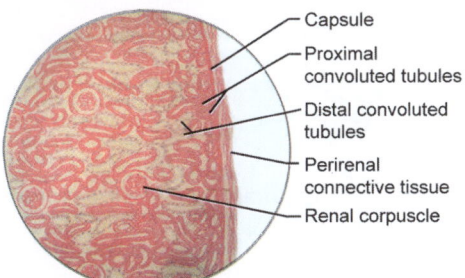

Fig. 21.3: Structure of kidney (cortex)

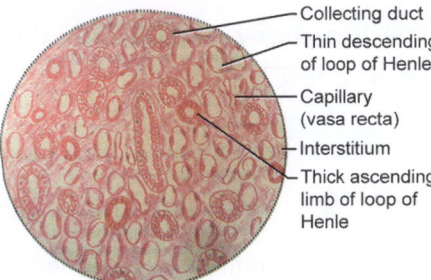

Fig. 21.4: Structure of kidney (medulla)

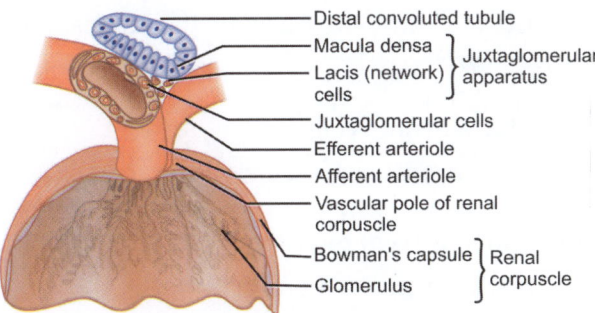

Fig. 21.5: Juxtaglomerular apparatus

5. Name the structures that are found in the cortical arch?

Ans. i. Medullary rays.
ii. Convoluted parts (cortical labyrinth).

6. What is a medullary ray?

Ans. A medullary ray is a striated light-colored conical mass, the apex of which is directed to the surface of kidney and the base is continuous with the striations of renal pyramid.

7. What do you mean by lobule of kidney?

Ans. Lobule of kidney is the part of cortical arch, the central axis of which is presented by a medullary ray and bounded on either side of the medullary ray by interlobular blood vessels.

8. What is a lobe of kidney?

Ans. A lobe of kidney is that part of the kidney which consists of a renal pyramid and its surrounding cortex.

9. How many renal lobes are there in a kidney?

Ans. About 10–18 lobes.

10. What is the origin of interlobular artery?

Ans. Interlobular artery is the branch of arcuate artery which is the branch of interlobar artery.

11. What is renal column of Bertin?

Ans. Renal column is the part of cortex lying between the adjacent pyramids, traversed by interlobar blood vessels.

12. What is cortical labyrinth?

Ans. Cortical labyrinth is the convoluted part of the cortical arch lying between the adjacent medullary rays.

13. What does the cortical labyrinth contain?

Ans. Cortical labyrinth contains renal corpuses and different convoluted portions of nephrons.

14. What are the presenting parts of renal medulla?

Ans. Renal medulla consists of about 8–18 renal pyramids.

Histology: Urinary System

15. What is a renal pyramid?

Ans. A renal pyramid is a pyramidal-shaped, striated pale and conical mass presenting a base, directed peripherally towards the surface and an apex projecting into a minor calyx of renal sinus.

16. What is renal papilla?

Ans. The apex of renal pyramid is known as renal papilla.

17. What are the structures that perforate renal papilla?

Ans. Renal papilla is perforated by 16–20 ducts of Bellini.

18. Where does the renal papilla end?

Ans. Renal papilla is received by minor calyx.

19. How many papillae are received by one minor calyx?

Ans. One minor calyx can receive 1 to 3 renal papillae.

20. What are the causes of striations of the pyramids?

Ans.
- i. U-shaped loops of Henle.
- ii. Collecting tubules and ducts of Bellini.
- iii. Arteriolae recti.
- iv. Venae recti.

21. What is renal sinus?

Ans. Renal sinus is a small cavity within the kidney, lined by renal capsule, communicates outside through the hilum.

22. What are the contents of renal sinus?

Ans.
- i. Excretory apparatus of kidney such as:
  - a. Minor calyx
  - b. Major calyx
  - c. Pelvis of ureter.
- ii. Perinephric fat.
- iii. Renal lymph vessels, blood vessels and nerves.

23. What are minor calyces?

Ans. Minor calyces are cup-shaped depressions to receive renal papillae.

24. How many minor calyces are there?

Ans. There are about 7–13 minor calyces.

25. How many renal papillae can be received by each minor calyx?

Ans. Each minor calyx can receive 1–3 renal papilla.

26. How are the major calyces formed?

Ans. Major calyces are formed by the union of minor calyces.

27. How many major calyces are there?

Ans. There are 2–3 major calyces.

28. What do the major calyces form?

Ans. The major calyces unite to form the pelvis of ureter.

29. What is the capacity of the pelvis of ureter?

Ans. About 5–7 mL.

30. What is the type of epithelium of these calyces and the pelvis of ureter?

Ans. Transitional epithelium.
Flowchart 21.1 shows the microscopic structure of kidney.

31. What are the main microscopic elements of kidney?

Ans. Uriniferous tubules.

Flowchart 21.1: Microscopic structure of kidney

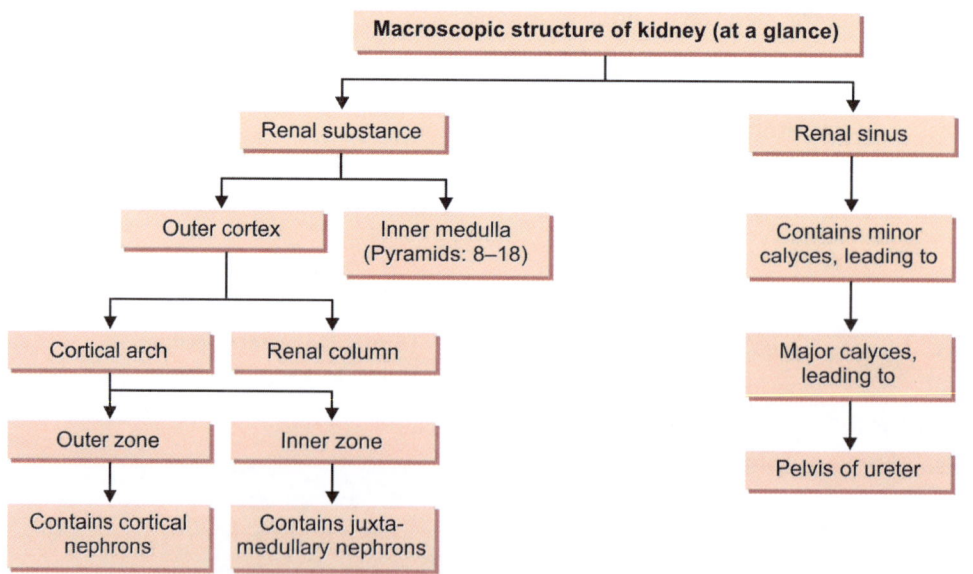

32. What are the parts of uriniferous tubules?

Ans. i. Secreting part which includes nephron.
ii. Collecting part which includes:
   a. Collecting tubules.
   b. Ducts of Bellini.

33. What are the parts of a nephron?

Ans. Two parts:
i. Renal corpuscles or malpighian body.
ii. Renal tubule.

34. What are the elements of a renal corpuscle?

Ans. A renal corpuscle consists of:
i. Glomerular plexus of capillaries.
ii. Bowman's capsule.

35. What are the parts of a renal tubule?

Ans. Each renal tubule consists of (Flowchart 21.2):
i. Proximal convoluted tubule (PCT).
ii. Descending loop of Henle (DLH).
iii. Loop of Henle (LH).
iv. Ascending loop of Henle (ALH).
v. Distal convoluted tubule (DCT).
vi. Junctional tubule (JT).

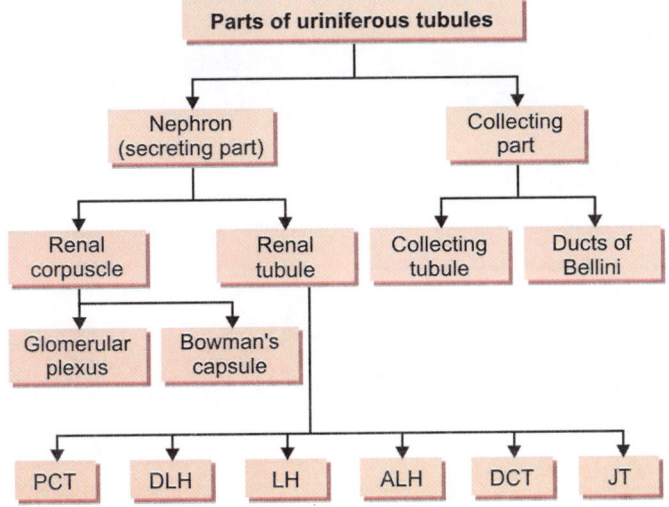

Flowchart 21.2: Parts of uriniferous tubules

36. How many nephrons are there in each kidney?

Ans. About 1–2 million.

37. What is the length of each nephron?

Ans. About 50–55 mm.

38. What are the functions of nephron?

Ans. i. Filtration. ii. Selective reabsorption.
iii. Secretion.

Nephron is called the structural and functional unit of kidney.

39. What is the developmental source of nephron?

Ans. Metanephric blastema.

40. Where are the nephrons distributed?

Ans. Nephrons are distributed in the cortex (cortical arch and renal column) and corticomedullary junction (juxtamedullary).

41. What are differences between cortical nephron and Juxtamedullary nephrons?

Ans. Table 21.1 shows the differences between cortical nephrons and juxtamedullary nephrons.

Table 21.1: Differences between cortical nephrons and juxtamedullary nephrons

	Cortical nephrons	Juxtamedullary nephrons
Location	Cortical arch and renal column	Corticomedullary junction
Loop of Henle	Short	Long
Number	More (85%)	Less (only 15%)
Function	Sodium reabsorption	Water reabsorption

42. What is a renal corpuscle?

Ans. A renal corpuscle is a small body consisting of glomerular plexus of capillaries invaginating into Bowman's capsule.

43. What is the other name for renal corpuscle?

Ans. Malpighian body.

Histology: Urinary System

44. What is the number and size of renal corpuscle?

Ans.   i. **Number:** 1–2 million in each kidney.
       ii. **Size:** 0.2 mm in diameter.

45. What is glomerular plexus?

Ans. Glomerular plexus is a lobulated tuft of convoluted capillaries projecting into the Bowman's capsule.

46. How many lobules are there in each glomerular plexus?

Ans. About 50 intercommunicating lobules are there.

47. What is the afferent and efferent arterioles of a glomerulus?

Ans. An afferent arteriole is one which carries blood into glomerulus and an efferent arteriole is one which carries blood out of the glomerulus.

48. Which of these two arterioles is wider?

Ans. Afferent arteriole is wider than the efferent arteriole. Due to this difference of caliber, the hydrostatic pressure of the glomerular plexus is increased.

49. What is the lining membrane of glomerular capillary?

Ans. Glomerular capillaries are lined by fenestrated flattened endothelium resting on a basement membrane. These capillaries are known as **fenestrated capillaries**.

50. What do mean by fenestrated endothelium?

Ans. Fenestrated endothelium are those endothelium which have apertures (pores) in their lining.

51. What is the pore size of glomerular capillary endothelium?

Ans. About 160 Å.

52. Name some organs where do you find fenestrated capillaries other than renal glomeruli.

Ans.   i. Pancreas.
       ii. Endocrine glands.
       iii. Intestinal villi.

53. What is the other type of capillary?

Ans. **Continuous capillary**. In this, the edges of the endothelial cells fuse completely with the adjoining cells forming a continuous wall.

54. What are the organs where do you get continuous capillaries?

Ans.
i. Lungs
ii. Brain
iii. Connective tissue
iv. Muscle
v. Skin.

55. What do you mean by glomerular basement membrane?

Ans. It is a thick membrane which intervenes between the capillary endothelium on one side and podocytes of the visceral layer of Bowman's capsule on other side.

56. How much is the thickness of this membrane?

Ans. About 300 nm.

57. What are the layers of this basement membrane?

Ans.
i. Inner electron-lucent lamina rara interna.
ii. Intermediate electron-dense lamina densa.
iii. Outer electron-lucent lamina rara externa.

58. What do you mean by vascular pole of renal corpuscle?

Ans. It is a point where the afferent and efferent arterioles of glomerular capillary lie close together.

59. What is mesangium?

Ans. Mesangium is the supporting element of glomerular capillary, which is made up of mesangial cells and surrounding noncellular mesangial matrix.

60. What are the functions of mesangial cells?

Ans.
i. Contractile properties control blood flow through the glomerulus.
ii. Phagocytic activity.
iii. Maintenance of glomerular basement membrane.

61. What is Bowman's capsule (glomerular capsule)?

Ans. It is a double-layered, blind, expanded end of renal tubule.

Histology: Urinary System

62. What are these two layers?

Ans. Outer or parietal layer, lined by squamous epithelium and inner or vascular layer, lined by discontinuous large polyhedral cells called podocytes.

63. What do you mean by capsular space?

Ans. The space between the inner visceral and outer parietal layer of Bowman's capsule is known as capsular space or urinary space.

64. What does this space contain?

Ans. The urinary space is filled with glomerular filtrate.

65. What are the podocytes?

Ans. Podocytes are large polyhedral cells which line the inner visceral layer of Bowman's capsule.

66. Why are they so named?

Ans. They are so named, because they possess foot-like processes.

67. What is the appearance of podocytes?

Ans. Podocytes and their processes present star-shaped appearance.

68. What are the podocyte processes?

Ans. The cell bodies of podocytes give rise to primary or major processes. These processes again give rise to a number of secondary or minor processes. These minor processes are attached to the basement membrane by footplates or pedicles or end feet.

69. What do you mean by glomerular filtration barrier?

Ans. It means the structures intervening between the blood of glomerular capillaries and the urinary space.

70. What are the structures that form the filtration barrier?

Ans.
    i. Fenestrated flattened endothelium of glomerular capillary.
    ii. Glomerular basement membrane.
    iii. Footplates of podocytes.

71. What do you mean by the neck of the renal tubule?

Ans. The narrow junction between the Bowman's capsule and PCT is referred to as neck of the renal tubule.

72. What are the parts of proximal renal tubule?

Ans.
i. Pars convoluta (PCT) which is the initial segment of renal tubule and continuous proximally with the Bowman's capsule.
ii. Pars recta (proximal straight tubule or PST) which becomes continuous with the descending loop of Henle.

73. Why is this part called proximal tubule?

Ans. This is because of its close proximity with the renal corpuscle.

74. Why are the sections of PCT usually more than DCT in a field of vision?

Ans. The length of PCT is several times greater than that of DCT. So, number of section of PCT is more than that of DCT.

75. What are the salient microscopic features of PCT?

Ans.
i. The epithelium is lined by low columnar or large cuboidal cells.
ii. The nucleus of these cells are central and euchromatic.
iii. Luminal surface presents brush borders due numerous microvilli.
iv. Cytoplasm of cells contains abundance of mitochondria which makes the cells strongly eosinophilic (acidophilic).

76. What is the diameter of PCT?

Ans. About 40–60 μm.

77. What are the functions of PCT?

Ans.
i. Active reabsorption of glucose, amino acids, sodium, chloride, phosphate, calcium, bicarbonate.
ii. Obligatory reabsorption of water.

78. What are the parts of loop of Henle?

Ans.
i. Thin segment of the loop.
ii. Thick segment of the loop.

79. What are the parts of thin segment of the loop?

Ans.
i. The descending limb of the loop.
ii. The loop itself.
iii. The proximal part of the ascending limb of the loop.

80. Which part of the loop is thick?

Ans. Distal part of the ascending limb of the loop is thick, which is continuous with the DCT.

81. What is the diameter of the loop of Henle?

Ans.
i. Thin segment—about 15 µm.
ii. Thick segment—about 30 µm.

82. Do all the loops of Henle extend up to the entire depth of the renal pyramid?

Ans. No, the loops of the cortical nephrons extend up to the basal region of the pyramid but the loops of the juxtamedullary nephrons extend up to the apex of the pyramid.

83. Do they possess microvilli?

Ans. The lining cells of the loop of Henle possess only a few small microvilli.

84. What is the other name for the loop of Henle?

Ans. Ansa nephroni.

85. What are the parts of distal convoluted tubule (DCT)?

Ans.
i. Initial straight part which is continuous with the thick part of the ascending limb of loop of Henle.
ii. Convoluted part.
iii. Terminal straight part, also called junctional or connecting tubule.

86. What is zig-zag tubule?

Ans. It is the short and irregular tubule which may lie between the DCT and JT.

87. What is the diameter of DCT?

Ans. About 20–50 µm.

88. Why are they called distal tubule?

Ans. This is because of the fact that these tubules are further away from the renal corpuscle.

89. How do you differentiate PCT from DCT?

Ans. Table 21.2 shows the differences between PCT and DCT.

Table 21.2: Differences between PCT and DCT

Parameters	PCT	DCT
Diameter	Wider (40–60 µm)	20–50 µm
Lumen	Smaller	Larger
Epithelium	Low columnar/large cuboidal	Cuboidal
Brush border	Present	Absent
Staining	Strongly eosinophilic	Lightly eosinophilic

90. **What are the collecting parts of uriniferous tubules?**

Ans. i. Collecting tubules.   ii. Ducts of Bellini.

91. **Where do they lie?**

Ans. They lie in the medullary rays of cortex and renal pyramids.

92. **How do you differentiate collecting tubules and convoluted tubules?**

Ans. Table 21.3 shows the differences between collecting tubules and convoluted tubules.

Table 21.3: Differences between collecting tubules and convoluted tubules

	Collecting tubules	Convoluted tubules
Diameter	Largest tubules may be as much As 200 µm. Smallest tubules are 40–50 µm	PCT: 40–50 µm DCT: 20–50 µm
Lumen	Larger and circular	Smaller and irregular
Lining epithelium	Cuboidal to columnar	PCT: Low columnar DCT: Cuboidal
Brush border	Absent	PCT: Present DCT: Absent
Staining	Light staining	PCT: Deep staining DCT: Light staining

93. **How are the duct of Bellini formed?**

Ans. Ducts of Bellini are the continuation of collecting tubules inside the renal pyramids.

94. **How are the collecting tubules formed?**

Ans. Collecting tubules are formed by the joining of junctional tubules from a number of nephrons.

95. **What is the developmental source of collecting tubules?**

Ans. Ureteric bud.

96. **Where do the ducts of Bellini open?**

Ans. They open at the summit of renal papilla.

97. **What do you mean by area cribrosa?**

Ans. Multiple perforations (8–18) at the summit of renal papilla (apex of renal pyramid) form a cribriform area, called area cribrosa. Tables 21.4 shows the distribution of different parts of uriniferous tubules in kidney.

Histology: Urinary System

Table 21.4: Distribution of different parts of uriniferous tubules in kidney

Cortex	Medulla (renal pyramid)	Medullary ray
i. Renal corpuscle	i. Loop of Henle	i. Collecting tubule
ii. PCT	ii. Collecting tubules	
iii. DCT	iii. Ducts of Bellini	
iv. Junctional tubule		

98. What is juxtaglomerular apparatus?

Ans. It is a complex of structures consisting of juxtaglomerular cells (JG cells) macula densa and Polkissen cells.

99. What are juxtaglomerular cells?

Ans. Juxtaglomerular cells are large and rounded cells in the tunica media of the afferent glomerular arterioles.

100. What do the juxtaglomerular cells secrete?

Ans. i. Renin (an enzyme) which maintains normal blood pressure.
 ii. Renal erythropoietin factor which helps in maturation of erythrocytes.

101. What is macula densa?

Ans. It is an area of densely-packed nuclei of the tubular epithelium of DCT, adjacent to the afferent glomerular arteriole at the vascular pole of renal corpuscle.

102. What are Lacis cells?

Ans. These are polyhedral clusters of cells between afferent and efferent arteriole and macula densa of DCT.

103. What are the other names for Lacis cells?

Ans. Polkissen cells/extraglomerular mesangial cells/Goormaghtigh cells.

104. What are the interstitial tissue of kidney?

Ans. Cortical interstitial space is occupied by lymphatics and blood vessels. Medullary interstitium contains interstitial cells, collagen fibers, protein and glycosaminoglycans.

LESSON 2: URETER

Key points for identification (Fig. 21.6)

1. Star-shaped lumen.
2. Mucous membrane is lined by transitional epithelium.
3. 2–3 layers of smooth muscle coat.

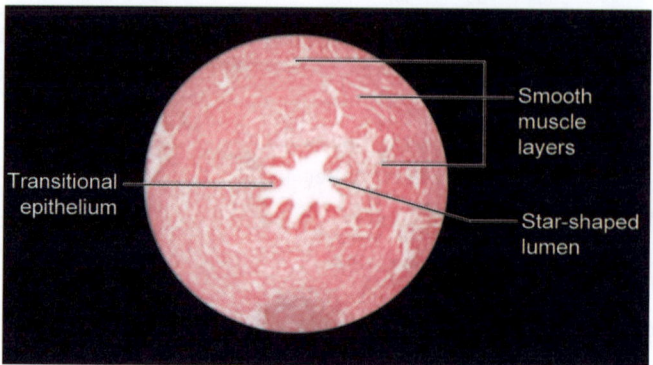

Fig. 21.6: Ureter

PROBABLE QUESTIONS AND ANSWERS

1. What are the layers of ureteric wall?

Ans. From outside inwards:
  i. Fibrous coat.
  ii. Muscular coat:
      a. Outer circular
      b. Inner longitudinal.
  iii. Mucous coat, lined by transitional epithelium.

2. Do you find two layers of muscle coat along the whole length of ureter?

Ans. **No,** two layers of muscle coat (outer circular and inner longitudinal) are found in the upper 2/3rd of the ureter. But in the lower 1/3rd the muscle coat consists of three layers (outer longitudinal, middle circular and inner longitudinal).

3. What is the arrangement of these muscles?

Ans. All these muscle layers are spirally arranged.

4. What is the milking action of circular muscle of ureter?

Ans. Upward continuation circular muscles surround the renal papillae of renal pyramids and exert action to squeeze urine from the ducts of Bellini. This action is called milking action.

5. Are the muscles of ureter continuous with the muscles of urinary bladder?

Ans. The outer two layers of ureteric muscles are continuous with the detrusor muscle of the urinary bladder and the inner longitudinal muscles blend with the mucosa of the internal trigone as the **muscle of Bell**.

Histology: Urinary System

6. What is the function of muscle of Bell?

Ans. It maintains the oblique coarse of the intravesical part of ureter and elevates the bladder neck to close the internal urethral orifice.

7. Why is the lumen of ureter star-shaped?

Ans. The mucous membrane of ureter is thrown into six longitudinal folds which gives a star-shaped lumen in cross-section.

8. Is the lumen is always star-shaped?

Ans. It is star-shaped only when the ureter is empty. In dilated ureter, the folds disappear and the lumen looks circular.

9. What are the layers of mucous membrane?

Ans.
i. Lining epithelium, lined by transitional epithelium.
ii. Lamina propria.
iii. **No muscularis mucosa.**

10. What do you mean by transitional epithelium?

Ans. Transitional epithelium is a type of tissue consisting of multiple layers of epithelial cells. These cells can contract and expand.

11. Why the epithelium is so named?

Ans. This epithelium is so named because of its function in the transition of degree of distension.

12. What are the layers of cells in transitional epithelium?

Ans. Three layers:
i. **Superficial** large polyhedral cells.
ii. **Intermediate** pear-shaped cells, the tips of which are directed towards the basement membrane.
iii. **Basal** smaller cells are columnar or cuboidal.

13. Why the transitional epithelium is also known as urothelium?

Ans. The transitional epithelium lines the mucous membrane of most of the parts of urinary system, hence called urothelium.

14. What are the sites of transitional epithelium as mucosal lining?

Ans.
    i. Minor calyces.
    ii. Major calyces.
    iii. Pelvis of ureter.
    iv. Ureter.
    v. Urinary bladder.
    vi. Proximal part of urethra in male (above the colliculus seminalis).

LESSON 3: URINARY BLADDER

Key points for identification (Figs 21.7A and B)

1. Mucosa is lined by transitional epithelium without mucous glands.
2. Mucosa is thrown into folds (in relaxed state).
3. Thick layer of smooth muscles arranged irregularly.

PROBABLE QUESTION AND ANSWERS

1. What are the layers of the wall of urinary bladder?

Ans. From outside inwards:
    i. Serous/adventitia coat.
    ii. Muscular coat.
    iii. Submucous coat.
    iv. Mucous coat.

2. What is the lining of the outermost coat?

Ans. Superior surface and upper part of the base of the bladder (in males) are lined by mesothelium of peritoneum and the rest of the wall is lined by loose connective tissue (adventitia).

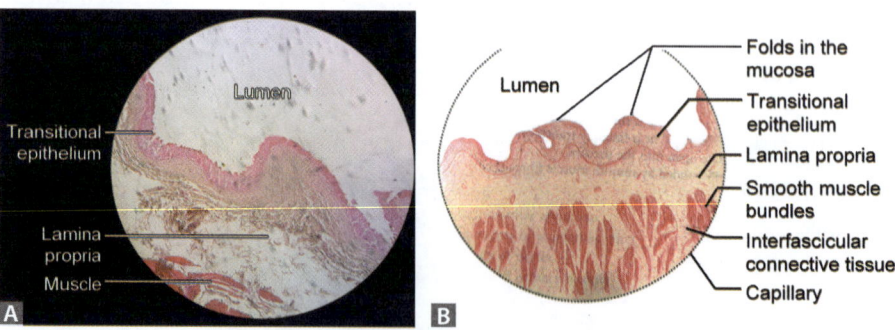

Figs 21.7A and B: Urinary bladder

Histology: Urinary System

3. What are the layers of muscular coat?

Ans. The thick muscular coat consists of three layers.
From outside inwards:
  i. Outer longitudinal fibers.
  ii. Middle circular fibers.
  iii. Inner longitudinal fibers.

4. Is the individual muscle layer well-defined?

Ans. No. The muscles are arranged in different planes with extensive interlacing of fibers of the different layers.

5. What are the extensions of the outer longitudinal layer?

Ans.
  i. Some fibers are continuous with the muscles of prostate.
  ii. Few fibers extend from the bladder to public bone forming pubovesicalis.
  iii. Few fibers extend from the bladder to the longitudinal muscle of rectum forming rectovesicalis.

6. How is the sphincter vesicae (sphincter urethrae) formed?

Ans. It is formed by the thickening of middle circular fibers of bladder around the internal urethral orifice.

7. What type of sphincter it is?

Ans. Sphincter urethrae (internal urethral sphincter) is involuntary in nature but external urethral sphincter, derived from sphincter urethrae muscle is voluntary.

8. What is the extension of inner longitudinal fibers?

Ans. There fibers are continuous with the muscle of prostatic urethra in male and along the entire length of urethra in female.

9. Why is the muscles of bladder known as detrusor muscle?

Ans. **Detrusor** (Latin) means a body part that pushes down. Contraction of this muscle pushes down urine for emptying of bladder.

10. Is the submucous coat present in bladder wall?

Ans. **Two views**:
  i. Some author says that the submucous coat is present all over the bladder wall except at the region of internal trigone (trigonum vesicae).
  ii. Other view is that the submucous coat is absent in the entire bladder wall because of absence muscularis mucosae in the mucous membrane.

11. What are the layers of mucous membrane?

Ans. i. Epithelium, lined by transitional epithelium. ii. Lamina propria.
(No muscularis mucosae)

12. Why are the superficial (luminal) or surface cells called umbrella cells?

Ans. These cells are large polyhedral and are connected to one another by tight junctions. These cells possess abundant cytoplasm and prominent nuclei. During distension of bladder, these cells are stretched and often shaped like an umbrella, hence called umbrella cells.

13. How does the stratified squamous epithelium differ from transitional epithelium?

Ans. Though both the types of epithelium are multilayered and their deepest cells are columnar, the surface cells of transitional epithelium are not squamous.

14. What are the specialities of transitional epithelium?

Ans.
  i. It is urine-proof.
  ii. Devoid of mucous glands and muscularis mucosae.
  iii. Power of stretchability of the surface cells (umbrella cells).

15. Are the mucous glands totally absent in mucous membrane?

Ans. Few mucous glands may be found close to the internal urethral orifice. The glands are known as subtrigonal and subcervical glands.

16. What is the thickness of lamina propria?

Ans. About 100–500 μm.

CHAPTER 22

Histology: Male Reproductive System

- Testis
- Vas Deferens
- Prostate

LESSON 1: TESTIS

Key points for identification (Figs 22.1A and B)

1. Presence of seminiferous tubules containing germ cells and Sertoli cells.
2. Interstitial cells of Leydig.

PROBABLE QUESTIONS AND ANSWERS

1. What are the coverings of testis?

Ans. From outside in words:
  i. Visceral layer of tunica vaginalis.
  ii. Tunica albuginea.
  iii. Tunica vasculosa.

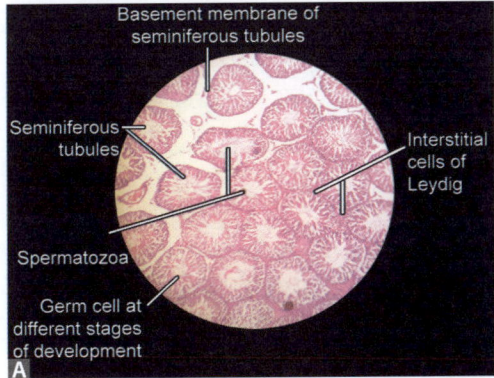
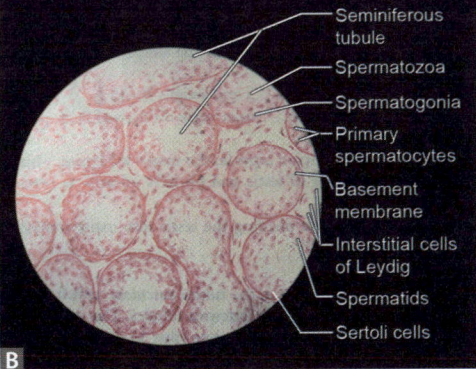

Figs 22.1A and B: Testis

2. What do you mean by a lobule of testis?

Ans. The interior of testis is divided by a number of fibrous septum radiating from the convex anterior surface of mediastinum testis into a number of cone-shaped compartments, called lobules of testis.

3. What is mediastinum testis?

Ans. It is an incomplete vertical partition within the interior of testis along its posterior border, formed by the inward projection of tunica albuginea.

4. How many lobules are there in each testis?

Ans. Each testis contains about 200–300 lobules.

5. What are the contents of a lobule?

Ans. Each lobule contains:
  i. 1–3 seminiferous tubules.
  ii. Interstitial cells of Leydig.
  iii. Loose connective tissue.

6. How many seminiferous tubules are there in each testis?

Ans. About 400–600.

7. What does the tunica albuginea layer contain?

Ans.
  i. Blood vessels.
  ii. Lymphatics.
  iii. Genital ducts.

8. What does the tunica vasculosa layer contain?

Ans. It contains plexus of blood vessels.

9. What are the layers of tunica vaginalis?

Ans. Two layers:
  i. Visceral layer.
  ii. Parietal layer.

10. Does the visceral layer cover the entire testis?

Ans. It covers the epididymis and testis except its posterior border.

Histology: Male Reproductive System

11. What are the parts of a seminiferous tubule?

Ans. Two parts:
   i. Convoluted or coiled part (in front).
   ii. Straight part or tubuli recti (behind).

12. What is the length of an uncoiled seminiferous tubule?

Ans. 70–80 cm.

13. What is its diameter?

Ans. 0.12–0.3 mm.

14. What is rete testis?

Ans. It is a plexiform network of straight part of seminiferous tubules within the mediastinum testis.

15. What are the efferent ductules?

Ans. The efferent ductules are those ductules which arise from the upper end of the rete testis and enter into the head of the epididymis.

16. How many efferent ductules are there?

Ans. About 12–20 ductules.

17. What is the canal of epididymis?

Ans. The canal of epididymis is a tubular structure which is formed by the union of efferent ductules.

18. What is the length of this canal?

Ans. About 20 feet.

19. What does the canal form?

Ans. It forms the body and tail of epididymis.

20. What are the parts of epididymis?

Ans. Three parts: i. Head   ii. Body   iii. Tail.

21. What is the continuation of the tail of epididymis?

Ans. Vas deferens is the continuation of tail of epididymis.

22. What are the layers of the wall of seminiferous tubule?

Ans. Two layers:
   i. Outer thick basal lamina.
   ii. Inner seminiferous epithelium.

23. What are the types of cells of seminiferous epithelium?

Ans. Two types of cells:
   i. Spermatogenic cells.
   ii. Sertoli cells (supporting cells).

24. What are the spermatogenic cells?

Ans.
   i. Spermatogonia.
   ii. Primary spermatocytes.
   iii. Secondary spermatocytes.
   iv. Spermatids.
   v. Spermatozoa.

25. What are the types of spermatogonia?

Ans. Two types:
   i. Type A:
      a. Dank A and
      b. Pale A.
   ii. Type B.

26. What is spermatogenesis?

Ans. The process of formation and maturation of spermatozoa from type A spermatogonia is called spermatogenesis.

27. What is spermiogenesis?

Ans. The process of metamorphosis of spermatid to spermatozoa is known as spermiogenesis.

28. What are the first cells for spermatogenesis?

Ans. Primordial germ cells: These cells arise from the endodermal wall of the yolk sac and migrate to the genital ridge. Differentiation of primordial germ cells forms spermatogonia. Flowchart 22.1 shows the steps of spermatogenesis.

Histology: Male Reproductive System

Flowchart 22.1: Steps of spermatogenesis

```
Primordial germ cells
         │
         │ Differentiation
         ▼
Spermatogonia (diploid chromosomes)
         │
    ┌────┴────┐
    ▼         ▼
  Type A    Type B
    │
    │ Mitosis
    │
  ┌─┴──┐
  ▼    ▼
Dark A  Dark A
  │
  ├──────────┐
  ▼          ▼
Dark A     Pale A
(Reserve cell)
             │
             │ Mitosis
          ┌──┴──┐
          ▼     ▼
        Pale A  Pale A
               (Reserve cell)
          │
       ┌──┴──┐
       ▼     ▼
     Type B  Type B
         │
         │ Repeated mitosis
         ▼
Four generations of Type B spermatogonia
         │
         │ Mitosis
         ▼
   Primary spermatocytes
         │
         │ 1st meiotic division
         ▼
Secondary spermatocytes (Haploid chromosomes)
         │
         │ Second meiotic division
         ▼
      Spermatids
         │
         │ Metamorphosis
         ▼
     Spermatozoa
```

29. Why are the primary spermatocytes always visible in section of seminiferous tubules?

Ans. The prophase of the first meiotic division is extremely prolonged (about 22 days).

30. Which one is the smallest cell of spermatogenic series?

Ans. Spermatid (about 10 mm is diameter).

31. What are the stages of spermatogenesis?

Ans. Three stages:
  i. Spermatocytosis.
  ii. Meiosis I and II.
  iii. Spermiogenesis.

32. What is the approximate time required for spermatogenesis?

Ans. About 64 days.

Spermatocytosis	—	16 days
Meiosis I	—	8 days
Meiosis II	—	16 days
Spermiogenesis	—	24 days
		64 days

33. What are the parts of a spermatozoa?

Ans. Four parts:
  i. Head (4 microns long).
  ii. Neck (0.3 micron long).
  iii. Body (middle piece) – 4 micron long.
  iv. Tail (principal piece) – 40 micron long.

Total length of a spermatozoon is about 50 microns.

34. What is the end piece of spermatozoa?

Ans. Terminal 5–7 microns of the tail is called the end piece which is devoid of fibrous sheath.

35. What are the factors which regulate spermatogenesis?

Ans.
  i. FSH of anterior pituitary.
  ii. Testosterone.
  iii. Vitamin E.
  iv. Low scrotal temperature.

36. What are the features of a Sertoli cell?

Ans.
  i. Elongated slender cells.
  ii. Extends from basement membrane to the lumen of the seminiferous tubule.
  iii. Possesses complex processes of plasma membrane which extend between the spermatogenic cells.
  iv. Nucleus is large, ovoid and lightly stained.
  v. Cytoplasm is acidophilic.
  vi. They are also called sustentacular cells or supporting cells.

Histology: Male Reproductive System

37. What are the functions of Sertoli cells?

Ans.
i. Mechanical and nutritive support for spermatogenic cells.
ii. Secrete Inhibin (a hormone) which inhibits production of FSH.
iii. Secrete androgen binding proteins that bind to testosterone.
iv. Secrete proteinaceous fluid into the lumen of seminiferous tubule for nutrition and transport of spermatozoa.
v. Form blood-testis barrier.
vi. Phagocytose residual bodies.
vii. Maintain cohesion between spermatogenic cells.
viii. Secrete Mullerian inhibitory substance (MIS) which suppresses the development of Mullerian duct in male fetus.

38. What are interstitial cells of Leydig?

Ans. These are polyhedral cells which lie in the interstitial tissue of testis.

39. Are they lie within the seminiferous tubules?

Ans. No. They lie outside the seminiferous tubules but within the lobules of testis.

40. What are the features of Leydig cells?

Ans.
i. Size—15–20 mm.
ii. Nucleus—large, round and often eccentric.
iii. Cytoplasm—finely granular and strongly acidophilic.
iv. Cytoplasm contains droplets of fat, vitamin C and sometimes rod-shaped "crystals of Reinke".
v. Richly supplied by capillaries.

41. What are the functions of Leydig cells?

Ans.
i. They synthesize and secrete testosterone which stimulates the growth of secondary sex characters in male.
ii. Their secretions help in the descent of testis.

42. What are the hormones for the proliferation and growth of Leydig cells?

Ans.
i. In fetal life: Placental gonadotrophic hormones.
ii. After puberty: Interstitial cells stimulating hormone of anterior pituitary gland.

43. What are the types of epithelium of the following parts?

Ans.
i. Rete testis: Flattened epithelium (squamous epithelium).
ii. Efferent ductules: Ciliated columnar epithelium.
iii. Canal of epididymis: Ciliated pseudostratified columnar epithelium.
(Note that these cilia are nonmotile, called stereocilia).

LESSON 2: VAS DEFERENS

Key points for identification (Figs 22.2A and B)

1. Mucosa lined by pseudostratified columnar epithelium.
2. Very thick muscular coat.
3. Stellate lumen.

PROBABLE QUESTIONS AND ANSWERS

1. What is the length of vas deferens?

Ans. About 45 cm.

2. What are the layers of the wall of the vas deferens?

Ans. From outside inwards:
- i. Adventitia (loose connective tissue layer)
- ii. Muscular coat:
    - a. Outer longitudinal fibers
    - b. Inner circular fibers
    - c. Innermost longitudinal fibers (found only in the proximal part of the duct).

3. What is the lining epithelium of vas?

Ans.
- i. In the proximal part: Nonciliated simple columnar epithelium.
- ii. In the distal part of the vas: Pseudostratified columnar epithelium.

Remember that the mucous membrane possesses lamina propria but the submucous coat is absent in the vas.

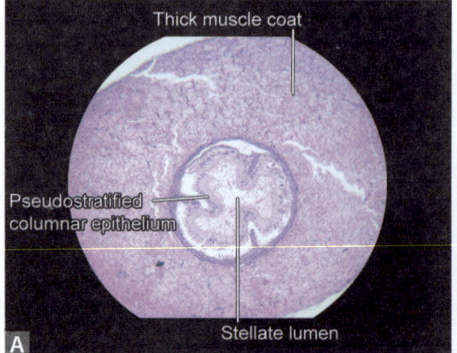 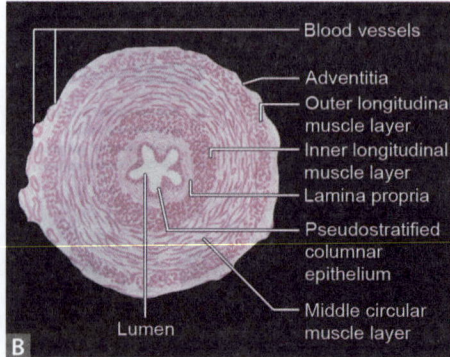

Figs 22.2A and B: Vas deferens

Histology: Male Reproductive System

4. Why the lumen of the vas is stellate shaped?

Ans. The mucous membrane is thrown into a number of longitudinal folds. So the lumen appears to be stellate in cross-section.

5. How do you differentiate the section of vas from ureter?

Ans.

	Ureter	Vas deferens
Lumen	Star-shaped	Small star-shaped lumen
Epithelium	Transitional	Pseudostratified columnar/nonciliated simple columnar
Muscle coat	Tick	Very thick

LESSON 3: PROSTATE

Key points for identification (Figs 22.3A to C)

1. Presence of irregular follicles of glandular tissue.
2. Glands are lined by simple columnar epithelium.
3. Glandular lumen may be filled with amyloid bodies.
4. Presence of fibromuscular stroma.

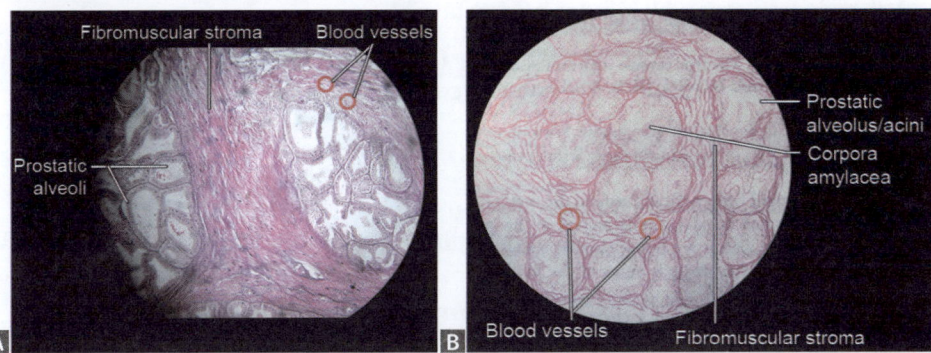

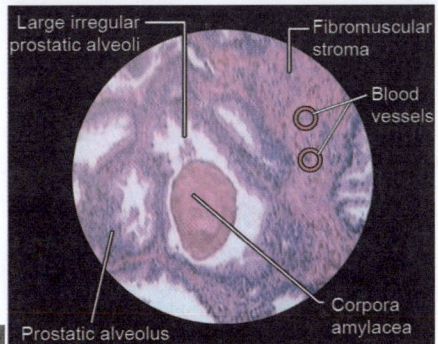

Figs 22.3A to C: Prostate

PROBABLE QUESTIONS AND ANSWERS

1. What are the structural components of prostatic gland?

Ans. It is a fibromusculoglandular organ. It consists of:
  i. Fibrous tissue—$\frac{1}{4}$ th part.
  ii. Muscular tissue—$\frac{1}{4}$ th part.
  iii. Glandular tissue—$\frac{1}{2}$ th part.

2. How the true capsule of the prostate is formed?

Ans. True capsule of the prostate is formed by the thickening of the fibrous tissue around the prostatic gland.

3. How are the muscle fibers arranged in the prostate?

Ans. The muscular tissue is arranged in two sheets:
  i. An outer sheet of muscle which lies deep to the true capsule.
  ii. An inner sheet of muscle which encircles the prostatic urethra.

4. What do you mean by lobules of prostate?

Ans. Numerous fibromuscular septae arise from the true capsule and extend into the gland forming indistinct lobules which contain glandular elements.

5. How many lobes of prostate are there?

Ans.
  i. Anatomically, three lobes: Median and two lateral.
  ii. Surgically, five lobes: Median, anterior, posterior and two lateral lobes.

6. What are the types of glands which form the glandular part of prostate?

Ans. Compound tubuloalveolar type.

7. How many glands are there in a prostate?

Ans. About 30–50 glands are there.

8. How do the glands look like?

Ans. They appear as irregularly shaped follicles.

9. What is the lining epithelium of these follicles?

Ans. Simple columnar epithelium.

Histology: Male Reproductive System

10. Where do these follicles drain?

Ans. These follicles drain into 12–20 excretory ducts which in turn open into the prostatic urethra.

11. What is the lining epithelium of these excretory ducts?

Ans. The excretory ducts are lined by double-layered epithelium; the superficial layer is columnar and the deeper layer is cuboidal.

12. What is corpora amylacea?

Ans. Corpora amylacea or amyloid bodies are condensed mass of glycoprotein and often calcified.

13. Where do you find these amylacea?

Ans. These bodies are seen within the follicles of prostatic glands.

14. Are they present in all age groups?

Ans. They are more abundant in older people.

15. What are the hormones that affect the growth of the follicles?

Ans. Testosterone stimulates and estrogen inhibits the growth of the follicles.

16. How are the glands of prostate arranged?

Ans. They are arranged in different zones on the basis of differences in the size and nature of the glands.
The zones are:
　i. Peripheral (outer zone): Contains main prostatic glands and their long ducts open into prostatic sinuses below the colliculus.
　ii. Internal (intermediate) zone: Contains submucous glands and their smaller ducts open into prostatic sinuses at the level of colliculus.
　iii. Innermost zone: Contains mucous glands and they directly open into the urethra.

17. What do you mean by central zone?

Ans. The internal and innermost zones together form the central zone.

18. What is the clinical importance of these zones?

Ans. Carcinoma usually affects the peripheral zone and benign hypertrophy affects the central zone.

CHAPTER 23

Histology: Female Reproductive System

- Uterus
- Uterine Tubes
- Ovary
- Mammary Gland

LESSON 1: UTERUS

Key points for identification (Figs 23.1A and B)

1. Mucosa, lined by simple columnar epithelium.
2. Lamina propria contains simple tubular uterine glands.
3. Thick muscle layer (myometrium).

PROBABLE QUESTIONS AND ANSWERS

1. What are the layers of uterine wall?

Ans. From outside inwards:
  i. Perimetrium (serosal layer).
  ii. Myometrium (smooth muscle layer).
  iii. Endometrium (mucosal layer).

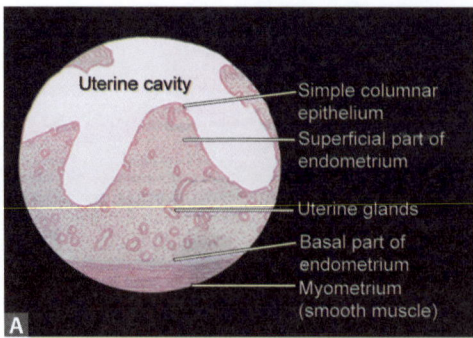 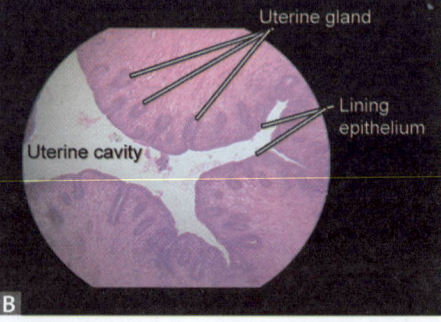

Figs 23.1A and B: Uterus

Histology: Female Reproductive System

2. What is perimetrium?

Ans. Perimetrium is the peritoneal covering of uterus. It is lined by a single layer of flattened mesothelial cells.

3. What are the nonperitoneal areas of uterus?

Ans.
i. Lateral borders of uterus.
ii. Anterior surface of supravaginal part of cervix.
iii. Vaginal part of cervix.

4. What are the layers of muscle coat (myometrium)?

Ans. Three ill-defined layers of muscles (from outside inwards):
i. Longitudinal layer.
ii. Circular layer.
iii. Reticular layer.

5. What is the other type of tissue found in the myometrium?

Ans. Dense connective tissue in the myometrium supports the interwoven bundles of smooth muscles of uterus.

6. What is endometrium?

Ans. It is the mucosal layer of uterus.

7. What is the lining epithelium of endometrium?

Ans.
i. **Before puberty:** Ciliated simple columnar epithelium.
ii. **After puberty:** Simple columnar epithelium.

8. Why is this change of epithelium?

Ans. After puberty (after menarche) cilia cannot grow due to cyclical destruction of superficial part of endometrium during menstruation.

9. What are the constituents of endometrium?

Ans.
i. Surface epithelium, lined by simple columnar epithelium.
ii. Subepithelial lamina propria.

10. What does lamina propria contain?

Ans. Lamina propria contains:
i. Simple tubular uterine glands.
ii. Blood vessels.
iii. Lymphatics.
iv. Nerves.

11. What do you mean by lamina propria?

Ans. Lamina propria is connective tissue layer of mucous membrane.

12. How are the uterine glands formed?

Ans. The uterine glands are formed by the invagination of the surface epithelium into the lamina propria.

13. What is lining epithelium of these glands?

Ans. Simple columnar epithelium (similar to surface epithelium).

14. What do these glands secrete?

Ans. Serous fluid (rich in glycogen).

15. What are the layers of endometrium?

Ans.
  i. Outer basal layer.
  ii. Inner functional layer.

16. What are the differences of features of these two layers?

Ans. Table 23.1 shows the differential features of outer basal layer and inner functional layer.

Table 23.1: Differential features of outer basal layer and inner functional layer

Outer basal layer	Inner functional layer
Deep layer	Superficial (luminal) layer
Thinner	Thicker
Unaltered in menstruation	Undergoes changes during menstruation
No shedding off	Shed off every month with menstrual blood
Contains tips of uterine glands	Contains main parts of uterine glands
Supplied by straight basal arteries	Supplied by spiral arteries
Helps in regeneration of denuded endometrium	Casting off this layer due to its chronic necrosis following vasoconstriction of spiral arteries

17. How do you differentiate proliferative endometrium from secretory endometrium?

Ans. Table 23.2 shows the differentiation of proliferative endometrium from secretory endometrium.

Table 23.2: Differentiation of proliferative endometrium from secretory endometrium

Endometrial features	Proliferative endometrium	Secretory endometrium
Thickness	2–3 mm	5–7 mm
Uterine glands	Straight	Convoluted and dilated with saw-tooth appearance

18. What do you mean by saw-tooth appearance of uterine glands?

Ans. In secretory phage of menstruation, the endometrial glands elongate, increase in diameter and become twisted on themselves. In section, these glands acquire a saw-tooth appearance because of this twisting.

LESSON 2: UTERINE TUBES

Key points for identification (Figs 23.2A and B)

1. Mucous membrane is lined by ciliated columnar epithelium.
2. Mucous membrane is thrown into numerous branching folds.
3. No submucosa coat.
4. Thin muscular layer.

PROBABLE QUESTIONS AND ANSWERS

1. What are the parts of uterine tube?

Ans. From medial to lateral:
   i. Intramural (1 cm).
   ii. Isthmus (3 cm).
   iii. Ampulla (5 cm).
   iv. Infundibulum (1 cm).

2. What are the layers of uterine tube?

Ans. From outside inwards:
   i. Serous (peritoneal) coat.
   ii. Muscular coat.
   iii. Mucous coat.

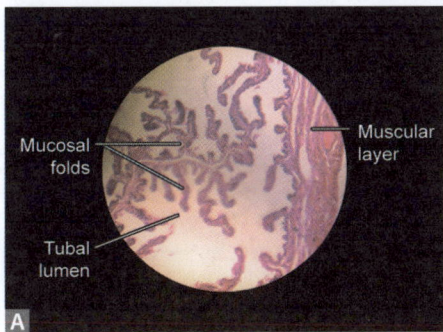
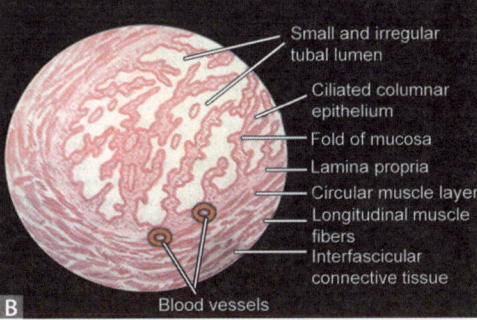

Figs 23.2 A and B: Fallopian tube

3. What is the lining cells of serous coat?

Ans. Mesothelial cells.

4. What are the nonperitoneal parts of uterine tube?

Ans.
  i. Lower border of the tube.
  ii. Intramural part of the tube.

5. What are the layers of muscular coat?

Ans.
  i. Outer longitudinal layer (thinner).
  ii. Inner circular layer (thicker).

6. Do you find any additional layer of muscle?

Ans. Sometimes, the intramural part of the uterine tube may possess innermost longitudinal layer.

7. In which part of the tube, the circular layer is thickest?

Ans. The circular muscle is thickest in the isthmus.

8. How many mucosal folds are there?

Ans. There are 3–6 primary longitudinal mucosal folds.

9. Why the lumen of the tube is irregular and small?

Ans. The primary mucosal folds give rise to secondary and tertiary folds which render the lumen small and irregular. These folds are conspicuous in the ampullary part of the tube.

10. What is the important function of the muscles of uterine tube?

Ans. Peristaltic contraction of the muscle helps to propel the oocyte from the peritoneal cavity towards the uterine cavity.

11. What are the different types of cells of epithelium?

Ans. Three types:
  i. Ciliated columnar cells.
  ii. Nonciliated secretory cells, also called peg cells.
  iii. Intercalary cells.

12. What is the function of ciliated cells?

Ans. These cells bear cilia which beat towards the uterus and help to propel oocyte towards the uterine cavity.

13. What is the function of secretory cells?

Ans. The secretion of these secretory cells provides nutrition to the fertilized ovum and spermatozoa. It also helps in capacitation of spermatozoa.

14. Why the surface lining is irregular?

Ans. Ciliated columnar cells are shorter than the nonciliated secretory cells. This difference of heights of these two types of cells make the surface lining irregular.

15. What are intercalary cells?

Ans. These are the variant of nonciliated secretory cells and are sandwiched between the basement membrane and the other two types of cells.

16. What is the importance of intercalary cells?

Ans. Probably they act as stem cells for the ciliated and secretory cells.

17. What does the lamina propria contain?

Ans. Lamina propria contains:
  i. Blood vessels.
  ii. Nerves.
  iii. Lymphatics.
  iv. Some smooth muscle fibers.

18. Name some organs that are devoid of submucous coat.

Ans.
  i. Gallbladder.
  ii. Uterus.
  iii. Uterine tubes.
  iv. Vocal folds.
  v. Dorsal surface of ventral ⅔rd of the tongue.
  vi. At the region of internal trigone of urinary bladder, etc.

LESSON 3: OVARY

Key points for identification (Figs 23.3A and B)

1. Surface, covered by germinal epithelium (simple cuboidal epithelium).
2. Outer cortex containing ovarian follicles in different stages of maturation.
3. Inner medulla containing stroma and blood vessels.

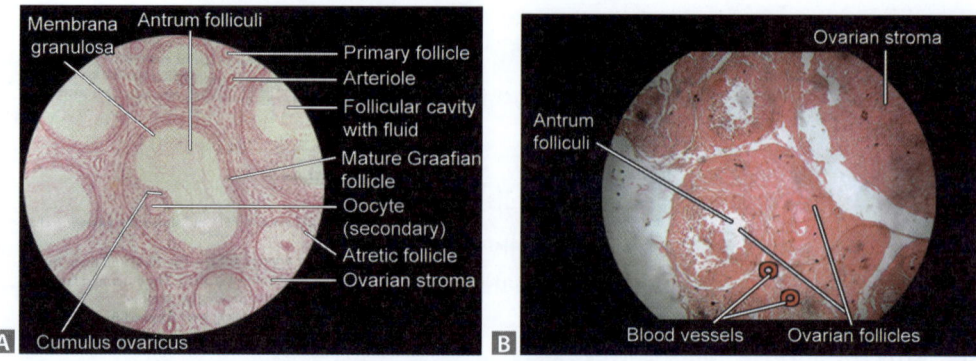

Figs 23.3A and B: Ovary

PROBABLE QUESTIONS AND ANSWERS

1. What is the lining epithelium of the surface of ovary?

Ans. The surface of the ovary is lined by a single layer of cubical epithelium, also called germinal epithelium.

2. What is tunica albuginea?

Ans. It is a layer of collagenous connective tissue lining just below the surface epithelium.

3. What are the parts of the substance of ovary?

Ans. i. Outer thick cortex.
 ii. Inner smaller medulla.

4. What are the contents of the cortex?

Ans. i. Stromal cells.
 ii. Ovarian follicles in different stages of development.
 iii. Corpus luteum.

5. What are the ovarian follicles that are found in different stages of maturity?

Ans. i. Primordial follicles.
 ii. Primary follicles.
 iii. Secondary follicles.
 iv. Mature or Graafian follicles.
 v. Atretic follicles.

6. What is the appearance of a primordial follicle?

Ans. It consists of a large oocyte surrounded by a single layer of flattened (squamous) follicular cells.

Histology: Female Reproductive System

7. Are they present throughout the life?

Ans. They are present at birth and remains as such till puberty.

8. What do you mean by the term 'primordial'?

Ans. Primordial is meant by something that exists at the beginning of time or something that is earliest.

9. Is the term germinal epithelium correctly named?

Ans. No, neither this epithelium gives rise to germ cells nor it helps in the development of pregranulosa cells of the follicles.

10. What is a primary follicle?

Ans. A primary follicle is one which consists of an oocyte (primary oocyte), surrounded by a single layer of cuboidal follicular cells (unlike squamous cells in primordial follicles).

11. What are the varieties of primary follicles?

Ans. Two varieties:
   i. Primary unilaminar follicles which consist of a single layer of cuboidal follicular cells.
   ii. Primary multilaminar follicles which consist of an oocyte surrounded by several layers of cuboidal follicular cells. These follicular cells proliferate to form granulosa cell layer. A rim of natural glycoprotein, the zona pellucida can be seen between the oocyte and the surrounding granulosa cells.

12. What is a secondary follicle?

Ans. A secondary follicle is one which consists of an oocyte surrounded by multiple layers of follicular (granulosa) cells in which a fluid filled space (antrum) can be seen.

13. What is the other name of secondary follicles?

Ans. Antral follicle, because an antrum can be seen in this follicle.

14. How is the antrum formed?

Ans. Flowchart 23.1 shows the formation of antrum.

15. What is theca folliculi?

Ans. Condensed stromal cells around the secondary follicle form a sheath, called theca folliculi.

Flowchart 23.1: Formation of antrum

```
┌─────────────────────────────────────────┐
│ Multiple small spaces appear within the │
│ multilayered follicular (granulosa) cells│
└─────────────────────────────────────────┘
                    ↓
┌─────────────────────────────────────────┐
│ Small spaces eventually coalesce to form│
│ a single large cavity, called antrum    │
│ folliculi                                │
└─────────────────────────────────────────┘
                    ↓
┌─────────────────────────────────────────┐
│ The follicular cells are separated into │
│ an outer layer (membrana granulosa) and │
│ an inner layer (cumulus ovaricus)       │
│ due to the formation of antrum folliculi│
└─────────────────────────────────────────┘
```

16. What are the layers of theca folliculi?

Ans. Two layers:
  i. Theca interna (inner layer).
  ii. Theca externa (outer layer).

17. What is the composition of theca interna?

Ans. It is composed of stromal cells of cortex (so it is a cellular layer).

18. What does the interna layer secrete?

Ans. The cells of theca interna secrete steroid precursors (estrogen).

19. How is the theca externa made up of?

Ans. Theca externa is made up of fibrous connective tissue.

20. What is its function?

Ans. This layer provides support for the developing follicles.

21. What do membrana granulosa cells secrete?

Ans. Progesterone.

22. Does this progesterone reach the circulation?

Ans. Though the estrogen (secreted by theca interna) can reach the circulation, the progesterone (secreted by the membrana granulosa) cannot reach the circulation due to the impermeable glass membrane.

23. What is glass membrane?

Ans. It is an a avascular membrane lying between the theca interna layer and the membrana granulosa layer.

24. What are the functions of estrogen?

Ans.
i. Stimulates proliferative phage of menstruation.
ii. Initiates libido.
iii. Helps in the development of secondary sex characters of female.

25. What do you mean by Graafian (tertiary) follicle?

Ans. It is a large sized mature ovarian follicle, almost occupying the whole thickness of ovarian cortex, containing a mature ovum.

26. What is the size of a Graafian follicle?

Ans. About 10 mm.

27. What is discus proligerus?

Ans. The layers of granulosa cells that attach the secondary oocyte to the wall of the mature ovarian follicle constitute the discus proligerus.

28. What is vitelline membrane?

Ans. It is the cell membrane of the secondary oocyte.

29. What is perivitelline space?

Ans. It is the space between the vitelline membrane and zona pellucida.

30. What is accommodated in the perivitelline space?

Ans. Polar body.

31. What do you mean by ovulation?

Ans. The shedding off ovum following rupture of a Graafian follicle is called ovulation.

32. What is stigma?

Ans. Stigma is an avascular and most convex point of the Graafian follicle just before it ruptures.

33. What are the stages of oogenesis?

Ans. Flowchart 23.2 shows the stages of oogenesis.

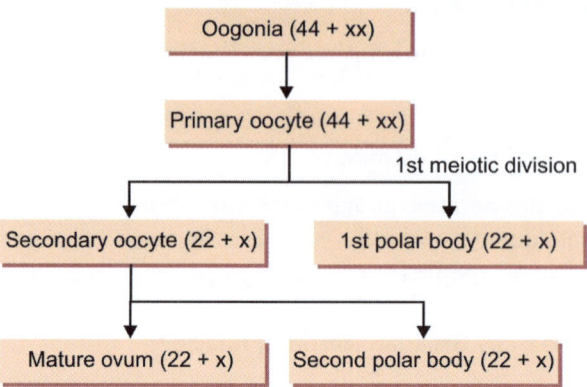

Flowchart 23.2: Stages of oogenesis

34. What are atretic follicle?

Ans. Except the Graafian follicle which ruptures and ovulates, the other developing follicles degenerate and undergo atresia.

35. What is corona radiata?

Ans. The first layer of granulosa cells (follicular cells) outside the zona pellucida is called corona radiata.

36. What is corpus luteum?

Ans. It is a solid mass or body formed by the remains of the Graafian follicle following ovulation.

37. How is the corpus luteum formed?

Ans. It is formed by the:
  i. Infolding and collapsing of the follicles.
  ii. Disintegration of glass membrane.
  iii. Invasion of blood vessels.

38. What does corpus luteum secrete?

Ans. It secretes progesterone.

39. What are the cells of corpus luteum?

Ans. Two types of cells:
  i. Luteal cells, derived from the cells of membrana granulosa.
  ii. Paraluteal cells, derived from theca interna.

40. What is the color of corpus luteum and why is this color?

Ans. Yellow in color due to the presence of lutein pigment (yellowish carotenoid pigment).

Histology: Female Reproductive System

41. What is the fate of corpus luteum?

Ans. Fate depends on fertilization:
  i. If fertilization occurs, the corpus luteum enlarges in size up to a diameter of about 2.5 cm and becomes corpus luteum of pregnancy.
  ii. If fertilization does not occur, the corpus luteum degenerates and gradually decreases in size (about 1 cm) to form a white fibrous nodule, called corpus albicans. This is called corpus luteum of menstruation.

42. What is the difference between primordial follicle and primary follicle?

Ans. In primordial follicle, the follicular cells are squamous.
In primary follicle, the follicular cells are cuboidal/columnar.

43. Points to remember:

Ans.
  i. Number of primordial/primary follicles at birth: About 2,00,000 in each ovary.
  ii. Follicles at puberty: About 40,000.
  iii. 5–12 primary follicles undergo a process of maturation in each menstrual cycle.
  iv. Only one follicle fully matures (Graafian follicle) and ovulates.
  v. From the beginning of formation of ovarian follicle up to the end of degeneration of corpus luteum is termed as ovarian cycle.

LESSON 4: MAMMARY GLAND

Key points for identification (Figs 23.4 to 23.6)

1. Lobules of glandular tissue are separated by considerable quantity of connective tissue and fat.
2. Presence of intralobular ducts, lined by cuboidal epithelium.

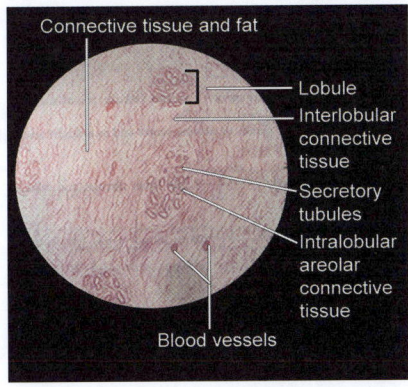

Fig. 23.4: Nonlactating mammary gland

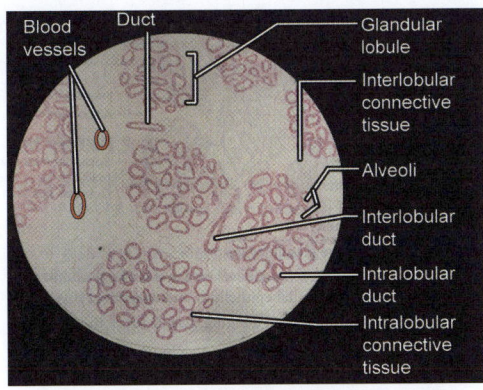

Fig. 23.5: Lactating mammary gland

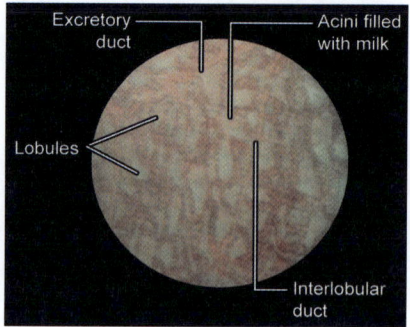

Fig. 23.6: Mammary gland

PROBABLE QUESTIONS AND ANSWERS

1. What are the structural components of breast?

Ans. Three components:
   i. Glandular tissue.
   ii. Fibrous tissue.
   iii. Interlobar fatty tissue.

2. What types of glands form the glandular tissue?

Ans. Tubuloalveolar types of glands.

3. What is a lobe of mammary gland?

Ans. It is a pyramidal-shaped area of mammary gland drained by a lactiferous duct and separated from each other by fibrous bands of connective tissue.

4. What is a lobule?

Ans. A lobule is a smaller unit of mammary gland drained by smaller branches of lactiferous duct and formed by a number of alveoli. Numerous lobules from a lobe.

5. What are alveoli?

Ans. Alveoli are the secretory element of mammary gland and are made up of alveolar cells. Each smallest branch of lactiferous duct leads to an alveoli.

6. How many lobes are there in a mammary gland?

Ans. About 15–20 lobes.

7. What is a lactiferous duct?

Ans. Lactiferous duct is the excretory duct of a lobe of mammary gland.

Histology: Female Reproductive System

8. How may lactiferous ducts are there?

Ans. As the number of lobes are 15–20 and each lobe is drained by a duct, the number of lactiferous ducts are 15–20.

9. What is lactiferous sinus?

Ans. Lactiferous sinus is the dilatation of lactiferous duct, which acts as a reservoir of milk.

10. Where do the lactiferous ducts open?

Ans. They converge towards the areola and open at the summit of the nipple.

11. What are the types of epithelial cells lining the ducts of mammary gland?

Ans.
 i. Smaller ducts are lined by simple cuboidal/columnar cells.
 ii. Larger ducts are lined by stratified columnar cells (2–3 layers of cells).
 iii. Near the openings of the ducts on the nipple—stratified squamous epithelium.

12. Do you find any other type of cell in the alveolar ductal system?

Ans. Myoepithelial cells resting on basal lamina.

13. What are the anatomical features of myoepithelial cells?

Ans.
 i. These cells rest on the basement membrane.
 ii. They occupy the position between the basement membrane and alveolar cells.
 iii. They are highly branched cells with radiating processes.
 iv. They contain contractile actin and myosin filaments.

14. What is the function of this myoepithelial cells?

Ans. The contraction of these cells helps to expel the secretions (milk) into the larger ducts. They can be stimulated by oxytocin of pituitary gland.

15. Do you observe any structural differences of mammary gland in different periods of life?

Ans. Yes, the changes vary with age, pregnancy and lactation.

16. What are the changes in different age groups?

Ans.
 i. From birth to puberty: No alveoli, though lactiferous ducts are present.
 ii. After puberty without pregnancy: Small alveoli filled up with spheroidal masses of cells forming solid mass. Branching of lactiferous ducts.
 iii. After puberty with pregnancy: Large alveoli with proliferation of central cells.
 iv. During lactation: Distended alveoli with fatty degeneration of central cells.
Alveoli are lined by a single layer of cuboidal epithelium in resting stage and columnar epithelium in lactational stage.

17. What is colostrum?

Ans. Milk secreted in the later part of pregnancy and following child birth is called colostrum. It is rich in fat, colostrum corpuscles and immunoglobulin.

18. What are the tissue components of stroma of mammary gland?

Ans.
  i. Fibrous tissue.
  ii. Fatty tissue.

19. How are the fibrous tissue organized?

Ans. Fibrous tissues are arranged loosely around the lobules to allow expansion of alveoli and lobules during pregnancy and lactation.

20. How is the suspensory ligament formed?

Ans. The suspensory ligament is formed by the condensation of fibrous tissue extending from the ducts and alveoli to the dermis of the skin.

21. What is the function of this ligament?

Ans. It keep the mammary gland in position.

22. What is the disposition of the fatty tissue?

Ans. They are interalveolar and interductular in position.

23. What are the function of these fatty tissue?

Ans.
  i. Maintains shape of the breast.
  ii. Acts as a cushion.
  iii. Regulates temperature.

24. What type of gland is the mammary gland?

Ans. It is a modified sweat gland. On the basis of mode of secretion, it is an apocrine and merocrine type.

25. Do you find fatty tissue everywhere in the breast?

Ans. No, fat is absent beneath the areola and nipple.

26. What are the cell types in connective tissue stroma?

Ans.
  i. Fibroblasts.
  ii. Mast cells.
  iii. Adipocytes.
  iv. Macrophages.
  v. Plasma cells.
  vi. T- cells.
  vii. Neutrophils and eosinophils.

CHAPTER 24

Histology: Nervous System

- Spinal Cord
- Cerebellum

LESSON 1: SPINAL CORD

Key points for identification (Figs 24.1A and B)

1. Inner H-shaped gray matter.
2. Central canal at the center of gray matter.
3. Outer white matter containing nerve fibers and neuroglia.

PROBABLE QUESTIONS AND ANSWERS

1. What is the extent of spinal cord?

Ans. It extends from the upper border of C_1 vertebra to the lower border of L_1 vertebra.

2. Where does the spinal cord lie?

Ans. It lies within the upper 2/3rd of vertebral canal.

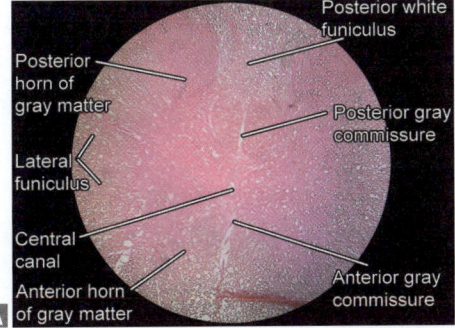 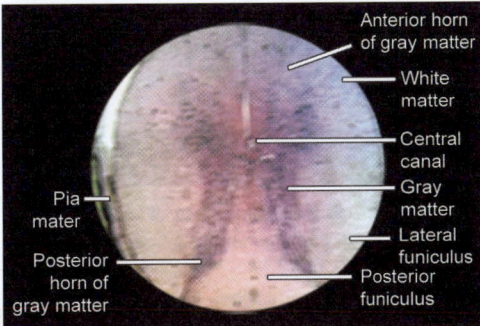

Figs 24.1A and B: Spinal cord

3. What is the shape of spinal cord?

Ans. It is somewhat cylindrical in shape with a tapering lower end, called conus medullaris.

4. Which parts of the spinal cord present enlargements?

Ans. Two enlargements of spinal cord:
  i. Cervical enlargement (C_4–T_2 segments).
  ii. Lumbosacral enlargement (L_2–S_3 segments).

5. Why such enlargements are there?

Ans. Cervical enlargement is formed due to more accommodation of motor neurons in gray matter of C_4–T_2 segments of spinal cord to supply the muscles of upper limb. Lumbosacral enlargement is formed due to more accommodation of motor neurons in gray matter of L_2–S_3 segments of spinal cord to supply the muscles of lower limb.

6. What is the length of spinal cord?

Ans. About 45 cm in adult males and 42 cm in adult females.

7. What is its weight?

Ans. About 30 g.

8. What are the coverings of spinal cord?

Ans. From outside inwards:
  i. Dura mater.   ii. Arachnoid mater.   iii. Pia mater.

9. What are the anatomical features that divide the spinal cord into right and left symmetrical halves?

Ans. A deep anterior median fissure and a shallow posterior median sulcus divide the spinal cord almost completely into two symmetrical halves.

10. How many cord segments (spinal segments) are there?

Ans. There are 31 spinal cord segments. So 31 pairs of spinal nerves are attached to spinal cord.

11. What do you mean by a spinal cord segment?

Ans. The portion of spinal cord which gives attachment to a pair of spinal nerves (right and left) is known as the spinal cord segment or neuromere.

12. What are the parts of a spinal cord segment?

Ans.
  i. Inner gray matter.   ii. A central canal inside the gray matter.
  iii. Outer white matter.

Histology: Nervous System

13. What do you mean by gray matter?

Ans. Gray matter means the substance which is composed of nerve cells (cell bodies of neurons), neuroglia and blood vessels.

14. How does it differ from white matter?

Ans. White matter is composed of nerve fibers (unlike nerve cells in gray matter), neuroglia and blood vessels.

15. Why does the area look white?

Ans. The whiteness is due to the myelination of the nerve fibers (nerve cells are devoid of myelin sheath).

16. What is the shape of the gray matter of spinal cord?

Ans. H-shaped.

17. What are the parts of gray matter?

Ans.
  i. A pair of anterior horns.
  ii. A pair of posterior horns.
  iii. An intermediate region between anterior and posterior horns.
  iv. Anterior gray commissure.
  v. Posterior gray commissure.

18. How do you differentiate anterior horns from the posterior horns?

Ans. Anterior horn is broad and short and does not reach the surface of the spinal cord, but posterior horn is narrow and elongated reaching the surface of the spinal cord.

19. What are the parts of anterior horn?

Ans.
  i. Head
  ii. Base.

20. Are the anterior horns uniform in shape throughout the entire length of spinal cord?

Ans. No, it is more broad in cervical and lumbosacral enlargements of spinal cord.

21. Why it is more broad in these two regions?

Ans. The anterior horns in cervical enlargement accommodate more motor neurons to supply the muscles of upper limb. In lumbosacral enlargement it accommodates more motor neurons to supply the muscles of lower limb.

22. What are the neurons of anterior horn?

Ans.
i. Motor neurons (lower motor neurons) which are multipolar.
ii. Interneurons (connector neurons).

23. How many types of motor neurons are there?

Ans. Three types:
i. α-neurons
ii. β-neurons
iii. γ-neurons.

24. What are the functions of these neurons?

Ans.
i. Alpha (α) neurons supply extrafusal fibers of skeletal muscles.
ii. Beta (β) neurons supply both extrafusal fibers of skeletal muscles and intrafusal fibers of muscle spindle.
iii. Gamma (γ) neurons supply intrafusal fibers of muscle spindle.

25. What are the subtypes of alpha neurons?

Ans. Two subtypes:
i. Phasic α neurons produce rapid contraction without any limit of shortening.
ii. Tonic α neurons maintain contraction of requisite length.

26. What are the subtypes of γ-neurons?

Ans. Two subtypes:
i. Dynamic γ-neurons supply nuclear bag type of intrafusal fibers.
ii. Static γ-neurons supply nuclear chain type of intrafusal fibers.

27. What are the differences between α and γ neurons?

Ans. Table 24.1 shows the differences between α and γ neurons.

Tble 24.1: Differences between α and γ neurons

	α-neurons	γ-neurons
Size	About 25 μm in dimensions	About 15–25 μm in dimensions
Myelination	Thickly myelinated	Thinly myelinated
Conduction velocity	About 15–120 m/sec	About 10–45 m/sec
Functions	a. Supply extrafusal muscle fibers	a. Supply intrafusal fibers of muscle spindle
	b. Produce shortening of muscle without any limit	b. Produce shortening of muscle to a predetermined length
Subtypes	a. Phasic α-neurons	a. Dynamic ($γ_1$) neurons
	b. Tonic α-neurons	b. Static ($γ_2$) neurons

28. What are interneurons?

Ans. Interneurons, also called connector or internuncial neurons are Golgi type II neurons which connect sensory neurons with motor neurons.

29. What do you mean by Golgi type II neurons?

Ans. The axons of these neurons are short and confined within the gray matter, whereas Golgi type I neurons possess long axons which form tracts of CNS or peripheral nerves.

30. What is the primary function of interneurons?

Ans. They are concerned with reflex activities.

31. Are they excitatory or inhibitory?

Ans. A particular interneuron may act as either excitatory or inhibitory neuron. A single neuron cannot produce both excitatory and inhibitory response.

32. Give an example of inhibitory interneuron?

Ans. Renshaw cells.

33. What is the neurotransmitter of these inhibitory neurons?

Ans. Glycine.

34. What are the parts of posterior horns?

Ans.
- i. Base
- ii. Neck
- iii. Head and
- iv. Apex.

35. What are the neurons of posterior horns?

Ans. Two types of neurons:
- i. Tract cells
- ii. Interneurons.

36. What is the role of these tract cells?

Ans. Tract cells are sensory neurons and their axons form ascending tracts which pass in the anterolateral white funiculi of the same side or opposite side or both sides.

37. What are the neurons of the lateral horns?

Ans. Lateral horns of thoracolumbar region (T_1–L_2) contain preganglionic neurons of sympathetic system. Lateral horns of sacral region (S_2–S_4) contain preganglionic neurons of sacral component of parasympathetic system.

38. How are the different nuclear groups arranged in different regions of spinal cord (Fig. 24.2)?

Ans. A. Nuclear groups in anterior gray columns (anterior horn):
  i. Medial group of motor neurons supplies trunk muscles and is arranged in two subgroups:
      a. Ventromedial nuclear group.
      b. Dorsomedial nuclear group.
  ii. Central group of neurons (C_3–C_5) supplies the diaphragm.
  iii. Lateral group of motor neurons supplies the limb muscles and is arranged in three subgroups:
      a. Ventrolateral group.
      b. Dorsolateral group.
      c. Retrodorsolateral group.
B. Nuclear groups in lateral horns (intermediate region of gray matter):
  i. Intermediolateral group of neurons.
  ii. Intermediomedial group of neurons.
     [Concerned with autonomic nervous system (visceral efferent)]
C. Nuclear groups in posterior gray columns (posterior horn)
   (from apex to base of posterior horns):
  i. Substantia gelatinosa of Rolando.
  ii. Nucleus proprius.
  iii. Nucleus dorsalis (Clarke's column).
  iv. Visceral afferent nucleus.

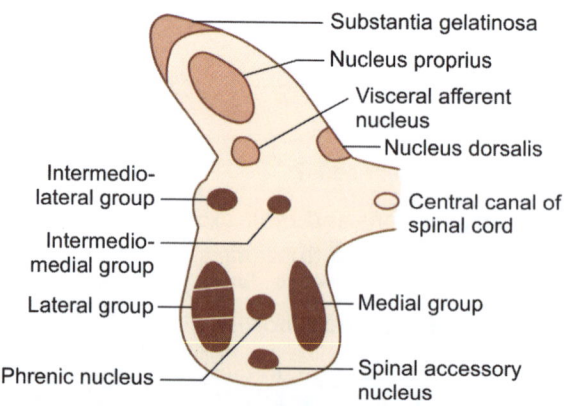

Fig. 24.2: Nuclei in spinal gray matter

Histology: Nervous System

39. What is substantia gelatinosa?

Ans. These are small and medium sized interneurons which form a cap-like appearance at the apex of the posterior gray column.

40. What is the extent of substantia gelatinosa in spinal cord?

Ans. It extends along the entire length of the spinal cord and it is continuous above with the nucleus of spinal tract of trigeminal nerve.

41. What is the composition and extent of nucleus proprius?

Ans. This nucleus is composed of interneurons and tract cells extending along the entire length of spinal cord, occupying its head and neck regions of posterior gray column.

42. What is nucleus dorsalis (Clarke's column)?

Ans. This nucleus is also composed of interneurons and tract cells, occupying the medial part of the base of posterior gray column extending from C_8 to L_2/L_3 segments of spinal cord.

43. What is the location and extent of visceral afferent nucleus?

Ans. This nucleus is located at the base of the posterior gray column and extends from T_1–L_2 and from S_2–S_4 segments of spinal cord.

44. What is the central canal of spinal cord?

Ans. It is an elongated canal within the gray matter of spinal cord.

45. What is the lining membrane of this canal?

Ans. Ependyma, made up of ciliated simple columnar epithelium.

46. What does this canal contain?

Ans. Cerebrospinal fluid (CSF).

47. What is the extent of this canal?

Ans. It is continuous above with the central canal of the medulla oblongata and below it extends into the proximal 4–5 mm of the filum terminale.

48. What is terminal ventricle?

Ans. The dilated central canal within the conus medullaris of spinal cord is known as terminal ventricle.

49. What is substantia gelatinosa centralis?

Ans. The neuroglial tissue surrounding the central canal is known as substantia gelatinosa centralis.

50. What do you mean by gray commissure?

Ans. The mass of gray matter which connects the right and left halves of gray matter is termed as gray commissure.

51. What is anterior and posterior gray commissure?

Ans. The gray commissure lying ventral to the central canal is called ventral or anterior gray commissure and the gray commissure lying dorsal to the central canal is called dorsal or posterior gray commissure.

52. Which important tract passes through the ventral gray commissure?

Ans. Spinothalamic tract (from right to left or left to right).

53. What do you mean by Rexed lamination of spinal gray matter?

Ans. The entire spinal gray matter is divided into ten laminae according to cytoarchitecture and packing-density of neurons.

54. What are the arrangements of these laminae?

Ans.

Laminae	Gray column
Laminae I to VI	Posterior gray column
Lamina VII	Intermediate region of gray matter
Lamina VIII	Medial part of anterior gray column
Lamina IX	Lateral part of anterior gray column
Lamina X	Gray matter around the central canal

55. How is the white matter arranged in the spinal cord?

Ans. The white matter of spinal cord is arranged in three columns:
  i. Anterior white column (anterior funiculus).
  ii. Lateral white column (lateral funiculus).
  iii. Posterior white column (posterior funiculus).

56. Are all these funiculi separated from each other?

Ans. Anterior and lateral funiculi are continuous with each other because of the short anterior gray column, called anterolateral funiculus.

Histology: Nervous System

57. What do you mean by ventral white commissure?

Ans. The band of white matter lying in front of the anterior gray commissure is called ventral or anterior white commissure.

58. What is dorsolateral tract of Lissauer?

Ans. It is a narrow strip of white fibers intervening between the apex of the posterior gray column and the posterior-lateral sulcus of the spinal cord, thus separating the posterior funiculus from the lateral funiculus.

59. Does the tract extend throughout the spinal cord length?

Ans. Yes. It extends along the entire length of spinal cord and dorsal nerve root fibers enter through this region.

60. What is the arrangement of fibers in anterior funiculi?

Ans.
  i. Superficial fibers are descending.
  ii. Intermediate fibers are ascending.
  iii. Deep fibers are both ascending and descending.

61. What is the arrangement of fibers in lateral funiculi?

Ans.
  i. Superficial fibers are ascending.
  ii. Intermediate fibers are descending.
  iii. Deep fibers are both ascending and descending.

62. What is the arrangement of fibers in dorsal funiculus?

Ans.
  i. Mostly ascending (fasciculus gracilis and fasciculus cuneatus).
  ii. Few fibers are descending.

63. Name some important ascending and descending tracts of spinal cord.

Ascending tracts	Descending tracts
Lateral spinothalamic tract	Corticospinal tract
Anterior spinothalamic tract	Rubrospinal tract
Anterior and posterior spinocerebellar tracts	Vestibulospinal tract
Fasciculus gracilis	Reticulospinal tract
Fasciculus cuneatus	Tectospinal tract

LESSON 2: CEREBELLUM

Key points for identification (Figs 24.3A and B)

1. Cortex consists of three layers:
   a. Outer molecular layer
   b. Middle Purkinje cell layer
   c. Inner granular layer.
2. Purkinje cells are arranged in a single row and flask-shaped.
3. Core of the leaf-like folium is formed by white matter.

PROBABLE QUESTIONS AND ANSWERS

1. What are the macroscopic parts of a section of cerebellum?

Ans. Two parts:
   i. Outer cortex or gray matter.
   ii. Inner medulla or white matter.

2. What are the distinctive features of cerebellar cortex?

Ans.
   i. Cerebellar cortex is entirely uniform (unlike cerebral cortex).
   ii. Cortical neurons and their processes are arranged in geometrical configuration.

3. What are the microscopic layers of cerebellar cortex?

Ans. From outside inwards:
   i. Molecular layer.
   ii. Purkinje cell layer.
   iii. Granular layer.

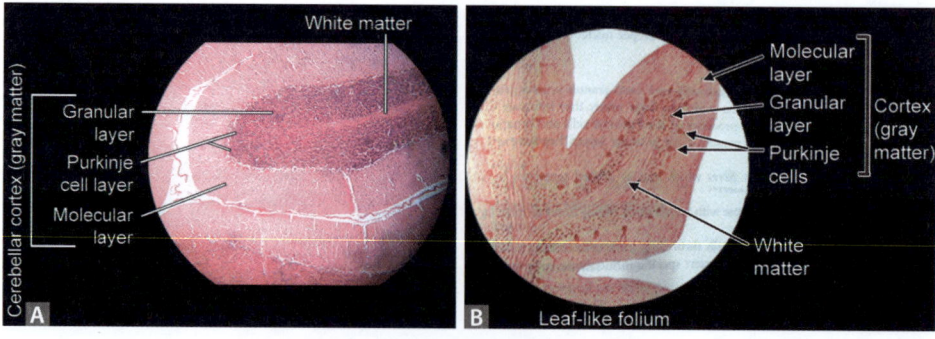

Figs 24.3A and B: Cerebellum

Histology: Nervous System

4. What is the thickness of molecular layer?

Ans. About 300–400 μm thick.

5. What are the neurons and nerve fibers of molecular layer?

Ans.
    a. Neurons of molecular layer are:
        i. Stellate cells.
        ii. Basket cells.
    b. Nerve fibers (unmyelinated) of molecular layer are:
        i. Axons of stellate and basket cells.
        ii. Parallel fibers of axons of granule cells.
        iii. Dendritic tree of Purkinje cells.
        iv. Dendrites of Golgi cells.
        v. Climbing fibers.

6. Where are these neurons located in the molecular layer?

Ans. Stellate cells are located near the surface of the cortex and the basket cells are located in the deeper ⅓rd of the molecular layer.

7. What type of neurons are they?

Ans. They are inhibitory interneurons.

8. How many basket cells are there in cerebellar cortex?

Ans. About 100 million. 1 mm thick cerebellar cortex contains about 600 basket cells.

9. What are the cells of Purkinje cell layer?

Ans.
    i. Mainly Purkinje cells.
    ii. Occasionally Bergmann glial cells and Golgi cells.

10. What is the shape of Purkinje cell?

Ans. Large flask-shaped.

11. How are the Purkinje cells arranged?

Ans. Purkinje cells are arranged in a single row and are evenly spaced.

12. What is the distance between the adjacent P-cells?

Ans. About 50 μm.

13. What are the poles of a Purkinje cell?

Ans. i. Superficial or outer pole.
ii. Deep or inner pole or basal pole.

14. What is the diameter of the basal pole?

Ans. About 30–35 μm.

15. What is the height of a P-cell?

Ans. About 50–70 μm.

16. How many Purkinje cells are there in cerebellar cortex?

Ans. About 30–50 million. 1 mm thick cerebellar cortex contains about 500 P-cell.

17. What are the features of dendrites of Purkinje cells (P-cells)?

Ans. The dendrites of P-cells show arborization (elaborate branching) in the molecular layer. 1st and 2nd orders of dendritic branches are nonspinous (smooth). The branches of 3rd order onwards are thick and provided with numerous spines.

18. How many spines are there in a Purkinje cell?

Ans. About 1,80,000 spines.

19. What are the synaptic contacts of the dendrites of Purkinje cells?

Ans. i. Nonspinous (smooth) dendrites synapse with the axons of stellate and basket cells.
ii. Spine of dendrites of Purkinje cells synapse with the parallel fibers of granule cells.
iii. Nonspiny areas of spiny branches (smooth inter spine) synapse with the branches of climbing fibers.
iv. **No input from Golgi cells.**

20. How many parallel fibers of Granule cells synapse with a single Purkinje cell?

Ans. Dendrites of single Purkinje cell synapse with 1,50,000–3,00,000 parallel fibers.

21. What do you mean by preaxon?

Ans. Initial narrow unmyelinated portion of axon of Purkinje cell is known as preaxon.

22. What is the length of preaxon?

Ans. About 30 μm.

Histology: Nervous System

23. Which cortical cell makes synapse with preaxon?

Ans. Axons of basket cells synapse with the preaxon of P-cells.

24. Where from the axons of P-cells arise?

Ans. The axons of P-cells arise from the center of the basal pole of the cell bodies.

25. Are the axons myelinated or unmyelinated?

Ans. Except the initial preaxon, the rest of the axon is myelinated.

26. Where do these axons terminate?

Ans. The axon of the Purkinje cells are the only efferent of the cerebellar cortex and they terminate in the deep cerebellar nuclei.

27. What type of influence is exerted by the P-cells?

Ans. P-cell axons exert inhibitory influence on the deep cerebellar nuclei and lateral vestibular nuclei.

28. What are deep cerebellar nuclei?

Ans. They are the output neurons of cerebellum and form final efferent pathway, located in the medullary core of white matter.

29. How many deep cerebellar nuclei are there?

Ans. There are four pairs of deep cerebellar nuclei.
These are (from medial to lateral):
  i. Nucleus fastigii.
  ii. Nucleus globosus.
  iii. Nucleus emboliformis.
  iv. Nucleus dentatus.

30. What do mean by nucleus interpositus?

Ans. Nucleus globosus and nucleus emboliformis are collectively called nucleus interpositus.

31. Phylogenetically, in which type of cerebellum these deep nuclei are found?

Ans. Nucleus fastigii (roof nucleus) belong to archicerebellum. Nucleus interpositus belong to paleocerebellum. Nucleus dentatus belongs to neocerebellum.

32. Which of these deep nuclei is the largest?

Ans. Dentate nucleus is the largest of all deep cerebellar nuclei.

33. What is the appearance of dentate nucleus?

Ans. It appears as a crenated mass with a hilum directing ventromedially.

34. Which important tract is formed by the axons of the dentate nucleus?

Ans. Dentatorubrothalamic tract which passes through the superior cerebellar peduncle.

35. What are the cells of the innermost granular layer?

Ans.
  i. Granule cells.
  ii. Golgi cells.
  iii. Brush cells.

36. How many granule cells are there in cerebellar cortex?

Ans. About 30–50 billion. 1 mm thick cerebellar cortex contains about 3,000,000 granule cells.

37. What is the size of a granule cell?

Ans. About 5–8 µm is diameter. Granule cell is the **smallest neuron** of the cerebellar cortex.

38. What is the fate of the axons of granule cell?

Ans. The axon of each granule cell extends into the molecular layer where it branches in the form of T-shaped manner and makes synaptic contacts with the dendrites of Purkinje and basket cells.

39. How many dendrites a granule cell possesses?

Ans. A granule cell possesses 3–5 (may be 1–7) dendrites, each of which is about 10–30 µm in length.

40. What do you mean by dendritic claw?

Ans. The enlargements at the terminals of the dendrites of granule cells are known as dendritic claws.

41. Are the granule cells excitatory or inhibitory?

Ans. Only the granule cells of the cerebellar cortex are excitatory, whereas the other cortical cells (Stellate, Basket, Purkinje and Golgi cells) are inhibitory.

42. What is the neurotransmitter of the excitatory granule cells?

Ans. Glutamate acts as a neurotransmitter.

43. What do you mean by cerebellar island?

Ans. The space within the granular layer, not occupied by the granule cells are called cerebellar islands.

44. What do these islands contain?

Ans. These islands are occupied by cerebellar glomeruli.

45. What is the composition of cerebellar glomerulus?

Ans.
i. In the center, terminal swelling of mossy fiber, called rosette.
ii. Dendritic claw of the granule cells surrounding the rosette.
iii. Axon terminals of Golgi cells.

46. What is the shape of a glomerulus?

Ans. Spherical or ovoid.

47. What is the size of a glomerulus?

Ans. About 20 µm in greatest dimension.

48. How many glomeruli are there?

Ans. About 600,000/cu mm.

49. What is the ratio between the number of granule cells and cerebellar glomeruli?

Ans. Granule cells: Glomeruli = 5:1 (3,000,000 : 6,00,000)

50. What are the largest neurons of the cerebellar cortex?

Ans. The Golgi cells (smallest being the granule cells).

51. How many Golgi cells are there in cu mm of thickness?

Ans. About 50/cu mm.

52. What is the thickness of the granular layer?

Ans. About 100 µm thick at sites of fissures. About 400–500 µm at the apex (foliar summit).

53. What are the nuclei in the medullary core of white matter?

Ans. Four pairs of deep cerebellar nuclei (discussed earlier).

54. What are the fibers in the white matter?

Ans.
i. Fiber propriae (intrinsic fibers).
ii. Projection fibers.
iii. Myelinated axons of Purkinje cells.
iv. Climbing fibers.
v. Mossy fibers.

55. What are the sensory inputs of cerebellum?

Ans. The climbing and mossy fibers are the sensory inputs of cerebellum (afferent fibers) and both these fibers are excitatory in nature.

56. What are the climbing fibers?

Ans. The climbing fibers are the afferent (sensory) fibers that are derived from the inferior olivary nucleus of opposite side. They (olivocerebellar fibers) pass through the granular and Purkinje cell layer to reach the outermost molecular layer of cerebellar cortex where they coil around the dendritic tree of a Purkinje cell and synapse with them.

57. What are mossy fibers?

Ans. All afferent fibers of the cerebellum other than the olivocerebellar (climbing) fibers are called mossy fibers.

58. What do you mean by rosette?

Ans. Each mossy fiber divides into about 30–40 terminal swellings in the granular layer of cerebellar cortex. These terminal swellings of mossy fibers are known as rosette.

59. Where do the mossy fibers terminate?

Ans. The mossy fibers terminate in the central part of cerebellar glomeruli as rosette in the in the granular layer.

60. How many granule cells are activated by a single rosette?

Ans. About 15 granule cells.

61. How many granule cells are activated by a single mossy fiber?

Ans. A single mossy fiber divides into about 30–40 rosette. A single rosette in turn activates about 15 granule cells. So, a single mossy fiber activates about (30–40) × 15, i.e. 450–600 granule cells.

Histology: Nervous System

62. How many Purkinje cells are excited by the parallel fibers of the axon of a single granule cell?

Ans. About 450–500 Purkinje cells.
Note that a climbing fiber excites a single Purkinje cell, whereas a mossy fiber excites thousands of Purkinje cells.

63. Which of these five sets of cortical intrinsic neurons are characteristic of cerebellum?

Ans. Purkinje cells are characteristic of cerebellum.

64. What are the layers of cerebral cortex?

Ans. Neocortex of cerebrum consists of six layers.
These are (from outside inwards):
  i. Molecular layer (lamina 1).
  ii. Outer granular layer (lamina 2).
  iii. Outer pyramidal layer (lamina 3).
  iv. Inner granular layer (lamina 4).
  v. Inner pyramidal layer (lamina 5).
  vi. Pleomorphic layer (lamina 6).

65. What are the cell types of cerebral cortex?

Ans. Four basic types of cells are there in cerebral cortex:
  i. Pyramidal cells (about 5.5 billion).
  ii. Stellate cells.
  iii. Cells of Martinotti.
  iv. Horizontal cells of Cajal.

66. What are Betz cells?

Ans. Large or giant pyramidal cells are also known as Betz cells which are found in the deeper zone of lamina 5 (inner pyramidal layer).

Number of cerebellar cortical cells at a glance:
  i. Stellate and Basket cells → About 100 million.
  ii. Purkinje cells → About 30–50 million.
  iii. Granule cells → About 30–50 billion.

Concentration of different types of cerebellar cortical cells:
  i. Basket cells → 600/cu mm.
  ii. Purkinje cells → 500/cu mm.
  iii. Granule cells → 3,000,000/cu mm.
  iv. Golgi cells → 50/cu mm.
  v. Synaptic cerebellar glomeruli → 6,00,000/cu mm.

CHAPTER 25

Histology: Integumentary System

- Skin

LESSON 1: SKIN

Key points for identification (Figs 25.1 to 25.6)

1. Skin consists of epidermis and dermis.
2. Epidermis is lined by keratinized stratified squamous epithelium.
3. Dermis contains hair follicles, sebaceous glands, sweat glands and arrector pili muscles.

PROBABLE QUESTIONS AND ANSWERS

1. What are the layers of skin?

Ans. Skin consists of two layers, from outside inwards:
   i. Epidermis
   ii. Dermis

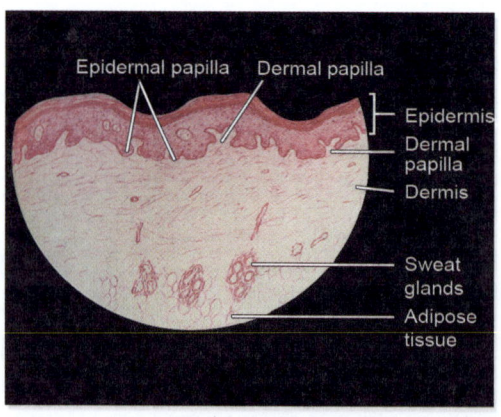

Fig. 25.1: Thick skin

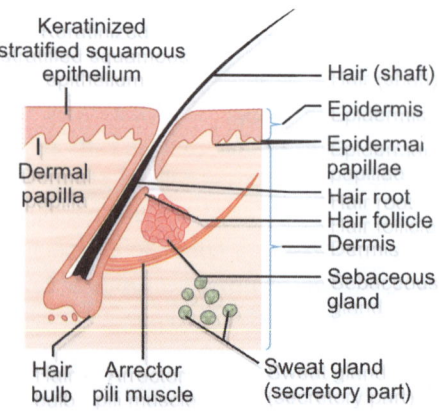

Fig. 25.2: Thin skin

Histology: Integumentary System

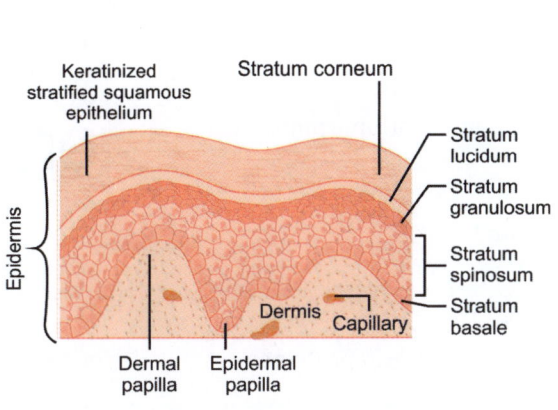

Fig. 25.3: Layers of epidermis

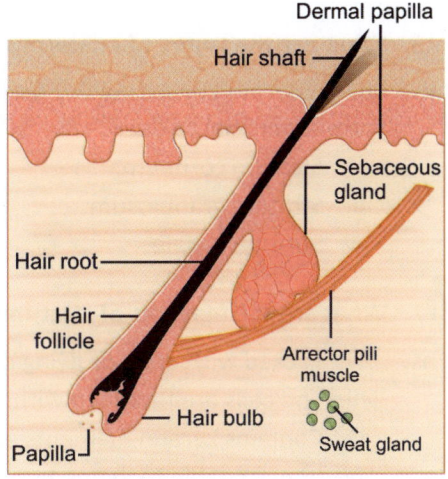

Fig. 25.4: Structure of a hair follicle

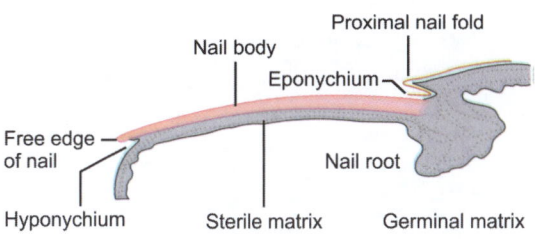

Fig. 25.5: Parts of nail

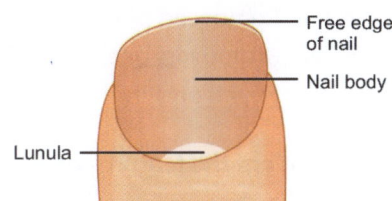

Fig. 25.6: Lunule of nail bed

2. What do you mean by hypodermis?

Ans. Hypodermis is the intervening tissue between the dermis and the underlying deep fascia and equivalent to subcutaneous fascia.

3. What is the composition of hypodermis?

Ans. The hypodermis is composed of fibroareolar and fatty tissue.

4. What is the thickness of skin?

Ans. The thickness of skin varies between 0.3 and 3 mm.

5. What is the lining epithelium of skin?

Ans. The epidermis of skin is lined by keratinized stratified squamous epithelium.

6. What are the layers of epidermis?

Ans. The epidermis consists of five layers:
From superficial to deep:
- i. Stratum corneum.
- ii. Stratum lucidum.
- iii. Stratum granulosum.
- iv. Stratum spinosum.
- v. Stratum basale.

7. What are the features of stratum corneum?

Ans.
- i. Packed with dead or dying cells.
- ii. The cells are flattened without nuclei.
- iii. Contains keratin filaments embedded in filaggrin proteins.
- iv. Impermeable to water, because of lipid content.

8. What are the features of stratum lucidum?

Ans.
- i. Appears homogeneous and clear; hence, the name lucidum.
- ii. Composed of non-nucleated cells.
- iii. Disappears as the cells fill with keratin.
- iv. Found only in thick glabrous skin.

9. What are the features of stratum granulosum?

Ans.
- i. Composed of several layers of cells (1 to 5 cell layers thick).
- ii. Cells are loaded with numerous keratohyalin granules.
- iii. Cells are flattened and nuclei are condensed.
- iv. Serves as waterproof layer due to presence of lipids in the intercellular spaces.

10. What are the features of stratum spinosum?

Ans.
- i. Composed of several layers of large polyhedral cells.
- ii. Cells show numerous cytoplasmic spine like processes (prickles), hence called prickle cell layer or stratum spinosum.
- iii. The cells show cytoplasmic basophilia.
- iv. The cells are attached to one another by desmosomes.

11. How do the cells of stratum spinosum produce spine like processes?

Ans. The cells are attached to one another by desmosomes. During tissue preparation, the cells retract and become separated from one another except at the sites of desmosomes. This produces spine like processes.

Histology: Integumentary System

12. What are the features of stratum basale?

Ans.
i. The cells of this layer are columnar and are arranged in a single layer. The cells are called keratinocytes.
ii. The cells rest on the basement membrane to which they are connected by hemidesmosomes.
iii. The cells show cytoplasmic basophilia.
iv. The cells serve as stem cells that undergo mitosis to give rise to new keratinocytes.
v. This stratum is also known as stratum germinativum.

13. What are the different types of cells of stratum basale (basal layer of epidermis)?

Ans. Most abundant cells are keratinocytes.
Other cells are:
i. Melanocytes.
ii. Langerhans cells.
iii. Merkel cells.

14. What is the function of keratinocytes?

Ans. These cells produce keratin through different stages of morphogenesis. This keratin is the most important structural protein of epidermis and renders the skin waterproof.

15. Why the epidermis of the skin is stratified?

Ans. The keratinocytes of the basal layer of epidermis undergo keratinization through different stages of morphogenesis which is responsible for the stratification of epidermis.

16. What is the source of origin of melanocytes?

Ans. Neural crest (ectoderm) → melanoblasts → melanocytes.

17. What does the melanocytes synthesize?

Ans. Melanocytes synthesize a brown pigment called melanin.

18. What is the process of melanin synthesis?

Ans. Flowchart 25.1 show the process of melanin synthesis.

19. How are the melanin pigments transferred to keratinocytes?

Ans. Exocytosis of the pigments from the melanocytes → exocytosed pigments are phagocytosed by the keratinocytes.

20. What are the different forms of melanin?

Ans. Two forms:
i. Eumelanin.
ii. Pheomelanin.

Flowchart 25.1: Process of melanin synthesis

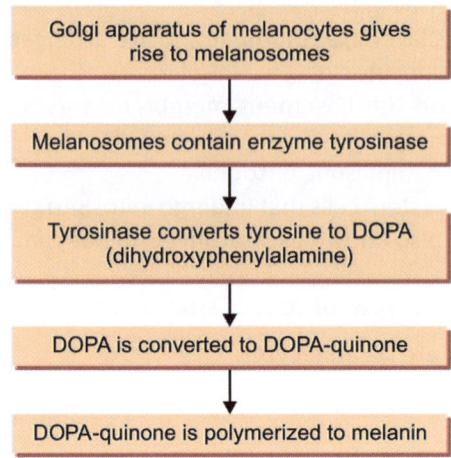

21. What are the functions of the two forms of melanin for hair color?

Ans.
    i. Eumelanin is responsible for dark brown color of hair.
    ii. Pheomelanin is responsible for reddish hair.

22. Why does the skin color differ from race to race?

Ans. This color difference among the different races are due to the size and number of melanosomes and their distribution in keratinocytes in basal layer of epidermis.

23. Where do you find the melanocytes?

Ans. They are found at the dermoepidermal junction or in the basal layer of epidermis.

24. What is the morphological appearance of melanocytes?

Ans. Melanocytes are irregular cells with many dendritic processes. These cells are connected to the basement membrane by hemidesmosomes (like keratinocytes of the basal layer).

25. What are the factors for melanin formation?

Ans.
    i. Genetic factor.
    ii. Melanocyte stimulating hormone (MSH).
    iii. Estrogen.
    iv. Progesterone.
    v. Exposure to ultraviolet (UV) rays.

26. What is the protective role of melanin?

Ans.
UV rays increase melanin secretion
↓
Increased melanin prevents the damaging effects of UV radiation of nuclear DNA on basal cells of epidermis, thus, acting as a protective role.

27. Why skin cancer is more in old age than in youngsters?

Ans. In old age, the melanin pigments decrease and thereby produce less protective role against UV radiation.

28. What are malignant melanomas?

Ans. Benign melanomas are the melanocytic moles, formed by the cluster of melanocytes in high densities. Chronic irritation causes transformation of these benign melanomas into malignant melanomas which are treated surgically.

29. What is albinism?

Ans. It is an autosomal recessive disorder characterized by total depigmentation of skin due to congenital absence of tyrosinase.

30. What is vitiligo?

Ans. It is the localized depigmentation of skin caused by failure of production of melanin or loss of melanocytes.

31. What are the features of Langerhans cells?

Ans.
a. **Source:** Bone marrow
b. **Location:**
   i. All layers of epidermis of skin
   ii. Oral mucosa
   iii. Esophageal mucosa
   iv. Vaginal mucosa
   v. Hair follicles
   vi. Around blood vessels in papillary layer of dermis.
c. **Morphology:**
   i. Dendritic cells
   ii. Cells contain membrane bound, elongated vacuoles, rod-like Langerhans bodies or Birbeck bodies
   iii. Nucleus is oval/reniform/irregular.
d. **Functions:**
   i. The cells are antigen presenting cells
   ii. Play role in immune system of skin
   iii. Protective role in epidermal carcinoma
   iv. Inactivated by prolonged exposure to ultraviolet light.

32. What are Merkel's cells?

Ans.
i. These are specialized touch receptors which lie in the stratum basale of the epidermis.
i. These cells are predominant in thick skin.

33. What is the blood supply of epidermis?

Ans. Epidermis is devoid of blood vessels (avascular), it receives its nutrition by diffusion from the capillaries of dermis.

34. What are the layers of dermis?

Ans. Two layers:
i. Outer papillary layer.
ii. Inner reticular layer.

35. What does the papillary layer contain?

Ans. Papillary layer contains:
i. Irregular network of collagen fibers.
ii. Loop of blood capillary.
iii. Tactile corpuscles (Meissner's corpuscles).

36. What do you mean by dermal papillae?

Ans. These are a series of conical projections of the papillary layer of dermis into the corresponding recesses of the epidermis.

37. Which type of skin possesses prominent dermal papillae?

Ans. Thick glabrous skin of palms and soles, which are subjected to mechanical stress.

38. What do you mean by epidermal papillae?

Ans. These are downward projections of epidermis into the dermis in the interval between the dermal papillae.

39. What are epidermal ridges?

Ans. These are elevations on the surface of the epidermis, particularly on the palms and ventral surfaces of the fingers of the hand and soles of the feet.

40. What is the importance of these epidermal ridges?

Ans. The ridges produce fingerprints which are highly specific for each individual.

Histology: Integumentary System

41. What are the contents of the reticular layer of the dermis?

Ans.
i. Irregular bundles of collagen and elastic fibers.
ii. Fibroblasts.
iii. Adipose tissue (fat).
iv. Sweat glands.
v. Blood vessels.
vi. Nerves.
vii. Macrophages, etc.

42. What is the developmental source of dermis?

Ans. Dermis is developed from mesoderm, whereas the epidermis is ectodermal in origin.

43. What are the appendages of skin?

Ans.
i. Hairs.
ii. Sweat glands.
iii. Sebaceous glands.
iv. Nails.
v. Modified sweat glands.

44. What are hairs?

Ans. Hairs are derived from invagination of epidermis (hence keratinized) and projected out from the body surface.

45. How are they distributed?

Ans. Distribution of hairs in the face is about 600/cm^2. Rest of the body surface (except face) is about 60/cm^2.

46. What are the important functions of hairs?

Ans.
i. Thermoregulation.
ii. Sensory function.

47. What are areas of absence of hairs?

Ans.
i. Palms and soles.
ii. Glans penis and clitoris.
iii. Prepuce.
iv. Labia minora and inner surface of labia majora.
v. Nipples.
vi. Lips.

48. What are the parts of hairs?

Ans.
i. **Shaft:** The part of hair projecting out of the body surface (visible part).
ii. **Root:** The part of hair lying within the skin (embedded part).

49. What do you mean by bulb of hair?

Ans. The expanded lower end of the root of the hair is called bulb.

50. What is hair papilla or dermal papilla?

Ans. Invaginated bulb by the part of the dermis is known as hair papilla.

51. What is hair follicle?

Ans. The tubular sheath around the root of the hair is termed as hair follicle.

52. Where from the hair follicle is derived?

Ans. The hair follicle is derived from the invagination of the epidermis into the dermis around the root of the hair. It means that the hair follicle is regarded as a part of epidermis.

53. What are the layers of the hair follicle?

Ans. From inside outward:
  i. Inner root sheath (derived from outermost cells of hair bulb).
  ii. Outer root sheath (derived from invagination of epidermis).
  iii. Connective tissue sheath (derived from the dermis).

54. What are the different zones of inner root sheath?

Ans. From outside inwards: Three zones:
  i. Henle's layer, made up of a single layer of keratinized cubical cells with flattened nuclei.
  ii. Huxley's layer, made up of 1–3 layers of partially keratinized flattened nucleated cells containing large eosinophilic trichohyalin granules.
  iii. Cuticle layer, made up of single layer of flattened keratinized cells with atrophied nuclei.

55. What are the segments or parts of hair follicle?

Ans.
  i. **Infundibulum:** The segment between the skin surface and the opening of the sebaceous duct.
  ii. **Isthmus:** The segment between the opening of the sebaceous duct and the attachment of arrector pili muscle.
  iii. **Inferior segment:** The part of the hair follicle below the attachment of arrector pili muscle up to the hair bulb.
  iv. **Fundus:** The lower expanded part of hair follicle (the lower expanded part of the root of the hair is called bulb).

Histology: Integumentary System

56. What is the structure of the shaft of the hair?

Ans. On cross-section, the hair shaft possesses **three zones from outside inwards**:
  i. Cuticle, a thin membrane formed by flattened keratinized cells.
  ii. Cortex, made up of keratin (acellular).
  iii. Medulla, made up of keratinized cells of irregular shape and it is present only in thick hairs.

57. What are arrector pili muscles?

Ans.
  i. These are small bands of smooth muscles.
  ii. **Extent:** From connective tissue sheath of a hair follicle to the papillary layer of the dermis or up to the dermoepidermal junction.
  iii. **Nerve supply:** Cholinergic sympathetic fibers.
  iv. **Disposition of muscles:** Diagonal, so that the muscle makes an obtuse angle with the skin surface.
  v. **Functions:**
     a. Contraction of the muscle causes the hair follicle to become almost vertical, i.e. erection of hair shaft.
     b. Contraction causes depression of the skin where the muscle is attached to the dermis producing goose flesh appearance.
     c. Contraction causes squeezing out of sebaceous secretion into the hair follicle because of the position of the sebaceous glands between the muscle and the hair follicle.

58. What are the areas where the arrector pili muscle are absent?

Ans.
  i. Hairs of face and axilla.
  ii. Hairs of eyebrows and eyelashes.
  iii. Hairs of nostrils and external auditory meatus.

59. What do you mean by holocrine mode of secretion?

Ans. It means that the secretion occurs following the complete destruction of cell cytoplasm.

60. Where are the glands located?

Ans. The sebaceous glands are located in the dermis at the angles formed by the hair follicles and the arrector pili muscles.

61. Where do the sebaceous ducts open?

Ans.
  i. Normally their ducts open into the hair follicle.
  ii. Some glands which occur independently of hair follicle open directly on the skin surface, e.g. lips and parts of external genitalia.

62. What are the areas where the sebaceous glands are abundant?

Ans.
i. Face
ii. Scalp
iii. Scrotum
iv. Ears
v. Nostrils
vi. Vulva
vii. Around the anus

63. What are the areas where the glands are absent?

Ans.
i. Palms of hands.
ii. Soles of feet.
iii. Flexor surface of the digits.

64. What is the lining epithelium of sebaceous ducts?

Ans. Keratinized stratified squamous epithelium.

65. What is the secretion of the sebaceous glands?

Ans. The secretion is known as sebum which is an oily fluid.

66. What is the composition of sebum?

Ans. It is a mixture of lipids including triglycerides, cholesterol, cholesterol esters and fatty acids.

67. What are the functions of sebum?

Ans.
i. Keeps the skin and hair soft.
ii. Prevents dryness of the sin.
iii. Weak antibacterial and antifungal action.

68. What do you mean by pilosebaceous apparatus?

Ans. The combination of arrector pili muscle, the hair follicle and the sebaceous gland form the pilosebaceous apparatus.

69. Where from the sebaceous glands are developed?

Ans. Most of the glands are developed as lateral outgrowths of the outer root sheath of the hair follicles.

70. What are the factors that control sebaceous secretion?

Ans.
i. Testosterone in males.
ii. Ovarian and adrenal androgens in female.
Note that sebaceous secretion is not under nervous control.

Histology: Integumentary System

71. What are modified sebaceous glands?

Ans. The secretion of these glands instead of opening into the hair follicle, directly opens on the skin surface.
Examples:
  i. Meibomian (tarsal) glands of eyelids.
  ii. Montgomery's tubercles in the areola.

72. What are the types of sweat glands?

Ans. Two types:
  i. Typical (eccrine/merocrine) sweat glands.
  ii. Atypical (apocrine) sweat glands.

73. What are the parts of a typical sweat gland?

Ans.   i. Body or fundus.   ii. Duct.

74. What is body (fundus) of a typical sweat gland?

Ans. The highly coiled secretory part of the tubular gland within the dermis is called body of the gland.

75. What are the parts of the ducts of the sweat glands?

Ans.   i. Part within the dermis (somewhat straight)
  ii. Part within the epidermis (spiral in course).

76. What is the shape of the opening of the duct on the skin surface?

Ans. Funnel-shaped opening.

77. What are the differences between typical and atypical sweat glands?

Ans. Table 25.1 shows the differences between typical and atypical sweat glands.

Table 25.1: Differences between typical and atypical sweat glands

	Typical sweat glands	Atypical sweat glands
Type	Simple tubular. The secretory part of the gland (body) is highly coiled	Branched tubular. The secretory parts branch and may form a network
Mode of secretion	Merocrine	Apocrine
Size	Smaller than atypical glands	Larger in size with a wide dilated lumen
Epithelium of secretory part	Cuboidal or pseudostratified	May be squamous or cuboidal or columnar
Opening of duct	On the skin surface	Into the hair follicle
Secretion	Clear, colorless, hypotonic fluid	Viscid, milky, containing proteins and odorless
Distribution	Most numerous on palms, soles, forehead, scalp	Axilla, areola, nipple, pubic and perineal region
Nerve supply	Cholinergic sympathetic nerves	Adrenergic sympathetic nerves

78. Are the secretions of atypical sweat glands always odorless?

Ans. No, they give off body odors after bacterial decomposition.

79. Name some modified sweat glands.

Ans.
  i. Mammary glands.
  ii. Ciliary glands (of eyelids).
  iii. Ceruminous glands (of external auditory meatus).

80. What do the nails represent?

Ans. The nails represent the plates of keratinized epithelial cells of epidermis.

81. What is the structural component of nails?

Ans. The structure of the body of the nails correspond to the stratum corneum of epidermis consisting of several layers of dead, cornified, anucleate cells filled with keratin.

82. What are the parts of a nail?

Ans. Three parts:
  i. Root which is implanted into groove on the skin.
  ii. Body (exposed part of the nail).
  iii. Free distal edge.

83. How many nail folds are there?

Ans.
  i. Two lateral nail folds of skin which overlap the lateral margins of the nails.
  ii. Proximal nail fold which covers the root of the nail.

84. What is lunule of nail?

Ans. The semilunar white area just distal to the proximal nail fold is known as lunule. It is most prominent in the thumb nail.

85. What do you mean by nail bed?

Ans. The nail bed is the tissue on which a nail rests.

86. Why do the nails look pink?

Ans. The tissue underlying the body of the nail is highly vascular which gives the nails its pink color.

87. Where from the nail substance is derived?

Ans. The nail substance is derived mainly from the proliferation of cells of the germinal matrix.

Histology: Integumentary System

88. What is germinal matrix?

Ans. The thickened germinative zone of the epidermis of skin (stratum spinosum and stratum basale) near the root of the nail forms the germinal matrix.

89. What is sterile matrix?

Ans. The thin germinative zone underlying the body of the nail has no role in forming the substance of the nail and is known as sterile matrix.

90. Is there any structural change of dermis underlying the sterile matrix?

Ans. The dermis underlying the sterile matrix does not possess dermal papillae but is highly vascular containing arteriovenous anastomosis and provided with numerous sensory nerve endings.

91. What do you mean by eponychium and hyponychium?

Ans.
i. **The eponychium** is the extension of stratum corneum from the deep surface of the proximal nail fold over the external surface of the body of the nail.
ii. **The hyponychium** is the reflection of stratum corneum of epidermis of the finger tip onto the under surface of the distal border of the nail.

92. What are the differences between thick and thin skin?

Ans. Table 25.2 shows the differences between thick and thin skin.

Table 25.2: Differences between thick and thin skin

	Thick skin	Thin skin
Epidermis	Stratum lucidum of epidermis is prominent	Stratum lucidum is absent. So, epidermal layers are four unlike 5-layered epidermis in thick skin
Dermis	a. Dermal papillae of papillary layer is more prominent b. Reticular layer of dermis is less thick	a. Dermal papillae are less prominent b. Reticular layer of dermis is thicker
Hair follicles	Absent	Present
Sebaceous glands	Absent	Present
Sweat glands	Less in number	More in number
Distribution	Palms and soles and flexor surface of the digits	All over the body except palm and sole

Chapter 26
Histology: Identification of Histological Slides At a Glance

Musculoskeletal System
1. **Compact bone:**
    i. Presence of Haversian system.
    ii. Lacunae containing osteocytes.
2. **Skeletal muscle:**
    i. Multinucleated muscle cells (muscle fibers).
    ii. Muscle fibers do not branch and possess peripherally placed nuclei.
3. **Cardiac muscle:**
    i. Interconnected cylindrical fibers by side branches with striations.
    ii. Centrally placed nuclei.

Blood-vascular System
4. **Artery:**
    i. Lumen is small, round and empty.
    ii. 3-layered wall with well-defined endothelial lining.
5. **Veins:**
    i. Wall is 3-layered but thin.
    ii. Larger quantities of collagen fibers and smaller quantities of elastic or muscle tissue in tunica media.

Gastrointestinal System
6. **Tongue:**
    i. Mucosa lined by nonkeratinized stratified squamous epithelium containing papillae.
    ii. Presence of serous and mucous glands in between muscles.
7. **Esophagus:**
    i. Mucosa lined by nonkeratinized stratified squamous epithelium with thick muscularis mucosa.
    ii. Presence of submucosal esophageal glands.
8. **Stomach:**
    i. Mucosa lined by simple columnar epithelium with tubular glands in lamina propria.
    ii. Three layers of smooth muscles.

9. **Duodenum:**
    i. Mucosa → presence of villi, lined by simple columnar epithelium with crypts of Lieberkuhn.
    ii. Presence of submucosal **Brunner's gland**.
10. **Ilium:**
    i. Mucosa → presence of villi, lined by simple columnar epithelium.
    ii. Presence of Peyer's patches.
11. **Appendix:**
    i. Poorly developed mucous membrane lined by simple columnar epithelium.
    ii. Presence of profuse numbers of lymphatic nodule in lamina propria.
12. **Rectum:**
    i. Mucosa are thrown into folds and lined by columnar cells without any villi.
    ii. Presence of numerous goblet cells.

Liver and Pancreas

13. **Liver:**
    i. Presence of hepatic lobule containing lamina of hepatic cells in radiating manner and central vein.
    ii. Portal triad consisting of interlobular branch of portal vein, hepatic artery and bile ductule.
14. **Pancreas:**
    i. Exocrine part containing serous acini in the lobule with intralobular ducts.
    ii. Endocrine part consists of group of cells that form Islets of Langerhans.

Salivary Glands

15. **Parotid gland:**
    i. Lobule contains serous acini.
    ii. Intra and interlobular ducts lined by columnar epithelium.
16. **Submandibular gland:**
    i. Mixed acini—both serous and mucous.
    ii. Mucous acini capped by serous demilunes.
17. **Sublingual gland:**
    i. Lobule contains mucous acini.
    ii. Intra and interlobular ducts lined by cuboidal to columnar epithelium.

Endocrine Glands

18. **Thyroid gland:**
    i. Made up of follicles lined by cuboidal epithelium containing pink staining colloid.
    ii. Presence of parafollicular cells.
19. **Suprarenal (adrenal) gland:**
    i. Cortex covered by capsule consists of three layers—zona glomerulosa, zona fasciculata and zona reticularis.
    ii. Inner medulla, containing chromaffin cells separated by sinusoids.

Lymphatic System

20. **Lymph node:**
    i. Outer cortex covered by thin capsule and presence of subcapsular sinus and lymphatic follicle with germinal center.
    ii. Inner medulla with medullary cords and blood vessels.
21. **Thymus:**
    i. Presence of lobulations with outer cortex and inner medulla.
    ii. Aggregation of lymphocytes in darker cortex and presence of Hassalls's corpuscle in inner medulla.
22. **Spleen:**
    i. Presence of thick capsule and trabeculae.
    ii. Splenic pulp consisting of white pulp and red pulp.
23. **Palatine tonsil:**
    i. Mucosa lined by nonkeratinized stratified squamous epithelium.
    ii. Subepithelial lymphoid nodules are present.

Respiratory System

24. **Trachea:**
    i. Mucous membrane lined by pseudostratified ciliated columnar epithelium.
    ii. Presence of hyaline cartilage and tracheal glands outside the mucous membrane.
25. **Lungs:**
    i. Presence of alveoli and septa lined by squamous epithelium.
    ii. Presence of capillaries, terminal and respiratory bronchiole.

Excretory System

26. **Kidney:**
    i. Presence of glomerulus, Bowman's capsule and tubules.
    ii. Parenchyma contains blood vessels and scanty connective tissue.
27. **Ureter:**
    i. Mucous membrane thrown into star-shaped folds.
    ii. Mucous membrane lined by transitional epithelium.
28. **Urinary bladder:**
    i. Mucous membrane lined by transitional epithelium without any submucous coat.
    ii. Thick muscular layer.

Male Reproductive System

29. **Testis:**
    i. Presence of seminiferous tubules containing germ cells and Sertoli cells.
    ii. Presence of interstitial cells of Leydig.
30. **Vas deferens:**
    i. Mucous membrane lined by pseudostratified columnar epithelium and profuse elastic fibers in lamina propria.
    ii. Very thick muscular coat.

31. **Prostate:**
    i. Presence of irregular follicles of glandular tissue lined by simple columnar epithelium with amyloid bodies in the lumen.
    ii. Presence of broad bands of fibromuscular tissue between the follicles.

Female Reproductive System

32. **Uterus:**
    i. Mucous membrane lined by simple columnar epithelium with tortuous numerous tubular uterine glands in lamina propria.
    ii. Presence of multiple layers of myometrium with blood vessels within them.
33. **Fallopian tubes:**
    i. Mucous membrane thrown into primary, secondary and tertiary folds lined by ciliated columnar epithelium without submucous coat.
    ii. Thin muscular layer.
34. **Ovary:**
    i. Outer cortex covered by germinal epithelium containing ovarian follicles.
    ii. Inner medulla contains stroma and blood vessels.
35. **Mammary gland:**
    i. Consisting of lobules of glandular tissue separated by considerable quantity of connective tissue and fat.
    ii. Presence of intralobular ducts lined by cuboidal epithelium.

Nervous System

36. **Spinal cord:**
    i. Surrounding the central canal, the inner grey matter is arranged into anterior and posterior horn.
    ii. Outer white matter consists mainly of nerve fibers and neuroglia without nerve cells.
37. **Cerebellum:**
    i. The core of leaf-like folium formed by white matter.
    ii. Outer cortex consists of outer molecular, intermediate single row of large flask shaped Purkinje cell layer and inner granular layer.

Integumentary System

38. **Skin:**
    i. Epidermis composed of keratinized stratified squamous epithelium traversed by hair follicle.
    ii. Dermis contains sweat and sebaceous glands.

SECTION 4

Radiological Anatomy

Section Outline

- ❖ Radiological Anatomy: Introduction
- ❖ Radiological Anatomy: Upper Limb (Superior Extremity)
- ❖ Radiological Anatomy: Lower Limb (Inferior Extremity)
- ❖ Radiological Anatomy: Abdomen
- ❖ Radiological Anatomy: Thorax
- ❖ Radiological Anatomy: Head and Neck

SECTION A

Radiological Anatomy

CHAPTER 27

Radiological Anatomy: Introduction

- General Considerations

GENERAL CONSIDERATIONS

PROBABLE QUESTIONS AND ANSWERS

Q.1. What do you mean by radiology?

Ans. **Radiology** is the science which deals with the use of radiant energy for the diagnosis and treatment of diseases.

Q.2. What is radiography?

Ans. **Radiography** is the procedure of obtaining radiographs of the internal structures of the body.

Q.3. What are X-rays?

Ans. **X-rays** are electromagnetic radiation of relatively high energy **photon** with very **short wavelength** (0.01–10 nm) which can penetrate structures within the body and create images of the internal structures on photographic film. X-ray is also called **Roentgen ray**.

Q.4. What is a radiograph?

Ans. Radiograph is the film produced by radiography.

Remember that wavelengths of X-rays are shorter than ultraviolet rays but longer than Y-rays. X-rays with high photon energies are called **hard X-rays**, whereas X-rays with lower energies are called **soft X-rays**.

Q.5. What are the important characteristics of X-rays?

Ans. Few characteristics of X-rays:
   i. They are electrically neutral.
   ii. They are invisible.
   iii. They travel in **straight lines**.
   iv. They travel at the speed of light in vacuum.
   v. They can be absorbed by the tissues in the body.
   vi. They can cause biologic and chemical damage to living tissues.

Q.6. Why do these X-rays can penetrate materials?

Ans. X-rays can penetrate materials because of their short wavelength.

Q.7. What is the fundamental principle of radiographic tests that employ X-rays?

Ans. A beam of X-rays passes through the exposed area of the body.
   i. X-rays are blocked or absorbed by the different body tissues in differing amounts depending on the density and compositions of the tissues (**radiopaque areas** i.e. looks white or near white on conventional X-ray films).
   ii. X-rays that are not absorbed pass through the area and appears black or dark on conventional radiographic films (**radiolucent areas**).

Q.8. Do the bones and muscles absorb X-rays equally?

Ans. No, the bones absorb X-rays well because of their higher density, whereas soft tissues like muscle fibers absorb fewer X-rays due to their lower density.

Q.9. Arrange following important tissues/materials in order of their decreasing radiopacity.

Radiopaque dense foreign bodies, bones, muscles, fatty tissue, air, enamel of teeth, radiopaque contrast media.

Ans.
   i. Radiopaque dense foreign bodies and radiopaque contrast media.
   ii. Enamel of teeth.
   iii. Bones (calcific tissues).
   iv. Soft tissues like muscles, heart, kidneys, etc.
   v. Fatty tissue.
   vi. Air (translucent areas, i.e. looks dark in plain X-ray films), as found in fundus of stomach, intestine, trachea, lungs, and paranasal air sinuses.

Radiological Anatomy: Introduction

Q.10. How does a long bone appear in an X-ray film?

Ans.
i. The peripheral **compact part** of the bone appears dense white owing to its high radiopacity.
ii. The **spongy substance** towards the end of the shaft presents interstices of soft tissue density.
iii. The **bone marrow** and **periosteum** are not distinguishable (both of them present soft tissue density).
iv. **Nutrient canal** may appear as an oblique radiolucent line traversing the compact part of the bone.
v. **Articular cartilage** is not distinguishable as such and presents a soft tissue density.

Q. 11. What do you mean by radiological joint space?

Ans. It is the interval between the radiopaque adjacent ends of the two articulating bones which is occupied almost entirely by their articular cartilages.

Q.12. What is the approximate width of this joint space?

Ans. About 2–5 mm in adults.

Q. 13. How do you identify a young (growing) long bone?

Ans. In an X-ray of a young bone the uncalcified epiphyseal plate appears as an irregular, radiolucent band termed as **epiphyseal line**. When the epiphysis and diaphysis are united or fused, the epiphyseal line is no longer seen and the bone is said to be an adult bone.

Q. 14. What do you mean by contrast radiograph?

Ans. It means that the radiographs are taken by using some contrast media.

Q. 15. What is the utility of using contrast media?

Ans. When the density of a structure is very similar to that of the adjacent structures, the contrast media is used to enhance or outline its contours.

Q. 16. What are the types of contrast media?

Ans.
i. Radiolucent media, e.g. air.
ii. Radiopaque media, e.g. barium or iodinated contrast media.

Q. 17. What are the routes of administration of these contrast media?

Ans.
  i. Oral, as in barium meal X-rays.
  ii. Intravenous, as in intravenous pyelography.
  iii. Intra-arterial, as in angiography.
  iv. Intrathecal.
  v. Intraluminal.
  vi. Into various cavities and organs.

Q. 18. What do you mean by double contrast X-rays? Give examples.

Ans. It means that radiography is done by using two contrast medias. Air is used together with barium sulfate for double contrast barium meals and barium enemas.

Q. 19. Where do you find the use of barium compound as a contrast media?

Ans. Barium is used in the form of barium sulfate to highlight the lumen of the gastrointestinal tract.

Q. 20. What are the common areas of using iodinated contrast media?

Ans.
  i. Intravenous pyelography (IVP).
  ii. Arteriograms or venograms.
  iii. Intravenous cholangiography.
  iv. Hysterosalpingography (HSG).

Q. 21. What are the different views used in taking plain radiographic images?

Ans. Common views are:
  i. Anteroposterior (AP view).
  ii. Posteroanterior (PA view).
  iii. Lateral view.
  iv. Oblique view.

Q. 22. How are the views named?

Ans. The views are named for the part of the body that is nearest the X-ray film.
  i. **The anteroposterior view means**: The X-rays are allowed to pass through the object (here part of the body) from front to back, i.e. the source of the X-rays is in front of the object and the film is behind.
  ii. **The posteroanterior view means**: The X-rays are allowed to pass through the object from back to front, i.e. the source of the X-rays is behind the object and the film is in front.

Radiological Anatomy: Introduction

Q. 23. What is the importance of these views?

Ans. The different views are chosen to highlight the particular areas or structures of the body being examined.

Q. 24. How do you describe an X-ray film?

Ans. A description of an X-ray film should include the following points:
  i. Type of X-ray (whether plain X-ray or contrast X-ray).
  ii. Region (which part of the body?).
  iii. View (AP/PA/lateral/oblique?).
  iv. Side (right/left?).
  v. Of whom? (Name of the person written on the X-ray film).
  vi. Date of exposure (written on the film).
  vii. Structures showing _____.

Q. 25. Give an example of description of a plain X-ray film.

Ans. Plain X-ray of abdomen (Fig. 27.1A):

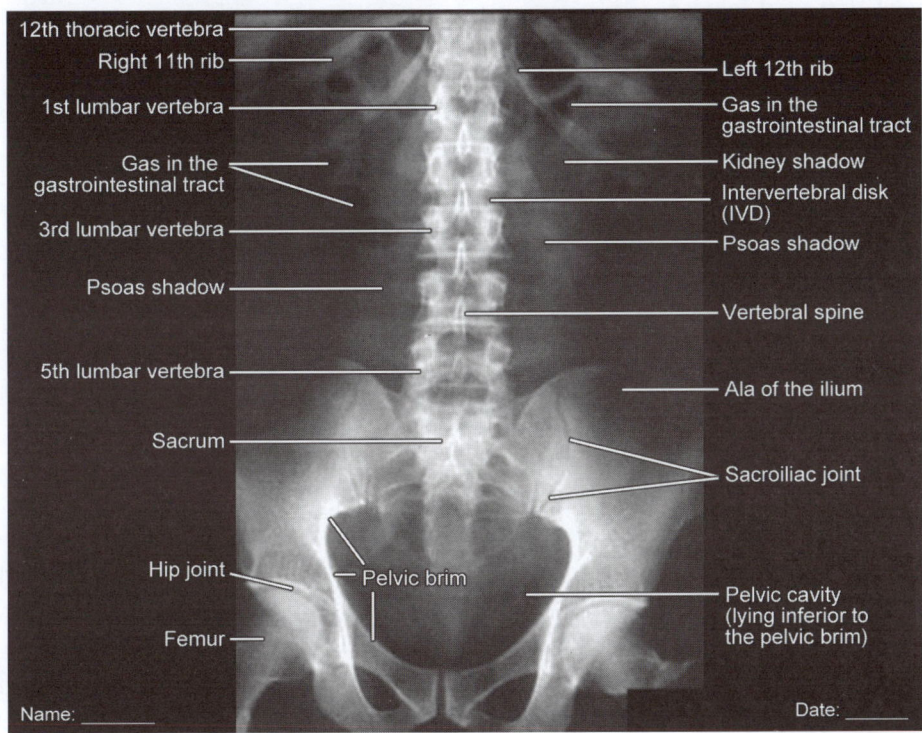

Fig. 27.1A: Plain X-ray of abdomen in anteroposterior (AP) view

Description

This is a straight/plain X-ray (**type of X-ray**) of abdomen (**region**) in anteroposterior view (**view**) of Mr SK Chakraborty (**name**) taken on 26.07.2016 (**date**), showing vertebral column, bony pelvis, lower ribs, sacroiliac joints of both sides, psoas shadow, kidney shadow and gas shadows (**structures**).

Q. 26. Give an example of a contrast X-ray film.

Ans. Barium-meal X-ray of stomach and duodenum (Fig. 27.1B)

Description

This is a barium-meal X-ray (**contrast X-ray**) of stomach and duodenum (**region**) in anteroposterior view (**view**) of Mr A Kulavi (**name**) taken on 23.04.2015 (**date**) showing:
  i. Stomach filled with barium.
  ii. Gastric curvatures.
  iii. Fundic gas shadow.
  iv. Incisura angularis.
  v. Pylorus.
  vi. Duodenal cap.
  vii. Vertebral column.
  viii. Intestinal gas shadow.
  ix. Lower ribs.

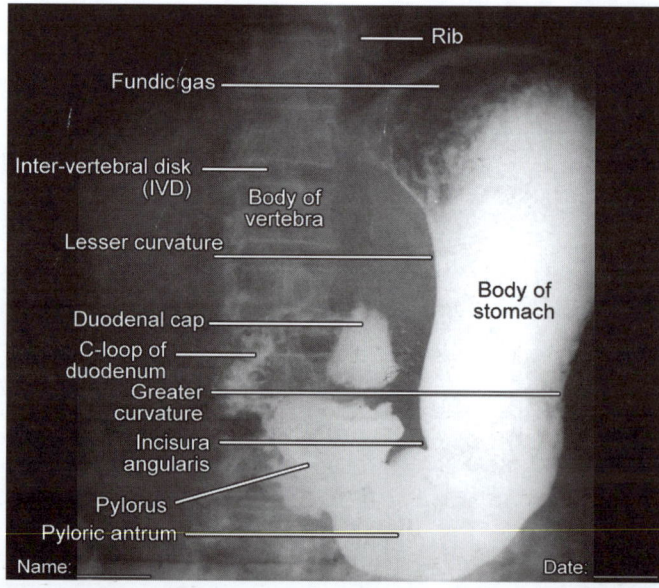

Fig. 27.1B: Barium-meal X-ray of stomach and duodenum in anteroposterior view

CHAPTER 28
Radiological Anatomy: Upper Limb (Superior Extremity)

- Shoulder Region
- Elbow Region
- Region of Wrist and Hand

LESSON 1: SHOULDER REGION

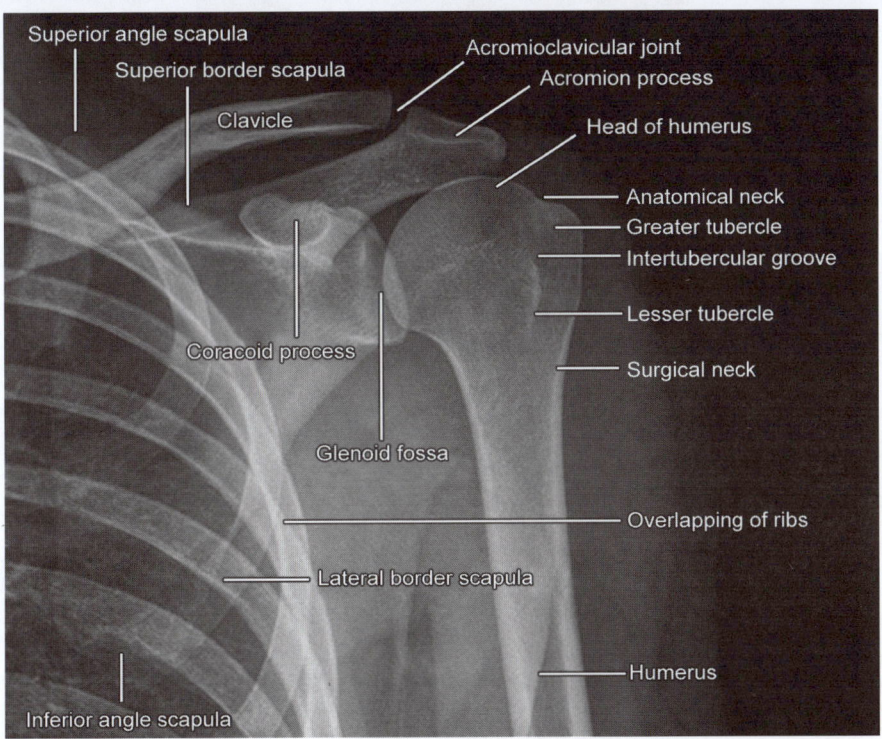

Fig. 28.1A: Shoulder region (anteroposterior view)

What is this film?

Ans. This is a plain (straight) X-ray of shoulder region in anteroposterior (AP) view.

Anteroposterior View

> **Radiographic Appearance—Anteroposterior View (Fig. 28.1A)**
> 1. Spherical head of the humerus is seen in the shallow glenoid fossa of scapula.
> 2. Coracoid process of scapula is seen superimposed on the acromial process.
> 3. Acromial end of the clavicle is seen clearly.
> 4. Acromioclavicular joint is visible.
> 5. Proximal end of the humerus is seen clear off the scapula.
> 6. Scapula and few upper ribs are also obvious in this film though the lateral border of scapula and its vertebral or medial border are superimposed on the ribs.

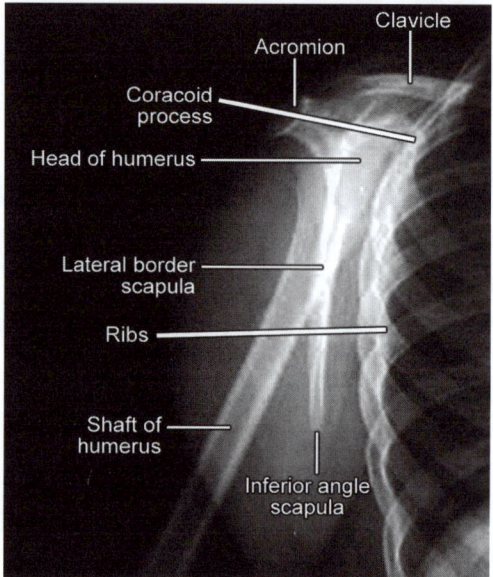

Fig. 28.1B: Shoulder region (lateral view)

 **What is this film?**

Ans. This is a plain X-ray of shoulder region in lateral view.

Lateral View

> **Radiographic Appearance—Lateral View (Fig. 28.1B)**
> 1. Scapula is seen clearly separated from the ribs.
> 2. Coracoid and acromion processes are seen away from one another on either side of the head of the humerus (they are superimposed in AP view).

Radiological Anatomy: Upper Limb (Superior Extremity)

Key points to be identified

1. Shoulder joint.
2. Scapula.
3. Proximal end of humerus.
4. Acromial end of clavicle.
5. Upper ribs.

PROBABLE QUESTIONS AND ANSWERS

1. What do you mean by plain X-ray?

Ans. It means that this particular film was taken without using any contrast media.

2. What do you mean by anteroposterior view?

Ans. It means that the X-rays traverse the patient from anterior to posterior (front to back).

3. What are the bones forming the shoulder joint?

Ans. It is formed by the spherical head of the **humerus** representing the **ball** and the pear-shaped small and shallow glenoid cavity of the **scapula** as the **socket**, hence called glenohumeral joint.

4. What type of joint it is?

Ans. It is a **synovial joint**; subtype being multiaxial **ball and socket variety**.

5. Can you see the joint space?

Ans. Yes, it looks as a small radiolucent (dark) area between the two participating parts of the bones (radiopaque).

6. Why the joint space looks dark (black)?

Ans. The articular cartilages of the either ends of the joint forming parts of the two bones are radiolucent. These cartilages actually represent the joint space.

7. What type of bone the scapula is?

Ans. It is a **flat bone**, triangular in shape.

8. How many ribs are overlapped by the scapula?

Ans. 2nd to 7th ribs.

9. Identify the three angles of scapula.

Ans.
  i. **Lateral angle**, bearing the glenoid cavity that forms the shoulder joint.
  ii. **Superior angle**, junction between medial and superior borders.
  iii. **Inferior angle**, meeting place of medial and lateral borders.

10. Why the inferior angle is considered as an important bony land mark?

Ans. The inferior angle of scapula overlaps the 7th rib and lies at the level of T_7 spine which is a useful guide for the counting of vertebral spine.

11. Identify the spine of scapula and what is its importance as a bony landmark?

Ans. Being a triangular bony projection, its base forms the posterior boundary of spino-glenoid notch and apex (medial border of spine) lies at the level of T_3 spine.

12. What are the other bony processes of scapula other than the spinous process?

Ans.
  i. **Coracoid process.**
  ii. **Acromial process.**

13. What are the parts of coracoid process?

Ans. Two parts:
  i. Vertical part.
  ii. Horizontal part.

14. What are the muscles that arise from its tip?

Ans.
  i. Short head of biceps brachii (laterally).
  ii. Coracobrachialis (medially).

15. What is acromial process?

Ans. It is a forward projection of a flattened plate of bone from the lateral end of the spine.

16. What do you mean by acromial angle?

Ans.
  i. It is the meeting place of lower lip of the spine and lateral border of acromial process.
  ii. It is an important subcutaneous bony landmark for surface anatomy.

17. Identify the different parts of the proximal end of humerus.

Ans.
  i. Head.
  ii. Greater tubercle.
  iii. Lesser tubercle.
  iv. Intertubercular sulcus.
  v. Neck.
  vi. Upper part of the shaft.

Radiological Anatomy: Upper Limb (Superior Extremity)

18. What type of epiphysis the head of the humerus is?

Ans. Pressure epiphysis.

19. What type of epiphysis the lesser and greater tubercles are?

Ans. Both these tubercles are examples of **traction epiphysis**.

20. At what age do these secondary centres of ossification arise?

Ans.
i. Head → 1st year.
ii. Greater tubercle → 2nd year.
iii. Lesser tubercle → 5th year.

21. Why the upper end of humerus is a compound epiphysis?

Ans. The above mentioned three secondary centers unite among themselves in the 6th year of age to form a **compound epiphysis** which then unites with the shaft at about the 20th year of age.
(**Remember** that in simple epiphysis the individual center unites with the diaphysis separately unlike the compound epiphysis.)

22. Which of these bony prominences forms the rounded contour of the shoulder?

Ans. Convex lateral surface of the greater tubercle.

23. Identify the neck.

Ans. The constriction succeeding the head is the **anatomical neck**.

24. Can you identify the surgical neck?

Ans. Surgical neck is the constriction between the expanded upper end and cylindrical shaft of the humerus.

25. Why it is so named?

Ans. It is a common site of fracture in the upper 1/3rd of the humerus. Surgical neck is embraced by the axillary nerve and posterior circumflex humeral vessels. The axillary nerve involvement in fracture of surgical neck of humerus may cause paralysis of deltoid muscle.

26. With what structure does the acromial end of the clavicle articulate?

Ans. It articulates with the acromial process of scapula forming **acromioclavicular joint**.

27. What type of joint it is?

Ans. Plane synovial joint.

28. What are bones of shoulder girdle?
Ans. Clavicle and scapula.

29. What are the joints of shoulder girdle?
Ans. Sternoclavicular and acromioclavicular joint.

30. Can you see the ribs clearly?
Ans. No, the ribs are overlapped by the scapula.

31. Can you observe the intercostal spaces?
Ans. Yes, the spaces are visible in between the ribs.

LESSON 2: ELBOW REGION

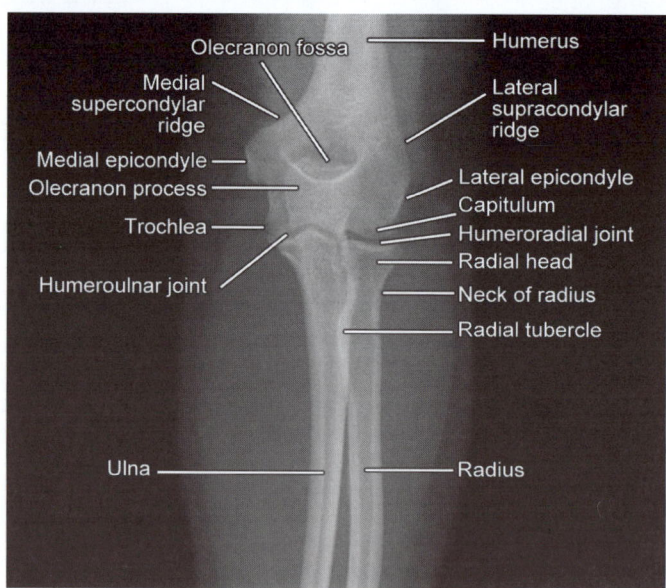

Fig. 28.2A: Plain X-ray elbow (anteroposterior view)

What is this film?
Ans. This is a plain X-ray of elbow region in anteroposterior view (Fig. 28.2A).

Anteroposterior View

> **Radiographic Appearance—Anteroposterior View (Fig. 28.2A)**
> 1. Distal end of the humerus with its both condyles are well seen.
> 2. Olecranon process of ulna is the superimposed on the lower end of humerus.
> 3. Upper end after ulna is seen separated from the upper end after radius.
> 4. Joint space of the elbow joint is well seen.
> 5. Superior radioulnar joint is also visible.

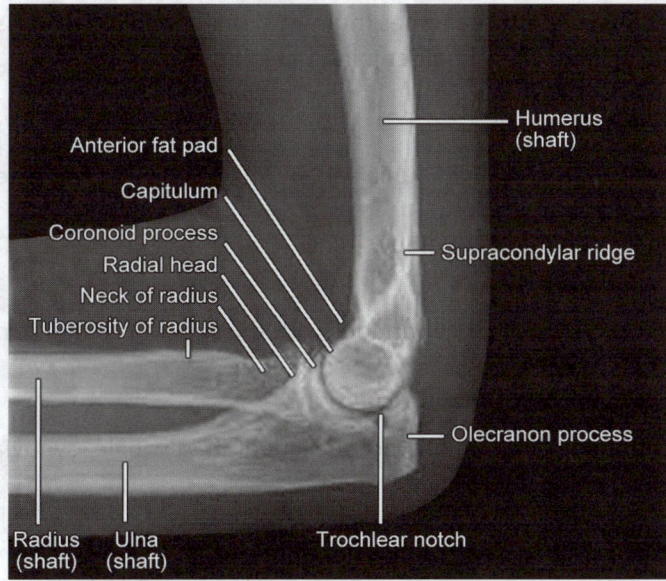

Fig. 28.2B: Plain X-ray elbow (lateral view)

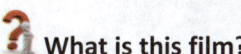 **What is this film?**

Ans. This is a plain X-ray of elbow region in lateral view.

Lateral View

> **Radiographic Appearance—Lateral View (Fig. 28.2B)**
> 1. Individual condyle of the humerus is not clearly seen as they are superimposed on each other.
> 2. Olecranon process of ulna is well seen.
> 3. Coronoid process of ulna is superimposed on the part of the head of the radius.

Key points to be identified
1. Elbow joint.
2. Superior radioulnar joint.
3. Distal end of humerus.
4. Proximal ends of radius and ulna.

PROBABLE QUESTIONS AND ANSWERS

1. Why the elbow joint is called a compound joint?

Ans. Because it is formed by three bones (humerus, radius and ulna). In a simple joint only two bones articulate, e.g. shoulder joint, hip joint, etc.

2. What type of joint the elbow is?

Ans. Type: Synovial; **Subtype:** Hinge (uniaxial), permitting flexion and extension.

3. What are the components of elbow joint?

Ans.
 i. Humeroulnar.
 ii. Humeroradial which is continuous with the superior radioulnar joint.

4. What do you mean by cubital articulation?

Ans. Above mentioned three joints are included in a single synovial envelop and collectively known as cubital articulation.

5. Identify the condyle and epicondyles of humerus.

Ans. The lower end of the humerus is known as the condyle. Medical epicondyle is a blunt projection from the medical aspect of the condyle and the lateral epicondyle is a less prominent projection from the lateral aspect of the condyle.

6. What are the articular parts of the condyle of the humerus?

Ans.
 i. **Capitulum** on the lateral part, articulates with upper surface of the head of the radius.
 ii. **Trochlea** on the medial part, articulates with the trochlear noch of the ulna.

7. What type of joint is the superior radioulnar joint?

Ans. Type: Synovial; **subtype:** Uniaxial pivot.

[Remember that superior and inferior radioulnar joints are synovial joints, whereas the middle radioulnar joint is a fibrous joint (syndesmosis)].

8. Identify the nonarticular parts of the lower end of humerus.

Ans. Nonarticular parts are:
  i. Two epicondyles—medial and lateral.
  ii. Three fossae—radial, coronoid and olecranon.
   - **Radial fossa** lies anteriorly, above the capitulum.
   - **Coronoid fossa** lies anteriorly, medial to the radial fossa.
   - **Olecranon fossa** lies posteriorly, above the trochlea.

But these three fossae cannot be demarcated properly due to their overlapping in the X-ray film.

9. What do you mean by carrying angle?

Ans. It is an obtuse angle (about 163°) between the long axis of the arm and forearm in fully-extended elbow and supinated forearm.

10. What are the anatomical causes in the formation of this angle?

Ans.
  i. Medial edge of the trochlea extends about 6 mm below the lateral edge.
  ii. Obliquity of the superior articular surface of coronoid process of ulna.

11. What is the relative position of the two forearm bones?

Ans. **Ulna** lies medial to **radius**.

12. Which of the leg bones correspond with the radius and ulna?

Ans. Radius corresponds with the tibia of lower limb (preaxial bone). Ulna corresponds with the fibula of lower limb (postaxial bone).

13. Identify the parts of the upper end of the radius.

Ans.
  i. Head.
  ii. Neck.
  iii. Radial tuberosity.

14. How do you identify the neck?

Ans. The constricted area succeeding the head is the neck.

15. How do you identify radial tuberosity?

Ans. The radial tuberosity lies in the lower part of the neck on its medial side.

16. When does the center of ossification appear for the head?

Ans. The secondary center for the head of the radius appears in the 5th year of age and unites with the shaft in the 18th year.

17. Identify the parts of the upper end of the ulna.

Ans.
i. **Olecranon process**—projecting upward and its beak-like summit fits in the olecranon fossa of the lower end of the humerus in extended elbow.
ii. **Coronoid process**—projecting forward and its tip comes in contact with the coronoid fossa of the humerus in flexed elbow.
iii. **Trochlear notch**—formed by the anterior surface of the olecranon process and superior surface of the coronoid process, articulates with the trochlea of the lower end of the humerus, thereby forming the **humeroulnar joint**.
iv. **Radial notch**—present on the lateral aspect, articulates with the medial part of the pheripheral margin of the radial head, thereby forming the superior radioulnar joint.

18. Can you identify the ulnar tuberosity?

Ans. Just below the anterior surface of the coronoid process, the elevated area is the ulnar tuberosity.

19. What are the muscular attachments of radial and ulnar tuberosities?

Ans.
i. In radial tuberosity—insertion of biceps brachii.
ii. In ulnar tuberosity—insertion of brachialis.

20. What is the time of appearance of secondary center of ossification for the upper end of ulna?

Ans. 10th year of age and it unites with the shaft in 16th year.

21. What are the epiphyseal centers (secondary centers) for the different parts of the lower end of the humerus?

Ans.
i. Medial epicondyle—5th year.
ii. Medial part of trochlea—10th year.
iii. Lateral part of trochlea with capitulum—2nd year.
iv. Lateral epicondyle—12th year.

Radiological Anatomy: Upper Limb (Superior Extremity)

 22. When do they unite with the shaft of the humerus?

Ans. Medial epicondyle unites with the shaft at about 16th year by a separate epiphyseal line. Other three centers unite themselves to form a compound epiphysis and then unite with the shaft by a separate epiphyseal line at about 18th year.

Remember that the lower end of the humerus possesses two separate epiphyseal lines with an intervening wedge-shaped diaphysis on which lies the ulnar nerve.

LESSON 3: REGION OF WRIST AND HAND

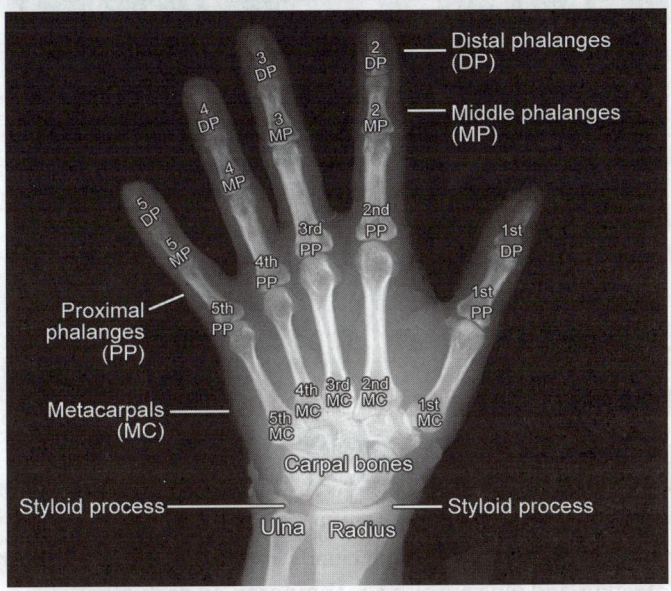

Fig. 28.3A: Wrist and hand (dorsipalmar view)

 What is this film?

Ans. This is a plain x-ray of the region of the wrist and hand in dorsipalmar view (Fig. 28.3A) and lateral view (Fig. 28.3B).

Radiographic Appearance—Dorsipalmar View (Fig. 28.3A)

1. Distal ends of radius and ulna are seen to lie side by side.
2. Very narrow wrist joint space is seen.
3. Carpals, metacarpals and phalanges are also seen.
4. Intercarpal, intermetacarpal, carpometacarpal, metacarpophalangeal and interphalangeal joints are visible.

Exam-Oriented Practical Anatomy

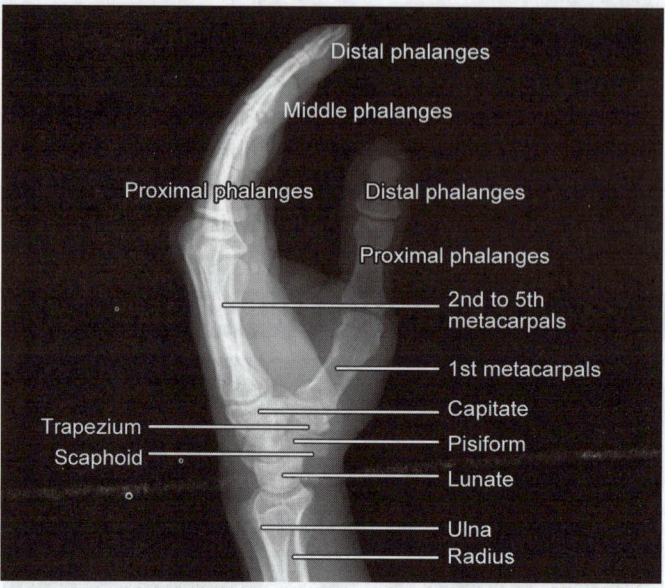

Fig. 28.3B: Wrist and hand (lateral view)

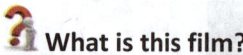 **What is this film?**

Ans. This is a plain X-ray of the region of wrist and hand in lateral view.

Radiographic Appearance—Lateral View (Fig. 28.3B)
1. Carpal bones are superimposed on each other.
2. Individual metacarpal bones are also not clearly visible due to superimposition.
3. Lower ends of radius and ulna are also superimposed on each other.

Key points to be identified
1. Wrist joint.
2. Lower ends of radius and ulna.
3. Carpals, metacarpals and phalangeal bones.

PROBABLE QUESTIONS AND ANSWERS

1. What type of joint the wrist is?

Ans. Type: Synovial; **Subtype:** Biaxial ellipsoid joint.

Radiological Anatomy: Upper Limb (Superior Extremity)

2. What are the participating bones to form this joint?

Ans. Proximally:
  i. Articular surface of the lower end of radius.
  ii. Articular disk of the inferior radioulnar joint **(ulna does not participate to form this joint).**

Distally:
  i. Scaphoid, lunate and triquetral bones.
  ii. Interosseous ligament connecting these three carpal bones.

3. Identify the styloid processes of radius and ulna.

Ans.
  i. The radial styloid process is the downward projection from the lateral surface of the lower end of radius.
  ii. The ulnar styloid process is a small downward projection from the posteromedial aspect of the head of the ulna.

4. Which of these two styloid processes lie at a lower level?

Ans. The radial styloid process lies about 1 cm below the ulnar styloid process.

5. How do you determine the age of a person from the radiological view of the lower ends of radius and ulna?

Ans.
  i. The secondary center of ossification for the lower end of radius appears in the 1st year and unites with the shaft in the 20th year. So, before 20th year (year of union), there would be an epiphyseal line (looks dark in X-ray film) between the diaphysis (shaft) and epiphysis (lower end) of the radius.
  ii. The secondary center for the lower end of ulna appears in the 5th year and unites with the shaft in the 15–16th year. So, before 16th year (year of union), there would be an epiphyseal line between the diaphysis and epiphysis of the ulna.

Remember that the lower end of the radius and ulna are their growing ends (**the law for growing end is**: The epiphyseal center which appears first, unites last with the shaft. The end of a long bone in which union occurs late is the growing end).

6. Identify the carpal bones.

Ans.
  i. Bones of **proximal row** (lateral to medial): Scaphoid, lunate, triquetral and pisiform.
  ii. **Bones of distal row:** (lateral to medial): Trapezium, trapezoid, capitate and hamate.

7. Which of these carpal bones ossify first and last?

Ans. The **capitate** (largest carpal bone) ossifies first and **pisiform** (smallest carpal bone) ossifies last.

Remember that all the carpal bones ossify after birth.

8. What is the chronological order of appearance of ossification centers of the carpal bones?

Ans.
  i. Capite (largest) → 2–3 months.
  ii. Hamate (wedge-shaped) → 2–3 months.
  iii. Triquetral (wedge-shaped) → 3rd year.
  iv. Lunate (semilunar in shape) → 4th year.
  v. Scaphoid (boat-shaped) → 5th year.
  vi. Trapezium (quadrilateral) → 6th year.
  vii. Trapezoid (boot-shaped) → 7th year.
  viii. Pisiform (pea-shaped) → 10–12th year.

9. What type of bones the carpals are?

Ans. The carpals are short bones.

10. How do you count the metacarpal bones?

Ans. The metacarpals are counted from lateral to medial side as 1st to 5th metacarpals.

11. What is the type of these bones?

Ans. All the metacarpals and phalanges are short (miniature) long bones.

12. What do you mean by short long bones?

Ans. The short long bones possess a diaphysis and a single epiphysis at one end of the bone.

13. Where do you find the epiphysis of these bones?

Ans. All the metacarpal bones possess the epiphysis towards their heads except the 1st metacarpal bone.

14. Where do you find the epiphysis of the 1st metacarpal bone?

Ans. The 1st metacarpal bone possesses the epiphysis at its base like that of all the phalanges. Therefore, the 1st metacarpal bone is called a **modified phalanx** by some authors.

15. How many phalanges are there?

Ans.
  i. In the thumb: 2 (proximal and distal).
  ii. In other fingers: 3 (proximal, middle and distal).
     Total: 2 + 4 × 3 = 14.

16. What is the position of the 1st metacarpal bone in relation to the other metacarpal bones?

Ans. 1st metacarpal bone lies in a more anterior plane than the others.

17. What is the axial-line of the hand?

Ans. The axial line represents a line which passes through the middle finger, 3rd metacarpal and capitate bone.

CHAPTER 29
Radiological Anatomy: Lower Limb (Inferior Extremity)

- Hip Region
- Knee Region
- Ankle and Foot

LESSON 1: HIP REGION

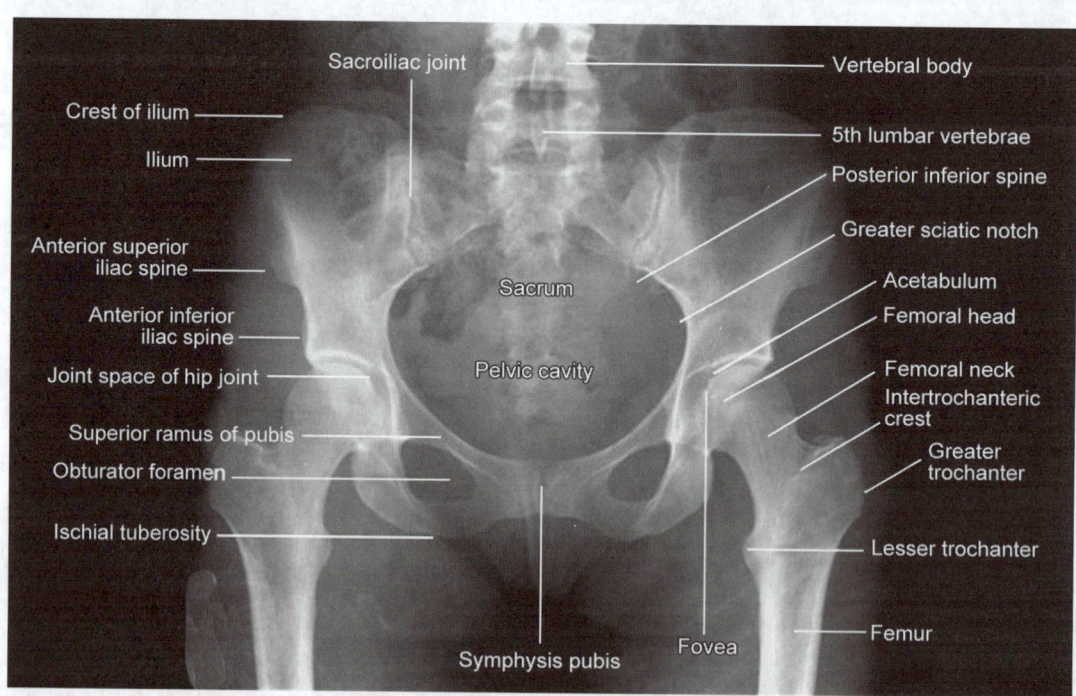

Fig. 29.1: Hip region (anteroposterior view)

Radiological Anatomy: Lower Limb (Inferior Extremity)

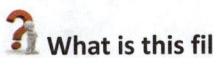 **What is this film?**

Ans. This is a plain (straight) X-ray of bony pelvis and upper end of femur in anteroposterior view.

Anteroposterior View

Radiographic Appearance—Anteroposterior View (Fig. 29.1)

1. Head of the femur is seen in the acetabulum of the hip bone (hip joint).
2. Three main parts of the hip bone (ilium, ischium and pubis) are seen separated from each other.
3. Head of the femur is partially superimposed on the acetabulum.
4. Joint space of hip joint (a curved dark area) is seen between the head of the femur and the acetabulum of the hip bone.
5. Neck, both trochanters and part of the proximal end of the shaft of the femur are well visible.
6. Sacrum is seen with its foramina.
7. Sacroiliac joints are seen separating the sacrum from the ilium of the hip bone.

Key points to be identified
1. Hip joints
2. Hip bones
3. Upper end of femur
4. Sacroiliac joints.

PROBABLE QUESTIONS AND ANSWERS

1. What are the bones forming the hip joint?

Ans.
  i. **Head of the femur**, representing the **ball**.
  ii. **Acetabulum** of the hip bone as the **socket**.

2. Can you show the joint space?

Ans. Yes, it is visible as a dark (black) area (radiolucent) between the two participating bones (radiopaque).

3. Why this space looks dark?

Ans. The articular cartilages of the two bones which represent the joint space are radiolucent *(all radiolucent areas look dark in an X-ray film)*.

4. What type of joint the hip is?

Ans. It is a **synovial** joint; subtype being multiaxial **ball and socket** variety.

5. What type of bone the hip is?

Ans. Hip bone is an irregular bone.

6. Identify the important parts of hip bone.

Ans.
  i. **Ilium** (upper expanded plate of bone).
  ii. **Ischium** (dorsolateral to the pubis).
  iii. **Pubis** (ventromedial to the ilium and ischium).
  iv. **Acetabulum** (the cup-shaped depression).
  v. **Obturator foramen** (below and in front of the acetabulum).

7. What is triradiate cartilage?

Ans. It is a **Y-shaped epiphyseal plate** between the ilium-ischium, ischium-pubis and pubis-ilium in the acetabulum which appears as a radiolucent area in the X-ray film.

8. When does the center of ossification for this cartilage appear?

Ans. **Two secondary centers** for the triradiate cartilage appear at the onset of puberty.

9. When this cartilage is replaced by bone?

Ans. Replacement of triradiate cartilage (acetabular fusion) occurs at 20–25 years of age.

10. What are the other secondary centers of ossification in hip bone?

Ans. Other than two centers for triradiate cartilage, other centers are:
  i. Two centers for iliac crest
  ii. One center for ischial tuberosity.
  iii. One center for anteroinferior iliac spine.
  iv. One center for symphyseal surface of pubis
All these centers fuse at 20–25 years of age.

11. When does this ischiopubic ramus fuse?

Ans. Ischiopubic ramus fuses with each other at 7–8 years of age.

12. How many primary centers of ossification are three in a hip bone?

Ans. Three centers:
   i. One for ilium (2nd month of IUL).
   ii. One for ischium (4th month of IUL).
   iii. One for pubis (4th–5th of IUL).

13. Identify the following parts of the hip bone.

Ans.
   i. Iliac crest and iliac spine.
   ii. Ischial spine.
   iii. Sciatic notches.
   iv. Ischial tuberosity.
   v. Ischiopubic rami.

14. What are the participating bones of bony pelvis?

Ans.
   i. Both hip bones (anteriorly and either sides).
   ii. Sacrum and coccyx (posteriorly).

15. Show the pubic angle and what is its importance?

Ans. The **pubic angle**, also called the subpubic angle is formed between the inferior ramus of the pubic bones of both sides. Its importance lies in the differentiation of sex in pelvic anatomy. In males, it is generally 50–60° and in females it is generally 80–90°.

16. How much is the radiological joint space in the symphyseal joint?

Ans. It is <5 mm.

17. What type of joint it is?

Ans. Secondary cartilagenous or symphyseal joint.

18. Show the sacroiliac joint and what is its type?

Ans. It is a plane synovial joint.

19. What is the importance of this joint?

Ans. The weight transmits through this joint from the trunk to the lower limb. So, it is more stable than mobile.

20. How much is the radiological joint space of this joint?

Ans. About 2–4 mm.

21. What is Shenton's line?

Ans. It is a curved imaginary line, drawn on a plain anteroposterior X-ray film of bony pelvis along the upper margin of obturator foramen and inferomedial border of the neck of the femur *(It is a continuous curved line)*. It was first described by an English radiologist **Edward Warren Hine Shenton.**

22. What is the clinical importance of this line?

Ans. Interruption of this line may indicate:
   i. Fractured neck of the femur.
   ii. Developmental dysplasia of the hip.

23. Identify the presenting parts of the upper end of femur.

Ans.
   i. Head.
   ii. Neck.
   iii. Greater trochanter.
   iv. Lesser trochanter.
   v. Intertrochanteric line and crest.

24. How many epiphyseal centers are there in the upper end and when do they fuse with the diaphysis (shaft)?

Ans. Three epiphyseal or secondary centers; one each for:
   i. Head—appears at about 1st year and fuses at about 18th year.
   ii. Greater trochanter—appears at about 4th year and fuses at about 17th year.
   iii. Lesser trochanter—appears at about 14th year and fuses at about 16th year.

Note that these three centers appear at different ages after birth and all are separated from the diaphysis by three separate epiphyseal plates. They fuse with the diaphysis separately at different ages of life. Therefore, these epiphyses are **simple epiphyses** unlike the compound epiphysis at the upper end of humerus.

25. What is the functional classification of these three epiphyses?

Ans. i. Head: **Pressure epiphysis** (weight is transmitted through the head).
ii. Both trochanters: **Traction epiphysis** (muscle pull at these points causes the appearance of trochanters).

26. What about the ossification of the neck of the femur?

Ans. Neck is the upward extension of the shaft (diaphysis) which is ossified from **one primary center** at about 7th–8th week of intrauterine life.

27. Identify the epiphyseal plate between the head and neck and what is its alignment?

Ans. This epiphyseal plate is horizontal in early part of the life but gradually becomes oblique by the age of 8–12 years.

28. What do you mean by neck-shaft angle (angle of inclination)?

Ans. It is the angle formed between the long axis of the neck and the long axis of the shaft of femur.

29. What is its normal value?

Ans. About 125° in adults and 160° in a child.

30. What is its importance?

Ans. In different pathological or clinical conditions this angle may be decreased or increased.

31. What is coxa vara or coxa valga?

Ans. The condition of diminished neck-shaft angle is known as coxa vara. The condition of increased neck-shaft angle is known as coxa valga.

32. What is calcar femorale?

Ans. It is a thin but dense vertical plate of bone which extends upward from the compact part of linea aspera to the interior of the neck.

33. What is its function?

Ans. It provides strength to the femoral neck. Its osteoporosis in old age makes the femoral neck susceptible to fracture even with minor trauma.

LESSON 2: KNEE REGION

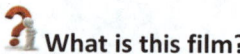

Fig. 29.2A: Plain X-ray of knee region (anteroposterior) view

What is this film?

Ans. This a plain X-ray of knee region in anteroposterior view (Fig. 29.2A).

Anteroposterior View

Radiographic Appearance—Anteroposterior View (Fig. 29.2A)

1. Lower end of femur is seen clearly.
2. Patella is superimposed on the lower end of femur.
3. Upper end of tibia and fibula are visible.
4. Joint space of knee is very clear.
5. Tubercles of intercondylar eminence of tibia are very clear in the film.
6. Upper end of fibula is partly superimposed on tibia at the superior tibiofibular joint.

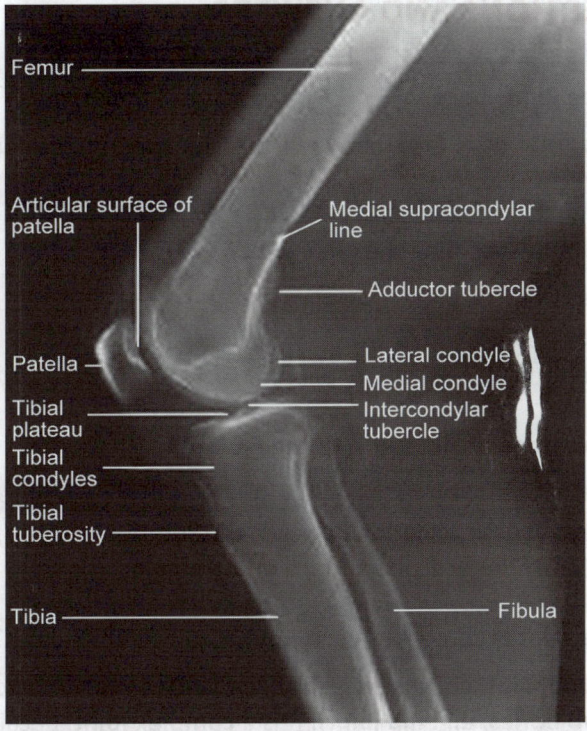

Fig. 29.2B: Plain X-ray of knee region (lateral view)

What is this film?

Ans. This is a plain X-ray of knee region in lateral view.

Radiographic Appearance—Lateral View (Fig. 29.2B)
1. Condyles of the femur are superimposed on each other.
2. Patella is clearly seen away from the femur with the femoropatellar joint space.
3. Upper end of tibia is superimposed on a small area of the upper end of fibula.
4. Knee joint space is visible.

Key points to be identified
1. Lower end of femur.
2. Upper end of tibia and fibula.
3. Patella.
4. Knee joint.
5. Superior tibiofibular joint.

PROBABLE QUESTIONS AND ANSWERS

1. Show the knee joint space.

Ans. It is the horizontal dark area between the condyles of the femur above and the condyles of the tibia below.

2. What ate the bones forming the knee joint?

Ans.
i. Condyles of the **femur**.
ii. Condyles of the **tibia**.
iii. Posterior articular surface of **patella**.

3. How many components are there in knee joint?

Ans.
i. **Condylar articulation** between the condyles of the femur and tibia.
ii. **Femoropatellar articulation** between the anterior articular surface of the femur and the posterior articular surface of the patella.

4. Why the knee joint is known as a compound and complex joint?

Ans. It is a **compound joint,** because it is formed by more than two bones (**Simple joint is formed by two bones, e.g. hip joint**). It is a **complex joint**, because the femorotibial articulation is subdivided incompletely by the menisci (semiulnar cartilage, a type of fibrocartilage).

5. What type of joint the knee is?

Ans. **Type**: Synovial; **Subtype:** Uniaxial condylar (modified hinge) joint.

6. Why the knee joint is a modified hinge joint?

Ans. A hinge joint is one where the flexion-extension movement occurs around a fixed transverse axis. But in the knee, the transverse axis is not fixed, rather it moves forward and backward during extension and flexion respectively. Moreover, a **conjunct rotation** of femur on tibia or tibia on femur occurs around a more or less vertical axis.

7. Which of the two condyles of the femur is more massive?

Ans. The lateral condyle is more massive than its medial counterpart, because it shares more in weight transmission from femur to tibia.

8. What type of bone is the patella?

Ans. It is the **largest sesamoid bone** in the human body.

9. What are the characteristics of a sesamoid bone?

Ans.
i. It ossifies from secondary center of ossification, i.e. after birth.
ii. It ossifies in a tendon.
iii. It is devoid of periosteum (so, no chance of union following fracture).
iv. It is devoid of Haversian system.

10. Can you remember other sesamoid bones in the body?

Ans.
i. **Fabella** (in the tendon of lateral head of gastrocnemius).
ii. Two sesamoid bones (unnamed) below the head of 1st metatarsal bone (in the tendon of flexor hallucis brevis).
iii. One sesamiod bone at the cuboid (in the tendon of peroneus longus).
iv. **Pisiform** in the hand (in the tendon of flexor carpi ulnanis).

11. Identify the epicondyles of femur.

Ans. Most projecting part on the lateral surface of lateral condyle is the **lateral epicondyle**. Most projecting part on the medial surface of medial condyle is the **medial epicondyle**.

12. Identify the adductor tubercle.

Ans. It is a small bony projection above the medial epicondyle of femur where the medial supracondylar line ends.

13. What are the importance of this tubercle?

Ans.
i. The epiphyseal line of the lower end of femur passes through this tubercle.
ii. It is an important land mark for surface anatomy.
iii. It is the site of attachment of ischial fibers of adductor magnus and tibial collateral ligament.

14. Why the tibia is more massive than fibula?

Ans. The tibia, representing the preaxial leg bone transmits much of the body weight than the postaxial leg bone fibula.

15. Identify the intercondylar area of the tibia.

Ans. It is an area between the articular surfaces of the condyles of the tibia, which presents an intercondylar eminence.

16. Identify the tibial tuberosity.

Ans. It is a bony projection from the anterior aspect of the upper end of the tibia, which is better seen in lateral view of X-ray.

17. What is its importance?

Ans. It belongs to the upper epiphysis as the upper epiphyseal line descends to include this tuberosity.

18. What is attached here?

Ans. Ligamentum patellae.

19. When do the epiphyseal centers appear at the upper end of tibia and lower end of femur?

Ans. Epiphyseal center (secondary center) one each for the lower end of the femur and upper end of tibia appears just before birth in 9th months of intrauterine life.

20. What is the importance of these epiphyseal centres ?

Ans.
  i. Though these centers appear from secondary centers of ossification, they appear before birth (exception to the law of ossification). **The law is:** Primary centers of ossification appear before birth and secondary centers appear after birth.
  ii. It bears a medicolegal importance to determine the maturity of the fetus.

21. What is the growing end of femur and tibia?

Ans. The lower end of femur and upper end of tibia, i.e. the ends around the knee are the growing ends, because fusion occurs late (at about 20 years) at these ends in comparison to the other ends (at about 17th to 18th year).

22. What are the joint forming parts of bones of superior tibiofibular joint?

Ans.
  i. Articular facet on the posterolateral part of the lateral condyle of **tibia** and
  ii. Articular surface of the head of fibula.

23. How many tibiofibular joints are there and what are their types?

Ans.
  i. Superior tibiofibular joint: Plane synovial type.
  ii. Middle tibiofibular joint: Syndesmosis type of fibrous joint.
  iii. Inferior tibiofibular joint: Syndesmosis type of fibrous joint.

Radiological Anatomy: Lower Limb (Inferior Extremity)

24. What is the anatomical angle of knee joint?

Ans. It is an obtuse angle of about 170–175°, formed between the long axis of the shaft of femur and tibia, which opens laterally.

25. What is genu varum or bow legs?

Ans. It is a deformity, wherein there is outward bowing of the legs at the knee and the anatomical angle exceeds 180°.

26. What is genu valgum or knock knee?

Ans. It is a deformity in which the anatomical angle is less than 165° and the knees touch one another when the legs are straightened.

27. What is genu recurvatum or knee hypertension or back knee?

Ans. It is a deformity in which excessive extension occurs in femorotibial joint. Normal range of extension of knee is up to about 10°. This deformity may lead to osteoarthritis of knee joint.

28. What is the clinical significance of these deformities?

Ans. These deformities are predisposing factors for osteoarthritis of knee joint.

29. What are the anatomical factors that prevent the inherent tendency of lateral displacement of patella?

Ans.
    i. The lateral condyle of the femur is more massive and projected forward.
    ii. The prolonged and lower attachment of vastus medialis along the medial border of patella exerts more medial traction on patella.

30. What is 'Q' angle?

Ans. It is the angulation, formed by the intersection of lines between the anterior superior iliac spine to the midpoint of patella and between the midpoint of tibial tuberosity to the centre of the patella extending upwards.

31. What is its normal value?

Ans. It is more in females (about 15°) due to wider pelvis than in males (about 10°).

32. What is the significance of Q angle?

Ans. Increased Q angle predisposes to lateral dislocation or subluxation of patella.

LESSON 3: ANKLE AND FOOT

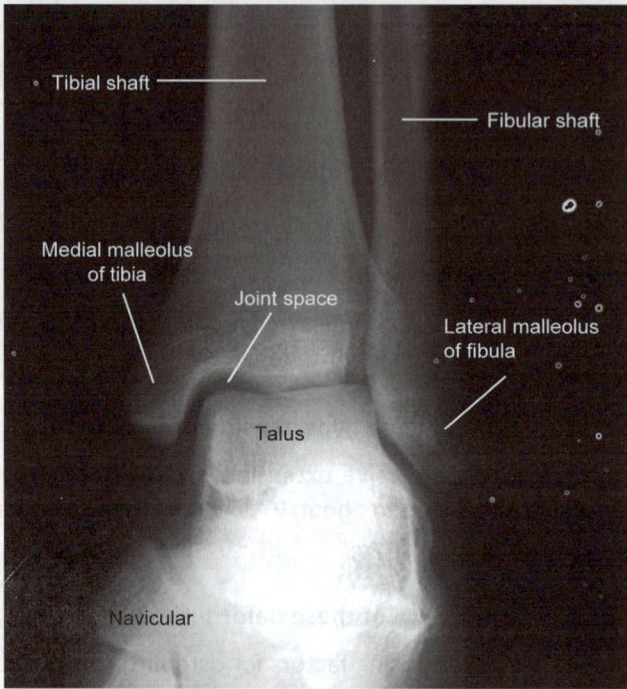

Fig. 29.3A: Ankle joint (anteroposterior view)

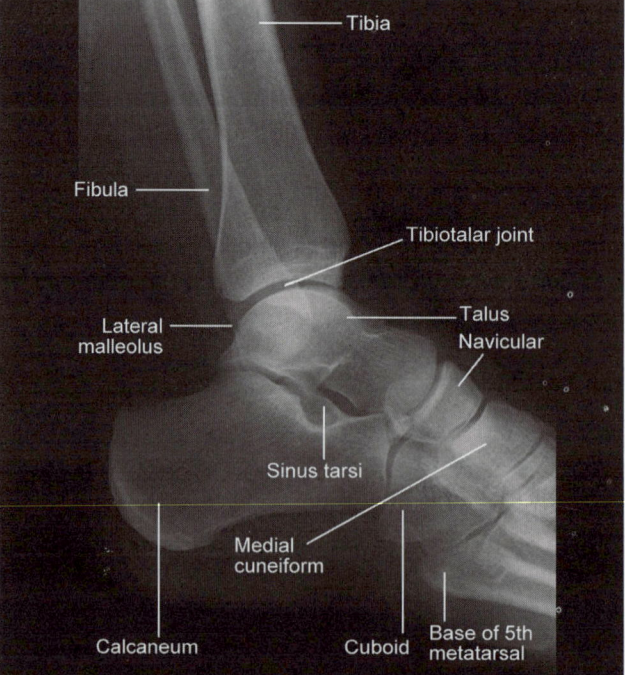

Fig. 29.3B: Ankle and proximal part of foot (lateral view)

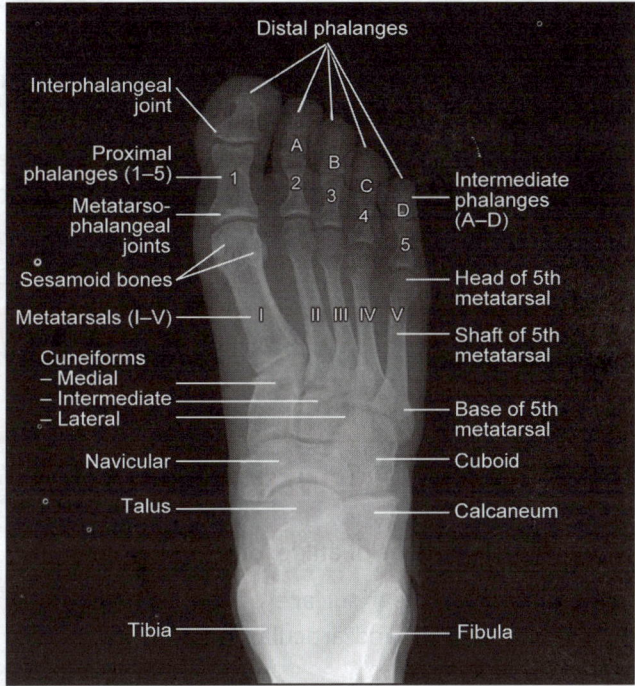

Fig. 29.3C: Dorsiplantar view of foot

❓ What are these films?

Ans. These are plain X-rays of ankle joint and foot in anteroposterior view (Fig. 29.3A), lateral view (Fig. 29.3B) and dorsiplantar view (Fig. 29.3C) of foot.

Anterolateral View

Radiographic Appearance—Anteroposterior View (Fig. 29.3A)

1. Lower ends of tibia and fibula are seen separated from each other except at the area of inferior tibiofibular joint.
2. Ankle joint space between the talus inferiorly and the tibia and fibula with their malleoli in the superior aspect and sides is clearly visible in the film.
3. Upper part of talus is seen separated from the other tarsal bones.

Lateral View

Radiographic Appearance—Lateral View (Fig. 29.3B)

1. Lower ends of tibia and fibula are seen superimposed on each other.
2. Body of the talus is partly superimposed on the extreme lower ends of the tibia and fibula.
3. Malleoli of tibia and fibula are superimposed on each other.
4. Calcaneum, head of the talus, navicular bone are well visible.

Dorsiplantar View

> **Radiographic Appearance—Dorsiplantar View (Fig. 29.3C)**
> 1. All the metatarsals and phalanges are clearly seen.
> 2. Tarsal bones are also seen clearly with partial superimposition among them.
> 3. In dorsiplantar oblique view, 1st and 2nd metatarsals are partly superimposed but other three MTS are clearly separated from each other.
> 4. Cuboid is seen clearly but the all three cuneiforms are superimposed on each other.

Key points to be identified
1. Lower ends of tibia and fibula
2. Ankle joint
3. Tarsals, metatarsals and phalangeal bones.

PROBABLE QUESTIONS AND ANSWERS

1. What are the participating bones of ankle joint?

Ans.
i. **Proximally,** inferior articular surface of the lower end and lateral surface of medial malleolus of **tibia** and articular facet on the medial surface of lateral malleolus of **fibula.**
ii. **Distally,** upper surface (trochlear surface) of the body of **talus.**

2. What is tibiofibular mortice?

Ans. **Mortice** means a recess designed to receive a corresponding projection so as to join or lock two parts together. Here, the tibiofibular mortice is formed by the articular surfaces of tibia and fibula to receive the body of the talus below.

3. Which of the malleoli lies at a lower level?

Ans. The tip of the lateral malleolus of fibula lies about 2 cm below the tip of the medial malleolus of tibia.

4. What is the level of epiphyseal line of these two malleoli?

Ans. The epiphyseal line of medial malleolus is above the joint space of ankle while the epiphyseal line of the lateral malleolus corresponds with the joint space.

5. What type of joint the ankle is?

Ans. It is a uniaxial hinge synovial joint.

6. When does the epiphyseal center appear for the lower end of fibula?

Ans. It appears at the 1st to 2nd year and unites with the shaft at the 15th to 17th year.

7. How the law of ossification is violated in fibula?

Ans. **The law is:** The epiphyseal center which appears first, unites last with the diaphysis. In fibula, the epiphyseal center for the upper end appears late (3rd to 4th year) and also unites late (17th to 19th year), whereas the center for the lower end appears first (1st to 2nd year) and unites early (15th to 17th year).

8. Why there is violation of law of ossification in fibula?

Ans. **As a rule,** pressure epiphysis appears earlier than the traction epiphysis. The lower end is a pressure epiphysis due to its property of weight transmission, whereas the upper end is a traction epiphysis due to the pull of biceps femoris.

9. When does the epiphyseal center appear at the lower end of tibia?

Ans. It appears at 1st to 2nd year and unites at the 17th to 18th year.

10. Identify the tarsal bones and how many are they?

Ans. **Proximodistally,** these are calcaneus, talus, navicular, there cuneiforms (medial, intermediate and lateral) and cuboid. They are **seven** in number, whereas carpals are **eight.**

11. When do the centers of ossification appear for the tarsal and carpal bones?

Ans. All the tarsal and carpal bones ossify in cartilage after birth except talus, calcaneus and cuboid which start ossification before birth.

12. What type of bones are they?

Ans. All the carpal and tarsal bones are **short bones.**

13. How do you count the metatarsals?

Ans. They are counted from medial to lateral as 1st, 2nd, 3rd, 4th and 5th metatarsals.

14. How many phalanges are there?

Ans. There are 14 phalanges (2 in the great toe and 3 in each of the other 4 toes).

15. What type of bones are the metatarsals and phalanges?

Ans. All metatarsals, metacarpals and phalanges are short long bones, because they possess a diaphysis at the center and a single epiphysis at one end only. In a typical long bone, epiphysis lies at its both ends.

16. Name the homologous tarsal and carpal bones.

Ans.

Tarsals	Talus	Posterior tubercle of talus	Calcaneus	Cuboid	Medial cuneiform	Intermediate cuneiform	Lateral cuneiform	Navicular
Carpals	Scaphoid	Lunate	Triquetral	Hamate	Trapezium	Trapezoid	Capitate	Tubercle of scaphoid

Note that pisiform (a carpal bone) does not possess homologous tarsal bone.

17. Which one is the largest tarsal bone?

Ans. Calcaneus.

18. Which of the tarsal bones is devoid of any muscle attachment?

Ans. Talus.

19. What are midtarsal joints?

Ans. i. Talocalcaneonavicular joint (restricted ball and socket variety of synovial joint).
ii. Calcaneocuboid joint (saddle variety of biaxial synovial joint)

20. What is subtalar joint?

Ans. It is the posterior talocalcaneal joint (synovial joint).

21. What are the movements taking place in midtarsal and subtalar joint?

Ans. Inversion and eversion of foot.

22. What are the bones forming the medial longitudinal arch?

Ans. *From behind forwards*: Calcaneus, talus, navicular, three cuneiforms, medial three metatarsals up to their heads.

23. What is the summit of this arch?

Ans. Trochlear surface of the talus.

24. What are the bones forming the lateral longitudinal arch?

Ans. *From behind forwards*: Calcaneus, cuboid, 4th and 5th metatarsals up to their heads.

Radiological Anatomy: Lower Limb (Inferior Extremity)

25. What is the most vulnerable part of the medial and lateral arches?

Ans.
  i. In medial arch: Head of the talus.
  ii. In lateral arch: Calcaneocuboid joint.

26. What is the commonest deformity of foot?

Ans. Flat foot or pes planus.

27. What is pes cavus?

Ans. It is the overarching of the longitudinal arch.

CHAPTER 30

Radiological Anatomy: Abdomen

- Plain X-rays
  1. Abdomen
  2. Lumbosacral Spine
- Contrast X-rays
  1. Barium Meal X-ray (Contrast)
2. Barium Follow Through
3. Barium Enema
4. Intravenous Pyelography
5. Hysterosalpingography
6. Choleycystogram

LESSON 1: PLAIN X-RAYS

1. ABDOMEN

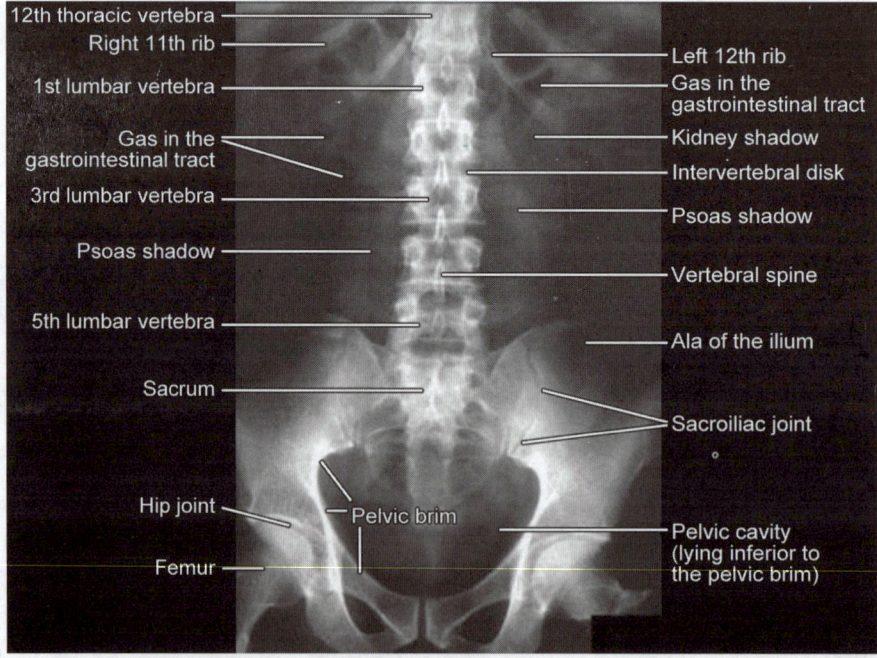

Fig. 30.1: Plain X-ray of abdomen (anteroposterior view)

Radiological Anatomy: Abdomen

What is this film?

Ans. This is a plain (straight) X-ray of abdomen in anteroposterior (Fig. 30.1) view.

Anteroposterior View

Radiographic Appearance—Anteroposterior View (Fig. 30.1)

1. **Bony structures** visible in this X-ray are bony pelvis, lumbar vertebrae, lower ribs.
2. **Soft tissue shadows** seen in this film are kindney shadow and psoas shadow.
3. **Joints** like intervertebral joints and sacroiliac joints are visible.
4. Gas-filled intestinal loops are also visible.

Key points to be identified

1. Lower ribs.
2. Lumbar vertebrae and sacrum.
3. Upper part of the hip bones (ilium).
4. Soft tissue shadows and gas shadow.
5. Sacroiliac joints.

PROBABLE QUESTIONS AND ANSWERS

1. How do you count vertebra from this film?

Ans. The lower most rib is the 12th rib. The vertebra which articulates with this rib is the 12th thoracic vertebra. So, the vertebra below T_{12} is 1st lumbar (L_1) vertebra and so on.

2. Do you know any other way to count vertebra?

Ans. The sacrum is easily identified in this film. The vertebra which articulates with the 1st piece of sacral vertebra is the 5th lumbar (L_5) vertebra and then count upwards as L_4, L_3 and so on.

3. Identify the Kidney shadow?

Ans. Kideny (renal) shadow is a soft tissue shadow which appears less dense due to lower radiodensity of kidney.

4. Is there any relation of this shadow with the ribs?

Ans. **Yes,** kidney shadow is superimposed on the lower ribs. The right kidney is superimposed on the 12th rib only, whereas, the left kidney is superimposed on the 11th and 12th ribs. This is due to the lower location of the right kidney because of the larger size of the right lobe of liver which pushes the right kidney downwards.

5. Can you see the psoas shadow?

Ans. **Yes**, it appears as an oblique, flat, low radiodense shadow running downwards and laterally from the sides of the upper lumbar vertebrae towards the ilium of the hip bone. This shadow represents the psoas major muscle.

6. Is there any relation of psoas major muscle with the kidney?

Ans. **Yes**, the psoas major muscle is related to the medial part of posterior surface of kidney.

7. What is the origin and insertion of psoas major?

Ans. **Origins:**
  i. Anterior surface and lower borders of transverse processes of L_1–L_5 vertebrae.
  ii. Sides of bodies of all lumbar vertebrae and their intervertebral disks.

Insertions: Anterior surface of the tip of the lesser trochanter of the femur.

8. What are the bones forming the bony pelvis?

Ans.
  i. Sacrum and coccyx behind.
  ii. Hip bones of both sides lie laterally and in front.

9. What are the joints that are involved in the bony pelvis?

Ans.
  i. Sacroiliac joints (plane synovial joint).
  ii. Sacrococcygeal joint (atypical intervertebral joint, a type of secondary cartilaginous joint).
  iii. Symphyseal joint (secondary cartilagenous Joint) between the public bones of both sides.

10. Outline the bony land marks of pelvic inlet.

Ans. **From behind forwards:** Sacral promontory → anterior margin of the ala of the sacrum → arcuate line → pectineal line of pubis (pecten pubis) → public tubercle → pubic crest → upper end of symphysis pubis. Join these landmarks of both sides to form a circular outline, representing **pelvic inlet.**

11. What is linea terminales?

Ans. Arcuate line + pectin pubis + pubic crest = Linea terminales.

12. Identify the sacroiliac joint.

Ans. It is the joint between the auricular surfaces of the sacrum and ilium of hip bone.

13. What type of cartilage covers the articular surface of sacrum and ilium in sacroiliac joint.

Ans. Sacral surface is covered with hyaline cartilage but that of ilium with fibrocartilage.

14. Why the sacroiliac joint is so stable?

Ans. The weight of the body is transmitted from the trunk to the lower extremities through these joints.

15. How do you identify the intestinal loops?

Ans. The loops of intestine contain gas which appears as diffuse dark (black) area throughout the film.

2. LUMBOSACRAL SPINE

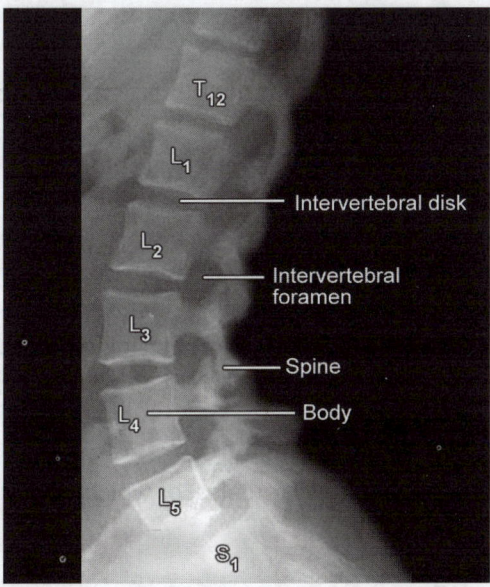

Fig. 30.2: Plain X-ray of lumbosacral spine (lateral view)

What is this film?

Ans. This is a straight X-ray of lumbosacral (LS) spine in lateral view.

Lateral View

> **Radiographic Appearance—Lateral View (Fig. 30.2)**
> 1. Part of the vertebral column is seen from its lateral aspect.
> 2. Intervertebral disks are well visible as radiolucent intervals between the adjacent vertebral bodies.
> 3. Bodies and spines of individual vertebra are well seen.
> 4. Intervertebral foramina are also clearly seen as small, more or less circular radiolucent (dark) areas behind the vertebral bodies in a vertical row.

Key points to be identified
1. Individual vertebra.
2. Vertebral column.
3. Intervertebral foramen and vertebral spines.
4. Intervertebral disk.

PROBABLE QUESTIONS AND ANSWERS

1. What do you mean by vertebral column?

Ans. It is the series of all the vertebrae and their intervertebral disks.

2. How many vertebrae are there in the vertebral column?

Ans. Total = 33 vertebrae of which:
  i. Cervical = 7
  ii. Thoracic =12
  iii. Lumbar = 5
  iv. Sacral = 5 (all of them are fused to form a single piece).
  v. Coccyx = 4 (all of them are fused to form a single piece).

3. What is the length of a vertebral column?

Ans. About 70 cm in males and 60 cm in females, of which:
  i. Cervical part = 12.5 cm.
  ii. Thoracic part = 27.5 cm.
  iii. Lumbar part = 17.5 cm.
  iv. Sacrococcygeal part = 12.5 cm.

4. How much of the vertebral column is contributed by intervertebral disks?

Ans. About 1/4th to 1/5th of the total length of the vertebral column is contributed by intervertebral disks.

Radiological Anatomy: Abdomen

5. What is the composition of intervertebral disk?

Ans.
    i. **Centrally:** Nucleus pulposus, **a remnant of notochord (ectodermal).**
    ii. **Peripherally: Annulus fibrosus,** composed of collagenous tissue and fibrocartilage **(mesodermal).**

6. Is the composition of disk always stationary?

Ans. No, there is diurnal variation of water content of the disk which may produce alterations in the height of the individual by 1 to 2 cm. Due to work load of the vertebral column during day time, the height of the individual may be diminished and the height may be increased in the morning after rest in lying down posture at night. Moreover, the water content of the disks is reduced in old age. So, the height of that individual is usually decreased in old age.

7. How many curvatures are there in the vertebral column?

Ans.
  a. **Two primary curvatures** (concave forwards):
    i. Thoracic.
    ii. Pelvic (sacrococcygeal).
  b. **Two secondary curvatures** (convex forwards):
    i. Cervical.
    ii. Lumbar.

8. What do you mean by primary curvature?

Ans. These curvatures develop **during fetal** life and persist in adult life.

9. What do you mean by secondary curvature?

Ans. These are compensatory curvatures and develop **after birth.**

10. When and why does the cervical curvature develop?

Ans. It develops at about 5–6 months of age when the child tries to raise the head.

11. When and why does the lumbar curvature develop?

Ans. It develops at about 1–1½ years of age when the child tries to start walking.
Remember that the forward convexity of the secondary curvatures is due to the greater thickness of the anterior part of the intervertebral disks.

12. What are the extents of these curvatures?

Ans.
i. Cervical: C_1 to T_2 vertebrae,
ii. Thoracic: T_2 To T_{12} vertebrae,
iii. Lumbar: T_{12} to L_5 vertebrae,
iv. Sacrococcyglal — From S_1 to C x 4 vertebrae.

13. What are the parts of a vertebra?

Ans.
i. Body.
ii. **Vertebral arch** which consists of:
   a. Pedicles
   b. Laminae
   c. Spine
   d. Transverse processes
   e. Superior articular processes
   f. Inferior articular processes.
iii. Vertebral foramen.

14. Can you identify clearly all the parts of a vertebra?

Ans.
i. Vertebral bodies are seen clearly in lateral view and appear as rectangular shadows in anteroposterior (AP) view.
ii. Spines are prominent in lateral view.
iii. Pedicles are seen in lateral view but in AP view they appear as **ring shadows** on the sides of the vertebral bodies.
iv. **Laminae** are superimposed on the vertebral bodies and intervertebral disks in AP view.
v. **Transverse processes** are well seen in AP view, but superimposed on lateral view.
vi. **Intervertebral disk spaces** are clear in lateral view.
vii. **Intervertebral** foramens are clearly seen in lateral view.

15. Which structure passes through the intervertebral foramen?

Ans. Corresponding spinal nerves.

16. How many ossification centers are there in a vertebra?

Ans.
i. **Primary centers:** 3 (one for body, one for each vertebral arch)
ii. **Secondary centers:** 5 (one for upper surface of the body, one for lower surface of the body, one for spine, one for each tranverse process).

17. What type of bone a vertebra is?

Ans. Irregular bone.

Radiological Anatomy: Abdomen

18. How do you explain vertebral body as a long bone?

Ans. From ossification point of view, the body is an example of long bone. Because, it is having one primary center for the body (diaphysis) and two secondary centers (epiphysis) one on upper and another on lower surface of the body.

19. What do you mean by lumbarization of sacral vertebra?

Ans. In this condition, S_1 vertebra fails to fuse with the other sacral vertebrae and behaves as the sixth (L_6) lumbar vertebra.

20. What do you mean by sacralization of lumbar vertebra?

Ans. In the condition, the L_5 vertebra is fused with the S_1 vertebra and the sacrum becomes a bone of six vertebrae and lumbar vertebra becomes four in number.

21. What is scoliosis?

Ans. It is the **lateral curvature** of the vertebral column.

22. What is kyphosis?

Ans. It is the increased posterior convexity of the thoracic vertebral column.

23. What is lordosis?

Ans. It is the increased anterior convexity of the lumbar vertebral column.

24. What do you mean by spondylosis?

Ans. It is a degenerative condition involving the intervertebral disks and the formation of osteophytes.

25. What do you mean by spondylolisthesis?

Ans. It is a condition in which the 5th lumbar vertebra along with it the whole vertebral column above slips over the S_1 vertebra below.

26. What is disk prolapse?

Ans. It is the herniation of intervertebral disk (nucleus pulposus) either posteriorly or posterolaterally affecting the spinal cord or spinal nerve roots respectively.

27. Which disk space is commonly involved in disk prolapse?

Ans. Between L_5/S_1 followed by L_4/L_5.

LESSON 2: CONTRAST X-RAYS

1. BARIUM MEAL X-RAY (CONTRAST)

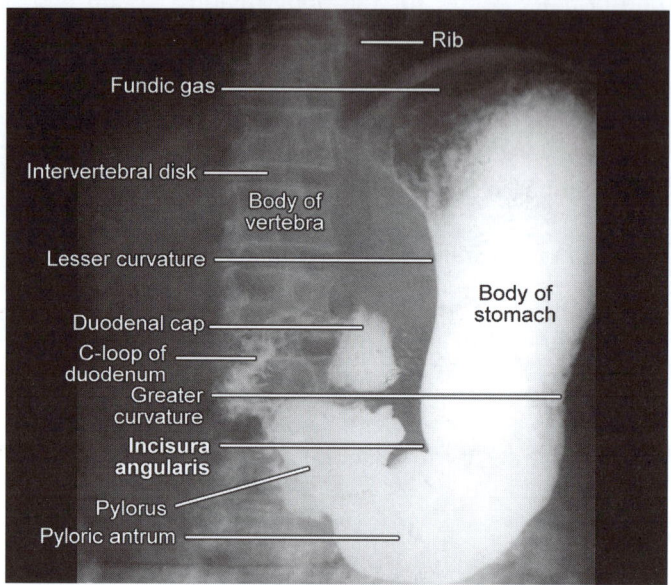

Fig. 30.3: Barium meal X-ray of stomach and duodenum (anteroposterior view)

What is this film?

Ans. This is a contrast X-ray (Barium meal X-ray) of stomach and duodenum in anteroposterior view.

Anteroposterior View

Radiographic Appearance—Anteroposterior View (Fig. 30.3)

1. **Stomach** filled with barium emulsion.
2. **Greater and lesser curvatures** of stomach are clearly outlined.
3. **Fundus** is easily identified by the presence of translucent gas bubble inside the fundus.
4. **Incisura angularis** at the lesser curvature is clearly pointed out.
5. The wide part of the **pyloric** region, called pyloric **antrum** is found distal to the incisura.
6. **1st part of duodenum** is identified by a radiopaque triangular upright mass, known as **duodenal cap**.
7. **2nd part** is floccular in appearance.
8. **3rd part** runs horizontally.
9. **4th part** is seen to be directed upwards and to the left.
10. **Duodenojejunal flexure** is usually not visible.

Radiological Anatomy: Abdomen

Key points to be identified
1. Stomach and its different parts.
2. Duodenum and its different parts
3. Other additional X-ray findings.

PROBABLE QUESTIONS AND ANSWERS

1. What is barium meal test?

Ans. It is a radiographic investigation of stomach and duodenum in which serial X-ray images are taken following oral administration of barium sulfate solution.

2. Why barium sulfate is used as a contrast media?

Ans.
  i. It has superior contrast qualities.
  ii. It is cheap.
  iii. It is safe (barium carbonate is poisonous).

3. How barium sulfate solution is prepared?

Ans. About 120–130 g of barium sulfate powder is mixed with about 180–200 mL of water to prepare the solution.

4. How the chalky taste of the solution can be overcomed?

Ans. Taste can be improved by using vanilla flavor and sweetened by using white saccharin.

5. What should be the PH of the solution?

Ans. It should have a PH of about 5.3 which makes the solution stable in gastric acid. Otherwise the barium particles would aggregate into clumps.

6. How do you prepare a patient for barium meal study?

Ans.
  i. **Nothing to be taken orally** or drunk for 6 hours prior to the test.
  ii. **No smoking** on the day of the examination, as it may increase gastric motility.
  iii. **No purgatives** should be given at night prior to the day of examination.
  iv. Chewing gums and medicines are not to be taken few hours prior to the test.

7. How much of barium solution to drink for barium meal test?

Ans. About 10–15 Oz (0.5 to 1 pint).

8. What are the timings of taking radiographs following oral administration of barium solution?

Ans.
  i. 1st film—immediately after the barium meal
  ii. 2nd film—at ½ hour of swallowing.
  iii. 3rd film—at 1 hour of swallowing.
  iv. 4th film—at 4 hours of swallowing.
  v. 5th film—at 6 hours of swallowing.

Remember: The stomach starts emptying within a few minutes after swallowing of barium suspension and the emptying is completed between 4–6 hours after meal. If barium-suspension remains in the stomach even after 6 hours, a 6th film is taken at 24 hours to ensure gastric outlet obstruction.

9. How do you identify pyloric canal in this film?

Ans. It is seen as a narrow column of barium of about 2–3 mm wide and 5–8 mm long running from the barium-filled wider pyloric antrum.

10. Can you see the duodenojejunal flexure?

Ans. No, it is hidden behind the gastric shadow.

11. What do you like to see in the barium-meal study of stomach?

Ans.
  i. Any filling defect.
  ii. Mucosal pattern.
  iii. Outlet obstruction.

12. What do you like to see in the duodenum?

Ans.
  i. Deformities of the duodenal cap.
  ii. Any filling defect.
  iii. Obstruction, if any.

2. BARIUM FOLLOW THROUGH

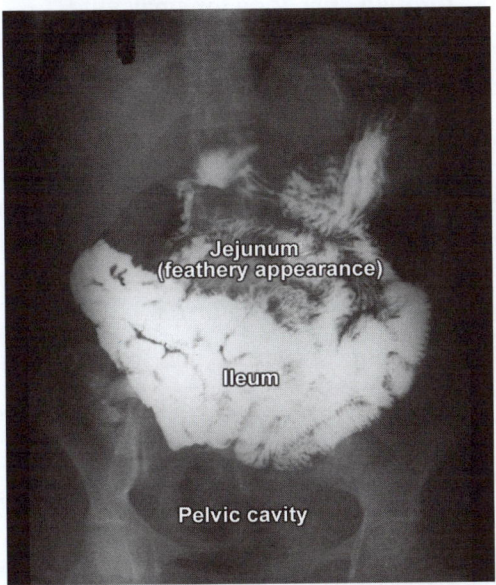

Fig. 30.4: Barium follow through (anteroposterior view)

What is this film?
Ans. This is a barium follow through contrast radiography in anteroposterior view.

Anteroposterior View

Radiographic Appearance—Anteroposterior View (Fig. 30.4)
1. Jejunum shows **feathery appearance** (valvulae conniventes).
2. Ileum shows homogenous shadow below the feathery jejunum.

Key point to be identified
1. Jejunum and ileum of small gut.
2. Other additional X-ray findings.

PROBABLE QUESTIONS AND ANSWERS.

1. When to take the radiographs for follow through?
Ans. Once the barium suspension crosses the duodenum which usually starts after ½ hour. of swallowing of contrast solution, radiographs are taken at hourly intervals up to 4 hours.

2. Is the barium follow through only done for small gut?

Ans. It can be done for proximal colon as well at about 5–6 hours following swallowing of barium solution (distal colon is better identified by barium enema).

3. How much of barium solution is required for follow through investigation?

Ans. About 300 mL (Note that the amount of solution may be reduced to 150 mL if it is done after a barium meal).

4. What are the lengths of Jejunum and Ileum?

Ans. Jejunum = About 8 feet; Ileum = About 12 feet. Total = 20 feet.

5. How do you identify proximal colon?

Ans. Haustrations (sacculations) are there.

6. What are the other X-ray findings in this film?

Ans.
  i. Vertebral column (partly overlapped by the small gut shadow).
  ii. Lower ribs.
  iii. Part of the bony pelvis.
  iv. Few soft tissue shadows.

3. BARIUM ENEMA

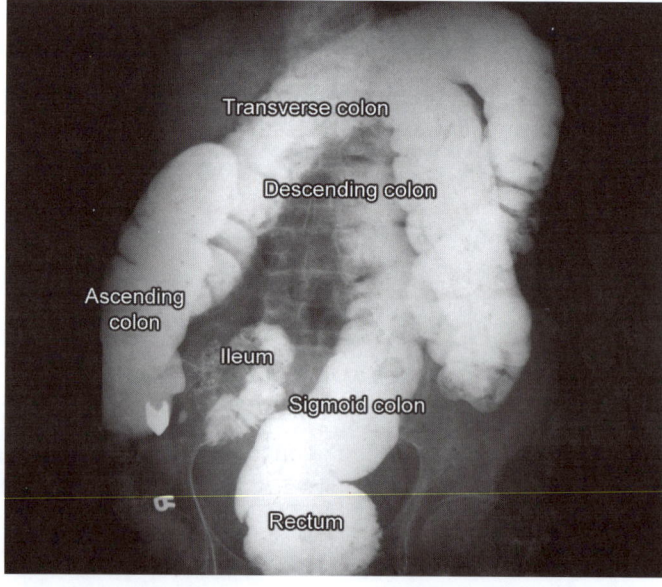

Fig. 30.5: Barium enema (anteroposterior view)

Radiological Anatomy: Abdomen

What is this film?

Ans. This is a contrast radiograph of barium enema in anteroposterior view.

Anteroposterior View

Radiographic Appearance—Anteroposterior View (Fig. 30.5)
1. Color is homogenously dense.
2. Regions of colic flexures are superimposed.
3. Haustrations of colon appear blunt.
4. Rectum looks most dense without haustrations and is found in front of the sacrum.

Key point to be identified
1. Parts of colon.
2. Rectum.
3. Additional X-ray findings.

PROBABLE QUESTIONS AND ANSWERS

1. What do you mean by enema?

Ans. An enema is the introduction of a liquid into the rectum through a small tube.

2. What is barium enema?

Ans. It is a radiographic procedure used to define the anatomy of the colon and rectum by introducing barium solution as a contrast media through the rectum.

3. What is double-contrast barium enema?

Ans. This procedure is done by using double contrast media (barium and air).

4. What is the advantage of double-contrast enema?

Ans. It is the method of choice to demonstrate mucosal pattern.

5. How do you prepare a patient for this test?

Ans.
  i. Low residue diet for three days prior to the test.
  ii. A suitable purgative is given two days prior to the test.
  iii. Only liquid diet is given 24 hours prior to the test.
  iv. A high colonic lavage by plain water or normal saline (without using any soap) on the day of examination.

6. What is the procedure for barium enema?

Ans. The barium sulfate solution is passed into the anus through a tube as high as possible and then serial X-rays are obtained to outline the rectum and colon.

7. How much of barium solution is required?

Ans. About 500 mL.

8. What are the other cardinal features of colon other than haustrations?

Ans.
i. *Taenia coli*.
ii. Appendices epiploica.
(*But these are not demonstrable in barium enema*).

9. What are the different parts of large intestine?

Ans.
i. Cecum (6 cm).
ii. Appendix (2–20 cm).
iii. Colon
    a. Ascending colon (15 cm).
    b. Transverse colon (50 cm).
    c. Descending colon (25 cm).
    d. Sigmoid colon (40 cm).
iv. Rectum (12 cm).
v. Anal canal (3.8 cm).

10. What are the differences between barium meal follow through and barium enema of large gut?

Ans. Table 30.1 shows the differences between barium meal follow through and barium enema of large gut.

Table 30.1: Differences between barium meal follow through and barium enema of large gut

Barium meal follow through	Barium enema
i. X-rays taken 5–6 hours after meal	i. X-rays taken almost immediately following enema
ii. Small bowel can be seen	ii. Small bowel—not seen
iii. Sacculations are normal	iii. Sacculations are blunt
iv. Rectum—least dense	iv. Rectum—most dense
v. Colon—not homogenous	v. Colon—homogenously dense

General Considerations on Contrast X-rays of Gastrointestinal System

1. **Most commonly used dye:** Barium sulfate ($BaSO_4$).
2. **Procedures:**
　i. **Oral route**
　　　a. Barium swallow for esophagus.

b. Barium meal for stomach and duodenum.
c. Barium follow through for small gut and partly proximal part of large gut.
ii. **Rectal route:** Barium enema (single and double contrast)
3. **Quantities of barium-sulfate solution required for:**
   i. Barium swallow—about 100 mL.
   ii. Barium meal—about 135–150 mL.
   iii. Barium follow through—about 300 mL.
   iv. Barium enema—about 500 mL.
4. **Indications:**
   i. To observe mucosal pattern.
   ii. To detect filling defect.
   iii. To detect outlet obstruction.
5. **Contraindications:**
   i. Past history of allergy to the contrast media.
   ii. Severely-ill patient.
   iii. Known case of intestinal obstruction.

4. INTRAVENOUS PYELOGRAPHY

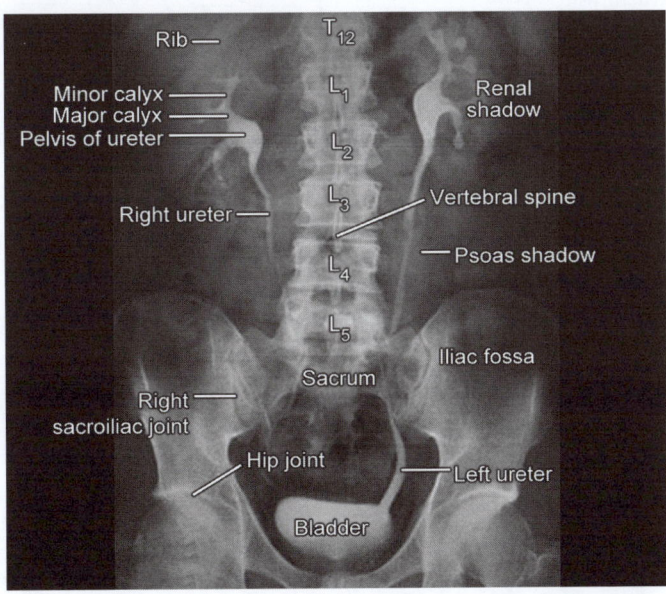

Fig. 30.6: Intravenous pyelography (anteroposterior view)

What is this film?

Ans. This is a contrast radiograph of intravenous pyelography (IVP) in anteroposterior view showing bilateral kidneys and ureters, urinary bladder and other adjacent structures.

Anteroposterior View

> **Radiographic Appearance—Anteroposterior View (Fig. 30.6)**
> 1. **Kidney** shadows are seen on both sides.
> 2. **Cup-shaped multiple minor calyces** are well visualized.
> 3. About **2–3 major calyces** are also seen in this film.
> 4. The pelvis of ureters appears funnel-shaped and joins with the ureters at an obtuse angle.
> 5. The **ureters** of both sides run downwards close to the tips of transverse processes of lumbar vertebrae and more distally they are passing in front of the sacroiliac joints.
> 6. The outline of urinary bladder is also seen.
> 7. Other structures like vertebrae, lower ribs, intestinal gas shadows are also visible in this X-ray film.

> **Key points to be identified**
> 1. Kidneys.
> 2. Major and minor calyces.
> 3. Pelvis of ureters and ureters.
> 4. Urinary bladder.
> 5. Other adjacent structures.

PROBABLE QUESTIONS AND ANSWERS

1. What do you mean by IVP or IVU?

Ans. Intravenous pyelogram (IVP) or intravenous urogram (IVU) is a radiological procedure used to visualize the kidneys, ureters and bladder by injecting iodinated contrast media into a vein.

2. What are the uses of this test?

Ans.
 i. To observe the anatomy of the urinary tract.
 ii. To determine the functional status of the kidneys.
 iii. To note any pathological change in the kidneys.

3. What is the contrast media used?

Ans. Conray 420 (preparation of choice). It is an **iodinated contrast media**.

4. What are the other contrast media that may be used?

Ans.
 i. Conray 280/480.
 ii. Urograffin 30/45/60/76.
 iii. Diatrizoate.

Radiological Anatomy: Abdomen

> **Note** that nowadays nonionic iodinated contrast media is used to increase the clarity of images. Few of these contrast medias are—Omnipaque (iohexol), Visipaque (iodixanol), Ultravist (iopromide), etc.

5. What is the usual dose of Conray 420 for IVP?

Ans. i. For adults: 50 mL.
ii. For children: 1–1.5 mL/kg of body weight.

6. What are the contraindications to perform this test?

Ans. i. Idiosyncracy to iodine. Therefore, iodine sensitivity test is a must.
ii. Poor renal function.

7. What are the timings of taking X-rays?

Ans. i. Immediately after the injection to see the outline of kidney (nephrogram).
ii. After 5 minutes: To see the major and minor calyces (pyelogram).
iii. After 10 minutes: To see the pelvis of ureter and ureters.
iv. After 30 minutes: To see the urinary bladder.

8. What are the other terms used for intravenous pyelography?

Ans. Excretion urography or descending pyelography.

9. What do you mean by retrograde urography or ascending pyelogram?

Ans. It is an imaging procedure where a physician injects a radiocontrast agent (Hypaque 45%) into the ureters by a cystoscopic guidance to visualize the bladder, ureters and kidneys by radiography.

10. What is the radiographic appearance of a retrograde urographic film?

Ans. i. Cystoscope is seen in the bladder.
ii. Ureteric catheter can be seen in the lumen of the ureter.
iii. Ureter, pelvis of ureter, major and minor calyces are seen filled with the contrast agent.

11. What is micturating cystourethrography?

Ans. It is a radiographic procedure of taking radiographs of urinary bladder and urethra while the patient voids urine.

12. What do you mean by antegrade pyelography?

Ans. It is an imaging procedure where a dilated renal calyx is punctured percutaneously and a contrast agent is injected.

5. HYSTEROSALPINGOGRAPHY

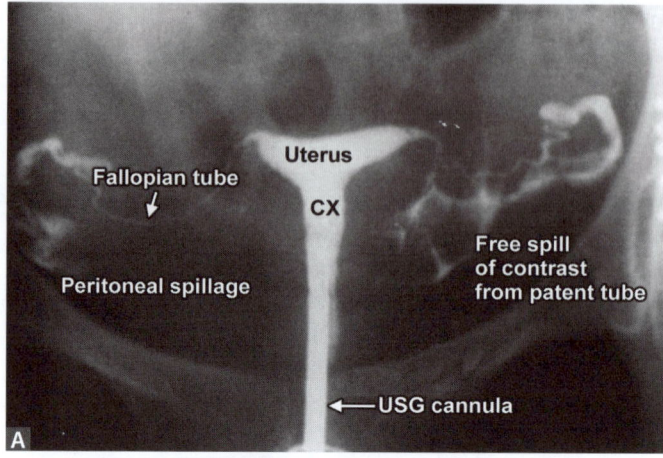

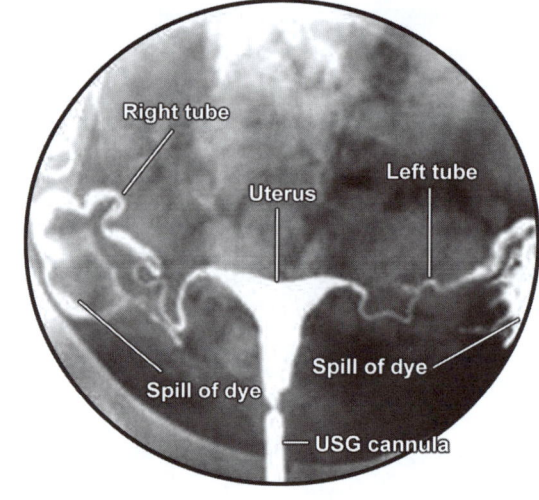

Figs 30.7A and B: Hysterosalpingography (anteroposterior view)

 **What is this film?**

Ans. This is a contrast radiograph of hysterosalpingography (HSG) in anteroposterior view showing uterus and uterine tubes.

Anteroposterior View

Radiographic Appearance—Anteroposterior View (Figs 30.7A and B)
1. Triangular shadow of uterine cavity is seen with the apex downwards.
2. Tortuous uterine tubes are seen passing from the angles of the uterus.
3. The caliber of the uterine tubes is wider at the lateral ends than at its medial ends.
4. Free spillage of contrast media into the peritoneal cavity.

Radiological Anatomy: Abdomen

Key points to be identified
1. Interior of the uterine cavity. 2. Fallopian tubes of both sides.

PROBABLE QUESTIONS AND ANSWERS

1. What is hysterosalpingogram?

Ans. It is a radiographic diagnostic procedure to study uterus and fallopian tubes.

2. What is the most common indication of doing this test?

Ans. It is most commonly done in the evaluation of **infertility.**

3. What is the dye used in this test?

Ans. Conray 420 (**Iothalamate sodium**).

4. How much dye is introduced?

Ans. About 15 mL.

5. How many X-ray exposures are taken?

Ans. Usually two:
  i. 1st film is taken immediately after the introduction of the dye into the cervical canal.
  ii. 2nd film is taken after 1–2 minutes.

6. Which time of the menstrual period is chosen for this test?

Ans. Between 7–14 days of the menstrual cycle.

7. What are the indications of HSG?

Ans.
  i. To visualize uterotubal canal.
  ii. To detect tubal block.
  iii. To detect hydrosalpinx (dilatation of fallopian tube).
  iv. As a follow-up study after tuboplasty operation.

8. What are the contraindications of this test?

Ans.
  i. Sensitivity to iodine.
  ii. Post-ovulatory period.
  iii. During menstruation.
  iv. Suspected pregnancy.
  v. Genital infection.
  vi. Abnormal uterine bleeding.

9. What does the peritoneal spillage indicate?

Ans. It indicates the patency of the fallopian tubes.

10. Can you see other structures in this film?

Ans. Part of the bony pelvis and lower lumbar vertebrae are also seen.

6. CHOLEYCYSTOGRAM

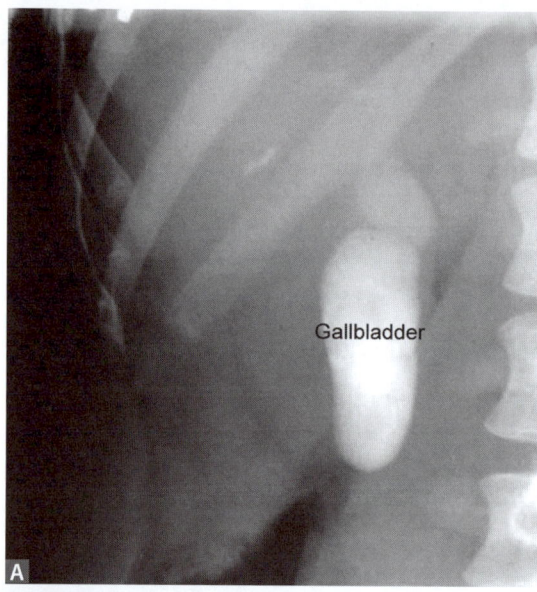

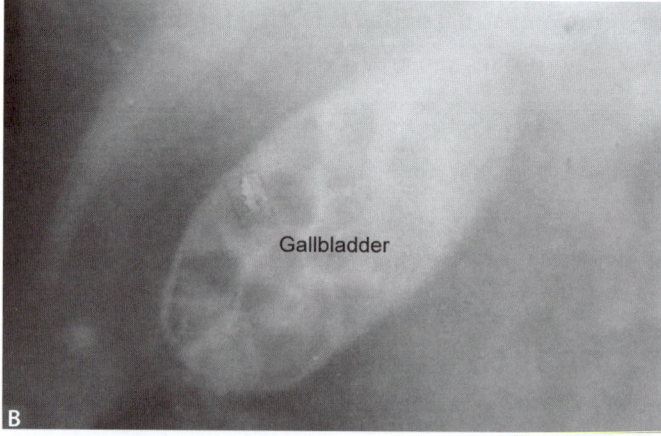

Figs 30.8A and B: Oral cholecystography (anteroposterior view)

What is this film?

Ans. This is a contrast radiograph of gallbladder in anteroposterior view.

Anteroposterior View

Radiographic Appearance—Anteroposterior View (Figs 30.8A and B)
i. **Gallbladder** is seen as a pear-shaped homogenous opaque mass.
ii. **Cystic duct** is also visualized.
iii. Lower ribs and part of the vertebral column are also seen in this film.
iv. Radiolucent gallstones are visible in Figure 30.8B.

Key points to be identified
1. Gallbladder.
2. Cystic duct.

PROBABLE QUESTIONS AND ANSWERS

1. What do you mean by cholecystogram?

Ans. It is the method of visualization of gallbladder by using a contrast agent.

2. What are the methods of cholecystography?

Ans.
  i. Oral cholecystography (OCG).
  ii. Intravenous cholecystography (ICG).

3. What is the principle of OCG?

Ans.

Oral administration of an iodinated contrast media
↓
Excretion of that contrast agent by the liver
↓
Contrast agent is concentrated in the gallbladder
↓
Gallbladder appears radiopaque in X-ray.

4. What is the commonly used contrast media?

Ans. **Iopanoic acid** (Telepaque).

5. How much of the contrast agent is to be taken for OCG?

Ans. Six tablets, each containing 500 mg are to be taken orally.

6. What is the alternative agent for OCG?

Ans. Oragrafin (**sodium ipodate**)—6 capsules, each containing 500 mg.

7. When the radiograph is taken following oral administration of the contrast agent?

Ans. About 12–14 hours after the ingestion of the tablets.

8. Do you need additional X-ray of the gallbladder?

Ans. After the exposure of the 1st film, the patient is allowed to take a fatty meal and a glass of milk. One hour after the intake of the fatty meal additional X-rays are taken to see the emptying capacity of the gallbladder due to its contraction after a fatty meal.

9. What are the contraindications of OCG?

Ans.
  i. Previous history of choleycystectomy (removed gallbladder).
  ii. Severe liver or kidney disease (hampers excretion).
  iii. Acute choleycystitis.
  iv. Dehydration.

10. When do you go for intravenous choleycystography?

Ans. It is done when the oral route is not useful.

11. In which condition oral route is not preferred?

Ans. In conditions where the absorption of orally administered contrast medium is interfered due to diarrhea, pyloric obstruction or other factors.

12. What is the contrast medium used in intravenous route?

Ans. Biligrafin (20% solution of sodium iodipamide).

13. How much contrast medium is required?

Ans. About 1–2 ampoules (20 mL per ampoule).

14. How the injection is given?

Ans. The injection is given very slowly over a period of 10 minutes.

15. When do you take an X-ray to visualize the gallbladder following intravenous injection?

Ans. 1½ to 2 hours after injection, because gallbladder fills very slowly after the ducts.

16. What are the modern techniques used to study the biliary tree?

Ans.
  i. Ultrasonography (USG)
  ii. Endoscopic retrograde cholangiopancreatography (ERCP)
  iii. Magnetic resonance cholangiopancreatography (MRCP).

CHAPTER 31

Radiological Anatomy: Thorax

- Plain X-rays
  Posteroanterior and Lateral View
- Contrast X-rays
  Barium Swallow of Esophagus

LESSON 1: PLAIN X-RAYS

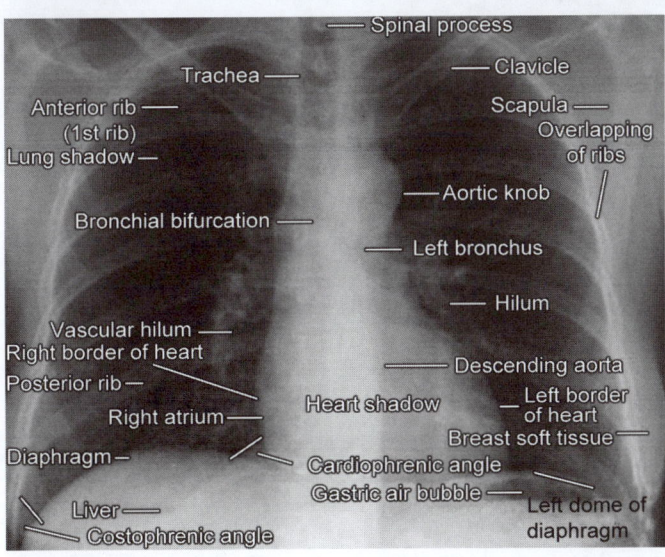

Fig. 31.1A: Plain X-ray of chest (posteroanterior view)

Exam-Oriented Practical Anatomy

 What is this film?

Ans. This is a plain X-ray of chest in posteroanterior (PA) view.

Posteroanterior View

Radiographic Appearance—Posteroanterior View (Fig. 31.1A)

1. Lung shadows are seen on either side of the cardiac shadow.
2. The ribs, medial halves of the clavicles and part of the scapula are superimposed on the lung shadows.
3. Sternum is superimposed on cardiac shadow.
4. The apices of both lungs are seen above the shadows of the clavicles.
5. Outline of both cupola of the diaphragm is clearly seen with convexity upwards.
6. The costophrenic and cardiophrenic angles are also seen.
7. Shadows of the heart and arotic arch are seen quite well.
8. Fundic gas shadow below the left cupola of the diaphragm is also seen.

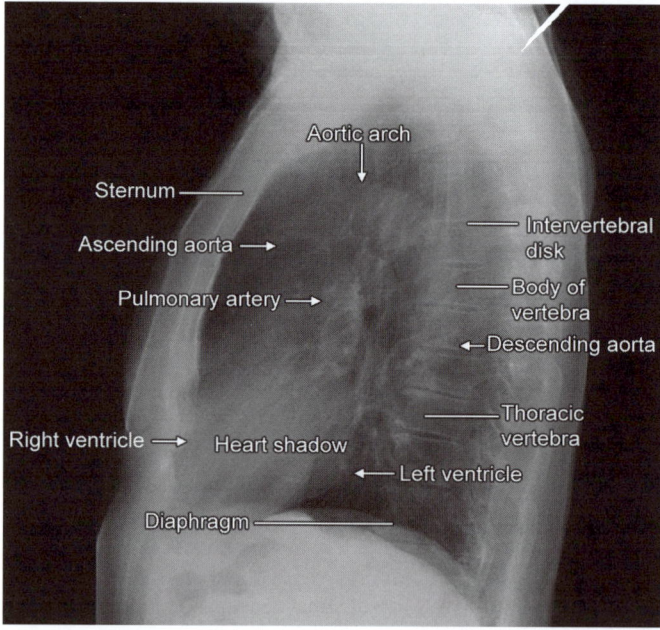

Fig. 31.1B: Plain X-ray of chest (lateral view)

What is this film?

Ans. This is a plain X-ray of chest in lateral view.

Lateral View

Radiographic Appearance—Lateral View (Fig. 31.1B)
1. Lung shadows are superimposed on each other.
2. Cardiac shadow and the shadow of aorta are superimposed on the central part of the lungs.
3. Shadows of the ribs are superimposed on the lungs.
4. The outline of the diaphragm is also visible.
5. Sternal shadow is seen anteriorly.
6. Thoracic vertebrae along with their intervertebral foramina and intervertebral disc spaces are clearly seen posteriorly.

Key points to be identified
1. Trachea.
2. Lungs.
3. Heart.
4. Mediastinum.
5. Hila.
6. Diaphragm.
7. Cardiophrenic and costophrenic angles.
8. Bones of the thoracic cage (ribs, sternum and thoracic vertebrae).

PROBABLE QUESTIONS AND ANSWERS

1. What are the bones forming the thoracic cage?

Ans.
    i. **Sternum** in front.
    ii. Bodies of **12 thoracic vertebrae** behind
    iii. **12 pair of ribs** on each side.

2. What is the boundary of thoracic inlet?

Ans.
    i. **In front:** Upper border of manubrium sterni.
    ii. **At the sides:** 1st rib and its cartilage of both sides.
    iii. **Behind:** Body of 1st thoracic vertebra.

3. Identify the bones of the thoracic cage.

Ans. Vertebral bodies and sternum are superimposed by the cardiac shadow in PA view but they are well visualized in lateral view. On the other hand, the ribs are well seen in PA view though these are superimposed by lung shadow.

4. **How do you count the ribs?**

Ans. The ribs are counted from above downwards, starting from the first rib which is seen passing behind the medial end of clavicle in downward and forward direction in the film.

5. **Can you see all the parts of the ribs clearly in the film?**

Ans.
  i. The anterior portions of the upper ribs are well seen.
  ii. The posterior ends of the ribs are not seen properly.
  iii. The vertebral portion of the ribs are obscured by the cardiac shadow, particularly on the left side.
  iv. The portions of the ribs in the mid-axillary line are superimposed on each other and seen as dense radiopaque shadow.

6. **How do you identify trachea?**

Ans. It lies in the midline in the upper part of the chest X-ray. Air inside the tracheal lumen looks dark in the film.

7. **What is the importance of tracheal localization?**

Ans. Trachea can be deviated from the midline in different pathological conditions of the lungs and mediastinal structures.

8. **How do you identify hilar shadow?**

Ans. Hilar shadows are lightly radiopaque and diffuse in appearance. These are seen on either side of the mediastinum.

9. **What are the structures that produce hilar shadows?**

Ans. Pulmonary vessels, lymphatics and the major bronchi.

10. **Do both the hila lie at the same level?**

Ans. The left hilum is about 2 cm higher than the right hilum.

11. **Show the lung fields.**

Ans. The radiolucent shadows of both the lung fields are seen on either side of the radiopaque shadow of heart and aorta.

12. **Why does the lung field show dark shadow?**

Ans. Lungs contain air which is radiolucent. All radiolucent areas look dark in a plain X-ray.

13. Identify the cardiac shadow.

Ans. It is a radiopaque area whose one-third diameter occupies the right side and two-third occupies the left side from the midline.

14. Identify the borders of the heart.

Ans. The right border is formed by the right atrium and the left border is formed by the left ventricle. Though these borders are well-defined, the inferior border which is formed by the right ventricle is not defined clearly because of superimposition with the shadows of liver and diaphragm.

15. How do you assess the size of the heart?

Ans. In a healthy individual, the transverse diameter of the heart shadow should be less than 0.5 (50%) the internal thoracic diameter (inner aspect of the ribs) at its widest point.

16. What do you mean by cardiothoracic ratio (CTR)?

Ans. It is the ratio between the maximum transverse diameter of the heart shadow and the maximum internal thoracic diameter. It should be less than 0.5 (50%) in a normal chest X-ray.

17. Name some pathological conditions where CTR is increased.

Ans.
  i. Cardiac failure.
  ii. Ventricular dilatation.
  iii. Pericardial effusion.

18. Can you tell any condition where cardiac shadow is magnified in a normal X-ray film?

Ans. Cardiac shadow looks enlarged in an anteroposterior view of X-ray film.

19. Show the cardiophrenic angle.

Ans. It is an obtuse angle between the shadows of heart and the diaphragm as seen in an X-ray film.

20. Which cardiophrenic angle is more clear?

Ans. The left one is more clear than the right-sided angle. The right cardiophrenic angle is obscured by the **cardiohepatic angle** (angle between the heart and liver).

21. What is the importance of this angle?

Ans. The cardiophrenic angle becomes acute in pericardial effusion.

22. What is aortic knuckle or aortic knob?

Ans. It is a convex bulging on the left side of the sternal angle representing the left lateral edge of the aorta.

23. In which condition the definition (contour) of aortic knuckle is lost?

Ans. In aneurysm of the arch of aorta.

24. Show the costophrenic angles.

Ans. The costophrenic angles are clearly visible in a normal chest X-ray as a well-defined acute angle between dome of each hemidiaphragm and the lateral chest wall.

25. What do you mean by costophrenic blunting?

Ans. Loss of the costophrenic angle is sometimes referred to as costophrenic blunting.

26. In which condition costophrenic blunting occurs?

Ans. Costophrenic blunting occurs in the presence of fluid or consolidation in this area.

27. Outline the diaphragm in the X-ray film.

Ans. The right hemidiaphragm is higher than the left one by 1–3 cm in healthy individuals.

28. Why the right diaphragm lies at a higher level?

Ans. It is due to the underlying large right lobe of the liver.

29. What lies underneath the left hemidiaphragm?

Ans. It is the stomach and it is best identified by the presence of fundic gas shadow (dark area).

30. What is silhouette sign?

Ans. Obliteration of normal air-soft tissue interface (i.e. air-filled lung shadow and radiopaque soft tissue shadow) is known as the silhouette **sign of Felson.**

31. Name the structures which are better visualized in lateral view.

Ans.
  i. Sternum.
  ii. Thoracic vertebrae with intervertebral foramina and intervertebral discs.
  iii. Structures of superior, anterior and posterior mediastinum, etc.

LESSON 2: CONTRAST X-RAYS

1. BARIUM SWALLOW OF ESOPHAGUS

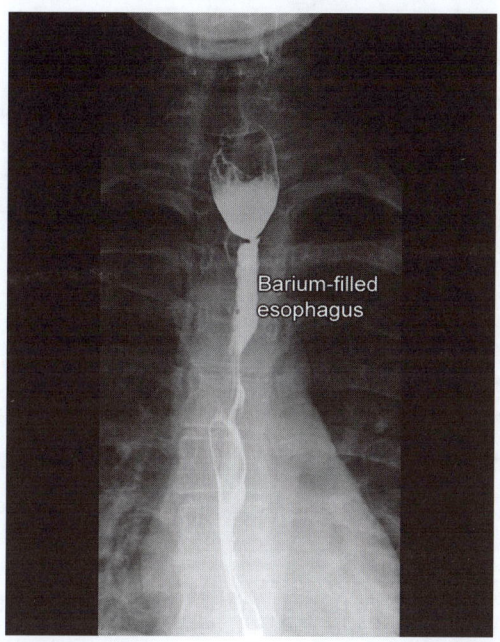

Fig. 31.2A: Barium swallow of esophagus (posteroanterior view)

What is this film?

Ans. This is a contrast X-ray (Barium swallow) of esophagus in posteroanterior (PA) view (Fig. 31.2A).

Posteroanterior View

Radiographic Appearance—Posteroanterior View (Fig. 31.2A)

1. Bony thoracic cage, pulmonary shadow, cardiac shadow, outline of the diaphragm are seen as usual.
2. A barium-filled tubular structure is seen in the median plane of the PA view of chest X-ray →the esophagus.

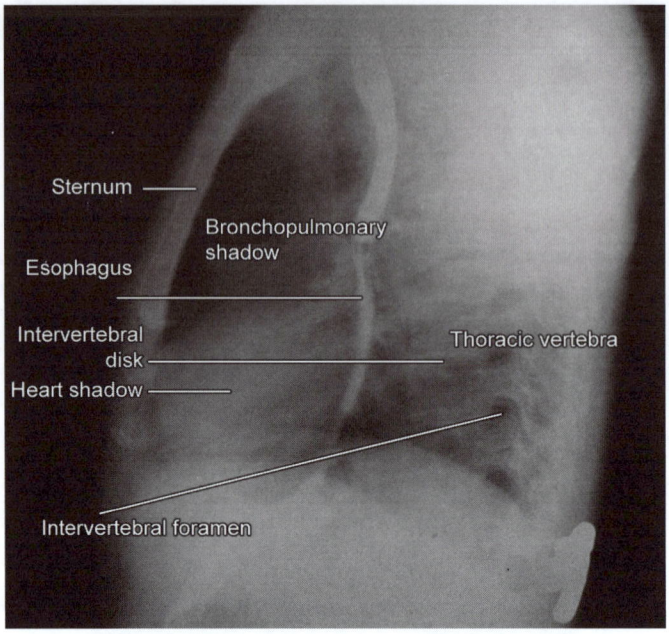

Fig. 31.2B: Barium swallow of esophagus (lateral view)

What is this film?

Ans. This is a barium swallow of esophagus in lateral view.

Lateral View

Radiographic Appearance—Lateral View (Fig. 31.2B)

1. Ba-filled esophagus in seen in the posterior mediastinum.
2. Heart shadow, lung shadows are seen, though the lung shadows are superimposed on each other.
3. Sternal shadow is seen very clearly.
4. Vertebral shadow and intervertebral foramina are also seen in the posterior part of the film.
5. Cupola of the diaphragm is seen, though the cupola of both sides is superimposed.

PROBABLE QUESTIONS AND ANSWERS

1. What do you mean by barium swallow of esophagus?

Ans. Barium swallow is an X-ray imaging test used to visualize the interior of the esophagus.

2. What is the procedure of doing this test?

Ans. Barium sulfate is mixed with water and a solution of barium sulfate is prepared. The X-ray images are obtained while the patient swallows this liquid barium.

3. What is the principle of barium swallow of esophagus?

Ans. The barium solution fills the esophagus and then coats its inside wall and appears white on the film so that it can diagnose the abnormalities.

4. Name few conditions where barium swallow is a useful investigation.

Ans.
i. Achalasia cardia.
ii. Carcinoma of esophagus.
iii. Pharyngeal pouch.
iv. Foreign bodies.

5. What is achalasia cardia?

Ans. Achalasia cardia is an idiopathic disorder of the cardiac end of the esophagus due to congenital absence of nerve cells in the wall of the lower esophagus. The lower end of the esophagus is grossly dilated followed by a tapering end which is termed as **rat's tail** appearance.

6. Where do you find bird's beak appearance?

Ans. In barium swallow of esophagus this appearance is also found in achalasia cardia.

7. What finding do you expect in carcinoma of esophagus?

Ans. Barium swallow shows filling defect' or rat's tail appearance.

8. Do you identify foreign body is esophagus in this contrast X-ray?

Ans. Only **radiolucent** foreign body can be visualized within esophagus in barium swallow study.

9. What is the extent of esophagus?

Ans. From C_6 vertebra to T_{11} vertebra.

10. Which part of a gastrointestinal (GI) tract is investigated by barium swallow other than esophagus?

Ans. Pharynx.

11. In which condition of esophagus do you find Cork-screw esophagus or nut crackers esophagus or curling esophagus in a barium swallow study?

Ans. In diffuse esophageal spasm.

Chapter 32

Radiological Anatomy: Head and Neck

- Plain X-rays
  Posteroanterior (Occipitofrontal) View
  Lateral View
  Occipitomental View

LESSON 1: PLAIN X-RAYS OF HEAD AND NECK REGION

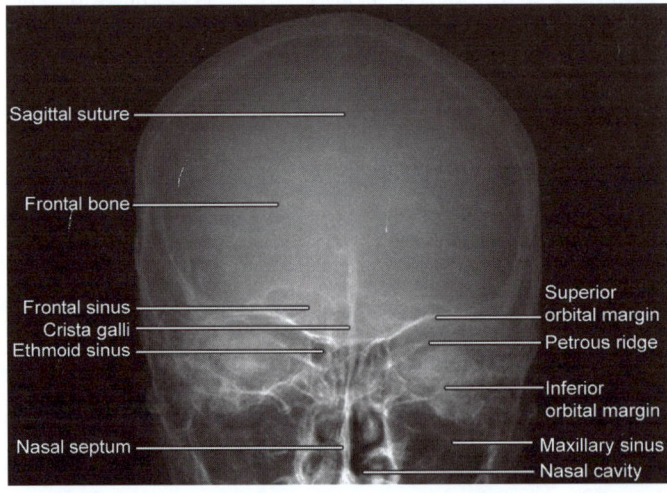

Fig. 32.1: Occipitofrontal view or posteroanterior view of skull

 **What is this film?**

Ans. This is a plain X-ray of skull in posteroanterior (occipitofrontal) view.

Occipitofrontal View or Posteroanterior View

> **Radiographic Appearance—Posteroanterior View (Fig. 32.1)**
> 1. The frontal bone and the sutures are well seen.
> 2. Facial bones (maxilla, zygomatic, mandible, etc.) are superimposed on the base of the skull.
> 3. Teeth of both upper and lower jaws are superimposed on the upper cervical spine.
> 4. The outline of both the orbits are well visualized
> 5. Rami of the mandible are seen.
> 6. Nasal septum and nasal cavities are well seen.
> 7. Maxillary, frontal and ethmoidal sinuses are clearly seen.

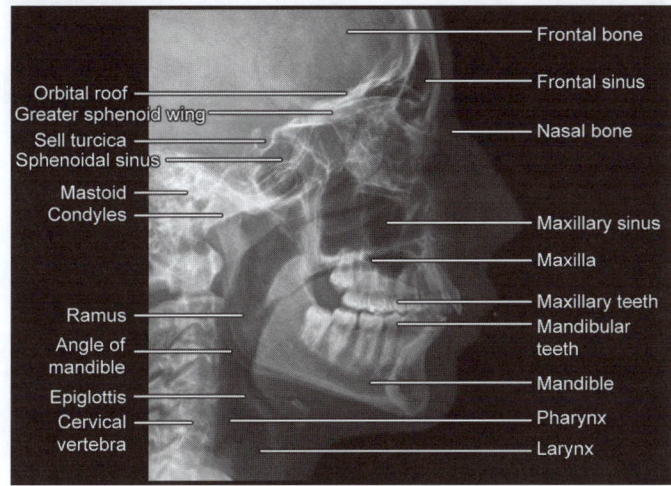

Fig. 32.2: Lateral view of skull

 What is this film?

Ans. This is a plain X-ray of skull in lateral view.

Lateral View

> **Radiographic Appearance—Lateral View (Fig. 32.2)**
> 1. Frontal, parietal and occipital bones and their sutures (coronal and lambdoid) are clearly seen.
> 2. Sella turcica appears as a small groove.
> 3. Anterior and posterior clinoid processes as well as dorsum sellae are clearly visualized.
> 4. Sphenoidal air sinus is seen just below the sella turcica.
> 5. Maxilla and mandible on the side nearest the X-ray film are well seen with the opposite side superimposed.
> 6. Facial bones are superimposed on each other.
> 7. Anterior arch of atlas and odontoid process of axis are also seen with other cervical spines.
> 8. Laryngeal gas shadow (dark) is also seen in the neck.

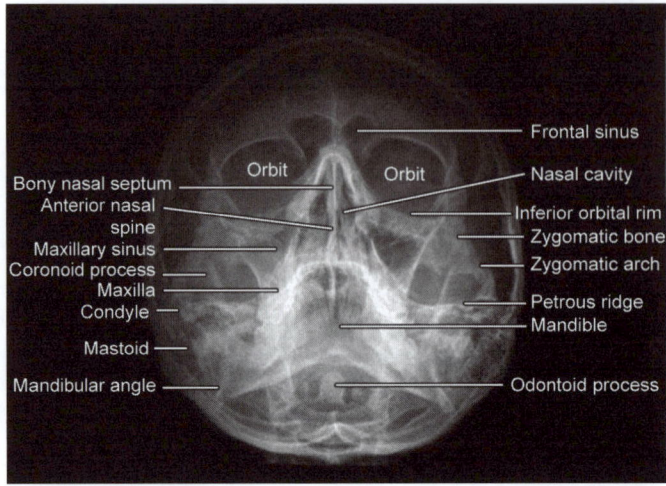

Fig. 32.3: Occipitomental or Water's view of skull

 What is this film?

Ans. This is a plain X-ray of skull in occipitomental view.

Occipitomental or Water's View (Fig. 32.3)

Key points to be identified
1. Frontal, parietal and occipital bones with their sutures. 2. Paranasal air sinuses. 3. Facial bones (maxilla and mandible). 4. Sella turcica of sphenoid and petrous part of temporal bone. 5. Bony orbits and nasal cavities. 6. Cervical vertebrae.

PROBABLE QUESTIONS AND ANSWERS

 1. How do you classify skull?

Ans. The skull is classified into **two parts:**
    i. **Calvaria** which forms a box to lodge and protect the brain.
    ii. **Facial skeleton** which again consists of two parts:
        a. Upper immovable part (fixed with calvaria).
        b. Lower movable part (mandible).

Radiological Anatomy: Head and Neck

2. What are the bones forming calvaria?

Ans. Total = 8
  i. Frontal (1)
  ii. Parietal (2)
  iii. Occipital (1)
  iv. Ethmoid (1)
  v. Sphenoid (1)
  vi. Temporal (2)

3. What are the bones of facial skeleton?

Ans. Total = 14.
  a. **Immovable = 13**
    i. Maxilla (2)
    ii. Zygomatic (2)
    iii. Lacrimal (2)
    iv. Nasal (2)
    v. Vomer (1)
    vi. Inferior nasal concha (2)
    vii. Palatine (2)
  b. **Movable = 1, i.e. mandible.**

4. What is the total number of skull bones in adults?

Ans. 8 + 13 + 1 = 22, of which only mandible can move but the other bones are connected with each other by sutures and hyaline cartilage. *Number of human skull bones during birth is 45.*

5. What do you mean by vault or roof of the skull?

Ans. It is the upper portion of the skull which caps over the brain, hence known as **skull cap**.

6. What are the bones that form the vault?

Ans. Frontal, two parietals, occipital, part of sphenoid and temporal bones.

7. What do you mean by base of the skull?

Ans. The inferior area of the skull on which the brain lies is the base of the skull.

8. What are bones forming the base?

Ans. Frontal, ethmoid, sphenoid, temporal and occipital.

9. How do the bones of the skull ossify?

Ans. The bones of the vault are ossified in **membrane,** hence called **membranons neurocranium.** The bones of the base of the skull are ossified in cartilage, hence called **chondrocranium.**

10. What is a suture?

Ans. A suture is a type of fibrous joint (synarthrosis) which is only found in the skull.

11. Name few important sutures of the skull in an infant?

Ans.
  i. **Metopic suture:** Supposed to close between 3 months and 9 months of age.
  ii. **Coronal, sagittal and lambdoid sutures:** Supposed to close between 22 months and 40 months of age.

12. What is metopic suture?

Ans. It is a midline suture of the squamous part of frontal bone which usually fuses between 3 months and 9 months of age.

13. What is craniosynostosis?

Ans. It is the premature fusion of one or more sutures and fontanelles of the skull.

14. What is the normal capacity of cranial box in adults?

Ans. 1300–1500 mL.

15. What is cephalic index?

Ans. **Breadth/length** x 100. Normally it is about 75–80 in Indian skulls.

16. How do you measure the breadth of the skull?

Ans. Breadth is measured from one parietal eminence to the other.

17. How do you measure the length of the skull?

Ans. It is measured from nasion in front to the inion behind.

Radiological Anatomy: Head and Neck

18. What do you mean by Frankfurt line/plane?

Ans. It is a plane which passes through the inferior border of bony orbit and the upper margin of the external auditory meatus.

19. What is its importance?

Ans. It is useful in orthodontic diagnosis and planning of treatment.

20. Identify the paranasal air sinuses?

Ans. These are air-filled spaces around the nasal cavities.

21. How do they appear in the X-ray film?

Ans. They appear as dark (black) area due to their air content (radiolucent area).

22. What are the different air sinuses?

Ans.
  i. Frontal (right and left).
  ii. Maxillary (right and left).
  iii. Ethmoidal (anterior, middle and posterior).
  iv. Sphenoidal (right and left).

23. Are they present at birth?

Ans. They are rudimentary or absent at birth.

24. How do you identify frontal air sinus in the film?

Ans.
  i. It lies within the frontal bone above the medial part of the bony orbit.
  ii. Its wall has a thin bony cortex which produces a sharp pencil-like outline on the radiograph.
  iii. Its roof is grooved by bony ridges which extend into the sinus.

25. Are both right and left sinuses equal in size?

Ans. Usually unequal in size. Congenital absence of one sinus is not uncommon.

26. What is its average size?

Ans. About 2.5 cm in all dimensions. However, they are better developed in males than in females.

27. When do these sinuses reach adult size?

Ans. They are rudimentary or absent at birth, well developed between 7 years and 8 years of age and reach adult size only after puberty.

28. Which of the paranasal sinus is the largest?

Ans. The maxillary sinus is the largest of all paranasal sinuses.

29. How do you identify the maxillary air sinus in a radiograph?

Ans.
  i. It lies within the body of the maxilla.
  ii. These are pyramidal-shaped dark air-filled spaces.
  iii. They lie below the orbits and inferolateral to the nasal cavities.

30. When do these sinuses develop? What is the other name for maxillary sinus?

Ans. They develop in the 4th months of intrauterine life (**first paranasal sinus to develop is the maxillary sinus**). Maxillary sinus is also known as **antrum of Highmore**.

31. What is the average adult size of the maxillary sinus?

Ans.
  i. Vertical length—3.5 cm.
  ii. Anteroposterior length—3.5 cm.
  iii. Transverse length (width)—2.5 cm.
  (Though the size is variable)

32. How do you identify sphenoidal air sinus?

Ans. Two sphenoidal air sinuses are seen below and anterior to the sella turcica and superimposed on each other on a **lateral view** radiograph.

33. Where do these sinuses lie? What is their average size?

Ans. They lie within the body of the sphenoid. Their dimensions are 2 cm (vertical) x 2 cm (anteroposterior) x 1.5 cm (transverse).

34. How do you identify ethmoidal air sinuses?

Ans. These are multiple (3–18) inter-communicating air-filled spaces and variable in size.
  i. On **lateral view radiograph**, they are seen behind the orbits and superimposed on each other.
  ii. On **occipitofrontal view**, they are seen on either side of the nasal cavities as radiolucent areas.

35. When do they develop?

Ans. They are present at birth and become mature during puberty.

36. Which of the skull bones contain these air sinuses?

Ans. Ethmoidal air sinuses lie within the labyrinth of ethmoid bone.

37. Where do all the paranasal air sinuses open?

Ans.
  i. **Frontal**—middle meatus of nose
  ii. **Maxillary**—middle meatus of nose.
  iii. **Sphenoidal**—sphenoethmoidal recess of nasal cavity.
  iv. **Anterior and middle ethmoidal sinuses**—middle meatus of nose.
  v. **Posterior ethmoidal sinuses**—superior meatus of nose.

38. Identify the sella turcica?

Ans. It is best seen in lateral view X-rays as a small grooved area, just above the sphenoidal air sinus.

39. What does it lodge?

Ans. It lodges the pituitary gland, hence called **pituitary fossa.**

40. What are the bony processes in front and behind the sella turcica?

Ans. Anterior and posterior clinoid processes.

41. Can you identify the mandible?

Ans.
  i. **In PA view,** the outline of whole of the mandible is seen except the coronoid processes and condyles which are superimposed on each other.
  ii. **In lateral oblique view**, the body, ramus and condyle of the mandible of the side nearest the film are clearly visible. The mandible of the opposite half is seen above the affected side but the area of chin is superimposed. In **true lateral view,** rami, condylar and coronoid processes of both sides are superimposed on each other.

42. Identify the angle of the mandible and how it is formed?

Ans. The lowest point of the posterior border of the ramus of the mandible is the angle. It is formed by the junction between the posterior and inferior border of the ramus of the mandible.

43. Identify mastoid process and mastoid air cells.

Ans. **Mastoid air** cells are well seen behind the external acoustic meatus in a lateral view radiograph.

44. Identify the nasal septum.

Ans. It is well visualized in PA view of skull as a white (radiopaque) midline shadow between the radiolucent nasal cavities of either side.

45. What are the principal bones forming the nasal septum?

Ans.
   i. The perpendicular plate of **ethmoid bone.**
   ii. The **vomer.**

46. Which view is suitable to identify all the paranasal sinuses?

Ans. **Occipitomental view** (Water's view).

47. Which view of the skull is better to locate the foramina of the skull?

Ans. Submentovertical or **'Hanging head'** position of the head.

48. What is this position?

Ans. The patient is supine, neck is extended as much as possible as the head hangs backwards.

49. Identify the temporomandibular joint.

Ans. Identify the condyle of the mandible and the external acoustic meatus. Between these two points, there is the temporomandibular (TM) joint.

50. What type of joint it is?

Ans. It is a condylar variety of synovial joint.

51. What are the bones forming TM joint.

Ans.
   i. Mandibular fossa of the temperal bone.
   ii. Head of the mandible (condylar process).

52. Identify the atlas and axis.

Ans. Anterior arch of atlas (C_1) and odontoid process of axis (C_2) are well seen in lateral view radiograph.

Radiological Anatomy: Head and Neck

53. Identify the black area in front of the cervical vertebra.

Ans. The black (dark) area is the air containing structure, the **larynx** (radiolucent).

54. Identify the cervical spines and intervertebral disk spaces.

Ans. The spines and disk spaces are well seen in the lateral radiograph.
   i. Spines look white (radiopaque, being a bony part).
   ii. Disc spaces look dark (radiolucent, being a cartilaginous part).

55. How do you classify cervical vertebrae?

Ans.
   i. Typical—C_3, C_4, C_5, C_6 vertebrae.
   ii. Atypical—C_1, C_2 and C_7 vertebrae.

56. What are the structural components of intervertebral disk?

Ans.
   i. Nucleus pulposus in the certer (remnant of notochord)
   ii. Annulus fibrosus (fibrocartilagenous) at the periphery.

57. What is the average length of cervical vertebral column in adult and what type of curvature it is?

Ans. About 12.5 cm. It has a secondary curvature, because it develops at about 6 months after birth during the holding up of the head.

RADIOLOGY WORKSHEET
(Identify and Label the Markings)

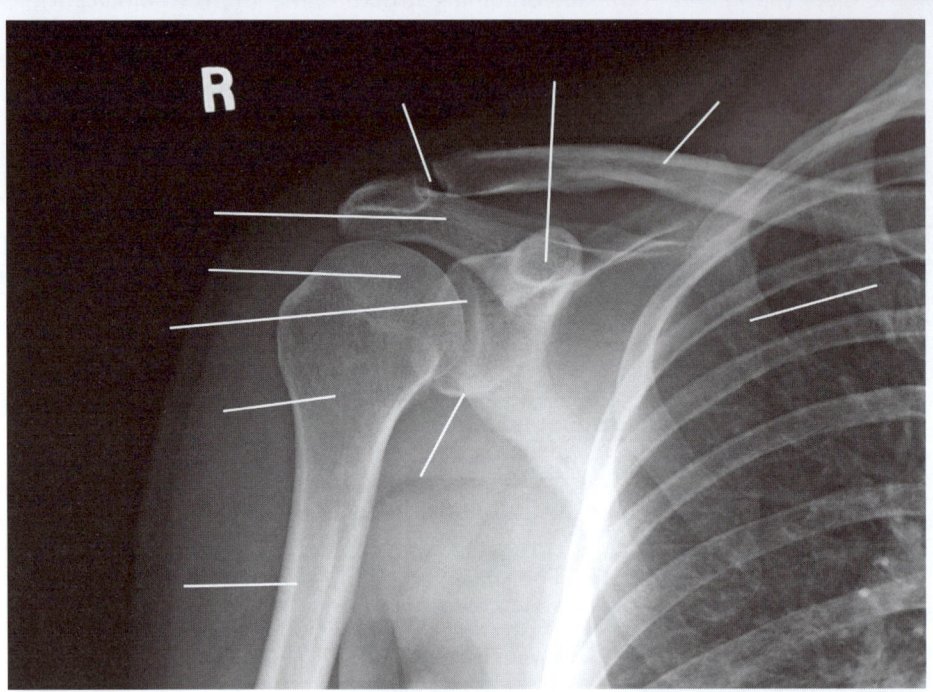

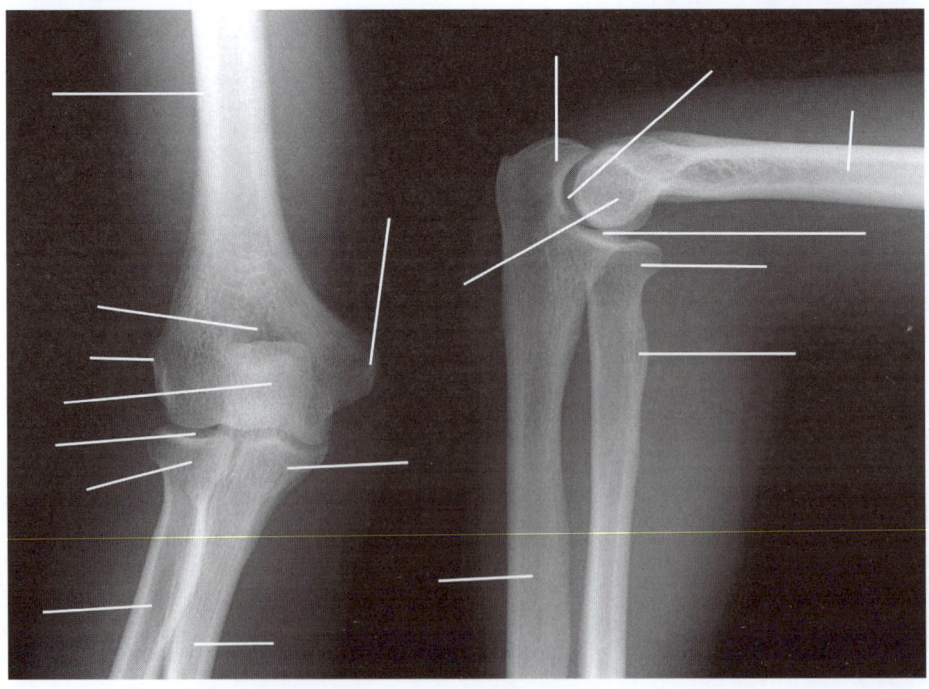

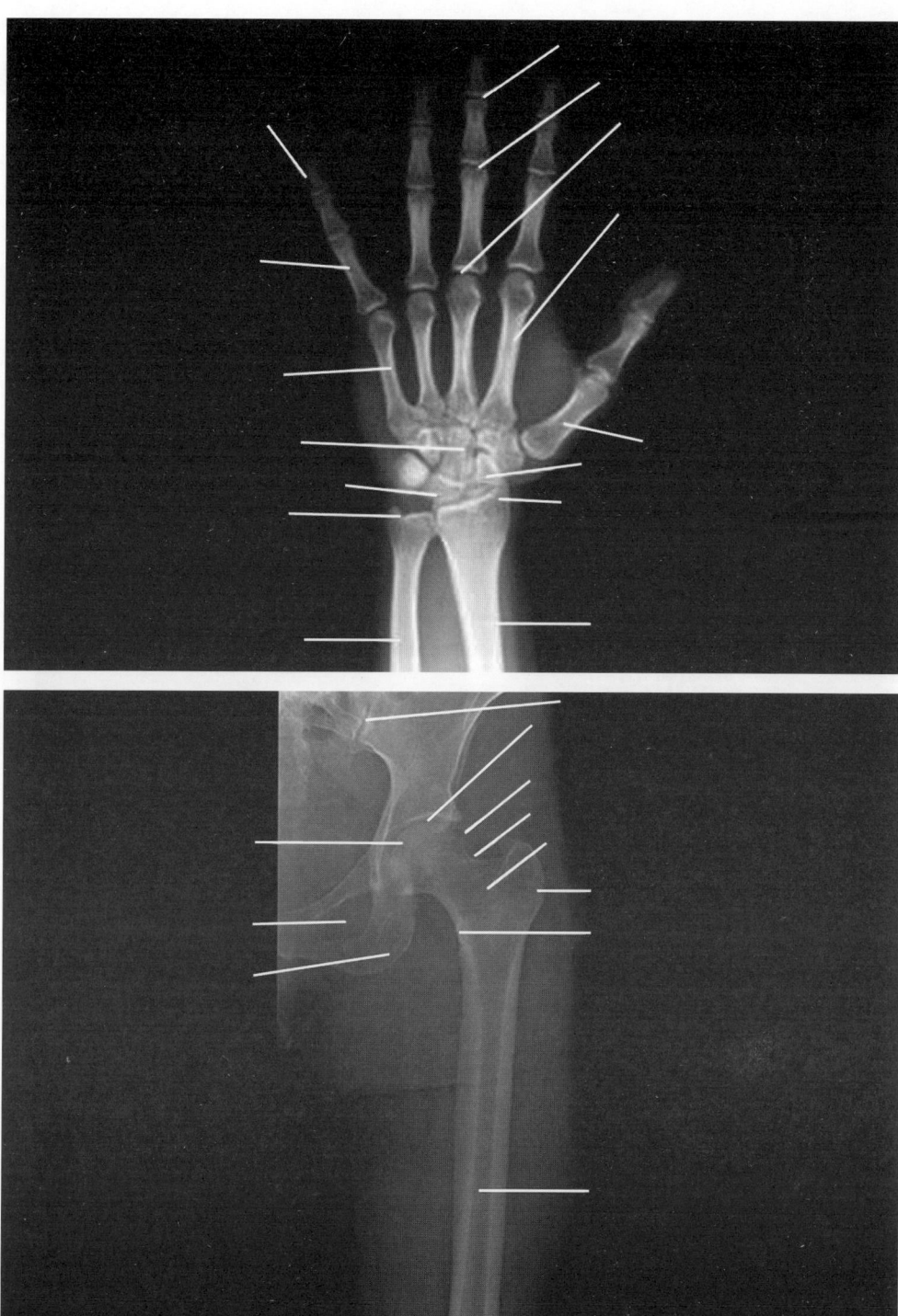

724 Exam-Oriented Practical Anatomy

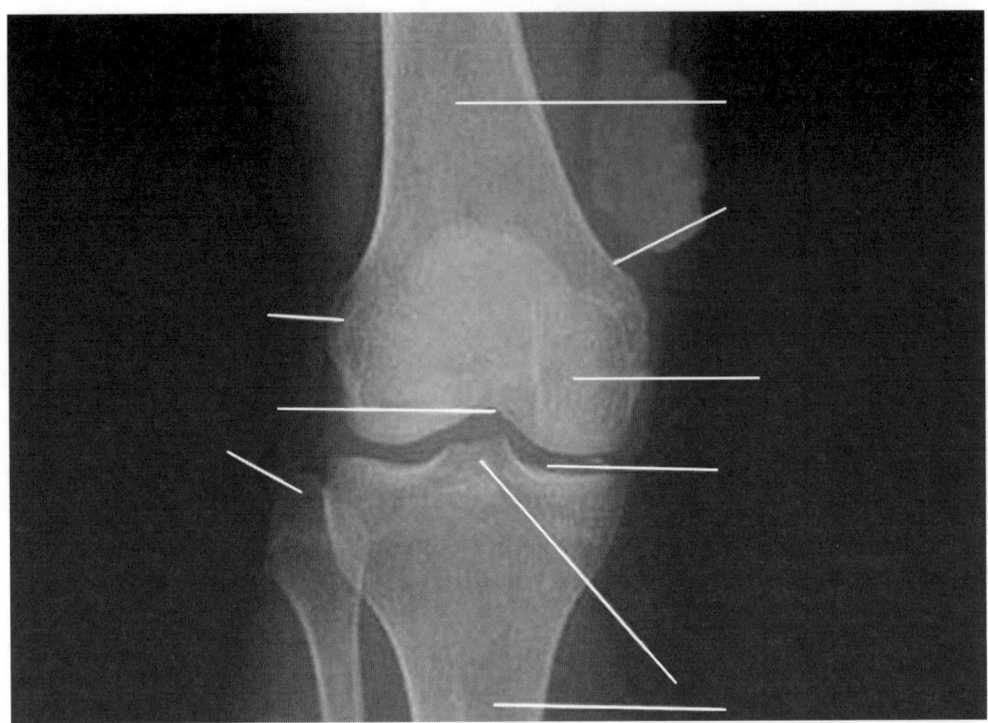

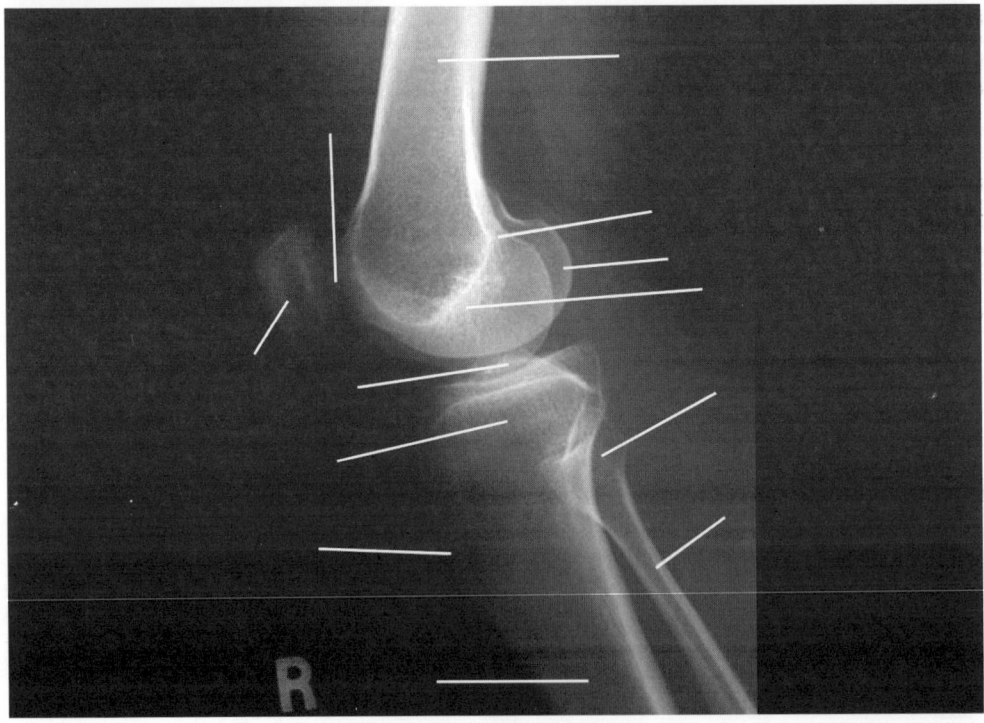

Radiological Anatomy: Head and Neck

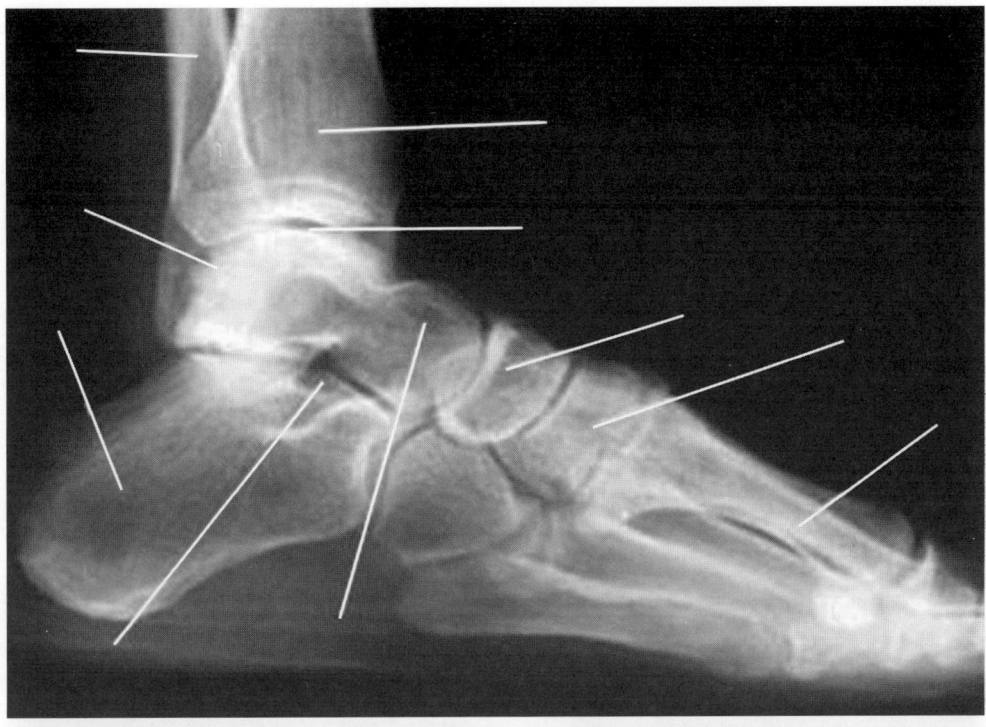

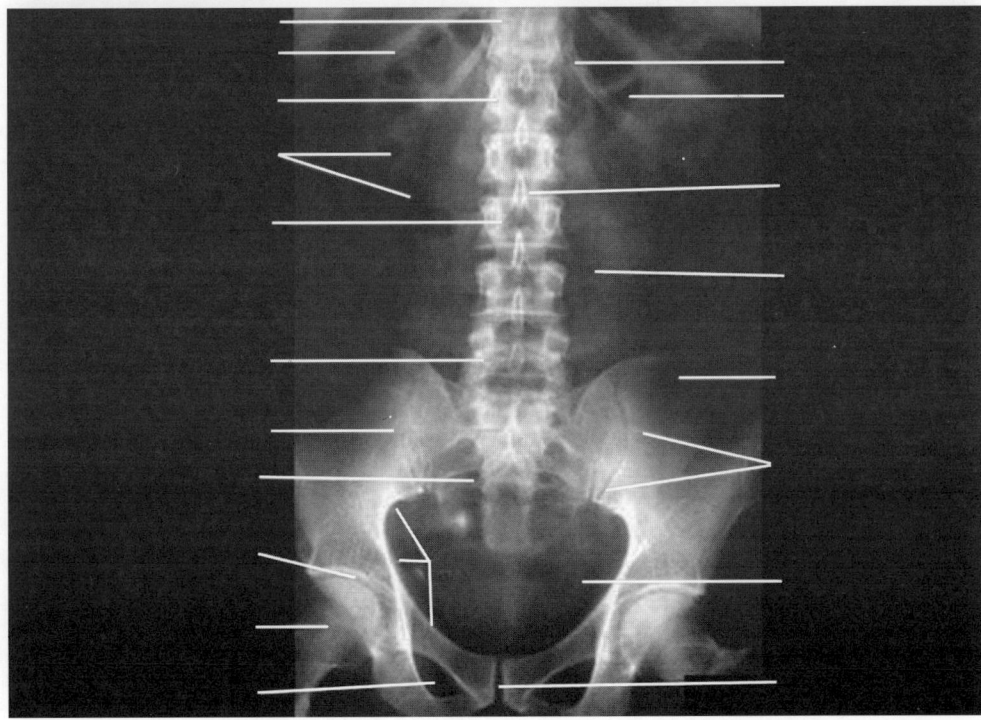

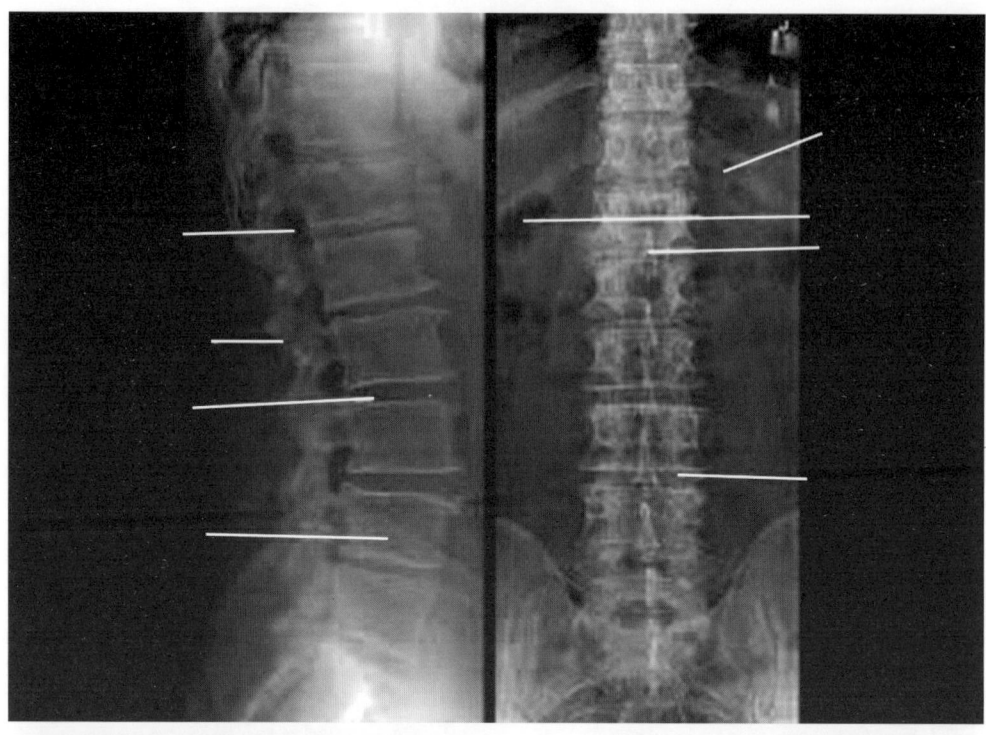
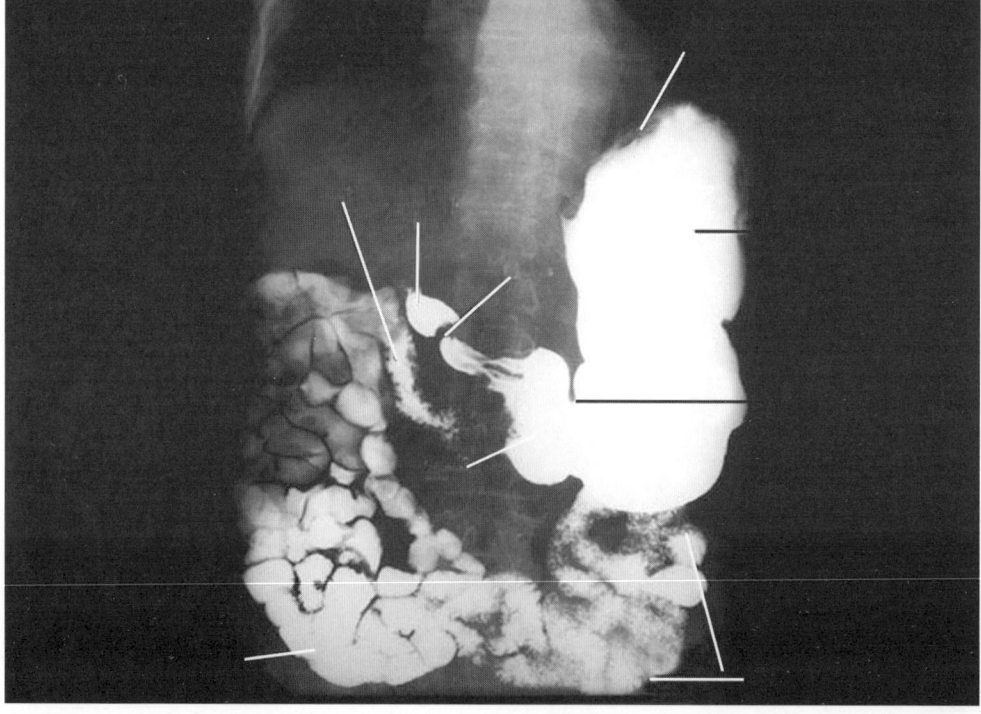

Radiological Anatomy: Head and Neck

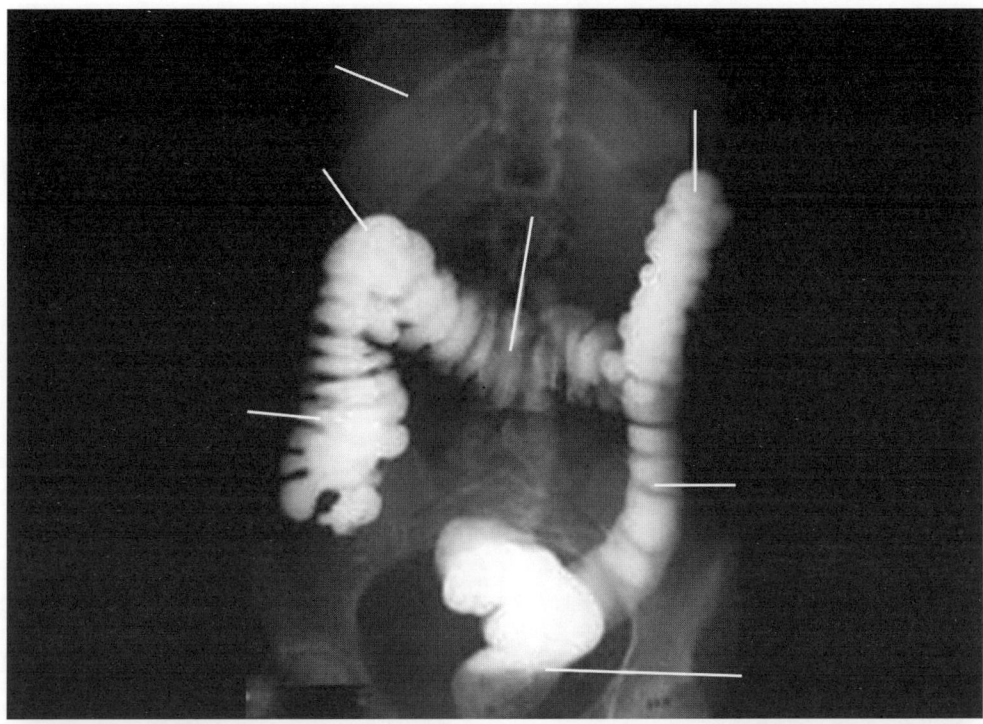

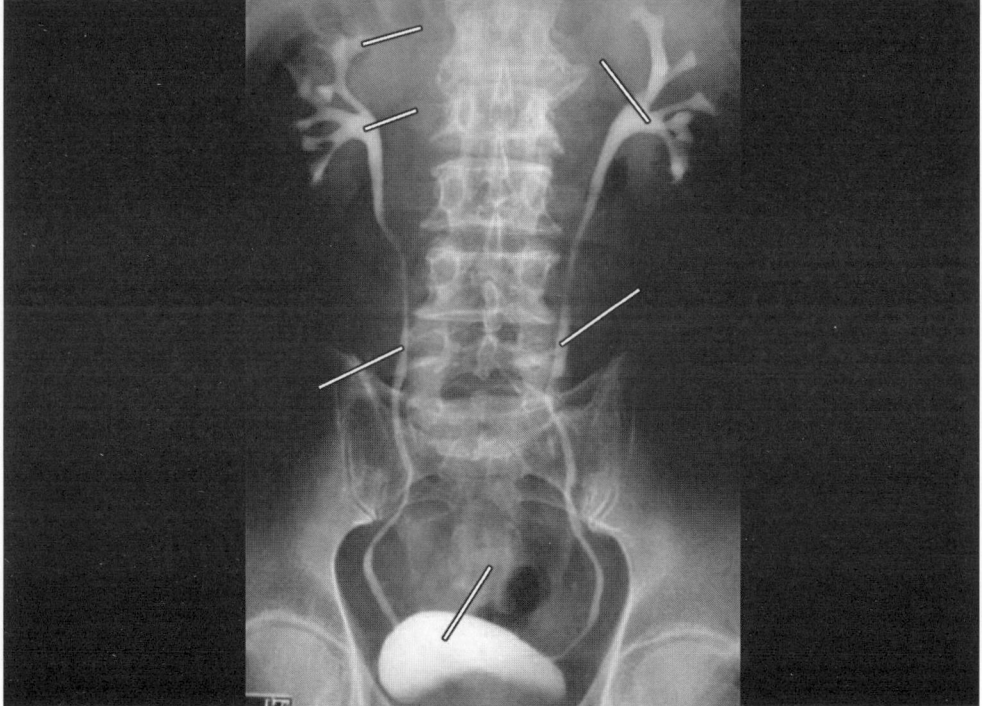

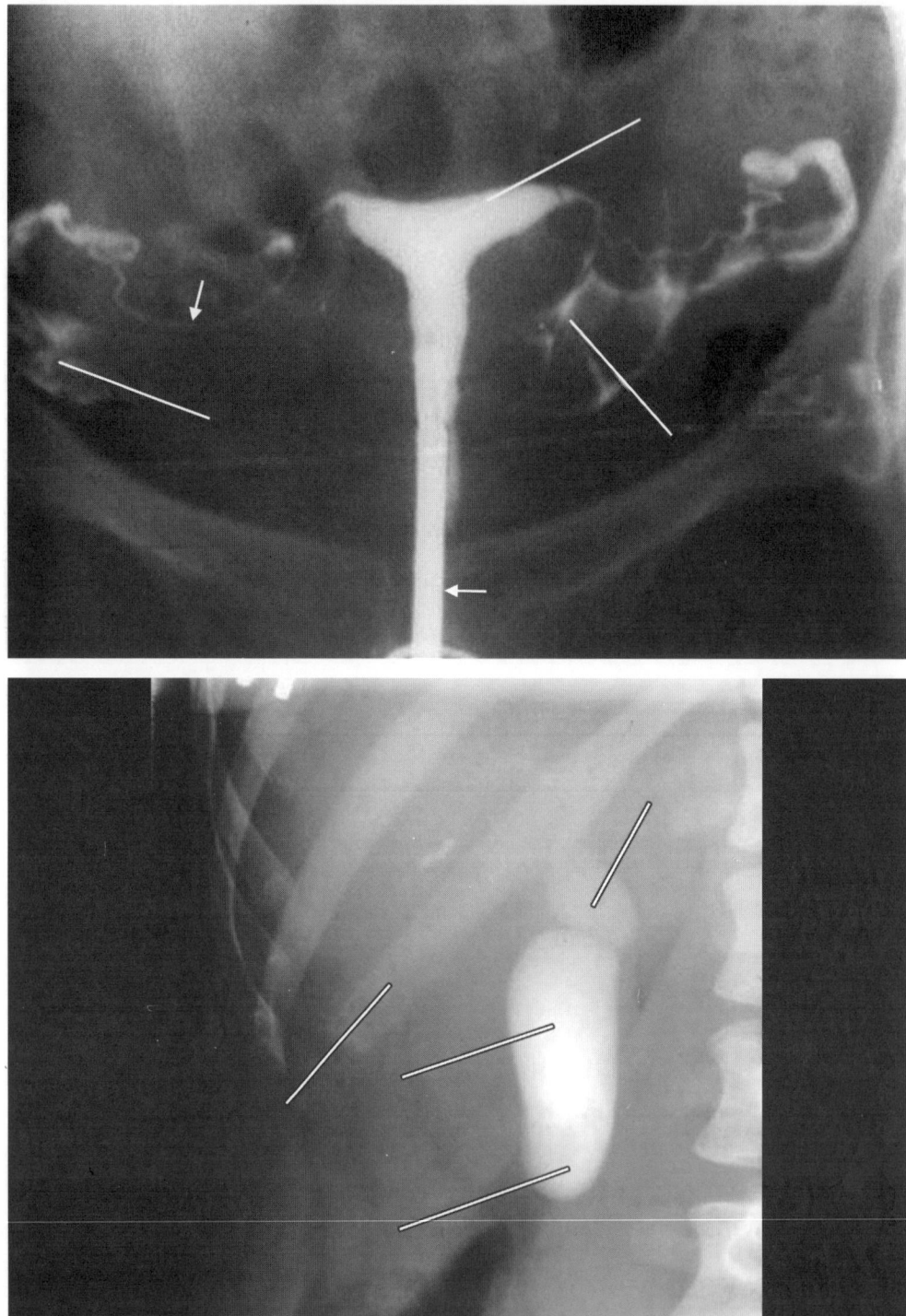

Radiological Anatomy: Head and Neck

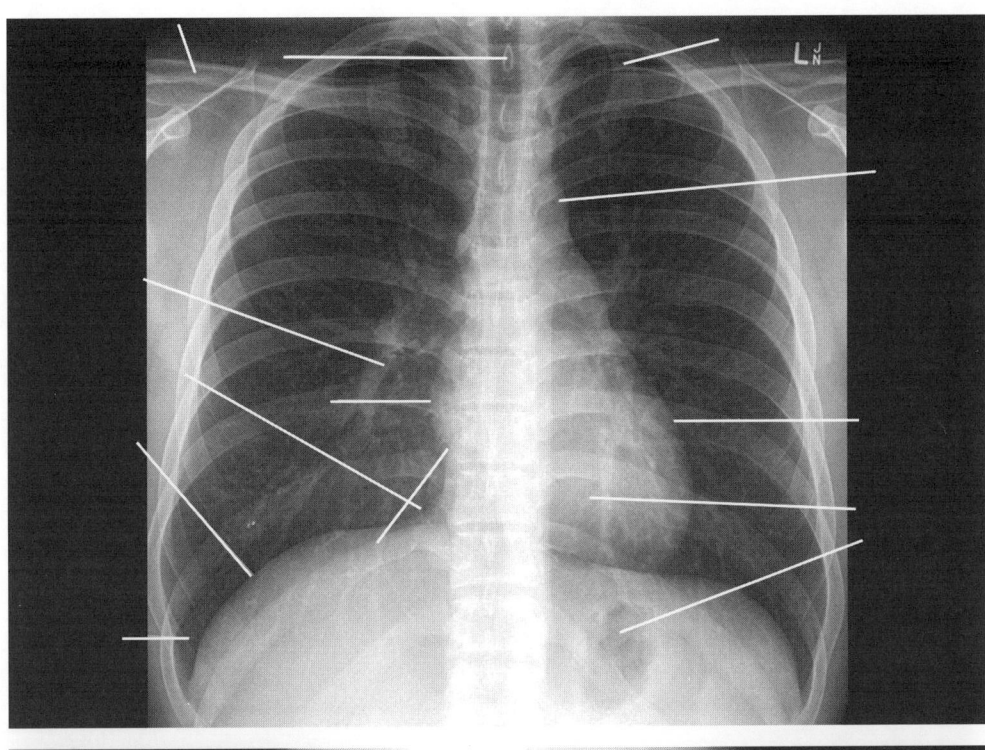

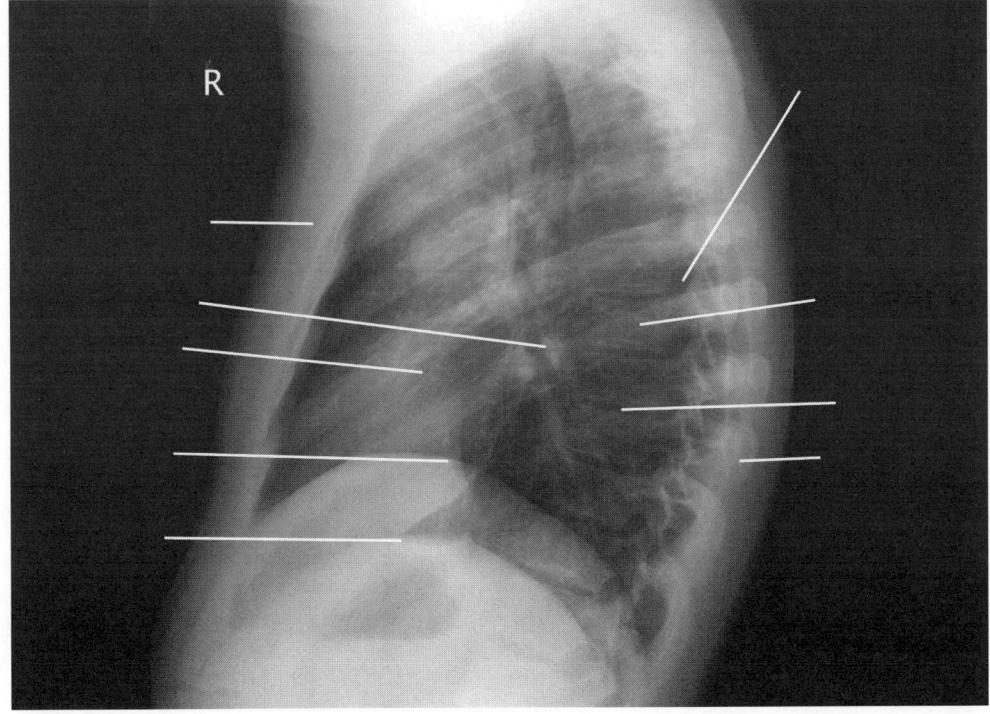

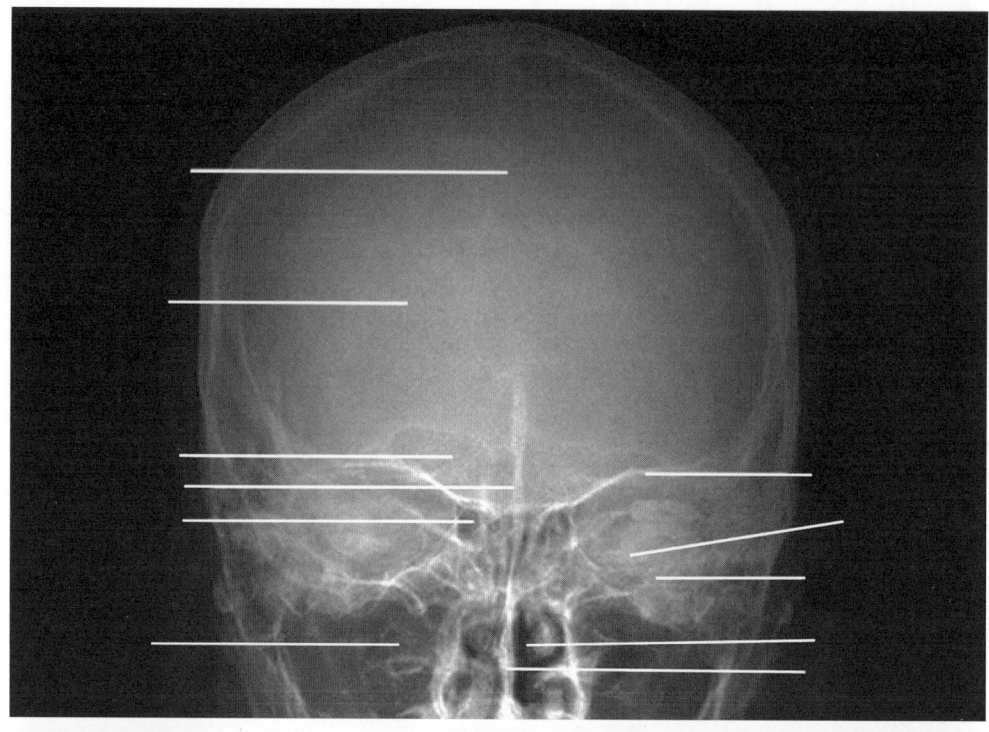
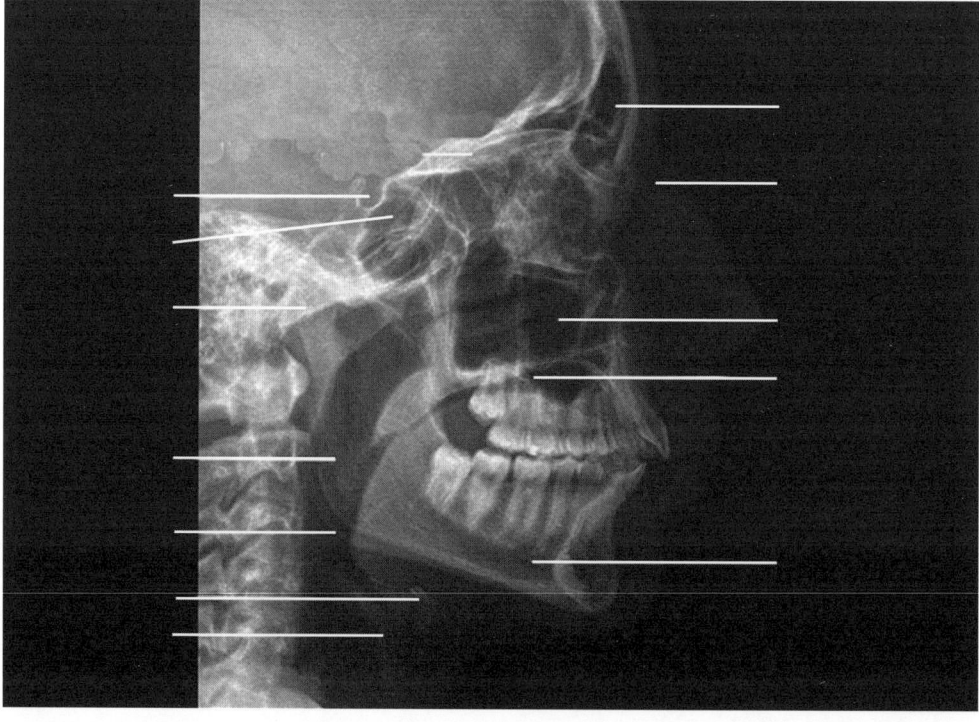

Index

Page numbers followed by *f* refer to figure and *t* refer to table

A

Abdomen 217, 357, 680
  boundaries of 217
  plain X-ray of 645, 680*f*
  viscera of 363
Abdominal wall, topographical divisions of 219, 219*f*, 220
Abduction 7, 98
Abductor digiti minimi 92, 94
Abductor hallucis 205
Abductor pollicis
  brevis 94
  longus 109, 112, 112, 335
Accessory nerve, spinal root of 424
Acetabulum 664
Achalasia cardia 711
Acini 502, 510
  types of 502, 510
Acinus of Rappaport 500
Adduction 7, 98
Adductor canal 143, 145-147
  boundaries of 146
  contents of 145*f*
  dissection of 143*f*
Adductor hallucis muscle 213
Adductor longus 132
  nerves supply of 143
Adductor magnus 147, 181, 183, 216
  ischial fibers of 182, 183
  muscle, hamstring part of 180
  parts of 183

Adductor minimus 148, 183
Adductor pollicis 93, 93*f*, 94
  muscle 92
Adductor tubercle 341, 342, 671
Adipose tissue 10
Adrenal cortex 525, 525*t*
Adrenal gland, regions of 521
Air blood barrier 557
  structure of 557
Air sinuses 717
Alveolar cells 556
Alveolar ductal system 601
Alveolar epithelium 555
Alveolar phagocytes 556
Alveoli 600
  epithelium of 555
  pulmonary 554
Alveolocapillary membrane 557
Ampulla 542
Amylacea 587
Anconeus muscle 106
  nerves supply of 115
Ankle 674, 674*f*
  joint 674*f*
  bones of 676
Annulus fibrosus 685
Ansa cervicalis 292, 293, 293*f*
Anserine bursa 201
Antebrachial fascia 60, 75, 76, 83
Anterior tibial artery
  branches of 156
  origin of 156

Anterior tibial nerve, origin of 155
Antrum 688
   formation of 596
Aorta
   arch of 381f, 393, 394, 414
   ventral branches of 367
Aortic knob 708
Aortic knuckle 394, 708
Apex
   beat 384
   direction of 383
Appendicular orifice 362f, 363, 364
Appendix 364, 365, 486, 487f
   arterial supply of 364
   inflammation of 365
   layers of 487
   length of 364
   parts of 364
Arch
   convexity of 338
   cortical 558, 560
Arcuate ligament
   lateral 243, 244
   medial 245
Areolar tissue 10
Argentaffin cells 481
Arm
   back of 102, 103f, 104f
   front of 60, 62f, 65
   lower part of 67
   middle of 65
   posterior compartment of 105
Arrector pili 8f
   muscle 9, 629
Arteria dorsalis pedis 156, 161f, 164, 209, 348, 350, 350f
   branches of 164
Arteria profunda brachii 330
   branches of 106
Arteria profunda femoris 135
   branches of 135
   fate of 135
Arterial wall
   coats of 14f
   layers of 460, 461

Arteries 14, 15f, 48, 78f, 110f, 196f, 259, 250f, 348, 349, 368, 427, 460
   acromiothoracic 37
   appendicular 364
   arcuate 351
   branches of 348
   concave side of 369
   convex side of 369
   cross-section of 14f
   elastic 14, 460f, 462t
   femoral 135-137, 147
   front of 333
   intercostal 251
   length of 346, 367
   lingual 282, 283f
   linguofacial 269
   metatarsal 351
   muscular 15, 462t
   musculophrenic 238
   peroneal 204
   pressure of 465
   size of 462
   subclavian 52, 303
   supraorbital 397, 404
   suprascapular 304
   tortuous 269, 422
   types of 462
   umbilical 226
Articularis cubiti 105
Articularis genu 140
Atavistic epiphysis 325
Atretic follicle 598
Auscultation, triangle of 45
Axilla 48
   base of 57
   medial wall of 55f
   muscles of 54
   shape of 49f
   skin incision of 49f
   suspensory ligament of 36, 57
   walls of 50f, 57
Axillary arch 56
Axillary artery 52, 59, 324f, 331-333
   branches of 52f, 332
   parts of 332

Axillary fascia 49
Axillary folds 57
Axillary nerve 48, 54
   branches of 47
Axillary sheath 51, 59
   content of 59
Axillary vein 52, 59

B

Barium enema 692-694
Barium meal 694
   study 689
   test 689, 690
Barium sulfate solution 689
   quantities of 695
Barium swallow study 711
Basal cells 473, 552
   role of 547
Basilic vein 60, 65, 69, 73, 76, 116
Basket cells 613
B-cells 530
Bell
   muscles of 572
   nerves of 49
Bertin, renal column of 560
Betz cells 619
Biceps
   brachialis 324
   brachii 13, 50, 56, 62, 65
      tendon of 73, 327
   femoris 181, 183, 184, 216
   long head of 324
   short head of 184
Bicipital aponeurosis 70, 71f, 74
Bile canaliculi 499
Biliary tree 702
Bird's beak appearance 711
Bladder, muscles of 575
Blind tube 365
Blood
   circulation 530
   nerve barrier 17
   pressure 467
   regurgitation of 420
   source of 496
   vascular system 13, 460, 634
   vessels 13, 23, 27, 28, 254
Blowing muscles 267
Blunt dissection 4
B-lymphocytes
   action of 530
   circulation of 530
Bones 3, 218, 649, 663, 670, 678, 715
   axial 30
   components of 443
   marrow, types of 448
   metacarpal 660, 661
   metatarsal 123
   shape of 322
   types of 322, 326, 443, 448, 649, 660, 664, 671, 677, 686
Bony landmark 342f, 355, 650, 682
Bony matrix
   composition of 445
   inorganic salts of 445
Bony pelvis 682
   bones of 665
Bow legs 673
Bowman's capsule 566, 567
Brachial artery 52, 63, 64f, 67, 70, 73, 75, 327, 333
   beginning of 327
   bifurcation of 74, 326
   branches of 68, 326, 327, 333
   terminal branches of 334
Brachial fascia 60, 66
Brachial plexus
   cord of 54, 59, 327, 329, 330
   largest branch of 328
   nerves of 53
   trunks of 304
   upper trunk of 302
Brachial vein 63, 64f
Brachialis 65, 73
   double nerve supply of 68
   muscle 62, 63f
Brachiocephalic vein, formation of 419
Brachioradialis 73, 114
Brain 255
Branchial arches 424
Breast 602
   structural components of 600

Brodie's bursa 201
Bronchial cartilage, types of 552
Bronchial muscle 551, 552
Bronchial wall, layers of 551
Bronchiole 554
Bronchus 554
Brunner's gland 483
    types of 482
Brush cells 552
Buccinator muscle 260, 267
Buccopharyngeal fascia 260, 273
Bugler's muscle 267
Bulkiest muscle 82
Burn, suprasternal space of 291f
Bursa 7, 20, 28

C

Cadaver, position of 34, 42, 69
Calcar femorale 667
Calvaria 714, 715
Canal, femoral 138, 139
Cancellous bone 447
Capillary 467, 496
    plexus 555
    types of 463
Cardiac glands 479
Cardiac muscle 456, 458, 459
    fibers 456f, 457
    microstructure of 456f
Cardiac myocytes 457, 458
    types of 458
Cardiac orifice 357, 358f, 359
    vertebral extent of 358
Cardiophrenic angle 392, 707
Carina 386
Carotid artery
    external 270, 281, 282, 297, 399, 415, 416, 418, 421
    internal 270, 281, 297, 399, 415, 416, 416f, 417, 418
Carotid body 284, 296, 399
    function of 399
Carotid sheath 282f, 292
Carotid sinus 296, 399, 414
    function of 414

Carotid siphon 418
Carotid triangle 281f
Carpal bones 30, 322, 323, 326, 660, 659, 678
Carpal tunnel syndrome 97, 98, 102, 340
Cartilage
    plates 552
    types of 385
Caval system 466
Cavity, abdominal 217
Celiac artery 367
    branches of 368
    origin of 367, 367f
Celiac plexus 368
Celiac trunk 368
    anomalous origin of 368
    largest branch of 368
Cells 506, 521, 523
    apical part of 491
    centroacinar 504
    intercalary 593
    intermediate 552
    layers of 573
    membranes 458
    mucous 503, 510
    types of 373, 472, 479, 480, 490, 505-507, 520, 532, 546, 552, 555, 580, 592, 601, 623
Cephalic index 403, 716
Cephalic vein 37, 38, 65, 69, 73, 116
Cerebellar
    cortex 612, 613, 614, 616, 617
        layers of 612
    cortical cells, types of 619
    glomerulus 617
Cerebellum 612, 612f, 618
    section of 612
Cerebral cortex, cell types of 619
Cervical
    curvature 685
    ganglia 312
    lymph nodes, superficial 311
    nerves 425
    plexus 258
    spines 721
    sympathetic ganglia 292
    vein, median 297

vertebra 721
vertebral column 721
Cervicoaxillary canal 48, 56, 57f
Chest, plain X-ray of 703f, 704f
Cholecystography
    intravenous 702
    methods of 701
    oral 700f
Chondroblasts 548
Chondrocranium 716
Chondroitin sulfate 548
Chromaffin cells 524
Ciliated columnar
    cells 552
    epithelium 439f
Circular fibers 478
Circular folds, functions of 483
Circumflex humeral artery 47, 48
Circumflex nerve 47
Circumflex scapular artery 48
Circumvallate papillae 471
Cirrhosis 498
Clara cells 552
Clarke's column 609
Clavicle 32f
Clavicular head 38, 54
Clavipectoral fascia 34, 34f-37f, 38, 39, 41
Cloquet's lymph node 143
Colloid, composition of 520
Colon 489
Colostrum 602
Common carotid artery 295, 398, 414, 415
    bifurcation of 296, 398
    branches of 399, 415
    right 414
    terminal branches of 414
Common facial vein 270
Common peroneal nerve 186, 192, 353-355
    branches of 155, 192, 354
Compact bone 442, 442f, 446, 447
Condylar articulation 670
Conheim's area 454
Coracobrachialis 50, 56, 62, 65
    origin of 324
Coracoid process, tip of 323, 324f

Cord
    medial 53
    medullary 532
    radial nerve 107
Cork-screw esophagus 711
Corona radiata 598
Coronoid fossa 655
Corpora amylacea 587
Corpus luteum 598, 599
    cells of 598
Cortex 544, 559f, 594
    parts of 534
    zones of 522
Costocoracoid ligament 41
Costophrenic angles 708
Coxa valga 667
Coxa vara 667
Cranial box, normal capacity of 716
Cranial cavity 266
Cranial nerve nuclei 423
Cranial root 424
    deep origin of 423
Craniosynostosis 716
Cribriform fascia 133
Cricoid cartilage 277, 400, 405
    arch of 400
    parts of 400, 401
    shape of 400
    vertebral level of 400
Crista terminalis 391
Cruciate anastomosis 176f
    formation of 176
Crural region, anterior 150f
Crypts of Lieberkuhn 483
Cubital articulation 654
Cubital fossa 68, 69f, 73, 74, 327, 336
    contents of 72f
    deep structures of 71f
    superficial structures of 69f
Cubital joint 318
Cubital tunnel syndrome 329
Cubital vein, median 70, 73
Cuboidal epithelium, simple 439f
Curling esophagus 711
Curvature, types of 721

Cutaneous nerve, medial 53, 62, 70
Cuvier, right duct of 395
Cystic duct 701
Cystic notch 376
Cystourethrography, micturating 697

D

Deep cerebellar nuclei 615
Deep cervical
  fascia 291
    layers of 290
    nodes 284
Deep fascia 10, 23, 25, 33, 34, 43, 49, 61, 70, 76, 87, 108, 127f, 133, 163, 260
  derivatives of 199
Deep inguinal
  lymph node 143
  ring 224, 226, 228
Deep nuclei 615
Deep palmar arch 117, 338
Deep peroneal nerve 155, 164, 350f, 355, 356
  branches of 356
  terminal branches of 164, 356
Deltoid ligament 344
Deltopectoral groove 40
Deltopectoral triangle 41
Dentate nucleus 616
Deodenojejunal flexure 362f, 369, 370f
Dermal papilla 8f, 626, 628
Dermatomes, overlapping of 18f
Dermis 8, 8f
  developmental source of 627
  layers of 626
Detrusor muscle 575
Diaphragma oris 294
Digastric muscle
  bellies of 288
  tendon of 288
Digastric triangle 286
Digestive tract, parts of 254
Digital sheaths 90f
Digitorum 204
Digits, abductor of 93
Dipalmitoyl phosphatidyl choline 556
Dissection, steps of 34, 42, 103

Distal tubule 569
Dorsal digital
  arteries 117
  expansion 120, 120f, 213
  veins 116, 162
Dorsal funiculus 611
Dorsal interossei 94, 95f, 209f, 213
  muscles 93
Dorsal metatarsal
  arteries 164
  vein 162
Dorsal subcutaneous space 116
Dorsal venous arch 159f, 162
  fate of 162
Dorsalis pedis artery 164
  pulsation of 164
Dorsum, intrinsic muscles of 161
Duct of Bellini 570
Duct
  basis of 509
  breadth of 413
  interlobular 511
  lactiferous 600, 601
  length of 413
  valve of 413
Duodenal mucosa 484
Duodenal wall, layers of 482
Duodenojejunal flexure 369, 688, 690
Duodenum 369, 370, 482, 482f, 485, 503, 649, 690
  barium meal X-ray of 688f
  muscles of 370

E

Ectoderm 5, 436
Elastic lamina 461
  internal 462
Elbow 68
  joint 654
    capsule of 105
    components of 654
    humeroradial part of 318
  plain X-ray of 652f, 653f
  region 652
Endocrine gland 254, 493, 517, 521, 635

Endoderm 436
Endometrium 589
   epithelium of 589
   layers of 590
Endomysium 11, 11f, 450f, 451
Endosteum 448
Endothelial cells 461
Endothelium 440
Enzymes, secretion of 505
Epidermal papillae 8, 8f, 626
Epidermis 8, 8f
   basal layer of 623
   blood supply of 626
   layers of 621f, 622
Epididymis
   canal of 579
   parts of 579
   tail of 579
Epigastric artery, superior 238, 239
Epigastric vessels
   superficial 126
   superior 232
Epimysium 11, 11f, 450f, 451
Epineurium 16
Epiphyseal line 342, 643, 660
Epiphyses
   compound 651
   simple 666
Epithelial cells 552
   types of 601
Epithelial tissue 436
Epitheliocytes, types of 537
Epithelium 436, 546
   characteristics of 436
   classification of 437
   simple 437, 438
   types of 437, 438t, 519, 562, 583
Eponychium 633
Erb's point 58, 59f
Erector spinae 241
Esophageal wall, layers of 475
Esophagus 475, 475f, 476, 711
   barium swallow of 709f, 710, 710f, 711
   carcinoma of 711
   length of 476
Estrogen, functions of 597

Ethmoid bone 720
Ethmoidal sinuses 719
   anterior 719
   posterior 719
Excretory system 636
Exocrine
   gland 493, 508, 509
   secretion, methods of 503
Extensor carpi
   radialis brevis 109
   radialis longus 109, 114
   ulnaris 109, 112, 321
Extensor digiti minimi 109, 112
Extensor digitorum 109, 112
   brevis 160, 161
      nerves supply of 163
   longus 151, 153, 160
Extensor hallucis
   brevis 161, 163
   longus 151, 153, 160
Extensor indicis 110, 113
Extensor pollicis
   brevis 109, 112, 335
   longus 110, 113, 335
Extensor retinaculum 118f
   function of 119
External carotid artery, branches of 294, 295f, 417
External jugular vein
   sinus of 420
   tributaries of 420
   valve of 309

F

Fabella 201
Face 255, 261
   deep fascia of 273
   muscles of 257f
   sensory supply of 264
   skin of 263
   veins of 271f
Facial artery 259, 260f, 269, 270, 272, 421, 422
   branches of 269, 294, 295f
   cervical part of 421
   origin of 269, 421
   parts of 421

Facial expression, muscles of 267
Facial nerve 259, 264
   branches of 265
   cervical branch of 277
   terminal branches of 265f
Facial skeleton 255, 714
   bones of 715
Facial vein 270, 270f, 272, 279, 422
Fallopian tube 591f
Fascia lata 133
   parts of 133
Fascia of Camper 221
Fascia of Scarpa 221
Fascia
   attachment of 36
   distribution of 36
   parotidomasseteric 260, 262, 273, 290
   superficial 10, 23, 25, 34, 43, 49, 60, 69, 76, 86, 107, 116, 126f, 221, 222f, 225, 231, 241, 262, 263, 273
   transversalis 226
Fasciculi 16
Fat 10
   paranephric pad of 241
Fatty tissue 602
Femoral artery
   branches of 135
   largest branch of 135
   last branch of 135
   origin of 134
Femoral nerve
   anterior division of 140
   branches of 139
   posterior division of 140
   root value of 139
Femoral ring 138
   boundaries of 139
Femoral sheath 129, 130f, 137
   compartments of 138
   function of 139
Femoral triangle 125, 131, 134, 136, 137
   boundaries of 125f, 132
   contents of 128f
   dissection of 125f
   floor of 131f
Femoropatellar articulation 670

Femur 123f
   epicondyles of 671
   head of 663
   upper end of 666
Fibers 611, 618
Fibrillary lamellae 447
Fibroelastic coat 551
Fibrous flexor sheath 90f
Fibula 123f
   lower end of 676
   neck of 348, 354, 355
Filiform 474
   papilla 470, 471
Flexion 7
Flexor carpi
   radialis 76, 81, 329
   ulnaris 77, 81, 331
Flexor digiti minimi brevis 92, 94
Flexor digitorum
   accessorius 212
   brevis 205
   longus 197
   profundus 77, 80, 82, 83, 331
   superficialis 76, 78f, 81
Flexor hallucis
   brevis 213
   longus 197
Flexor muscle
   layers of 84
   superficial 85
Flexor pollicis
   brevis 92, 94
   longus 80, 82, 88, 92
   tendon 91
Flexor retinaculum 88, 88f, 89f, 97, 338, 339f, 340
   attachment of 326, 339
   opening of 89f
Fluteal region 172
Foliate papillae 473
Follicle
   lymphatic 528, 531
   primary 595, 599
   primordial 594, 599
   secondary 595
   size of 518
   structure of 519
   types of 543

Follicular cells 519, 520, 521
Follicular epithelium, types of 519
Foot 674
    bones of 124*f*
    commonest deformity of 679
    dorsiplantar view of 675*f*
    dorsum of 158, 158*f*-160*f*, 161, 163, 165, 356
    eversion of 345
    layer of 206f-208*f*
    proximal part of 674*f*
    sole of 204, 210
    transverse arches of 214
Foramen of Morgagni 239
Forearm
    back of 107, 107*f*, 108*f*, 110*f*, 112
    front of 75, 81
    internervous line of 82
    posterior compartment of 109, 109*f*, 115
Fossa 7
    femoral 139
    roof of 188
Frankfurt line/plane 717
Frontal air sinus 411, 411*f*, 412
Fundic gland tube, parts of 480
Fundus
    capacity of 361, 372
    direction of 363
    vertebral level of 363
Fungiform papillae
    location of 471
    shape of 471

G

Gallbladder 362, 701
    additional X-ray of 702
    arterial supply of 363
    capacity of 362
    fundus of 361, 362, 362*f*
    parts of 362
Gamellus
    inferior 171
    superior 171
Ganglia 16
Gastric
    epithelium 482
    gland 477*f*, 479
        types of 479
    mucosa 479
    pit 479
        depth of 479
Gastrocnemius 197, 200
    muscle 194
        transection of 195*f*
    part of 201
Gastrohepatic ligament 374
Gastrointestinal
    system 469, 634
    tract, part of 711
Gemelli muscles, nerves supply of 175
Genicular nerve 353
Geniohyoid 284
Genital organs, external 10
Genu
    recurvatum 673
    valgum 673
    varum 673
Germinal
    epithelium 595
    matrix 633
Glands 509, 515, 586, 590
    activity of 519
    adrenal 521
    endocrine types of 492
    exocrine types of 492
    medial surface of 409
    mucous 576
    parathyroid 410
    pyramidal lobe of 410
    substance, parts of 527
    types of 476, 493, 502, 509, 600
    weight of 409
Glandular tissue 600
Glass membrane 597
Glisson's capsule 494
Glomerular
    basement membrane 566
    capillary
        endothelium 565
        membrane of 565
    capsule 566
    filtration barrier 567
    plexus 565

Glomerulus
  arterioles of 565
  shape of 617
  size of 617
Glucagon 507
Glucocorticoids 523
  function of 523
Gluteal aponeurosis 167f, 168f, 173
Gluteal nerves, superior 169f, 174
Gluteal region 165, 170
  dissection of 166f
  gateway of 173
  largest muscle of 173
Gluteus maximus 170, 173, 174, 176, 177
  nerves supply of 173
Gluteus medius 170, 178
Gluteus minimus 170, 178
Goblet cell 484, 490, 491, 546, 552
  function of 491
  secretion of 490
Goiter 410
  lactational 410
  physiological 410
  retrosternal 410
Golgi cells 614, 617
Graafian follicle 597
Gradual dehydration 434
Granule cell 616, 618
  fibers of 614
Gravity
  center of 122f
  line of 122f
Great auricular nerve 302, 310
Great saphenous vein 136, 137, 162, 344
  valves of 137
Greater sciatic foramen 174
Greater tubercles 651
Gustatory cells 472
  function of 472

H

Hair 8f
  bulb of 628
  follicle 8f, 628
    parts of 628
    structure of 9f, 621f

functions of 627
papilla 628
Hallucis 204
Hamate bone, shape of 326
Hamstring muscles 182
  origins of 181f
Hand
  dorsum of 115, 116f, 117f, 118, 118f
  retinaculum of 340
  skeleton of 33f
  spaces of 96
Hassall's corpuscles 537, 544
Haversian canal 446
Haversian system 446, 447
  components of 446
Head and neck 254, 396
  radiological anatomy 712
Heart 13, 14, 390
  apex of 381f, 383, 384
  borders of 390, 707
  chamber of 383
  failure 556
  inferior border of 392
  left border of 391
  peripheral 202
  right border of 389
Hemidiaphragm 708
Hemolymph organ 538
Hepatic artery, radicles of 496
Hepatic cells 495
  surfaces of 499
Hepatic lacunae 495
Hepatic lamina 495
  cells of 495
Hepatic lobule 494
Hepatic lymph 498
Hepatic sinusoids 495
  cells of 497
Hepatocyte 494
Herings, canal of 500
Hernia, femoral 139, 143
Hesselbach's triangle 230
Heterocrine gland 514
Hiatus muscularis 365, 487
Hilton's law 47

Hip
  bone 664
    parts of 664, 665
  joint 663, 670
Homocrine gland 514
Hormones 520
  parathyroid 521
Humerus 32f
  condyle of 654
  epicondyles of 654
  lower end of 655, 656
  proximal end of 650
  surgical neck of 48
Hyaline cartilage 547, 548, 552
Hybrid muscle 142, 183
Hydration 435
Hyoid
  bone 277, 406
    ossification of 407
    shape of 406
  greater cornu of 406, 406f
  parts of 406
  vertebral level of 406
Hypodermis 621
  composition of 621
Hyponychium 633
Hypothenar muscles 92f, 94
Hysterosalpingogram 699

I

Ileum 485, 485f, 486
  length of 692
Iliac artery, external 136
Iliac crest 218
Iliac spine, anterosuperior 221
Iliocecal orifice 364
Ilioinguinal nerve 223, 224f, 228
  branches of 126f
  root value of 228
Iliolumbar ligament 243
Iliopubic tract 235
Iliotibial tract 134, 167f
Ilium 664
Immunity, types of 530
Incision 103f, 107, 107f

Incisor teeth 358
Incisura angularis 374, 688
Incisura apicis cordis 392
Indirect hernia 230
Inferior epigastric
  artery 224, 227, 233, 238
  vessels 232
Inferior extensor retinaculum, function of 163
Inferior gluteal
  nerves 169f
  vessels 169f
Inferior labial artery 421
Inferior radioulnar joint, types of 319
Infrahyoid muscles 280, 285, 286f, 294, 298, 407
Infraorbital artery, origin of 404
Infraorbital foramen 398, 403
Infraorbital nerve 403, 404
  branches of 404
  fate of 404
Infraorbital plexus 404
Infratrochlear nerve 404
Inguinal canal 221, 224, 225, 225f, 226, 228, 230
  dissection of 222f
Inguinal fossa, medial 230
Inguinal hernia 230
Inguinal ligament 135, 142, 223
  parts of 142
Inguinal ring, superficial 227f
Inlet venules 496
Insertion of sartorius 141
Insulin 507
Integumentary system 620, 637
Intercostal arteries
  anterior 251
  posterior 251
Intercostal muscle
  external 249
  internal 249
Intercostal nerve, branches of 252
Intercostal space 249, 250f
  dissection of 248
  incision of 249f
  types of 251
Intercostal veins, anterior 252
Intercostalis intimus 249, 253
Interlobular artery, origin of 560

Interlobular ducts, epithelium of 511
Intermuscular septum, medial 66
Internal carotid artery
    branches of 417
    origin of 415
Internal jugular vein, drainage of 296
Interneurons, primary function of 607
Interossei muscles 94, 95f, 213
    nerves supply of 214
Interosseous artery
    anterior 80
    posterior 80, 84, 110, 115
Interosseous nerve
    anterior 80, 84
    posterior 79, 329
Intersinusoidal pressure 497
Interstitial lamellae 447
Intestine, large 488
Intralobular ducts, types of 511
Intrinsic muscles 96, 210, 474
Ischiopubic ramus 664
Ischiorectal fossa 10
Ischium 664
Islets of Langerhans 501f
Ito cells 498
    role of 498

J

Jejunum 485, 485f, 486
    length of 692
Joints 2, 20, 28, 325f, 405, 665, 672, 681
    acromioclavicular 651
    cartilaginous 20
    classification of 21
    complex 670
    compound 654, 670
    fibrous 20
    humeroulnar 656
    lower limb 124f
    midtarsal 678
    movements of 29
    shoulder girdle 652
    simple 670
    space 663
    subtalar 678
    symphyseal 665
    types of 28, 649, 651, 654, 658, 664, 665, 670, 676, 720
Jugular foramen 424
Jugular vein
    anterior 277, 297
    external 270f, 297, 300, 304, 308, 309, 416, 419, 420
    internal 281, 296, 416, 418, 419
Jugular venous arch 297
Juxtaglomerular apparatus 559f, 571
Juxtaglomerular cells 571
Juxtamedullary nephrons 564, 564t

K

Keratinized stratified squamous epithelium 440f
Keratinocytes 623
    functions of 623
Kidney 242, 243, 377, 378f, 558, 564, 681, 696
    back of 379
    coverings of 379
    cut-section of 559f
    exposure of 240
    interstitial tissue of 571
    lobe of 560
    lobule of 560
    macroscopic parts of 558
    microscopic
        elements of 562
        structure of 562
    structure of 559f
Knee joint 670
    anatomical angle of 673
    space 670
Knee region 668
    plain X-ray of 668f, 669f
Kocher's vein 409
Krause's membrane 454
Kupffer's cells 497
Kyphosis 687

L

Labial artery, superior 421
Laborer's nerve 329
Lacunae 444

Lamellae, types of 447
Lamellar bone 445
Lamina propria 487, 547, 589, 593
   thickness of 576
Langer's line 8, 24
Langerhans cells 625
Large intestine, parts of 694
Laryngeal cartilages 400
Lateral plantar nerve, branches of 215
Latissimus dorsi 55
Laughing muscle 268
Leg
   anterolateral compartment of 148, 153
   back of 192
   compartments of 123*f*, 149*f*, 154, 193*f*
   deep transverse fascia of 202
   lateral compartment of 151*f*
   posterior compartment of 199
   retinaculum of 156
Lesser omentum, layers of 374
Levator glandulae thyroideae 410
Leydig cells 583
   functions of 583
   growth of 583
Ligament of Treitz 370
Ligamentum denticulatum 424
Ligamentum nuchae 401
Ligamentum teres 376
   femoris 376
   hepatis 376
Linea alba 238
Linea aspera 342
Linea semilunaris 238, 380
Lingula 389
Lissauer, dorsolateral tract of 611
Liver 492-494, 500, 635
   bile, source of 499
   composition of 494
   coverings of 493
   injury 498
   lower border of 374, 375*f*
   size of 493
Lobule of Kiernan 494
Loop of Henle 569
   parts of 568
Lordosis 687

Lower limb 341
   bones of 124*f*
   radiological anatomy 662
Lumbar
   curvature 685
   fascia, layers of 244, 379
   spine 366*f*
   vertebra, sacralization of 687
Lumbocostal ligament 244
Lumbosacral spine, plain X-ray of 683*f*
Lumbricals, nerves supply of 213
Lung 549
   anterior border of 388, 389
   elastic recoil of 557
   layers of 550
   lobule 550
Lymph capillaries 16
Lymph node 16, 142, 254, 526, 526*f*, 527, 528
   capsule of 527
   components of 527
   lobules of 527
   parts of 532
   peripheral part of 532
   section of 526*f*
   supraclavicular 304
Lymph vessels 13, 15, 38
Lymphatic follicle, secondary 543
Lymphatic nodule, primary 528
Lymphatic system 526, 636
Lymphocyte 529, 530, 535, 552
   types of 529-531, 540, 544
Lymphoid follicles, types of 528
Lymphoid tissue, mass of 427

M

Macula densa 571
Malignant tumors, metastasis of 467
Malleolar fossa 345
Malleoli lies 676
Mammary gland 599-602
   duct of 601
   lactating 599*f*
   lobe of 600
   nonlactating 599*f*
Mandible, angle of 719

Marginal vein, medial 162
Mast cells 552
Mastication, muscles of 262, 267
Mastoid air cells 720
Mature bone 443
Maxillary sinus 718, 718f
McBurney's point 364, 365, 365f
M-cells 486
    function of 486
Medial circumflex femoral artery, branches of 136
Median nerve
    branches of 68, 75
    medial root of 53
    root value of 328
Mediastinum testis 578
Medulla 531, 544
    adrenal 524
Melanin
    formation 624
    role of 625
    synthesis 623, 624
Melanocytes 623, 624
    origin of 623
Melanomas, malignant 625
Membrana granulosa cells 596
Merkel's cells 626
Mesangial cells, functions of 566
Mesenteric artery, superior 367f, 368
Mesentery
    borders of 377
    breadth of 377
    contents of 377
    functions of 377
    root of 370f, 376, 377
Mesoderm 436
Mesothelium 440
Metopic suture 716
Microscope, types of 431f, 432
Mineralocorticoids, role of 522
Morris parallelogram 242, 243f, 378
Motor nerves 259, 259f, 424
    types of 606
Mucosa, layers of 483
Mucous acini 510, 510t, 511f
Mucous membrane 576
    composition of 441
    layers of 478, 546, 573, 576

Muscles 3, 11, 23, 38, 45, 58, 65, 73, 81, 94, 112, 131, 153, 161, 170, 181, 182, 210, 218, 239, 251, 261, 284, 304, 324, 343, 356, 400, 450, 572
    action of 12, 26
    antigravity 83
    coat, layers of 589
    composite 183
    deep 76, 83, 108f, 112
    extrinsic 96, 210, 474
    fasciculi of 451
    fiber 451, 452, 454, 456, 586
        development of 455
        types of 459
    hamstring group of 183f
    hyoglossus 298
    intercostal 250f
    kinds of 26
    layer 575, 592
    lumbrical 212
    paradoxical action of 27
    peroneal 157
    proteins of 453
    soleus 201
    sources of 451
    superficial 76, 78f, 83, 108, 108f, 112
    suprahyoid 284, 285f, 294, 407
    tissue, components of 451
    transection of 62f
    types of 450
Muscularis externa 476
Muscularis mucosae 488
Musculoaponeurotic canal 225
Musculoskeletal system 442, 634
Musician's nerve 99, 329
Myasthenia gravis 537
Mylohyoid 284, 298
Myoepithelial cells 512, 514, 601
Myometrium 589
Myotome 18, 28

N

Nail
    components of 632
    lunule of 721f, 632
    parts of 621f, 632

Nasal artery, lateral 421
Nasal septum 720
Navicular bone 343
   surface of 343
Neck
   anterior triangle of 274, 275f, 276f, 284
   front of 398f
   muscles of 288
   posterior triangle of 298, 299f, 300f, 301f, 304, 424, 425
Nephron 564
   cortical 564, 564t
   developmental source of 564
   functions of 564
   parts of 563
Nerves 3, 16, 49, 65, 110f, 196f, 250f, 258, 289, 328
   accessory 423, 424
   femoral 139, 140
   hypoglossal 279, 282
   intercostal 239, 251, 252
   intercostobrachial 58
   lingual 279
   median 53, 60, , 62, 67, 70, 72f, 73, 74, 79, 81, 327, 327f, 328
   musculocutaneous 53, 59, 63f, 67
   peripheral 27
   peroneal 353f
   supraclavicular 302
   supraorbital 397, 404
   suprascapular 58, 302, 304
Nervi hesitans 356
Nervous system 603, 637
   classification of 19
Neurocranium 716
Neurons 613
   structure of 16f
   types of 613
Neurovascular hilus 12
Nodules, lymphatic 528
Nonciliated simple columnar epithelium 439f
Nonkeratinized stratified squamous epithelium 439f
Nose, root of 403
Nuclei 16
Nucleus
   dorsalis 609

   interpositus 615
   proprius 609
Null cells 531

O

Obturator
   externus 171
   foramen 664
   internus 171
   nerve 183
Olecranon fossa 655
Omohyoid 280, 285
   fascia 308
   inferior belly of 308
   muscle 300
   superior belly of 293
Oogenesis, stages of 597, 598
Opponens
   digiti minimi 92, 94, 99
   pollicis 92, 94
      muscle 99
Orbicularis oculi 261, 268
   muscle 257f
Orbicularis oris, parts of 268
Organs
   lymphoid 544
   types of 533, 538
Orifices 358, 363
Osteoblasts cells 444
Osteoclasts cells 444
Osteocytes cells 444
Osteoprogenitor cells 443, 444
   types of 444
Ovarian follicles 594
Ovary 593, 594f
   gubernaculum of 230
   substance of 594
   surface of 594
Oxyntic cells 480

P

Palatine tonsil 416, 425, 427, 542, 542f
   epithelium of 543
   parts of 426
Palm, deep muscles of 93f

Palmar
　　aponeurosis 77, 87, 87f, 212
　　arch, superficial 337, 337f, 338
　　carpal arch 117
　　fascia 87
　　ganglion 100
　　interossei 93, 94, 95f
Palmaris brevis 94
　　muscle 86
Palmaris longus 76, 81
　　parts of 77
Pancreas 492, 501, 501f, 505, 635
　　exocrine
　　　　part of 502
　　　　system of 504
　　functional components of 501
　　parts of 505
Pancreatic
　　acinar cells 503
　　cell, types of 502
　　serous cells 503
Paneth cells 484, 486
Panniculus adiposus 10, 25
Panniculus carnosus 25, 97, 261, 276, 289
　　remnant of 86
Papillae
　　lingual 470
　　types of 470, 472, 474t
Paracrine gland 514
Parafollicular cells 521
Parallelogram, level of 243
Paranasal air sinuses 412, 717-720
　　functions of 412
Paratonsillar vein 426
Parenchyma 517
Parietal cells 480
Parotid duct 260, 273, 411f, 412, 413, 513
　　bents of 274f
　　emergence of 273
　　structural layers of 414
Parotid fascia 509
Parotid gland 260, 413, 507-510, 516t
　　accessory 260
　　development of 414
　　microstructure of 508f
Patella, lateral displacement of 673

Patellar plexus 148
Pectineus supply muscle 140
Pectoral fascia 34, 40
Pectoral nerve, lateral 37
Pectoralis major 54
Pectoralis minor 38, 54, 332
Pelvic inlet 682
Pelvifemoral space 218
Pelvis 219f
Peptic ulceration 374
Pericardium, fibrous 395
Perimetrium 589
Perimysium 11, 11f, 450f, 451
Perineurium 16
Periosteum, layers of 448
Peritoneal spillage 700
Perivitelline space 597
Peroneal artery, branches of 203
Peroneal nerve, superficial 155, 355
Peroneus brevis 151-153
Peroneus longus 151-153, 354
Peroneus tertius 151, 153
Pes cavus 679
Petineus, medial part of 140
Petoralis major 38
Peyer's patch 486
　　length of 486
Phalanges 123, 661, 677
Phospholipid, principal constituent of 556
Phrenic nerve 304, 311
Pilosebaceous apparatus 630
Pisiform 323
　　bone 321
　　ossify 323
Pituitary fossa 719
Plantar aponeurosis 205, 206f, 209, 210-212
　　parts of 211
Plantar arterial arch 215
　　branches of 215
Plantar arteries 209, 216
Plantar fascia 211
Plantar interossei muscles 214
Plantar nerves 216
Plantaris 200
　　muscle 200
Platysma 276, 277f, 289, 300, 304, 305

morphological basis of 289
muscle 289, 308
nerves supply of 308
Pneumocytes 555, 556
Podocytes 567
Popliteal artery 136, 189, 190, 203, 345, 346, 346f, 347
branches of 347
genicular branches of 189
length of 189
origin of 346
terminal branches of 190, 349
Popliteal fossa 184, 188-190, 192, 347, 351, 352, 352f
boundaries of 185f, 188
contents of 186f
dissection of 185f
floor of 187f
Popliteal nerve, lateral 191
Popliteal vein 190, 347
Popliteus muscle 353
Porta pedis 204
Portal acinus 493f
Posterior cerebral artery, origin of 417
Posterior crural region 192, 196f, 203
dissection of 193f
Posterior tibial artery
branches of 349
origin of 349
Posterior triangle, floor of 307f
Potato tumor 298, 399
Prepyloric vein of Mayo 360
Pressure epiphysis 667
Primordial germ cells 580
Progesterone 596
Pronator quadratus 80, 82
Pronator teres 73, 75, 76, 81
muscles 329
Prostate 585f, 586
capsule of 586
glands of 587
glandular part of 586
lobe of 586
Prostatic gland, components of 586
Proteins 453
Proximal colon 692

Proximal tubule 568
Pseudoganglion 165, 356
Pseudostratified ciliated columnar epithelium 439f
Psoas major muscle 682
Pterygoid
lateral 262
medial 262
Pubic angle 665
Pubic tubercle 226
Pubis 664
Pudendal arteries, external 136
Pulp space 96f, 101f
Purkinje cell 613, 614, 619
shape of 613
Pyelogram, ascending 697
Pyelography
antegrade 697
intravenous 695f, 697
Pyloric canal 690
Pyloric constriction 359
Pyloric glands 481
Pyloric orifice 358f, 359, 360
position of 359
vertebral level of 360
Pyloric sphincter 359, 360
Pyramidalis, action of 239

Q

Q angle 673
significance of 673
Quadrangular space 42, 43f, 46
Quadratus lumborum 241
Quadratus plantae 212
Quadriceps femoris 13, 140
nerves supply of 140

R

Radial artery 63, 79, 334, 334f, 335
beginning of 334
branches of 330, 335
position of 335
unusual origin of 335
Radial bursa 96, 100
Radial fossa 655

Radial head 317
Radial nerve 52, 53, 59, 63, 67, 74, 79, 81, 115, 327, 328-330, 335, 427
    injury 328
    branches of 106
    root value of 107, 328, 330
Radial notch 656
Radial tuberosity 655
Radiating septa 101f
Radiological joint space 643, 665, 665
Radioulnar joint 318, 654
Radius 32f
    head of 317, 319
    lower end of 320, 659
    styloid process of 320
    upper end of 655
Raphe 7
Rectal wall, coats of 488
Rectum 488, 488f, 489
Rectus abdominis 233f
    lateral border of 380
    muscle 232, 239
Rectus sheath 231, 234, 238, 239
    dissection of 232f
    formation of 234f, 237f
    posterior layer of 232
    posterior wall of 233f, 236f
Rectus sternalis muscle 55
Recurrent genicular nerve 192, 355
Renal corpuscle 564
    elements of 563
    vascular pole of 566
Renal cortex, parts of 558
Renal
    glomeruli 565
    lobes 560
    medulla, parts of 560
    papilla 561
    parenchyma, parts of 558
    pyramid 561
        depth of 569
    sinus 561
    tubule
        neck of 567
        parts of 563

Reproductive system
    female 588, 637
    male 577, 636
Respiratory system 545, 636
Rete testis 579
Retinaculum 7, 25
Retromandibular vein, formation of 271
Rhisorius muscle 268
Ribs 30
    parts of 706
Rider's bone 142
Risus sardonicus 268

S

Sacral vertebra, lumbarization of 687
Sacroiliac joint 665, 682, 683
Saliva, functions of 509
Salivary corpuscles 544
Salivary glands 254, 508, 509, 511, 512, 515, 516, 635
    duct of 512
    lobules of 514
Saphenous opening, dimensions of 133
Saphenous veins 162, 163
Sarcolemma 452
Sarcomere, length of 455
Sartorius
    muscle 141
    nerves supply of 141
Satellite cells 456
Saturday night palsy 330
Scalenovertebral triangle 312
Scalpel 4
Scapula 32f
    three angles of 650
Sciatic nerve 184, 354
    distribution of 175
    root value of 175
    terminal branches of 351, 354
Scoliosis 687
Sebaceous
    cyst 102, 210
    ducts 630
    gland 8f, 630, 631

Sebum
  composition of 630
  functions of 630
Secretion
  apocrine method of 490
  function of 472
  method of 490
  types of 504
Secretory cells, functions of 593
Secretory endometrium 590, 590t
Sella turcica 719
Semimembranosus muscle 189
Seminiferous epithelium 580
Seminiferous tubule 583
  parts of 579
  section of 581
  wall of 580
Sensory nerve 258, 424
Septum, femoral 138
Serous acini 510, 511f
  cells of 503
Serous cells 503, 510, 552
Serous coat, cells of 592
Serous demilunes 514, 516
Serratus anterior 55, 55f
Sertoli cells, functions of 583
Sesamoid bone 322, 671
Sex hormones, role of 523
Shenton's line 666
Shoulder girdle 31
  bones of 652
Shoulder joint 649
Sign of Felson 708
Silhouette sign 708
Single lymph node 527
Sinus
  lactiferous 601
  medullary 532
  subcapsular 527, 544
  venarum 391
Sinusoid 465, 467, 468, 496
  lumen of 496
  structure of 496
  walls of 468
Skeletal muscle 11, 11f, 12, 449, 452, 458, 459
  cell 452

  component parts of 11f
  fiber 449f, 451
  microstructure of 449f
  nomenclature of 12
  parts of 26
Skin 7, 9, 20, 620
  appendages of 24, 627
  epithelium of 25, 621
  functional aspect of 8
  incision 22f, 34, 43, 43f, 60f, 69, 69f, 75, 76f, 86, 103, 107, 116, 125f, 143f, 149f, 158f, 166f, 179f, 185f, 193, 205f, 221, 225, 256f, 313, 314
  layers of 620
  major layers of 24
  source of 24
  structure of 8f
  thick 620f, 633t
  thin 620f, 633t
Skull
  bones 719
    total number of 715
  breadth of 716
  cap 715
  foramina of 720
  lateral view of 713f
  length of 716
  ossify, bones of 716
  posteroanterior view of 712f
  roof of 715
  Water's view of 714f
Smooth muscle 459
  fiber 459f
Sodium ipodate 702
Soft tissue shadows 681
Sole
  chief
    arteries of 214
    nerves of 215
  deep fascia of 211
  muscles of 211
Soleus
  nerves supply of 201
  origin of 201
Spermatic cord 136, 228, 229, 229f
  coverings of 229f

Spermatogenesis 580
　stages of 581, 582
Spermatogenic cells 580
Spermatogonia, types of 580
Spermatozoa, parts of 582
Sphenoidal air sinus 718
Sphincter
　types of 575
　urethrae 575
　vesicae 575
Spinal accessory nerve 301f, 310, 422, 423f, 424, 425
Spinal cord 603, 603f, 604, 608-610
　central canal of 609
　descending tracts of 611
　gray matter of 605
　length of 604
　parts of 604
　shape of 604
Spinal nerve, divisions of 18f
Spinal root, deep origin of 424
Spine 401
Spiral cord 255
Spleen 538, 538f
　coverings of 539
　lobules of 539
　trabeculae of 539
Splenic circulation 538f, 541
Splenic parenchyma, red pulp of 540
Spondylolisthesis 687
Spondylosis 687
Spongiocytes 523
Spongy bone 447
Squamous epithelium 576
　simple 439f, 555
Stave cells, function of 541
Stem cells 529
　function of 481
Sterile matrix 633
Sternal angle 359, 382, 382f
　level of 383
Sternocleidomastoid 305
　muscle 291
Sternocostal head 38, 54
Sternocostalis 253
Sternohyoid 280, 285

Sternothyroid 280, 285
Stigma 597
Stomach 361, 371, 477, 646, 688
　barium meal study of 690
　borders of 373
　capacity of 361
　curvatures of 373
　fundus of 363, 371, 371f, 477f
　lesser curvature of 371f, 373
　orifices of 358
　parts of 371
　shape of 361
　situation of 361
　wall of 477
Strap muscles 298
Stratified squamous epithelium, function of 475
Stratum basale 9, 623
Stratum corneum 9, 622
Stratum granulosum 9, 622
Stratum lucidum 9, 622
Stratum spinosum 9, 622
　cells of 622
Stroma 517, 518
　tissue components of 602
Stylohyoid 284
　muscle 279
Styloid process 321
Stylomandibular ligament 290
Stylomastoid foramen 259
Subclavian artery, branches of 311
Subclavius muscle 40
Subcostal nerve, branches of 126f
Subcostalis 253
Sublingual gland 515, 515f, 516t
　duct of 516
Submandibular
　acini, types of 514
　duct 280
　ganglion 279
　gland 278, 512, 514, 516t
　　location of 513
　　parts of 513
　　system of 514
　lymph nodes 279
　salivary gland 513f
　triangle 279f

Sub-sartorial plexus 146, 147
Subscapular artery 58
Subscapular nerve 50, 51, 54, 58
Subscapularis 43, 55
Substantia gelatinosa 609
    centralis 610
Subtrapezoid plexus 310
Sulcus
    intertubercular 68
    terminalis 391
Superficial fascia
    exposure of 22f
    membranous layer of 222f
Superficial palmar arch, branches of 338
Superficial ring, boundaries of 227
Superior gluteal nerve, root value of 174
Superior mesenteric artery, origin of 368
Superior vena cava 382f, 383, 394
Supinator longus 75
Supinator muscle 329
Supraclavicular triangles 308
Supraorbital foramen 397, 404
Supraorbital notch 396-398, 403
Suprarenal cortical tissue 525
Suprarenal gland, microstructure of 522f
Supraspinous ligament 401
Supravesical fossa 230
Sural nerve 186
    fate of 353
Surface epithelium, types of 484
Sweat gland 8f, 631, 632
    atypical 631, 631t
    types of 631
    typical 631, 631t
Symphysis pubis 226, 235f, 236
Synovial joint 20, 649
Synovial sheath 7, 88

T

Taenia coli 489
Tarsal bones 123, 343, 677, 678
Taste
    bud 470f, 471-473
        types of 473
    pore 473
    types of 473

T-cells 530
Temporal bone, petrous part of 417
Temporomandibular joint 720
Tendinous intersections 233f, 236, 237
Tensor fascia lata 170
Teres major 45, 55, 63
Teres minor 45
Testis 577, 577f
    coverings of 577
    lobule of 578
Theca folliculi 595
    layers of 596
Theca interna, composition of 596
Thenar eminence 86
Thenar muscles 92f, 94, 98
Thigh
    back of 178, 179f, 180, 180f182
    compartments of 123f
Thoracic aperture, superior 254
Thoracic cage 246, 247f, 705
    bones of 705
Thoracic inlet 247f, 254
    boundaries of 705
Thoracic nerve 49, 58, 59, 302, 304
Thoracic spinal nerves 20
Thoracodorsal nerves 54
Thoracolumbar fascia 241f, 379
    anterior layer of 243
    layers of 243
    middle layer of 241, 244
    posterior layer of 241, 244
Thorax 246, 380
    abdomen 10
    inlet of 246
    outlet of 248
    radiological anatomy 703
Thymic lobule, regions of 534
Thymic lobules 533
Thymic lymphocytes 535
Thymic nurse cells 537
Thymus 532, 534
    cortex of 534
    coverings of 533
    functions of 537
    lymphocyte of 534
    role of 537

Thyroglobulin, function of 520
Thyrohyoid 280, 285
Thyroid 405, 517, 518
   angle 405
   artery, superior 282
   cartilage 277, 405
   follicle 518
      parts of 521
      structure of 519f
   gland 409, 410, 517, 518f
      isthmus of 407, 408f
      lateral lobe of 408
      parts of 408
   hormones 520
   lingual 410
   lobule 518, 520
   prominence 404
   venous plexus 410
Thyroidectomy, partial 411
Tibia 123
   intercondylar area of 671
   lower end of 677
Tibial artery
   anterior 151, 156, 344f, 347, 348
   posterior 203, 346f, 348, 349
Tibial nerve 347, 349, 351, 352f
   branches of 190, 352
   fate of 353
   root value of 351
   termination of 215
Tibial tuberosity 672
Tibial veins, anterior 348
Tibial vessels
   anterior 356
   posterior 194
Tibialis 204
   anterior 151, 153, 216
   posterior 196f, 198, 216
Tibiofibular joints 672
Tibiofibular mortice 676
Tissue 436
   collection of 433
   forcep 4
   lymphatic 547
   lymphoid 487, 538
   stroma 602

   trabeculae 518
   types of 436, 589
T-lymphocytes 529, 530, 544
   action of 530
   types of 531
Tongue 469
   lining epithelium of 470
   microstructure of 469f
   muscles of 474, 474t
   parts of 470
Tonsil
   abdominal 365, 487
   lymphatic follicle of 543
   origin of 427
   principal artery of 427
Tonsillar
   crypts 543
   fossa, boundaries of 425
   pits 543
   sinus 425
Torticollis 311, 425
Trabeculae 447, 527
Trabecular septa 534, 539
Trachea 384, 545, 706
   breadth of 384
   cartilaginous rings of 385
   cross-section of 545f
   divisions of 387
   layers of 545
   length of 384
   microstructure of 545f
   posterior aspect of 548
   vertebral extent of 384
Tracheal
   bifurcation 384, 385, 385f, 386, 386f, 387
   cartilages 547
   epithelium 546
   localization 706
   muscle 552
Trachealis muscle, contraction of 548
Tracheobronchial tree 550, 551
Traction epiphysis 667
Transitional epithelium 440f, 573
Transpyloric plane of Addison 360
Transtubercular plane 221

Transverse
  cervical
    artery 302, 303f, 304, 311
    nerve 276, 302
  facial artery 260
  intermuscular septum 194
Transversus thoracis 253
Trapezius 305
Traube's space 372
  boundaries of 372
Triangular ligament 375, 376
Triceps brachii 45
Triceps surae 200
Tricipital tendon 177
Trigeminal nerve 264, 266, 266f
Triradiate cartilage 664
Trochlear notch 656
Tubercle of Lister 114
Tuberosity 343
Tunica adventitia 14
  composition of 461
Tunica albuginea 578, 594
Tunica intima 15, 462
  components of 461
Tunica media 14, 462
  composition of 461
Tunica vaginalis, layers of 578
Tunica vasculosa 578

U

Ulna 32f
  head of 319
  lower end of 659
  styloid process of 320
  upper end of 656
Ulnar artery 63, 72f, 74, 77, 84, 331, 335, 336, 336f, 337, 338
  beginning of 336
  branches of 331, 337
  largest branch of 337
  terminal branches of 338
Ulnar bursa 88, 96, 100
Ulnar head, shape of 319
Ulnar nerve 52, 53, 63, 74, 77, 81, 84, 116, 324f, 327, 329, 331, 337
  deep branch of 93

Ulnar styloid 320
Ulnar tuberosity 656
Umbilical vein, right 376
Umbrella cells 576
Upper limb 30, 317, 647
  back of 318f
  front of 321f
  postaxial borders of 31f
Upper subscapular nerve 51
Upper triangular space 42, 46
Ureter 571, 572f, 696
  circular muscle of 572
  length of 572
  lumen of 573
  pelvis of 562
Ureteric wall, layers of 572
Urinary bladder 574, 574f
  fundus of 363
  muscles of 572
  wall of 574
Urinary system 558
Uriniferous tubules 571t
  parts of 563, 570
Urogastrone 485
Urography, retrograde 697
Urothelium 573
Uterine
  glands 590, 591
  tube 591
    layers of 591
    muscles of 592
    nonperitoneal parts of 592
    parts of 591
  wall, layers of 588
Uterus 588, 588f
  fundus of 363
  ligament of 223, 230
  nonperitoneal areas of 589

V

Vagus nerve 282
  branches of 297
Vallate papilla 469f, 471
Valve of Gerlach 364
Valve of Kerkring 483
Vas deferens 229, 230, 584, 584f
  wall of 584

Vasa nervorum 16
Vasa vasorum 15, 28
Vastus medialis 147
Veins 27, 260, 270, 465-467
    acromiothoracic 37
    femoral 134, 136, 137, 147
    intercapitular 119
    intercostal 250f, 251
    large 465f
    subclavian 303
    superficial 289
    tributary of 27
Venae comitantes 15, 27, 203, 348, 349
    function of 27
Venae nervorum 17
Venous sinusoid, structure of 541
Venous system, types of 466
Vertebra, parts of 686
Vertebral arch 686
Vertebral column 684
    length of 684
Vessels 3, 13, 65, 289
    lymphatic 254, 527
Villi 483
    density of 484
Villus
    core of 484
    length of 484
Vincula 90, 92f
    brevia 90
    longa 90

Vitelline membrane 597
Vitiligo 625
Volkmann's canals 446

W

White pulp 539, 540
    germinal center of 540
Workman's nerve 329
Woven bone 447, 448
Wrist
    band 10, 25
    joint 319
Wry neck 311

X

Xylene 434

Y

Yellow marrow 448
Y-shaped epiphyseal plate 664

Z

Zig-zag tubule 569
Zona fasciculata 523
    cells of 522
Zona glomerulosa, cells of 522
Zona reticularis 523
Zymogen cells 480